The Atmospheric General Circulation

The Atmospheric General Circulation

JOHN M. WALLACE
University of Washington

DAVID S. BATTISTI
University of Washington

DAVID W. J. THOMPSON
Colorado State University

DENNIS L. HARTMANN
University of Washington

CAMBRIDGE
UNIVERSITY PRESS

Shaftesbury Road, Cambridge CB2 8EA, United Kingdom

One Liberty Plaza, 20th Floor, New York, NY 10006, USA

477 Williamstown Road, Port Melbourne, VIC 3207, Australia

314–321, 3rd Floor, Plot 3, Splendor Forum, Jasola District Centre,
New Delhi – 110025, India

103 Penang Road, #05–06/07, Visioncrest Commercial, Singapore 238467

Cambridge University Press is part of Cambridge University Press & Assessment,
a department of the University of Cambridge.

We share the University's mission to contribute to society through the pursuit of
education, learning and research at the highest international levels of excellence.

www.cambridge.org
Information on this title: www.cambridge.org/highereducation/isbn/9781108474245

DOI: 10.1017/9781108563857

First published 2023

Printed in the United Kingdom by TJ Books Limited, Padstow, Cornwall

A catalogue record for this publication is available from the British Library.

A Cataloging-in-Publication data record for this book is available from the Library of Congress.

ISBN 978-1-108-47424-5 Hardback

Additional resources for this publication at www.cambridge.org/agc.

Contents

Preface

How the book came into being

The inspiration for this book dates back to JMW's years as a graduate student at MIT from 1962 to 1966. Beginning in the early 1970s, he taught a graduate-level general circulation course at the University of Washington. In the early years it closely paralleled the one that Victor Starr had taught at MIT, but as time went on, more and more new material was incorporated, some of it inspired by research collaborations with James Holton on stratospheric phenomena, and later adding a section on the zonally varying, extratropical tropospheric general circulation, which was the focus of his own research at that time. The revisions and additions also reflected breakthroughs in our understanding of wave-mean flow interactions in the general circulation dating back to the late 1970s and 80s. As he approached retirement about ten years ago, Chih-Pei Chang, one of Holton's earliest students and research collaborators, urged him to polish up the notes and publish them as a monograph.

From its inception, the book was envisioned as summarizing what the community has learned about the atmospheric general circulation since the mid-twentieth century, when it emerged as an important intellectual focus within the field of dynamical meteorology. It didn't become clear until the project was underway that surveying the voluminous literature in the field, deciding what to include and what not to include in the book, and organizing the contents into a coherent narrative, complete with illustrations and a supporting mathematical framework would require a team effort. DSB, the first to join in, took the lead in converting the fragmentary formal derivations scattered through the text and appendices of early versions of the book into a more complete and orderly development of the governing equations, supported by the relevant scaling considerations. DWJT made major contributions to Chapters 8 and 9 and to designing and creating graphics for the book. DLH made available the collection of global maps and meridional cross sections that he had created for his own book and oversaw the creation of additional figures for this book. Section 2.4 is based on his work.

The title

One of the primary aims of atmospheric dynamics is to develop an understanding of the macroscopic behavior of the global circulation – i.e., its structure and evolution, and the ways in which it mixes and disperses trace constituents, mediates radiative transfer, cloud microphysics, and microscale turbulence, interacts with the oceans, the biosphere, and the cryosphere, and imparts a regional character to climate change. In pursuit of this goal, the study of the atmospheric general circulation has assumed a role in atmospheric sciences curricula somewhat analogous to that of statistical mechanics in physics curricula, interpreting the macroscopic behavior of the global circulation in terms of the dynamics and statistical properties of ensembles of waves and vortices of which it is made up. Just as synoptic meteorology is concerned with the various dynamical entities that shape and mediate day-to-day weather, the general circulation is concerned with the entities that shape and mediate climate. When first introduced into atmospheric sciences curricula almost 70 years ago, general circulation courses were considered specialty courses, but they have since become mainstream, connecting atmospheric dynamics with the other main subdisciplines within the field and elucidating the behavior of the global atmosphere as a system. It thus seems an apt title for a book designed to facilitate the teaching of such courses.

Consistent with the book's title, the image on the cover of the book is global in scope and it ranges in scale from ~10 km to planetary. The atmospheric circulation is rendered in terms of the distribution of clouds that it produces. The oceans, the cryosphere, and the terrestrial biosphere, with which it interacts, are also visible. The image is a frame from an animation – not an extraordinary one, but a typical one. Rather than featuring a single storm or event the term "general circulation" connotes many different kinds of weather systems on many different space and timescales, as discussed in the book.

The scope

The book summarizes the major advances in our understanding of the general circulation during the past 70 years, and offers a brief history of how they came about. In contrast to most books and review articles, it places greater emphasis on the achievements of the broader community in advancing the science and less on individual contributions. We regret that we have space to cite only a small fraction of the thousands of published papers that have contributed to these advances.

Compared to early textbooks on the general circulation, ours is less focused on how the atmosphere fulfills the balance requirements for the conservation of mass, angular momentum, and energy, and offers a more holistic treatment of the underlying dynamics. Though not a self-contained dynamics textbook in its own right, it does attempt to explain atmospheric phenomena on a fundamental level, starting from the equations of motion. In comparison to other dynamics textbooks, it is more oriented toward the interpretation and explanation of phenomena and less toward the application of general principles. It is also more oriented toward the visual learner, with about a thousand panels of graphics, which greatly exceeds the number of equations. It touches on a wide range of atmospheric phenomena, including many that are not mentioned in other textbooks. The focus is on understanding the general circulation as it exists today, rather than on how it was different in the distant past or on how it is changing on timescales of decades or longer. Our book can be viewed as relating to climate dynamics, and planetary atmospheres, but compared to other books in these fields it is more focused on the circulation and on the departures from the vertical profile of globally averaged temperature than on the vertical profile itself.

The content

The book is divided into six parts, each consisting of a set of sequentially ordered chapters with numbered sections and subsections. Many of the sections are followed by exercises, some of which have well-defined answers and others intended to open a line of inquiry that the reader or instructor might wish to pursue. Each chapter contains numerous figures, most of them multi-panel, The main text is followed by Appendices A–F, a bibliography, and an index. Supporting materials made available on the companion web page include supplementary figures, a *Solutions Manual* for most of the exercises, and an extensive set of animations that illuminate the phenomena described in the text.

Part I is a two-chapter introduction to the book. The first chapter traces the development of the observing system from the mid-twentieth century onward and it introduces the reader to the "building blocks" of the general circulation – the jet streams, the Hadley cell, the storm tracks, the intertropical convergence zones, etc. – through an extended sequence of graphics reminiscent of a "mini-atlas." The formalism for decomposing the flow into time means and transients and into zonal means and eddies (or waves) is presented in Appendix A. It shows how the set of governing equations that is used generally in fluid dynamics can be simplified through the use of the hydrostatic and geostrophic approximations to derive simpler sets of equations that provide insight into the various phenomena that will be considered in later chapters. The chapter ends with a brief history of general circulation modeling. The second chapter is designed to give the reader an intuitive feel for the processes that will be described in greater detail and with more mathematical rigor in subsequent parts of the book. It describes and reflects upon a set of "thought experiments" and numerical simulations that explore the roles of differential heating, frictional dissipation, and gravity waves in the development, maintenance, and "spin-down" of the circulation and the role of rotation in shaping it. The book also describes how the general circulation can be represented in terms of a thermodynamic cycle consisting of a sequence of reversible processes –i.e., a heat engine.

The four chapters of Part II (Chapters 3, 4, 5, and 6 of the book) parallel Starr's graduate-level general circulation course at MIT during the 1960s, which was framed in terms of the balance requirements for angular momentum, total energy, mechanical (kinetic and potential) energy, and water vapor. It is more straightforward to cover these topics now than it was back then, because it does not require an extensive error analysis or elaborate justifications for neglecting some of the small terms that could not be evaluated on the basis of the data available back at that time. The treatment in the book is enriched by the inclusion of observational material that was virtually unimaginable at the time Starr taught his course – precise measurements of length of day and net radiation through the top of the atmosphere; reliable estimates of the geographical distribution of precipitation and surface energy fluxes; and global fields of vertically integrated water vapor and a host of other trace substances from satellites. As in Starr's course, the horizontal wind field is decomposed into an axially symmetric component consisting of the zonally averaged zonal winds and the mean meridional circulations and an eddy (or wave) component defined as the departure from the zonal mean, which is capable of transporting mass, zonal momentum, and energy in the meridional plane, analogous to the way in which turbulence produces down-gradient eddy diffusion. These eddy transports are represented by zonally and time-averaged covariance terms, which appear in the conservation equations, as elaborated in Appendices A and B.

Part III (Chapters 7 and 8) on the dynamics of the zonal mean flow is more advanced conceptually and some of it requires more mathematical manipulation than the reader will encounter in Part II. Chapter 7 transcends the limitations inherent in the balance requirement approach, in which angular momentum conservation and energy conservation are self-contained topics. It treats the general circulation more holistically, in terms of the set of equations that governs the evolution of the zonally symmetric flow. In his Planetary Circulations course at MIT, which he taught during the 1960s, Charney included an extensive section on this topic based on Eliassen's (1951) paper on the dynamics of an axisymmetric vortex. In our book, it is presented in a simplified form. The subject of Chapter 8 is wave–mean flow interaction. Jule Charney talked about it at length in his course, with specific reference to baroclinic waves, but he lamented that he didn't know how to represent the eddy transports after the waves grow to finite amplitude. It was not until the late 1970s that the newly developed transformed Eulerian mean (TEM) formalism made it possible to go beyond such questions as "What maintains the trade winds and the midlatitude westerly wind belts?", and explain why they exist in the first place. The nonlinear life cycle of baroclinic waves, as simulated in a primitive equation model, is used as a concrete illustration of the two-way wave–mean flow interaction that had previously eluded general circulation theorists. Another example, discussed briefly at the end of Chapter 8, is the annular modes.

Part IV (Chapters 9 and 10) focuses on the stratospheric general circulation. Chapter 9 discusses the global, time-varying Brewer–Dobson circulation, stratosphere–troposphere exchange, planetary waves in the winter stratosphere, and sudden warmings. In all these phenomena, the breaking of Rossby waves dispersing upward from below in middle and high latitudes plays a central role in permitting air parcels to flow poleward under the constraint of the conservation of angular momentum. Chapter 10 zooms in on the tropical stratosphere, where the breaking of inertio-gravity waves dispersing upward from directly below is instrumental in driving the remarkable quasi-biennial oscillation in zonally averaged zonal wind. The reader is introduced to the family of equatorially trapped planetary waves and to the linear theory based upon the shallow water wave equations that accounts for their existence. Comparison of two-sided wavenumber–frequency spectra and theoretically derived dispersion curves provides a graphic example of how linear theory can be of use in interpreting observations.

Up to this point in the book, the treatment is framed entirely in terms of zonally averaged means, variances, and covariances. In effect, the atmosphere is treated as if the Earth were an aquaplanet, with no significant longitudinal variations in climate. The distinguishing characteristic of Part V (Chapters 11, 12, and 13) is that it treats the extratropical general circulation as zonally varying. In the interest of brevity, coverage is limited to the Northern

Hemisphere wintertime circulation. Chapter 11 focuses on zonally varying features of the (time) mean climate such as the jet streams, the stationary waves, and the subpolar lows and the subtropical highs. Numerical experiments with a state-of-the-art general circulation model are used to discern the roles of orography, zonally varying diabatic heating, and tropical forcing in maintaining these longitudinally dependent features, taking into account their nonlinear interactions with the zonally averaged zonal wind field. Chapters 12 and 13 are devoted to the zonally varying transients with periods shorter and longer than about 1 week, respectively, defined on the basis of time-filtered data.

Part VI (Chapters 14–21) is devoted to the zonally varying tropical general circulation, with emphasis on the troposphere. Chapter 14 documents and interprets the year-round features, such as the Intertropical Convergence Zones and the equatorial dry zones. Chapter 15 illustrates the importance of deep convection in the tropical energy balance and explains why tropical convection is organized more by moisture gradients than by temperature and potential vorticity gradients. Chapters 16–18 document and interpret the variability of the tropical circulation observed in association with low frequency, planetary-scale phenomena: the climatological mean annual cycle, ENSO, and the Madden–Julian oscillation, respectively. Chapter 19 focuses on higher frequency, more regional scale phenomena, including equatorially trapped waves, easterly waves, and midlatitude baroclinic waves dispersing into the tropics at upper tropospheric levels. Chapter 20 discusses the dynamics of warm core tropical vortices including intense tropical cyclones. The final chapter (Chapter 21) ties together the various sections of the book in which inertio-gravity waves with periods shorter than a day are discussed in a more coherent way, and it discusses atmospheric tides and more regional diurnal variations in tropical rainfall.

To help readers navigate back and forth between the text and the diagrams, we have assigned titles to most of the numbered figures. Captions are short and place greater emphasis on what the figures show than on how they were produced. Further details on data sources and methodology used in constructing the figures are presented in Appendix F, and URLs to websites detailing the various datasets are provided on the companion web page, where they can be updated as needed.

The exercises are intended to help readers to assimilate the content of each section of the book, to think about it more deeply and from different points of view, and to relate it to influential papers in the literature. Some exercises open lines of inquiry that could not be pursued in the text because of space constraints. The more open-ended exercises, for which we do not provide complete solutions, are indicated by asterisks. These are intended to propose possible topics for class discussions, projects, or term papers. For most of the open-ended exercises, the *Solutions Manual* offers hints, examples, and/or suggested references.

In the online *Solutions Manual*, readers are introduced to online tools that will enable them to access gridded fields of atmospheric and oceanic variables, to display them, and to perform various kinds of analyses on them, ranging from time and zonal averaging and compositing to more complex operations, such as empirical orthogonal function (EOF) analysis. Relying on these online resources, instructors can assign "hands on" exercises of their own to enhance the learning experience.

Appendices A–F at the end of the printed text provide more detailed information on analysis techniques (A and C), mathematical derivations (B and D), notation and terminology (E), and on the data and analysis techniques used in creating the various figures that appear in the book (F). Additional information and resources are provided online, including high resolution PDFs of all the figures, a "mini-atlas" that extends the one presented in Chapter 1 of the printed text, additional supplementary figures, an animations library, and a compilation of URLs for the sources of the various datasets used in creating the figures.

The illustrations in the book are derived from a variety of sources, the most important of which are global reanalyses. The book's ~10-year "gestation period" saw important advances in the state of the art of data assimilation. Most of the figures that appear in the book are based on products derived from the ERA Interim Reanalysis, which was discontinued in 2019, replaced by the more advanced ERA5. Virtually all of the figures relating to gravity waves and many of those in Part VI documenting the tropical general circulation are based on ERA5. The reader can be assured that, unless otherwise noted, the figures that appear in the book based on the earlier ERA interim are not appreciably different from their counterparts based on ERA5.

The community

The general circulation research community dates back to the late 1940s. Pioneers in the field included Carl G. Rossby, Victor Starr, Jule Charney, Eric Eady, Edward Lorenz, Arnt Eliassen, and Alan Brewer. Other major contributors who came on the scene in the 1960s include Syukuro Manabe, Taroh Matsuno, and James Holton. In the early years, the field was populated by their former students and postdocs, many of whom made important contributions in their own right. With the passage of time, the community has become progressively more diverse with regard to nationalities and institutional affiliations, and in recent years, many more women are holding faculty and research scientist positions. Vigorous and sustained efforts will be needed to attract more minorities into geosciences from the existing pool of qualified applicants and to increase the size of the pool. Improving the quality and accessibility of kindergarten-to-12th grade science, technology, math, and engineering (STEM) education will not be enough. Students from those groups need role models, who can inspire more of them to enjoy and value science and math, and convince them that they can master it, starting in their early years. Some of our current students are contributing to this effort through departmental outreach programs to K-12 schools.

Using the book

The book is intended as a text for advanced undergraduate and graduate level courses on the general circulation and as a reference for atmospheric dynamists, synopticians, and those who work in the field of climate dynamics. It is written at a level that should be comprehensible to students who have successfully completed at least one atmospheric dynamics course.

The book is sufficiently long that an instructor using it as required reading for a course will need to make some choices. Instructors teaching classes that include students with minimal backgrounds in atmospheric dynamics will have to forgo Part III and present the TEM formulation in an abbreviated form or avoid talking about it altogether, taking care to omit subsequent sections that are heavily dependent on it. Conversely, in a more advanced course to be offered to students with strong backgrounds in atmospheric dynamics, a cursory treatment of Part II on the balance requirements should be sufficient. Either option offers the instructor quite a lot of latitude in choosing which of the chapters and sections in Parts IV, V, and IV to include. For example, a course could focus on the middle atmosphere (Part IV), the extratropical troposphere (Part V), or on the tropical troposphere (Section 10.2 and Part VI). If Part III is included in the course syllabus, the instructor might consider assigning or at least recommending a more conventional dynamics text such as *Atmospheric and Oceanic Fluid Dynamics*, by G.K. Vallis for the benefit of the student who learns more effectively by studying equations than by studying graphics.

Acknowledgments

We are thankful to many colleagues who helped us to assemble the many pieces woven into the fabric of this book – the diagrams, the explanations, the online supplementary material – and who, at the request of the publisher, anonymously critiqued preliminary versions and offered suggestions for improving the presentation.

Ángel Adames, Grant Branstator, Xianyao Chen, George Kiladis, Hamid Pahlavan, and Rachel White are chapter co-authors, and most of them contributed beyond their own chapters. Their involvement enabled us to widen the scope of the book beyond our own areas of expertise.

Relatively few of the diagrams that appear in the book are reproductions of figures that have previously appeared in journal articles. Most of them have been created by colleagues at our request and adapted to a format designed to match other figures in the book. The two major contributors are Ying Li, who created many of figures, and Marc Michelson, who created the three-panel climatologies that appear mainly in Chapter 1, as well as some of the figures in Chapters 3 and 5, and most of the figures in the online mini-atlas. Other contributors are Kallista Angeloff, David Bonan, Eliza Dawson, G. Neljon Emlaw, Jinhyuk Kim, Jonathan Martinez, Yumin Moon, and Simchan Yook.

Our department colleague, Christopher Bretherton, "test drove" an early version as the assigned text in his general circulation course. He and the students and postdocs who attended his classes offered a number of comments and suggestions for making the explanations more robust and understandable. More polished versions of the book were used in departmental general circulation courses in 2020 and 2021, with students leading the classroom discussions. Further edits were motivated by their input. We thank Brian Hoskins for many stimulating conversations over the years and more specifically for his advice on how to improve the treatment of some of the more difficult topics.

Among the colleagues with whom we consulted on specific topics are Panos Athanasiadis, Thomas Birner, C.-P. Chang, Clara Deser, Dargan Frierson, Qiang Fu, Kevin Hamilton, Peter Haynes, Isaac Held, Robert Houze, Lyatt Jaegle, Daniel Jaffe, Ian James, Daehyun Kim, N.-C. Lau, Joy Monteiro, Hisashi Nakamura, Noboru Nakamura, Paul Newman, William Randel, Richard Rosen, David Salstein, Takatoshi Sakazaki, Jai Sukhmate, Kevin Trenberth, Geoffrey K. Vallis, V. Venugopal, Katrina Virts, Justin Wettstein, and Fuqing Zhang.

Part I Background

Part I consists of two chapters. The first describes the observational basis for general circulation, documents its salient features, and introduces the reader to the kinds of models that are being used to simulate it. The second chapter describes a set of experiments designed to give the reader an intuitive feel for some of the concepts that are the basis for our understanding of the general circulation. Neither chapter goes into great depth: their purpose is to provide background for the more comprehensive treatment of this material that follows. The centerpiece of Part 1 is an abbreviated general circulation atlas. Most of the figures that appear in this section show *reanalysis products* that are derived by fitting observations to forecast fields based on the dynamic and thermodynamic equations that govern the state of the atmosphere. Though model dependent to some degree, these products are arguably much closer to reality than any interpolated version of the pure observations.

The narrative in Chapter 2 is framed in terms of a series of experiments, several of which might actually be performed in the laboratory, several others based on numerical simulations of the atmospheric circulation on a so-called *aquaplanet*, whose bottom boundary consists of a prescribed zonally symmetric sea surface temperature field, and one based on a historical event.

1 Atmospheric Observations and Models

The term *atmospheric general circulation*, as used in this book, connotes a statistical representation of the three-dimensional, time varying flow in the global atmosphere, including the cycling of zonal momentum, energy, water vapor, and other trace constituents. The circulation may be considered in its totality or it may be partitioned into time means versus transient (for a prescribed averaging interval), zonal mean versus zonally varying, or vertically integrated versus vertically varying components. It encompasses motions and physical processes on all resolvable space and timescales. What is considered "resolvable" is in reference to the current state-of-the-art global observing system and the numerical models used in assimilating the observations into a space–time matrix. With the passage of time, the term has become more inclusive of smaller space scales, higher frequencies, and higher levels of the atmosphere and, as such, has come to encompass an increasingly wide range of phenomena.

To set the stage for an in-depth discussion of the atmospheric general circulation, the reader who is not already familiar with atmospheric observations and models may benefit from reviewing the material presented in this chapter. Some background on the observing system is presented in the first two sections, the first relating to meteorological observations and the second to measurements of trace gases and aerosols. The third section presents global maps and pole-to-pole meridional cross sections of climatological means for selected variables. It contains a mini-atlas that introduces readers to most of the features and phenomena that will be discussed in subsequent chapters. The fourth section provides an orientation to the types of numerical models that are used in general circulation research, including the simplifying assumptions inherent in their design, and the final section offers further specifics relating to the history and uses of general circulation models.

1.1 THE GLOBAL OBSERVING SYSTEM

In fields such as Earth and planetary sciences, observations drive the research agenda. The availability of new observational data used in research on the general circulation has closely paralleled the development of the global observing system in support of weather prediction. A century ago, that system was largely based on ground-level measurements at land stations and aboard ships of opportunity. Systematic upper air measurements began in the 1930s with pilot balloons carrying instruments, including aneroid barometers, thermistors for monitoring temperature, and sensors for measuring relative humidity. The balloons were tracked optically using theodolites, yielding vertical profiles of the horizontal wind vector. Radar developed during World War II made it possible to track the balloons as they ascended to the jet stream level (~10 km) and beyond. The radar-tracked balloons were referred to as *rawinsondes*. Soon after the war ended in 1945, the balloon-based observing network had become sufficiently comprehensive to support systematic hemispheric analysis of winds and temperatures at the jet stream level, which provided a basis for investigating the meridional transport of angular momentum and energy. It is remarkable that despite the tensions related to the Cold War and the isolationist tendencies in some quarters, the nations of the world have never abandoned the policy of the free exchange of weather data in support of the network.

A breakthrough in balloon technology in the early 1950s made it possible for rawinsondes to reach the 10 hPa (30 km) level on a routine basis, opening up the lower stratosphere as a new research frontier. The International Geophysical Year (IGY) 1957–1958 provided an impetus for expanding the network into the tropics and the data-sparse Southern Hemisphere, as shown in Fig. 1.1 Almost coincidentally with the IGY, the United States military augmented the radiosonde network that it was operating in the western Pacific in support of nuclear weapon and missile testing in the Marshall Islands. High-quality observations from these stations have been used extensively in studies of tropical waves.

During the 1960s and 1970s, technological advances in remote sensing of the atmosphere using satellite-borne instrumentation radically transformed the observing system. The earliest satellite-based products were crude images based on reflected solar radiation, in which highly reflective cloud tops could be distinguished from the much darker ocean and land surfaces. The visible imagery was soon followed by infrared imagery, which was available 24 hours a day and in which the colder tops of deep convective clouds could be distinguished from those of the warmer low clouds. From the standpoint of general circulation research, the most important contribution of satellite-based remote sensing was the vertical profiles of temperature and humidity, which were retrieved from radiances in a suite of different infrared

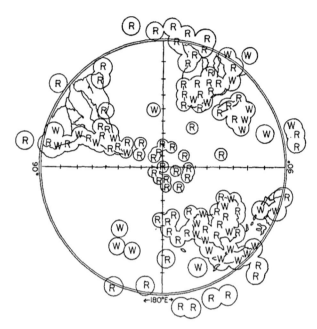

Figure 1.1 Locations of Southern Hemisphere sounding stations operating during the IGY (calendar year 1958). The South Pole is at the origin and the equator is along the circumference. R refers to pilot ballon stations, at which balloons were tracked optically using theodolites, and W to rawinsonde stations at which balloons were tracked using radar. Note the sparseness and the unevenness of the observations, bearing in mind that pilot balloons cannot often be tracked as high as the jet stream level. From Fig. 1 of Obasi (1963). Published (1963) by the American Meteorological Society.

wavelength bands ("channels") with differing emissivities and was hence emitted by air molecules at different atmospheric levels. These infrared soundings were initially used to supplement the radiosondes in data-sparse regions such as the Southern Hemisphere, but as their accuracy improved, they became the backbone of the global observing system.

The gridding of meteorological observations from land stations and ships of opportunity began during the 1960s, before the advent of satellite-based remote sensing. Dealing with the large data gaps was a formidable challenge, especially over the Southern Hemisphere oceans as shown in Fig. 1.1. Through the use of so-called *optimal interpolation* and *data assimilation* techniques, it was possible to glean much more information from the limited available observations than would have been possible using conventional interpolation schemes. In optimal interpolation a grid-point value X_{ij} is a weighted average of the values of that variable at the surrounding stations and adjacent levels. The weights are a function of the latitudinal and longitudinal distances between the stations and the grid point, as determined empirically for each meteorological variable. Observations that were considered to be more reliable were also assigned larger weights than those considered less reliable. In the more advanced schemes the hydrostatic and geostrophic approximations described in Section 1.5 were applied so that wind, temperature, and geopotential height observations could be combined in a *multivariate analysis*, rather than

being analyzed separately. Rather than performing a brand new analysis at each time step, the fields from the previous analysis (12 hours back) were used as the "first guess," which was updated or "corrected" by the interpolation scheme to bring it into closer alignment with the current observations.

The transition from *optimal interpolation* to *data assimilation* came when, rather than using the previous analysis as a first guess field, the operational centers began to use the 12-hour forecast fields derived from the numerical weather prediction (NWP) models. Using the forecast fields assured a higher level of dynamical consistency between the fields for different variables and more realistic time continuity than was possible using statistical relationships alone, resulting in smaller incremental corrections when the current observational data were assimilated. The analysis benefited from the data assimilated in the previous time step and from data assimilated in previous cycles extending back over a week or longer. Through this "bootstrap process," surface observations and remotely sensed temperature and humidity observations from space-borne instrumentation provided useful quantitative information on the flow throughout the troposphere. By the time of the Global Weather Experiment in 1979 the quality of gridded datasets derived from "four-dimensional (space and time) data assimilation" had improved to the point where the analyses could be considered to be truly global.

The quality of the gridded data provided by the operational centers for numerical weather condition has continued to improve due to numerous innovations such as the direct assimilation of radiances for temperature and water vapor and measurements of surface winds over the oceans based on microwave radiation. Horizontal and vertical resolution has also increased dramatically over the past 40 years. To address the lack of stationarity of their operational analysis products that they make available to researchers, several of the leading operational NWP centers have begun performing retrospective *reanalyses* from time to time, applying the current state-of-the-art NWPs and data assimilation techniques to archived observational data. The model used in the most recent reanalysis, the European Centre for Medium Range Forecast (ECMWF) Reanalysis v5 (ERA5),[1] has a spatial resolution of 31 km, 137 levels from the surface to 80 km, and produces gridded fields at hourly intervals. Most of the figures in this book are based on ERA5, released in stages from 2017 to 2019, and its immediate predecessor, the ERA Interim Reanalysis (ERA-I) released in 2006. The successor to ERA5 will likely be based on a cloud-resolving model like the one that was used to produce the image that appears on the cover of this book.

The model-dependent reanalysis datasets have not entirely supplanted the pure observations. A notable example is the high resolution, gridded precipitation dataset that makes use of microwave measurements from NASA's *Tropical Rainfall Measurement Mission* (TRMM) launched in 1997 and continuing under the name of NASA's *Global*

---

[1] Hersbach et al. (2020).

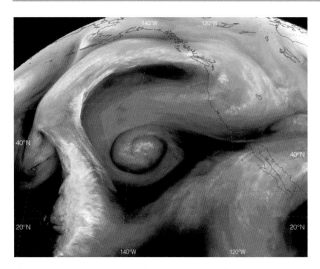

Figure 1.2 Satellite imagery over the eastern North Pacific, February 4, 2021, based on infrared radiation at wavelengths ∼7.3 microns, most of which is emitted by water vapor molecules. The lighter gray shades are indicative of higher column-integrated water vapor, and middle and high cloud decks are also discernible in the image as richly textured white patches. The bands wrapping around the cyclone at 40°N, 140°W reflect the Lagrangian character of the moisture field. Air parcels in the black streamer have a recent history of subsidence, which has thinned the moist layer and reduced the relative humidity. A moist, light gray streamer originating in the tropics can be traced as it flows northward in the central Pacific, curves eastward, and makes landfall along the west coast of Canada. NOAA GOES imagery.

Precipitation Mission (GPM), based mainly on microwave satellite imagery but also using station measurements for purposes of calibration. Other global or near-global datasets for meteorological fields that exist independently of the reanalyses include:

- temperature of atmospheric layers with depths on the order of 10 km from the microwave sounding unit (Microwave Sounding Unit, MSU, and Advanced Microwave Sounding Unit, AMSU),
- surface winds over the ocean, as inferred from measurements relating to capillary waves by satellite-borne scatterometers (NASA Scatterometer, NSCAT),
- column-integrated water vapor based on infrared satellite imagery (see the example shown in Fig. 1.2),
- radiative fluxes at the top of the atmosphere from the Earth Radiation Budget Experiment (ERBE) and the follow-on project entitled Clouds and the Earth's Radiant Energy System (CERES),
- fractional areal coverage of clouds in various height ranges (Cloud-Aerosol Lidar and Infrared Pathfinder Satellite Observation, CALIPSO, and CloudSat),
- temperature profiles with high vertical resolution derived from the Global Positioning System Radio Occultation (GPS RO) data including the Constellation Observing System for Meteorology, Ionosphere, and Climate (COSMIC) satellite programs,
- total column ozone starting with the Total Ozone Monitoring System (TOMS) and continuing with the Ozone

Monitoring Instrument (OMI), and vertical profiles of ozone mixing ratio from the Microwave Limb Sounder (MLS),
- lightning flashes inferred from electromagnetic waves (called sferics) detected and tracked by the ground-based World Wide Lightning Location Network (WLLNN).

Although many of these observations are assimilated directly into the reanalyses, they are of substantial value to the research community in their own right. The univariate datasets derived from them shed light on systematic errors in the reanalyses, such as the tendency to underestimate precipitation, and they are essential for monitoring climate change.

1.2 IMAGERY

The aim of the global observing system described in the previous section is to produce gridded digital datasets describing the evolving state of the atmosphere on a horizontal scale of tens of kilometers or longer and a time resolution of hours or longer. The imagery provided by some of the sensors carried aboard satellites and by ground-based radar have space/time resolutions an order of magnitude higher than that, enabling them to resolve phenomena such as deep cumulus convection and small-scale gravity waves, which still need to be parametrized in the models. The quality and quantity of imagery has improved dramatically with (i) the increasing (electromagnetic) spectral range and resolution of the data, (ii) advances in computer technology that have made it possible to store the virtual "firehose" of high resolution imagery data and make it available to the community, and (iii) the development of algorithms for extracting useful information on rain rate and other variables from radiance data. The imagery is valuable in its own right, providing detailed information across a range of horizontal scales of up to four orders of magnitude. It is only in recent years that researchers have had access to the computer resources required to enable them to process much of it digitally.

Imagery falls into three broad wavelength categories: visible, infrared (IR), and microwave (including the sensing of reflected radiation by ground-based weather radar). Dating back to the mid-1960s, visible imagery has been used to map the distribution of highly reflective clouds, and it continues to be widely used, especially for animations based on imagery from the geostationary satellites and for documenting the fine structure of low level, stratiform cloud decks. It is virtually unlimited in spatial resolution and it has an appealing familiarity. Drawbacks are that it is available only during the daylight hours and in the still images it is sometimes difficult to distinguish between cloud decks at different levels. Because of the broad extent of low, but optically thick, nonprecipitating cloud decks, the brightness of visible images is not highly correlated in space and time with precipitation rate.

In the infrared part of the spectrum, the opacity of clear air is highly wavelength dependent. At wavelengths ranging from 10.3 to 12.6 μm, cloud-free air is nearly transparent. In this "window," the radiance reaching a satellite-borne sensor is in accordance with the Stefan–Boltzmann law, where the "blackbody temperature" corresponds to the cloud-top temperature or, in cloud-free regions, to the surface temperature. The images are often displayed using a grayscale, with high radiances at the dark end so that the Earth's surface is rendered nearly black and cloud-tops at the tropopause level are at the other end of the scale, making for a clear distinction between relatively warm (gray) low cloud decks and the cold (white) anvils of deep convective clouds. False color is sometimes used to emphasize the features of interest, such as the coldest cloud-tops. Radiances may be labeled in energy units or in terms of the inferred equivalent blackbody temperature, referred to as the *brightness temperature*.

At wavelengths shorter than about 9 μm, an appreciable fraction of the upward radiance is emitted by water vapor molecules. In satellite imagery based on channels within this range of wavelengths, deep clouds are visible by virtue of their low radiance, but elsewhere the radiance is determined by the temperature at the level of unit of optical depth, which is located in the upper troposphere. The deeper the moist layer, the higher the level of unit optical depth, the lower the temperature at which the water vapor molecules emit radiation to space, and the lower the radiance. Figure 1.2 shows an example, with interwoven moist (lighter) and dry (darker) bands created and sustained by the vertical velocity field.

Infrared imagery (also referred to as outgoing longwave radiation – OLR – imagery) is a much better indicator of precipitation rate than visible imagery, but it does not distinguish between the extensive stratiform cloud shields created by the spreading anvils of deep convective clouds and the convective cores within those shields, in which most of the precipitation is concentrated. Hence, for the purpose of monitoring the precipitation rate, it is better to rely on ground-based weather radar or satellite-borne microwave sensors that are capable of distinguishing between the larger droplets and ice crystals lofted by the updrafts and the smaller ones in the spreading anvils. A remote sensing capability for microwave radiation has taken longer to develop than the visible and IR imagery described in the previous paragraphs because the emitted radiation is much weaker. Thus far it is available only in rather narrow swaths along the tracks of the polar orbiting satellites.

Much of the satellite imagery used in general circulation research is merged or blended products derived from the radiances in more than one channel, tailored for specific purposes, such as monitoring precipitation rate or aerosols and trace species, or ocean color and land vegetation. The data are interpolated onto a regular grid, sacrificing resolution space/time continuity. As explained in Section 1.1, the loss of high resolution information can be minimized by assimilating the data into a state-of-the art numerical weather prediction model. For example, the NASA Global Modeling and Assimilation Office (GMAO) makes extensive use of the GEOS-5 model for this purpose. The image that appears on the cover of this book was produced using that model.

> **Exercise**
>
> 1.1 Show examples of satellite imagery that illustrate the differences between the information derived from visible and infrared window channels.

1.3 MEASUREMENTS OF CHEMICAL TRACERS

The global observing system described in Section 1.1 that was developed in support of operational numerical weather prediction is inherently *Eulerian*, with wind, temperature, and other meteorological fields defined at a fixed array of grid points at successive time steps. In contrast, information derived from measurements of chemical tracers can be used to infer the *Lagrangian trajectories* of air parcels. Such measurements have been particularly helpful in revealing the pathways for the exchange of air between the troposphere and stratosphere and the mechanisms responsible for the mixing of air within the stratosphere. The earliest measurements of stratospheric water vapor revealed that the air was remarkably dry, from which it was deduced that it enters the stratosphere by way of the *tropical cold point tropopause*, where the ambient temperature and hence the dewpoint of saturated air is on the order of $-80°$C.

An important pathway by which stratospheric air re-enters the troposphere was revealed by radioactive aerosols that penetrated into the stratosphere in the US and USSR nuclear weapons tests circa 1960 and remained there long after the tropospheric aerosols were scavenged out by precipitation processes. Some of this long-lived debris was shown to re-enter the troposphere in discrete "tropopause folding events." In effect, the lower stratosphere was acting as a reservoir for the radioactive debris, and episodic intrusions of stratospheric air into the troposphere were bringing it down to the Earth's surface in concentrations high enough to be considered a health hazard.

In the early 1970s, one of the US aircraft companies proposed building a fleet of supersonic aircraft that would cruise at altitudes in the lower stratosphere, emitting copious amounts of trace gases that were believed to have the potential to cause the catalytic destruction of ozone. The scientific community was asked to assess the level of risk. It was, in part, because of its cautionary response that the company decided not to move forward with the project. A decade later, a similar question was posed with respect to the emergence of the so-called *ozone hole* over the Antarctic during the spring months, which was becoming more extensive with each passing decade. The culprit was found to be long-lived chlorofluorocarbon (CFC) trace gases used in refrigerants, as a propellant in aerosol cans, and in a variety of industrial applications. The chemical reactions that led to the ozone destruction were found to be mediated by atmospheric transport processes within the stratosphere.

There now exists a plethora of global, gridded datasets relating to the concentrations of trace gases and aerosols. Here we will mention just a few of them. The existing network of ground-based measurements of total column ozone dates back to the early twentieth century. Ground-based measurements of CO_2 concentrations at the Mauna Loa Observatory date back to 1958. The ground-based measurements have subsequently been augmented to include a suite of trace gases, including isotopic forms, and are now being taken at a network of stations operated by the National Oceanic and Atmospheric Administration's (NOAA) Global Monitoring Laboratory. Together with measurements of total column CO, CH_4, and CO_2 from satellite-borne *nadir-viewing* instruments, they have been used to document how the spread of tracers depends upon the sources and residence time of the species. For example, space-based measurements of total column ozone dating back to 1978 have been instrumental in documenting the spread of the ozone hole and relating it to the planetary-scale circulation. Vertical profiles of ozone, water vapor, and a number of other trace species are provided by the *microwave limb sounder* (MLS), in which the sensor is aimed at the limb of the Earth and consequently samples the atmosphere along a long, slanting path. MLS records date back into the 1990s.

The research community owes a huge debt of gratitude to the providers – federal agencies and laboratories operated by international consortia – for making the datasets described in this section and in the previous section available free or at a nominal cost. Had they been treated as commodities, available to customers at a "market price," it is questionable whether the remarkable scientific achievements documented in this book could have been realized.

1.4 THE OBSERVED GENERAL CIRCULATION

In this section, we present a brief survey of the atmospheric general circulation. All but a few of the figures are based on reanalysis products. Further documentation of the datasets used in these figures is presented in the online supplemental material.

In principle, the general circulation can be decomposed into (i) the circulation that would exist on an idealized Earth-like planet with a zonally (i.e., longitudinally uniform bottom boundary under annual mean or perpetual equinox insolation), (ii) the response to the zonally varying land–sea distribution and orography, and (iii) the response to seasonal variations in solar declination. Formally partitioning the circulation into these components on the basis of observations is precluded by the fact that they are inextricably interrelated. Nevertheless, examining the zonally symmetric circulation (i.e., the longitudinally averaged circulation around latitude circles) is illuminating, especially when both Northern and Southern Hemispheres are taken into consideration and

the zonally symmetric features are viewed in the context of the zonally varying circulation.

Beyond pointing out the features of interest in the figures, we will offer little, if anything in way of dynamical interpretation at this point, reserving that for later chapters, from which we will be referring back to some of these same figures. Most are arranged in a three-panel format with annual means at the top and means for the boreal winter (December–February, DJF) and boreal summer (June–August, JJA) below it. In the printed book, we will present a comprehensive set of world maps on a Hammer projection, most of them centered on the Americas (90°W), and an abridged set for the Eurasian sector centered on 90°E. A more complete set of maps centered on the Greenwich Meridian, 90°W and 90°E, and for each of the four seasons is available for most of these fields on the companion website.

This section is divided into three subsections: the first shows climatological mean maps of temperature, geopotential height, zonal wind, precipitation, diabatic heating rate, and other selected variables; the second introduces the zonally symmetric component of the climatological mean general circulation; and the third shows the statistics of departures from climatological (time) means and zonal means. We apply the methodology described in Appendix A to partition these departures into three categories depending on whether they are zonally varying time means, time-varying zonal means, or departures from both zonal means and time

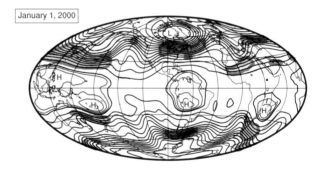

January 1, 2000

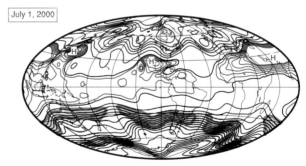

July 1, 2000

Figure 1.3 Contours showing the geopotential height of the 200 hPa surface Z_{200} on arbitrarily chosen dates in contrasting seasons, January 1 and July 1, 2000. Contour interval 100 m for the black contours and 15 m for the blue contours. Red Hs denote positive extrema. Orange shading indicates the westerly jet streams, also implied by the tight spacing of the Z_{200} contours. The threshold for the shading is $u > 40$ m s^{-1}.

means. Throughout this book we will use overbars to denote the time mean of a field variable such as temperature (i.e., $\overline{T}$) or zonal wind ($\overline{u}$), and square brackets (i.e., $[T]$ or $[u]$) to denote averages around latitude circles, referred to as *zonal averages*. Throughout Part I, *time mean* denotes an annual mean or seasonal mean, averaged over many (typically ~40) years.

1.4.1 The Time-Averaged General Circulation

Figure 1.3 shows the distribution of geopotential height Z at the jet stream level (200 hPa) on two arbitrarily chosen days, the first in the boreal winter and the second in the boreal summer. The wind field can be inferred from the geopotential height field using the geostrophic equation. Westerly jet streams are highlighted by orange shading. The meanders in the Z_{200} field in the extratropics are indicative of waves or eddies embedded in a westerly background flow. The prominence of the meanders is directly proportional to the eddy amplitude in the Z field and inversely proportional to the strength of the westerly background flow. The longitudes of the wave axes change from day to day. The wavy features that cancel in the computation of the time-averaged "climatological mean," are referred to as *transient eddies*. The residual (time-averaged) departures from zonal symmetry are referred to as the climatological mean *stationary waves*. Geographically anchored, zonally asymmetric features in the climatology are also referred to as *standing eddies* in

$$\overline{u}_{200}, \overline{Z}_{200}$$

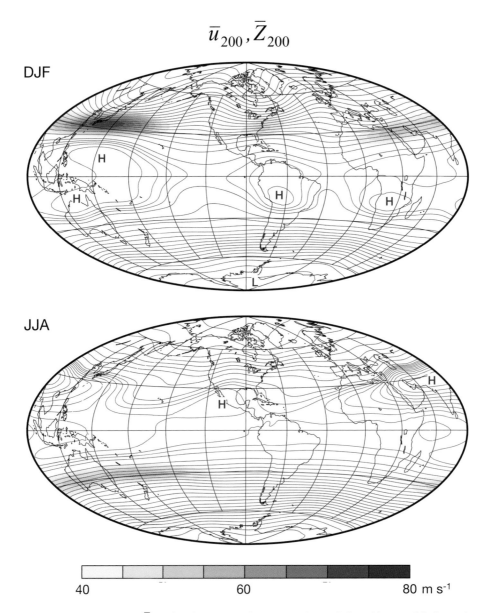

Figure 1.4 Contours showing the climatological mean $\overline{Z}_{200}$ in boreal winter DJF and summer JJA. The period of record for most of the figures shown in this chapter is 1979–2018. Contour interval 75 m, smaller than in Fig. 1.3, with lower values toward the poles; 15 m for blue contours in the tropics. Orange shading where $u > 40$ m s^{-1} indicates the westerly jet streams. The red H and L symbols denote positive and negative extrema in the height field, respectively.

the early general circulation literature, but the use of the term "standing" has come to be restricted to time-dependent variations with fixed nodes and antinodes.

Figures 1.4 and 1.5 show the boreal winter (DJF) and boreal summer (JJA) climatologies, computed by averaging the 200 hPa geopotential height Z_{200} and zonal wind u_{200} fields over the respective seasons and over all the years in the record (the 40 year period 1979–2018 for most of the figures shown in this chapter). That the stationary waves are stronger in the Northern Hemisphere than in the Southern Hemisphere reflects the stronger forcing by the larger continents and the higher and more extensive mountain ranges. The Northern Hemisphere extratropical stationary waves are particularly strong during DJF, when they are marked by prominent troughs in the upper tropospheric geopotential height field over eastern Asia and eastern North America.

The extratropical waves are driven mainly by orographic forcing (i.e., variations in terrain height, dominated by the major mountain ranges). East–west thermal contrasts in the lower troposphere also contribute.

The more prominent low latitude features in the seasonal climatologies of Z_{200} are related to the monsoons, which are driven by the continent–ocean thermal contrasts. The most striking of these features is the *Tibetan anticyclone* in JJA, which is clearly discernible even in the instantaneous July 1 field shown in the bottom panel Fig. 1.3. The annual mean Z_{200}^* climatology is dominated by the *equatorial stationary waves*, which are driven by the thermal gradients between the warmer continents and cooler oceans and between the warmer eastern hemisphere (the Greenwich Meridian, GM, eastward to the Date Line) and the cooler western hemisphere

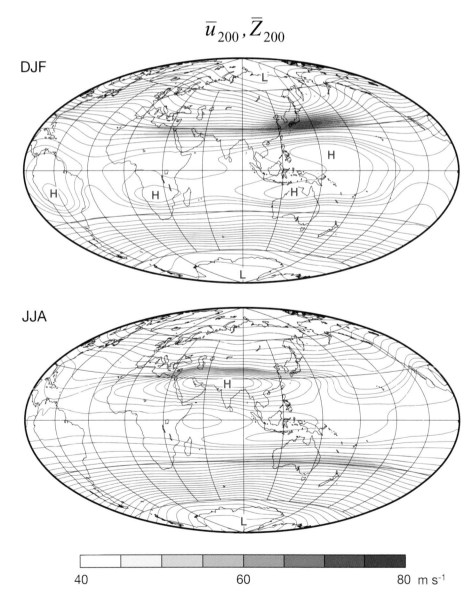

Figure 1.5 As in Fig. 1.4, but for the eastern hemisphere centered on 90°E.

Skin temperature

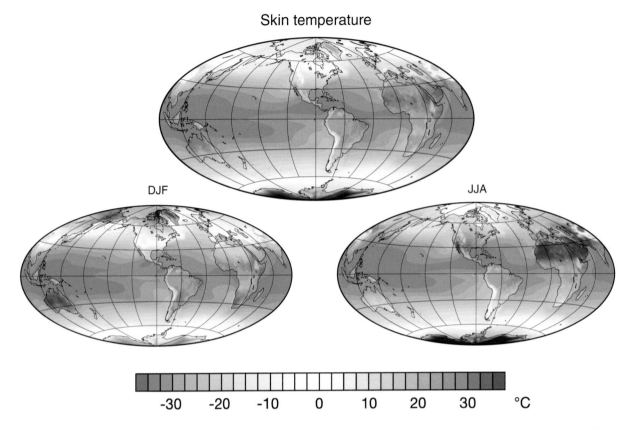

Figure 1.6 Climatological mean satellite-derived skin temperature in units of °C. In this and subsequent figures, the top panel represents the annual mean and the bottom panels seasonal means as indicated.

Skin temperature

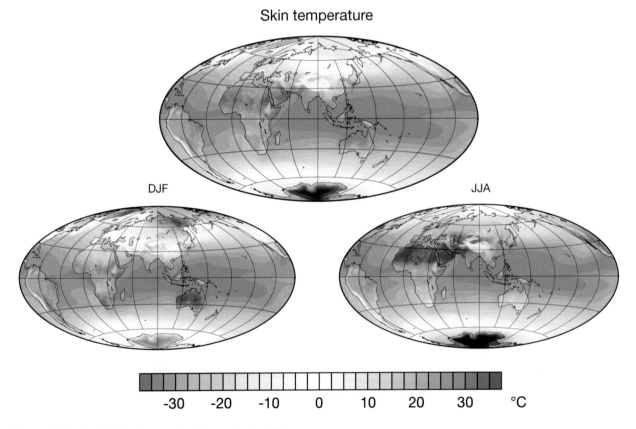

Figure 1.7 As in Fig. 1.6, but for the eastern hemisphere centered on 90° E.

SST

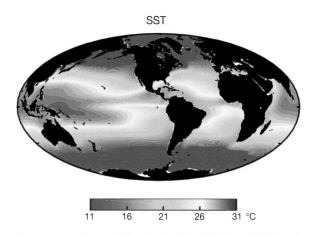

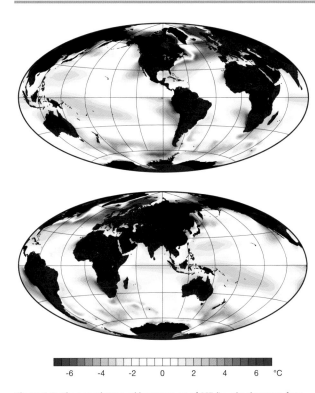

-6   -4   -2   0   2   4   6   °C

Figure 1.8 The annual mean eddy component of SST (i.e., the departure from the zonal mean on each latitude circle), considering only ocean grid points. (Top) Western hemisphere and (bottom) eastern hemisphere. Restricting the field to the eddy component reveals the zonally varying features more clearly. Land is not included because of the complicating influences of topography and ground cover.

Figure 1.9 Annual mean SST plotted using color scale limited to the high end of the temperature range, which emphasizes the subtle contrasts in the tropics and subtropics. The Indo-Pacific warm pool and the equatorial cold tongues in the eastern Pacific and Atlantic are more clearly apparent in this version. Courtesy of Ángel Adames.

of the tropics (the Date Line eastward to the GM). They are clearly discernible throughout the year except during JJA.

The next series of figures shows the climatological mean temperature distributions at the Earth's surface (Figs. 1.6–1.9) and at the top of the boundary layer. (Figs. 1.10 and 1.11). The first group shows the satellite-derived temperature field at the Earth's surface, referred to as the *skin temperature*. Over the oceans, satellite-derived sea surface temperature (SST) differs slightly from SST derived from in situ measurement (bucket and condenser intake temperature) because it is determined by a thin film of water molecules right at the air–sea interface. Land skin temperature is inferred from the surface energy balance, making use of satellite measurements of the emitted radiation.

Figures 1.6 and 1.7 show global distributions of skin temperature. The dominant high latitude features are the Antarctic and Greenland ice sheets, which are cold year-round, and the high latitude Northern Hemisphere continents, which are much colder during winter than the oceans at comparable latitudes. Both Eurasia and North America tend to be colder toward their eastern sides during winter. In this representation, the pattern is dominated by the equator-to-pole temperature gradient, which tends to be concentrated at high latitudes, leaving the low latitudes rather bland and featureless. Some of the more subtle but dynamically important features in the surface temperature climatology are

more clearly visible in Fig 1.8, which shows departures from the zonally averaged temperature at each latitude. Restricting the display to SST eliminates the complicating influences of topography and ground cover. In this representation, the SST signature of the Gulf Stream is clearly discernible. Other prominent features are the patches of relatively cold SST at subtropical latitudes off the west coasts of Africa, the Americas and Australia, a total of five of them. There is also a hint of equatorial "cold tongues" off the west coasts of South America and Africa. The same features are discernible in the (total) SST field, when it is plotted on a color scale that is limited to the high end of the temperature range, as shown in Fig. 1.9. In this representation, the equatorial cold tongues are more prominent and the five subtropical cold patches less so. The so-called "Indo-Pacific warm pool" is also evident, encompassing the western Pacific and Indian Ocean sectors. The warm water extends eastward all the way to the coast of South America in a narrow band along 7°N.

Figure 1.10 shows the climatology of 850 hPa temperature in the same format as Fig. 1.6. The 850 hPa level is representative of conditions at or near the top of the *boundary layer*: the layer in which the air is relatively well mixed by small-scale turbulence and thereby directly influenced by the fluxes of momentum, heat, and moisture at the Earth's surface. The 850 hPa temperature patterns at high latitudes are similar to their counterparts in the field of skin temperature shown in Fig. 1.6 but the isotherms are much smoother and as such, they bear a stronger resemblance to the geopotential height contours shown in Fig. 1.3: along 60°N the 200 hPa troughs over the east coasts of Asia and North America overlie the lowest 850 hPa temperatures, and the ridges overlie the highest 850 hPa temperatures. These features are reflections of the Northern Hemisphere winter-time stationary waves, which extend through the depth of the troposphere. The warmth of the deserts is more clearly discernible at the top of the boundary layer than at the Earth's

$$\overline{T}_{850}$$

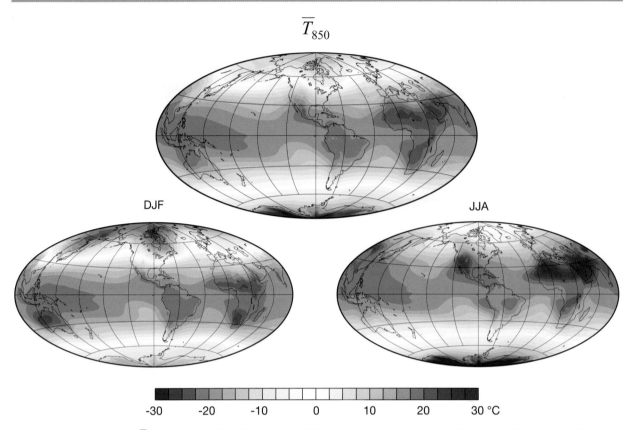

-30   -20   -10   0   10   20   30 °C

Figure 1.10 Climatological mean $\overline{T}_{850}$ representative of conditions at the top of the boundary layer. The patterns are similar to those in Fig. 1.6, but coastlines and topographic features are less prominent and there are indications of planetary-scale, wavelike features, especially in the seasonal plots.

surface and it is apparent that the tropical continents are warmer than the oceans in the annual mean and even during winter. The low latitude features show up even more clearly when temperature is represented on a limited color scale in Fig. 1.11. The oceanic features are not as pronounced at the top of the boundary layer as they are at the Earth's surface. The equatorial cold tongues are barely discernible and the five cold patches on the eastern sides of the subtropical oceans are much weaker and displaced to the west of their counterparts in SST. These are evidently shallow features in the temperature field, but we will see in Section 14.3 that they are the footprints of deeper overturning circulations.

Figures 1.12 and 1.13 show the climatological mean sea level pressure and the wind field at the 10 m level, $\mathbf{V}_{10m}$. The extratropical surface westerly wind belts and the tropical trade-wind belts are clearly evident as year-round features. The westerly belts are bounded on their poleward sides by subpolar low pressure belts comprising the Northern Hemisphere *Icelandic* and *Aleutian lows* and the Southern Hemisphere low pressure belt that encircles Antarctica. On their equatorward side, they are bounded by the subtropical highs, which are most prominent during summer, when they are all centered over the eastern oceans. The westerlies and the subpolar lows are stronger in the winter hemisphere. In both seasons and in the annual mean, the westerlies are substantially stronger in the Southern Hemisphere, inspiring the monikers "Roaring Forties" and "Furious Fifties."

The equatorward flow around the eastern flanks of the subtropical anticyclones, which overlies the five cold patches in Fig. 1.8 spreads out downstream to form the trade-wind belts. It drives the cool ocean currents and the upwelling along the west coasts of the continents. The southeast trades flow across the equatorial Pacific from the cold tongue in SST toward the Indo-Pacific warm pool, down the SLP gradient.

The surface wind field in the western hemisphere (Fig. 1.12) is dominated by the extratropical westerlies and tropical trade winds over the oceanic sectors. In contrast, the surface winds over the eastern hemisphere, dominated by the Eurasian sector (Fig. 1.13), are driven by north–south land–sea thermal contrasts, which drive a strong cross-equatorial monsoonal flow from the winter hemisphere into the summer hemisphere. This seasonally reversing flow is referred to as the *monsoon*. During JJA the northward flow assumes the form of the low-level Somali jet, which crosses the equator along the east coast of Africa and then curves eastward across southern India and Indochina, toward the Philippines. In contrast, during DJF, the circulation over east Asia is dominated by a weaker dry, northeasterly outflow from the cold interior of the continent.

The climatological mean precipitation field shown in Figs. 1.14 and 1.15 exhibits dramatic contrasts between western and eastern hemispheres. The dry zones over the tropical Atlantic and Pacific sectors of the western

$$\overline{T}_{850}$$

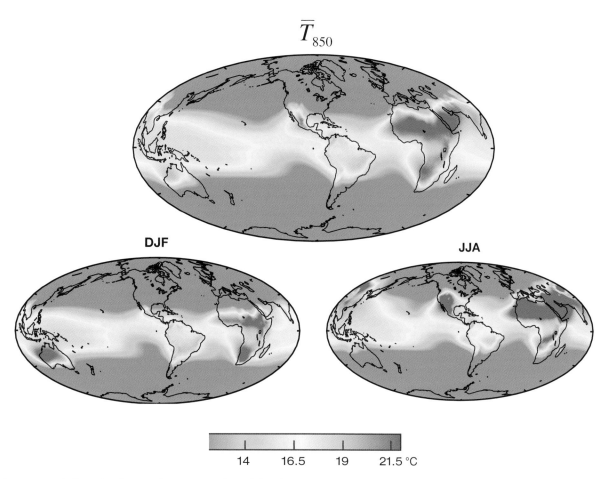

DJF

JJA

14	16.5	19	21.5 °C

Figure 1.11 As in the previous figure, 850 hPa temperature, but with the color gradations restricted to the high end of the temperature range to emphasize the subtle low latitude features. Note that the tropical continents tend to be warmer than the oceans year-round. Courtesy of Ángel Adames.

$$\text{SLP, } \overline{V}_{1000}$$

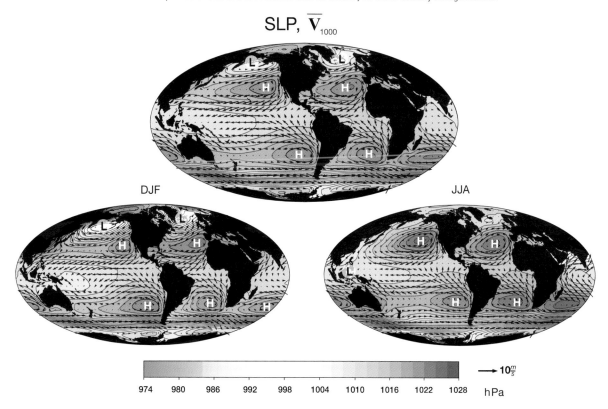

DJF

JJA

$$\longrightarrow 10\tfrac{m}{s}$$

974	980	986	992	998	1004	1010	1016	1022	1028	hPa

Figure 1.12 Climatological mean sea level pressure and 10 m wind. The westerly wind belt lies poleward of the subtropical highs (white Hs) and the trade winds lie equatorward of them. The black Ls denote centers of the subpolar lows.

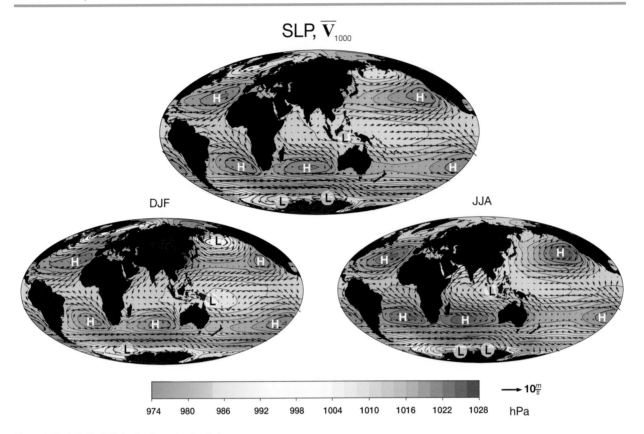

Figure 1.13 As in Fig. 1.12, but for the eastern hemisphere.

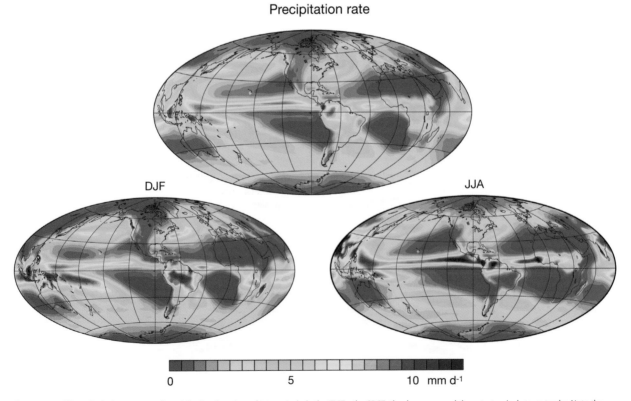

Figure 1.14 Climatological mean rate of precipitation. Prominent features include the ITCZs, the SPCZ, the dry zones, and the extratropical storm tracks. Note the resemblance of the tropical features to the SST field at low latitudes shown in Fig. 1.9.

Precipitation rate

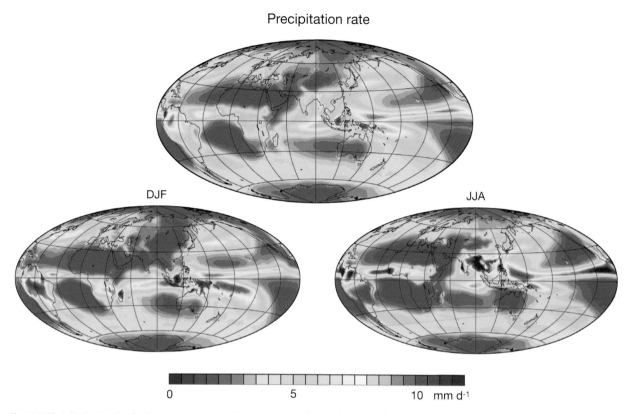

Figure 1.15 As in Fig. 1.14, but for the eastern hemisphere. Prominent features include the Australasian and African summer monsoons.

hemisphere correspond to the regions of divergent outflow from the subtropical anticyclones shown in Fig. 1.12. Both Atlantic and Pacific basins exhibit a pair of them: a Northern Hemisphere subtropical dry zone and a more extensive *equatorial dry zone* centered in, but not limited to, the Southern Hemisphere. The equatorial dry zones are bounded on their northern sides by the *intertropical convergence zones* (ITCZs), regions in which water vapor carried by the converging Northern and Southern Hemisphere trade winds rises and condenses in narrow bands of heavy rainfall. Some authors use the term ITCZ quite inclusively to denote the bands of confluence between climatological mean low-level flows originating in the Northern and Southern Hemispheres, irrespective of how distinct they are. We will use it only in reference to these well-defined features in the Pacific and Atlantic sectors. It is notable that both Atlantic and Pacific ITCZs are located not on the equator, but well to the north of it year-round, at a latitude of about 7°N in the annual mean. They overlie the bands of high SST in Fig. 1.9 The equatorial dry zones are bounded on their poleward flanks by the South Pacific convergence zone (SPCZ), and a weaker South Atlantic convergence zone (SACZ). The SPCZ is a year-round feature but it is stronger during the austral summer. Both the SPCZ and the SACZ overlie regions of relatively high SST.

Tropical rainfall over the eastern hemisphere (Fig. 1.15) is dominated by the summer monsoons. The strongest of these features is the Asian summer monsoon during the boreal

summer (JJA), in which the convergent, westerly boundary layer flow extending from the Arabian Sea eastward to the Philippines (Fig. 1.13) is marked by belts of heavy rainfall. Bands of relatively heavy rainfall are also observed over the southern tropics of the eastern hemisphere during DJF and over North and South America during their respective summer seasons. Heavy year-round rainfall is observed over Borneo and other large islands in the archipelago stretching from Southeast Asia to the north coast of Australia, referred to as the "Maritime Continent".

Other notable features in Figs. 1.14 and 1.15 are the zonally elongated precipitation maxima over the extratropical western Atlantic and Pacific and the Southern Ocean. These are mainly wintertime features: the *extratropical storm tracks* located over and extending downstream of bands of strong gradient in the underlying SST distribution.

The next two figures show the lower tropospheric diabatic heating rate fields – Fig. 1.16 in the free atmosphere (850–500 hPa) and Fig. 1.17 in the boundary layer (1000–850 hPa). (As opposed to *adiabatic* warming or cooling rates associated with compression or expansion of an air parcel, the *diabatic* heating rate Q expresses the time rate of change in the energy of an air parcel per unit mass.) In the free atmosphere the zones of heavy precipitation are marked by strong diabatic heating, while radiative cooling is prevalent in the dry zones. In contrast, the boundary layer heating bears little relation to the precipitation pattern. It is strongest over "hot spots" – such as the western boundary currents

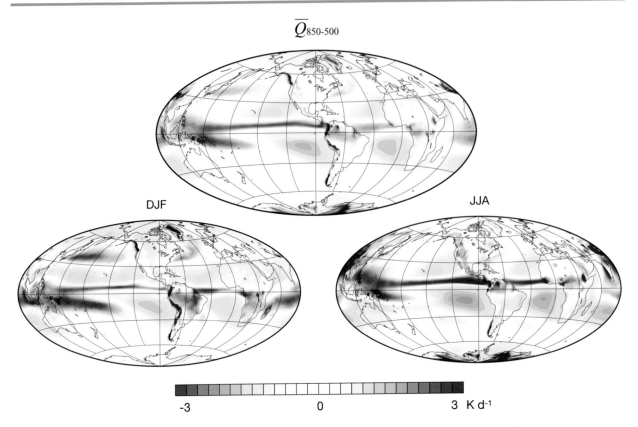

Figure 1.16 Climatological mean, vertically averaged 850–500 hPa diabatic heating rate. Here and in figures throughout the book, the diabatic heating rate is denoted by Q in the title and it is expressed in units of K d^{-1}. Strictly speaking, what is shown is Q/c_p.

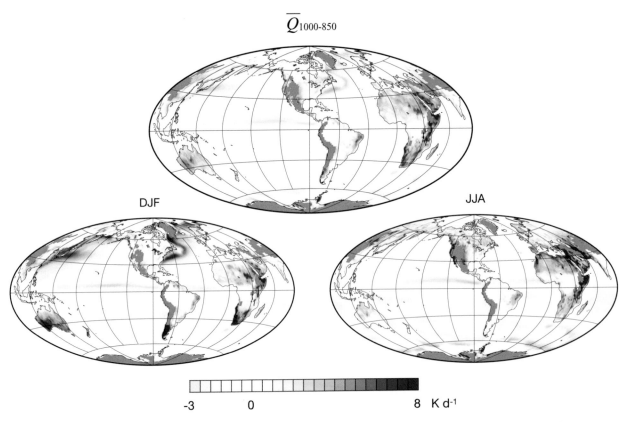

Figure 1.17 Climatological mean, vertically averaged 1000–850 hPa diabatic heating rate. Gray shading indicates terrain higher than the 850 hPa surface. The continental maxima correspond to arid and semiarid regions. The marine maxima in DJF lie over the western boundary (ocean) currents.

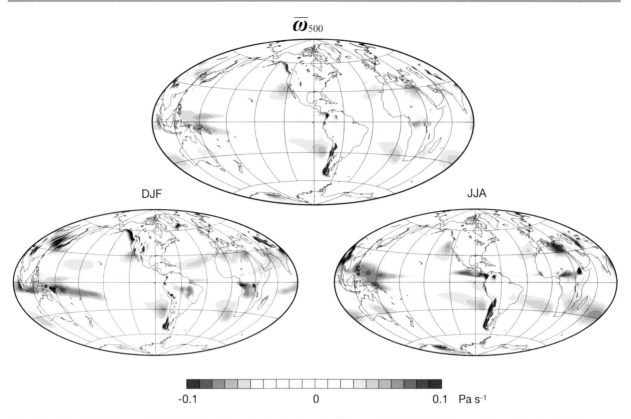

Figure 1.18 Climatological mean vertical velocity $\overline{\omega}_{500}$. Warm colors denote descent and blue ascent. This is the first of numerous figures in this book to show patterns of vertical velocity, whose units in pressure (x, y, p) coordinates are Pa s^{-1}, which is roughly equivalent to 1000 hPa d^{-1}. Hence, the largest vertical velocities in this figure are on the order of 100 hPa d^{-1}. At any given pressure level, $\omega/p = w/H$, where w is geometric vertical velocity expressed in m s^{-1}, and H is the scale height, of order 8 km. At the 500 hPa level, 100 hPa d^{-1} $\sim$1.6 km d^{-1} $\sim$2 cm s^{-1}. Note the resemblance to the diabatic heating distribution in the 850–500 hPa layer shown in Fig. 1.16.

in the winter hemisphere and the deserts in the summer hemisphere, where the fluxes of sensible heat at the bottom boundary are strongest.

The 500 hPa climatological mean vertical velocity field in pressure coordinates $\omega \equiv dp/dt$, shown in Fig. 1.18, mirrors most of the prominent features of the precipitation climatology such as the ITCZs and the regions of heavy summer monsoon rainfall. It is notable that the extratropical storm tracks are not marked by time-mean ascent: they are rectified signatures of transient disturbances, as will be explained in Section 1.4.3. Strong subsidence is observed over each of the five patches of low SST over the subtropical eastern oceans in Fig. 1.8.

1.4.2 The Zonally Averaged General Circulation

Early studies of the atmospheric general circulation were mainly concerned with its zonally averaged component. Except where it is obvious from the context, such zonally averaged variables are indicated in this book by square brackets [ ]. In this section, we document some of these distributions and relate them to the zonally varying patterns described in the previous section. Most of these sections in this and later chapters extend up to the 10 hPa level in the middle stratosphere. The vertical coordinate is log pressure,

which is "height-like" (i.e., nearly proportional to geometric height), so visual impressions concerning the strength of vertical gradients in the sections should be interpreted as relating to $\partial/\partial z$, not $\partial/\partial p$. When interpreting these diagrams it should be borne in mind that the 100–10 hPa layer, which occupies the top half of them, accounts for only $\sim$9% of the atmospheric mass.

The distributions of zonally averaged zonal wind [u], shown in Fig. 1.19, are dominated by two pairs of jets: one near the 200 hPa level and the other at the top of the section. The lower pair, referred to as *tropospheric jet streams*, are both westerly. The upper pair consists of a so-called westerly *polar night jet* that encircles the wintertime polar cap and a subtropical, summertime easterly jet. The Antarctic polar night jet in JJA is noticeably stronger than its Arctic counterpart in DJF.

At the Earth's surface the zero-wind line, which marks the poleward limit of the subtropical trade-wind belt, is located near 30°N/S throughout the year. The Northern Hemisphere jet is stronger during winter than in summer, reflecting the stronger equator-to-pole thermal gradient, and it is displaced equatorward by $\sim$12 degrees of latitude. In the Southern Hemisphere, the annual mean configuration is characterized by a pair of jets: one centered $\sim$30°S that looks like a mirror image of its Northern Hemisphere counterpart and the other,

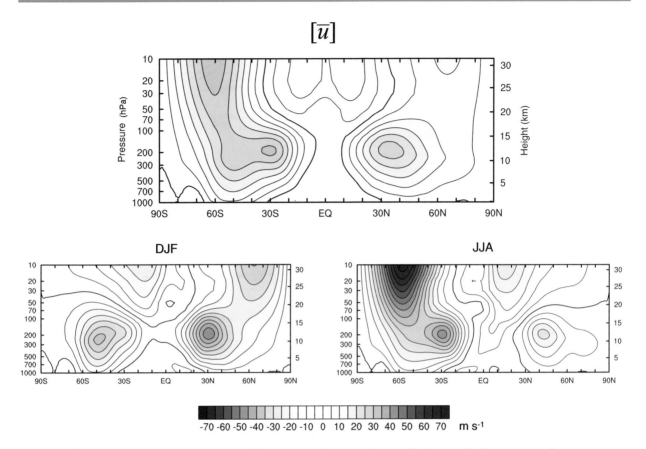

Figure 1.19 Climatological mean zonally averaged zonal wind [$\overline{u}$] Warm colors indicate westerlies and cold colors easterlies. The contour interval is 5 m s^{-1} and the zero contour is thickened. The tropospheric jet streams and the stratospheric polar night jets are clearly visible.

not quite as strong, centered ~50°S. The former is largely confined to the upper troposphere, while the latter extends downward all the way to the Earth's surface. During the austral winter JJA, the subtropical jet is dominant, while during summer DJF, the high latitude jet is dominant.

The zonal flow in the tropics is dominated by weak easterlies. The low-level easterly maxima in the DJF and JJA plots are related to the trade winds, which are stronger in the winter hemisphere. In contrast, the DJF and JJA patterns in the tropical upper troposphere exhibit stronger seasonality. In DJF, weak westerlies extend across the equator, whereas in JJA an easterly jet with wind speeds of up to 10 m s^{-1} is centered ~5°N, close to the latitude of the easterly jet on the equatorward flank of the Tibetan anticyclone. The low latitude zonal flow in the stratosphere is dominated by easterly jet streams centered at ~15° latitude of the respective summer hemispheres, increasing in amplitude with height all the way to the top of the sections and into the mesosphere. In the annual mean, these features are reflected as a pair of easterly jets, symmetric about the equator.

The corresponding section for temperature [T] is shown in the online supplementary materials. Here in the text, we show it decomposed into sets of global mean temperature profiles in Fig. 1.20 and *departure cross sections* (i.e., [T] at each latitude minus the global mean temperature at that level) in Fig. 1.21. The annual mean, global mean profile

shown in the left panel of Fig. 1.20 is made up of a lower layer extending from the Earth's surface up to 250 hPa (the 10 km level), in which the *lapse rate* $\Gamma \equiv -dT/dz$ is ~6.5 K km^{-1}, an upper layer above 100 hPa in which temperature increases with height, and a middle (250–100 hPa) layer in which temperature decreases with height, but not as rapidly as it does below 250 hPa. In the middle panel of Fig. 1.20, separate temperature profiles are shown for the tropics and extratropics. The tropical profile exhibits a distinct temperature minimum at 100 hPa which marks a sharp break in the lapse rate. The layer below the break is the tropical troposphere, the layer above it is the tropical stratosphere and the break itself is referred to as the tropical cold point tropopause. In the extratropical profile, the break in the lapse rate is not as pronounced and it occurs at ~250 hPa. As shown more clearly in Fig. 1.21, throughout the depth of the troposphere the tropics are warmer than the extratropics, but within the layer extending from ~200 to ~25 hPa the tropics are colder. The right panel of Fig. 1.20 contrasts DJF and JJA means for the Northern Hemisphere extratropics. The winter profile is colder at all levels but the seasonal contrast is about twice as large at the Earth's surface as it is in the upper troposphere, as reflected in the smaller wintertime lapse rate.

Figure 1.22 shows time series of near-equatorial temperature at selected levels in the upper troposphere and lower

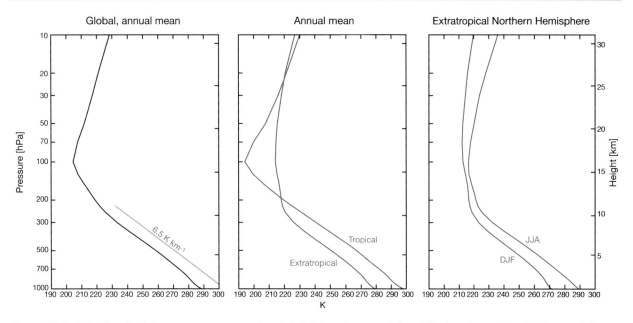

Figure 1.20 Vertical profiles of global mean temperature averaged over latitude belts and seasons as indicated. "Tropical" denotes 30°N–30°S, "Extratropical" denotes 30–90°N/S. Courtesy of Simchan Yook.

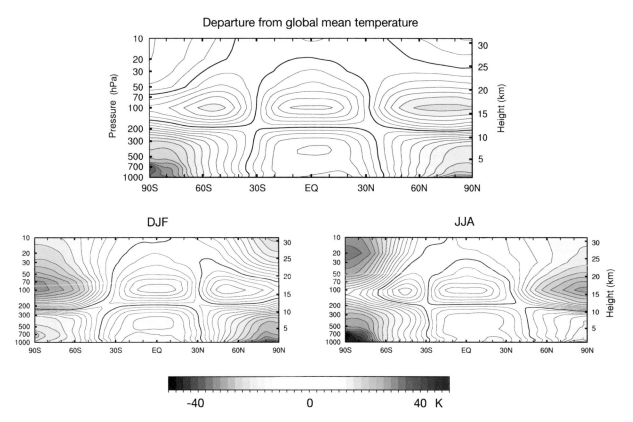

Figure 1.21 Meridional cross sections of zonally averaged temperature, departures from the global mean at each level, contour interval 2.5 K, zero contour bolded. The distinctive quadrupole pattern in temperature centered on the tropospheric jet stream reflects the constraint of thermal wind balance between the zonal wind and temperature fields. Courtesy of Simchan Yook.

stratosphere based on radio occultation measurements. At 20 km and above, it exhibits strong nonseasonal fluctuations with a period slightly in excess of 2 years, the so-called *quasi-biennial oscillation* (QBO), which will be discussed

in Sections 10.1 and 10.4. At 18 km, just above the tropical cold point, the variability is dominated by the annual cycle, with a maximum in August and a minimum in February. The amplitudes of these two phenomena are compared

$$[T]\,10°N–10°S$$

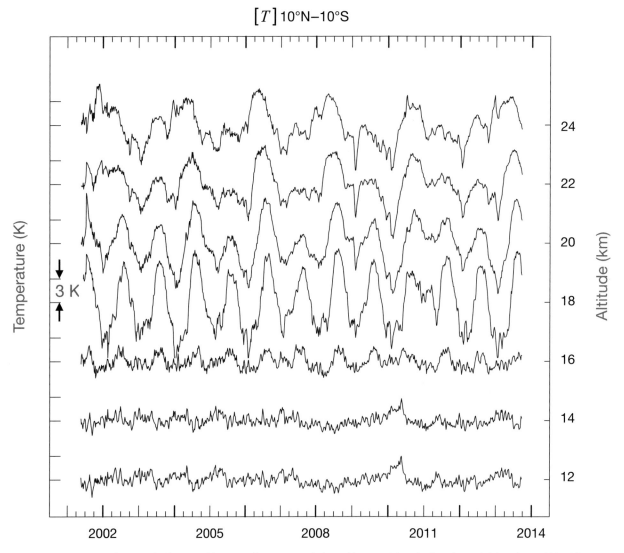

Figure 1.22 Time series at 2 km intervals of equatorial (10°N–10°S) temperature. The interval between tick marks along the y axis is 3 K. The quasi-biennial oscillation is clearly evident at the uppermost levels (22 and 24 km). At and just above the tropical cold point tropopause (~17 km) the variability is dominated by the annual cycle. From Randel and Wu (2015). © American Meteorological Society. Used with permission.

in Fig. 1.23. The existence of the QBO is evidence that the atmosphere is capable of generating near oscillatory variability, independent of the periodic forcing related to the Earth's orbit around the Sun. The pronounced annual cycle in cold point temperature is evidence that differences between the Northern and Southern Hemisphere boundary conditions exert a strong influence on the global general circulation.

The meridional cross section for potential temperature [θ], shown in Fig. 1.24, provides further insight into the structure and seasonality of the temperature field. Before describing the figure, we present the definition and a brief interpretation of potential temperature and static stability and the relationship between them for the benefit of the reader who might not already be familiar with these terms. Potential temperature is the temperature that an air parcel would

attain if it were adiabatically compressed to a reference pressure p_o:

$$\theta = T\,(p_o/p)^\kappa, \tag{1.1}$$

where T and p are the original temperature and pressure of the air parcel, κ is the ratio of the gas constant R to the specific heat c_p of dry air, which is 2/7.[2] The reference pressure p_o is usually taken as 1000 hPa. The distribution of θ is simpler than that of T because its vertical derivative does not change sign, as that of temperature does in the tropical lower stratosphere. The tropopause is clearly marked by a rapid change in its vertical gradient, $\partial\theta/\partial z$, with much higher values in the stratosphere than in the troposphere. It can be shown that the restoring force on an air parcel

[2] Holton and Hakim (2012), p. 52.

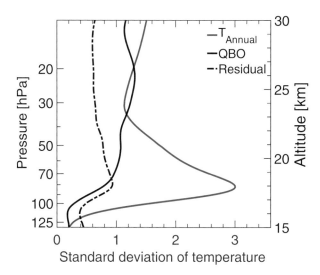

Figure 1.23 Vertical profiles of r.m.s. amplitude of zonally averaged, monthly mean temperature perturbations in units of K, averaged from 10°S to 10°N attributable to the QBO, the climatological mean annual cycle, and the residual perturbations after these two components are removed, as indicated in the legend. Courtesy of Hsiu-Hui Tseng and Qiang Fu.

that is adiabatically displaced a vertical distance δz from its equilibrium level in the atmosphere is $-N^2 \delta z$, where

$$N^2 = g \, \frac{\partial \ln\theta}{\partial z} = \frac{g}{T}(\Gamma_d - \Gamma), \qquad (1.2)$$

where $\Gamma_d \equiv g/c_p = 9.8 \text{ K km}^{-1}$ is the dry adiabatic lapse rate and Γ is the actual lapse rate.[3] N is the frequency of a buoyancy oscillation (referred to as the Brunt Väisälä frequency) and is thus a measure of the *static stability*. When pressure is used as a vertical coordinate static stability assumes the form

$$s = \left(\frac{\kappa T}{p} - \frac{\partial T}{\partial p} \right). \qquad (1.3)$$

The transition from tropospheric air, marked by weak static stability, to stratospheric air, marked by much stronger static stability, is evident at all latitudes in Fig. 1.24. Consistent with Fig. 1.20, the altitude of the tropical tropopause is about twice as high as that of the winter polar tropopause. At 200 hPa, the cruising level for commercial jet aircraft, the air on the poleward side of the jet stream is stratospheric, whereas the air on the equatorward side is tropospheric.

Figures 1.25 and 1.26 show distributions of the zonally averaged meridional and vertical wind components, $[v]$ and $[\omega]$. They are dominated by dipoles in $[v]$ with centers of opposing sign in the boundary layer and in the upper troposphere. The pattern in $[\omega]$ is dominated by a narrow "chimney" of ascent centered at $\sim 7°$N in the annual mean, flanked by broader bands of descent centered in the subtropics. The patterns of $[v]$ and $[\omega]$ are indicative of circulation cells spanning the depth of the troposphere. They satisfy the two-dimensional continuity equation

$$\frac{1}{R_E \cos\phi} \frac{\partial}{\partial \phi}([v]\cos\phi) + \frac{\partial[\omega]}{\partial p} = 0, \qquad (1.4)$$

from which it follows that they can be written as the gradient of a streamfunction ψ:

$$[v]\cos\phi = \frac{\partial\psi}{\partial p} \qquad (1.5a)$$

$$[\omega] = \frac{-1}{R_E \cos\phi} \frac{\partial\psi}{\partial\phi}. \qquad (1.5b)$$

where ϕ is latitude and R_E is the radius of the Earth. The ψ field is obtained by integrating Eq. (1.5a) upward from the Earth's surface or downward from the top of the atmosphere, subject to the boundary condition that $\psi = 0$ at the poles, at the Earth's surface, and at the top of the atmosphere.

Figure 1.27 shows the *mean meridional circulation* represented in terms of the ψ field. Values of ψ at the centers of the respective circulation cells are measures of the mass flux around those cells. If the contour interval is uniform, an equal amount of mass flows through each "channel" between pairs of adjacent contours. Hence, it is important to bear in mind that ψ is the streamfunction for the mass flux $(\rho v \cos\phi, \omega)$, not the geometric velocity (v, w) and it is thus referred to as the *mass streamfunction*. Because density decreases exponentially with height, so does the amplitude of the features in the ψ field. Above the 100 hPa level they are too weak to be discernible with the contour spacing used in Fig. 1.26. Hence, the top of the domains in Figs. 1.25, 1.26, and 1.27 is placed at 100 hPa, rather than 10 hPa. The stratospheric mean meridional circulation will be documented using other metrics.

The mean meridional circulations shown in Fig. 1.27 are superimposed upon the diabatic heating field, which takes into account radiative transfer, the release of latent heat when water vapor condenses in clouds, and the convergence of the vertical heat fluxes on horizontal scales smaller than the grid of the model used in the reanalysis. The mean meridional circulation is dominated by the tropical *Hadley cells*,[4] with an ascending branch in the equatorial belt and descending branches in the subtropics, and whose lower branch is the trade winds. The equatorward flow in the lower branch is warmed slightly by the flux of sensible heat and moistened by evaporation at the bottom boundary. The air in the rising branch is warmed by the release of latent heat in deep cumulus convection, which was acquired in the lower branch of the cell, while the air in the poleward and sinking branches is cooled by radiative transfer.

The zonally averaged Hadley cells shown in Fig. 1.27 reflect the superposition of separate features in the "western" and "eastern" hemispheres, as illustrated in Fig. 1.28 for the annual mean climatology. The former, extending westward from the Greenwich Meridian to the Date Line, is dominated by a narrow chimney centered along 7°N and only about

---

[3] Wallace and Hobbs (2006) Eqs. (3.75) and (3.76).

[4] After George Hadley, who was the first to recognize the importance of the Earth's rotation in accounting for the existence of the trade winds (Hadley, 1735).

$$[\bar{\theta}], [\bar{u}]$$

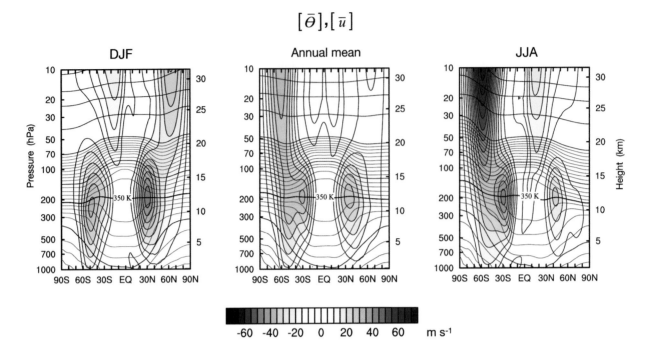

Figure 1.24 Climatological mean zonally averaged potential temperature $[\bar{\theta}]$ (contour interval 10 K, bolded contours 50 K) and zonal wind $[\bar{u}]$ (thick contours with colored shading). The vertical spacing of the contours, called *isentropes*, is a measure of the static stability. The middle panel is the annual mean. The 300, 400, 500... K isentropes are thickened. For values above 500 K only thickened contours are shown. Contour interval 5 m s^{-1} for zonal wind. The zero $[u]$ contour is omitted.

$$[\bar{v}]$$

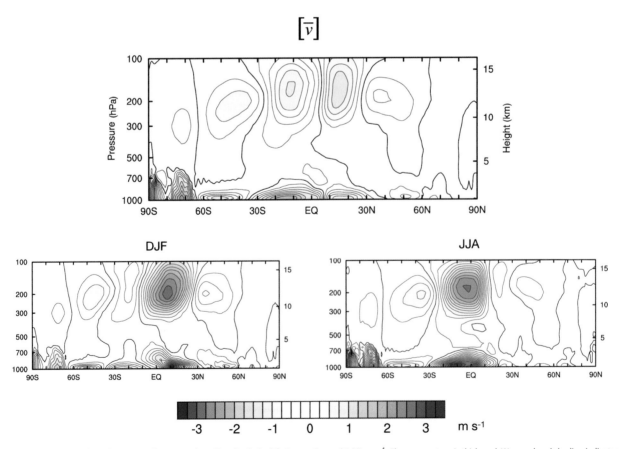

Figure 1.25 Climatological mean zonally averaged meridional velocity $[\bar{v}]$. Contour interval 0.25 m s^{-1}. The zero contour is thickened. Warm colored shading indicates northward and blue shading indicates southward flow. Note that the domain of this and the next three figures extends upward only to 100 hPa.

$[\bar{\omega}]$

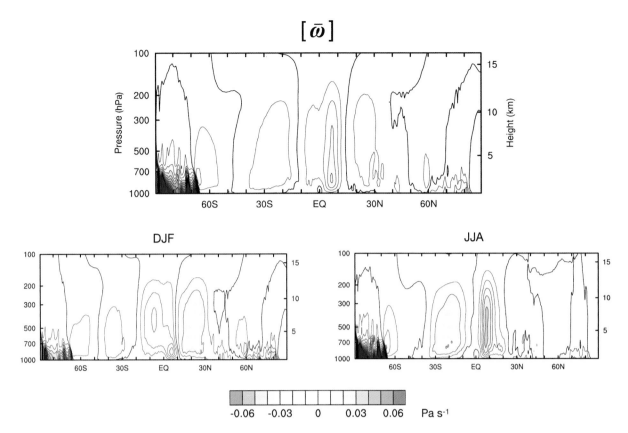

Figure 1.26 Climatological mean zonally averaged vertical velocity $[\bar{\omega}]$. Contour interval 0.01 Pa s^{-1}. The zero contour is thickened. Blue shading indicates ascent and warm colored shading indicates descent.

$\bar{\psi}$

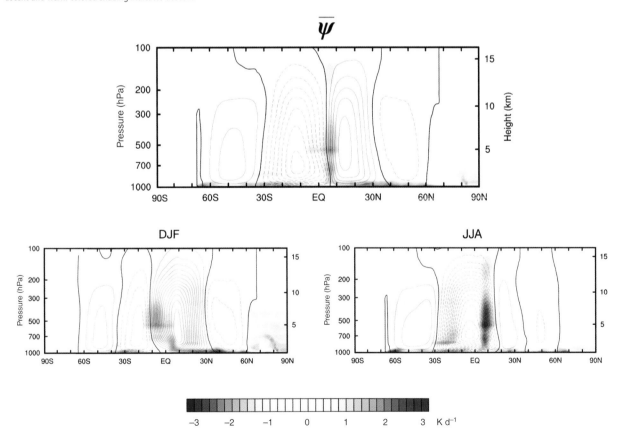

Figure 1.27 Climatological mean meridional mass streamfunction $\bar{\psi}$ (contour interval 15 $\times 10^9$ kg s^{-1} (15 Sv)) and diabatic heating rate $\bar{Q}$ in units of K per day (colored shading). The zero contour is thickened. The circulation is clockwise around the maxima and counterclockwise around the minima in the ψ field. The Hadley cells flank the band of tropical ascent, marked by the narrow maximum in Q. They are flanked by much weaker Ferrel cells centered in midlatitudes.

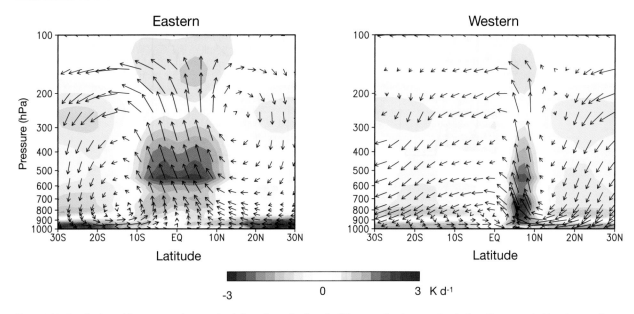

Figure 1.28 Contributions of the eastern and western hemispheres (extending from the GM eastward and westward to the Date Line, respectively) to the annual mean pattern in Fig. 1.27. The streamfunction cannot be computed for partial zonal averages, so the overturning circulation is represented by vectors representing the average of $\overline{v}$ and $\overline{\omega}$ over the respective hemispheres. The longest horizontal component of the arrows corresponds to 4 m s^{-1} and the longest vertical components corresponds to roughly 1 cm s^{-1} or 100 hPa d^{-1}. The narrow chimney of ascent in the western hemisphere corresponds to the ITCZ. Courtesy of Ying Li.

5 degrees of latitude in width, whereas the latter, extending from Greenwich eastward to the Date Line exhibits a much broader band of ascent extending almost all the way from 10°N to 10°S. The low-level convergence feeding into the regions of ascent is deeper and the level of strongest ascent is noticeably higher in the eastern hemisphere than in the western hemisphere. Comparing these features with the regional rainfall patterns shown in Figs. 1.14 and 1.15, it is evident that the narrow "chimney" of ascent in the western hemisphere corresponds to the Pacific and Atlantic ITCZs, while the more diffuse area of ascent in the eastern hemisphere reflects the superposition of the western end of the Pacific ITCZ, the SPCZ, and the belt of heavy year-round rainfall over the Maritime Continent that extends westward into the Indian Ocean south of the equator. The rising branch of the zonally averaged Hadley cell is thus an amalgam of a number of distinct regional features.

The DJF and JJA mean meridional circulation cells shown in Fig. 1.27 are about twice as strong as the annual mean cell. They straddle the equator, with ascent and diabatic heating in the summer hemisphere and cooling and descent in the winter hemisphere. Most of the mass flux in these overturning circulations is in the eastern hemisphere and is associated with the Australasian monsoon. The seasonally varying, climatological mean meridional circulation can thus be viewed as being dominated by an annual mean, equatorially symmetric component, and a stronger, seasonally reversing, equatorially antisymmetric component. The term *Hadley cell* may refer to either or to the total, depending upon the context.

The Hadley cells in the annual mean circulation are flanked by much weaker circulation cells in the opposite

sense, referred to as *Ferrel cells*,[5] with poleward flow in their lower branches, coincident with the belts of surface westerlies. Radiative cooling prevails at low levels over the polar cap regions, stronger over the Antarctic. In the global average, diabatic heating is prevalent in the boundary layer and cooling is prevalent in the free atmosphere and is particularly strong in the upper troposphere. Hence, in a mass-weighted sense, the diabatic heating is acting to warm the atmosphere at the bottom and cool it higher up, thereby raising its center of mass and reducing its static stability.

To conclude this section, Fig. 1.29 shows the zonally averaged zonal wind, the temperature departure, and mean meridional circulation fields together on the same plot. Here the circulation represents an estimate of the zonally averaged motions, not at a set of fixed grid points in the meridional plane, but following the centroids of hypothetical clouds of tagged air parcels initially located at those grid points. It is thus an estimate of the *Lagrangian* as opposed to the *Eulerian* mean meridional motions. It corresponds closely to the mean meridional motions in isentropic coordinates, that is zonally averaged along constant potential temperature (i.e., constant internal plus potential energy) surfaces rather than pressure surfaces. The methodology for making this estimate, which is referred to as the *transformed Eulerian mean* (TEM) circulation, is described in Section 8.2. Other distinctions between this representation and the one in Fig. 1.27 are: (i) the mean meridional circulation is represented by a vector field rather than as the gradient of a

---

[5] Named after William Ferrel, who showed that the Earth's rotation mediates the atmospheric circulation, not by the conservation of (zonal) momentum as postulated by Hadley (1735), but through the tendency for the conservation of angular momentum (Ferrel, 1856).

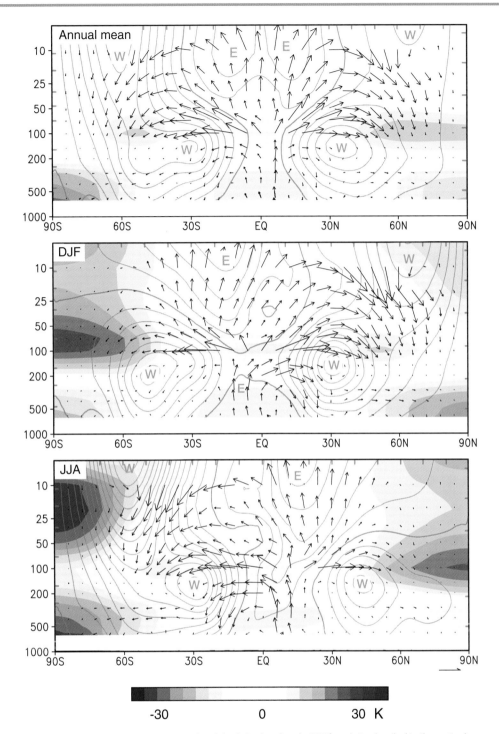

Figure 1.29 Vectors indicate the estimated Lagrangian mean meridional circulation based on the TEM formulation described in Chapter 8, whose stratospheric component is also referred to in many places in this book as the Brewer–Dobson circulation, superimposed upon the distribution of the temperature departures from the global mean at each pressure level. The colored shading indicates the temperature departure field, as in Fig. 1.21, and the contours represent the zonally averaged zonal wind. Velocities are multiplied by the cosine of latitude and the scaling of the arrows is different in the troposphere (T, 1000–150 hPa) and stratosphere (S, 100 hPa and above). In T the longest vertical arrows correspond to 1 cm s^{-1} and the longest horizontal arrows to 4 m s^{-1}; the corresponding values for S are 0.1 cm s^{-1} and 0.4 m s^{-1}. Courtesy of Ying Li.

streamfunction, and (ii) in order to make the relatively weak stratospheric mean meridional circulations more clearly visible, the vectors at 100 hPa and above are lengthened relative to the vectors below that level.

In response to the stronger equator-to-pole gradient of absorbed insolation in the winter hemisphere, the meridional temperature gradient at tropospheric levels is substantially stronger than in the summer hemisphere. The strong

$$\sigma(\overline{u}^*) \qquad\qquad \sigma(\overline{v}^*)$$

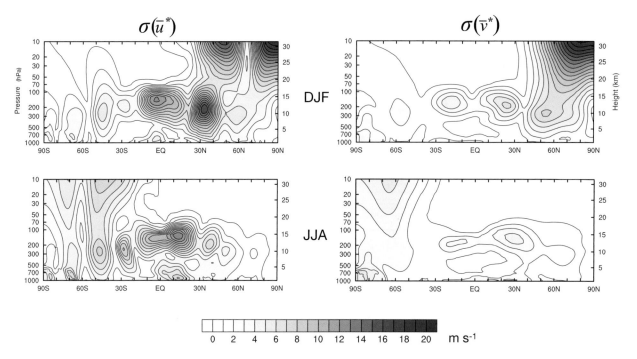

Figure 1.30 Climatological mean longitudinal standard deviations of (left) $\overline{u}^*$ and (right) $\overline{v}^*$ averaged around latitude circles in the stationary waves for (top) DJF and (bottom) JJA. The zonally average reference field for computing the standard deviation in each panel is the seasonal mean, not the annual mean.

cross-equatorial mass flux from the summer hemisphere to the winter hemisphere in the upper branch of the Hadley cell can be viewed as the zonally symmetric component of the monsoon circulations. In this representation, it is clear that mass flows from the summer to the winter hemisphere in the stratosphere as well. In this deep circulation cell most of the ascent is in the tropics and subtropics of the summer hemisphere, and the descent is in the polar cap region of the winter hemisphere. Were it not for the heating of this descending air due to adiabatic compression, the polar night region would be even colder than observed. The mass circulation in the lower stratosphere ($\sim$50 to 100 hPa) is more equatorially symmetric, with poleward flow in both winter and summer hemispheres above the tropospheric jet streams, fed by tropical upwelling, which is responsible for the remarkable coldness of the tropical lower troposphere.

A notable difference between the estimate of the Lagrangian mean meridional circulation shown in Fig. 1.29 and the Eulerian mean meridional circulation shown in Fig. 1.27 is that in the former there are no Ferrel cells in the midlatitude troposphere. At all but the lowest latitudes, the mass flux in the upper troposphere is poleward and it is stronger in the winter hemisphere than in the summer hemisphere. The Lagrangian climatological mean circulation that ventilates the stratosphere – upward at low latitudes, meridionally directed toward the winter pole, and downward in the wintertime polar night region – is referred to as the Brewer–Dobson circulation because long before reliable stratospheric wind measurements became available, its existence was inferred on the basis of Brewer's measurements of stratospheric water vapor and Dobson's measurements of stratospheric ozone.

1.4.3 Eddies and Transients

As noted at the beginning of this chapter, departures from zonal means are referred to as *eddies* and designated by asterisks; for example, the T^* field is the *eddy component* of the T field: $T^* = T - [T]$. In a similar manner, departures from time means are referred to as *transients* and designated by primes ( )'; for example, the T' field is the *transient component* of the T field: $T' = T - \overline{T}$, where $\overline{T}$ is the climatological (time) mean temperature. Using these definitions, we can decompose a field x that varies with longitude and time into mean and departure terms as follows:

$$x = \underbrace{[\overline{x}]}_{\text{zonal/time mean}} + \underbrace{\overline{x}^*}_{\text{stationary wave}}$$
$$+ \underbrace{[x]'}_{\text{transient zonal mean}} + \underbrace{x'^*}_{\text{transient eddy}} . \qquad (1.6)$$

The first term on the right-hand side represents the average over both time and longitude. The second represents the departure of the time mean at that grid point from $[\overline{x}]$, the third represents the departure of the zonal mean at that time from $[\overline{x}]$, and the fourth term represents the residual (i.e., the departure from both time and zonal means). In Appendix A, it is shown that the same four-term decomposition can be applied to variances and to covariances. In this section, we will show standard deviations, denoted by the symbol σ, a measure of the amplitude of the departure fields in each of these categories.

Stationary Waves
The top panels of Fig. 1.30 provide a global survey of the amplitude of the DJF stationary waves in the climatological

Figure 1.31 Temporal standard deviations of (top) $[u]'$ and (bottom) $[v]'$ based on daily data for all calendar months. The climatological mean reference fields used in computing the variance are smoothly seasonally varying, so the variability represented here is *nonseasonal*. Note the difference in scales between top and bottom panels.

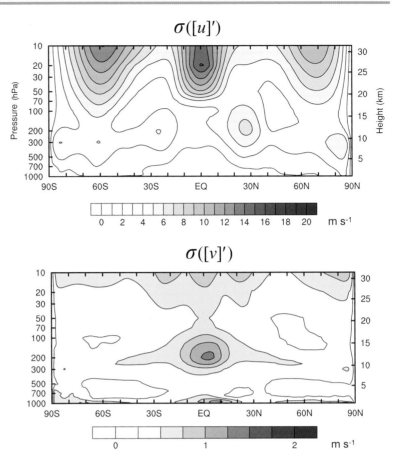

$\sigma([u]')$

$\sigma([v]')$

mean zonal and meridional wind components $\bar{u}^*$, and $\bar{v}^*$ on latitude circles. The meanders in the 200 hPa height contours in Figs. 1.4 and 1.5 over high latitudes of the Northern Hemisphere in winter are reflected in the maximum in $\sigma(\bar{v}^*)$ at 52°N, and the zonally localized westerly jets at the base of the wave troughs in Z_{200} contribute to the maximum in $\sigma(\bar{u}^*)$ around 32°N. The prominent equatorial maximum in $\sigma(\bar{u}^*)$, flanked by much weaker $\sigma(\bar{v}^*)$ maxima near 20°N/S, is the signature of the equatorial planetary waves identified in Figs. 1.4 and 1.5. Their upper tropospheric maxima are centered at the 150 hPa level, substantially higher than the corresponding features in the extratropical stationary waves. Also evident in Fig. 1.30 is a winter subpolar stratospheric maximum in $\sigma(\bar{v}^*)$ flanked by $\sigma(\bar{u}^*)$ maxima. These features are signatures of geographically anchored, meanders of the wintertime polar night jet.

The corresponding pattern for the JJA stationary waves are shown in the bottom panels of Fig. 1.30. The extratropical stationary waves are substantially weaker in the summer hemispheres than in the winter hemispheres, particularly in the Northern Hemisphere. The tropical features are not as equatorially symmetric as they are in DJF: the most prominent features are the maxima in $\sigma(\bar{u}^*)$ at ~150 hPa and in the lower troposphere, which are mainly attributable to the Australasian summer monsoon. The Southern Hemisphere polar night jet exhibits zonal asymmetries analogous to those in the Northern Hemisphere winter, albeit much weaker.

Transient Zonally Symmetric Variability

Seasonal variations in the climatology of the zonally symmetric flow were documented in Sections 1.4.1 and 1.4.2. Large nonseasonal variability is also observed, particularly in the zonal wind component. Figure 1.31 shows the standard deviations of $[u]'$ and $[v]'$ about their respective seasonally varying climatological means. They are computed on the basis of daily data and thus include variability on timescales ranging from days to years, but since the reference state is seasonally varying, it specifically excludes the variance associated with seasonality. Note that in general, $\sigma([u]') \gg \sigma([v]')$: just as the kinetic energy of the climatological mean zonally symmetric circulation is dominated by the zonal wind component, the same is true of the variations about that climatology. That $\sigma([v]')$ is so small reflects the inhibition of the meridional motion of zonally symmetric rings of air on a rotating planet, a constraint imposed by the conservation of angular momentum. The most prominent feature in $\sigma([u]')$ is the maximum in the equatorial stratosphere with an r.m.s. amplitude of ~20 m s^{-1}, which is associated with the QBO. The equatorial stratospheric maximum is flanked by maxima centered 60°N/S, which closely match the location of the wintertime polar night jets in Fig. 1.19. By consulting the corresponding figures on the companion web page, the reader can verify that the QBO is present year-round, whereas the high latitude maxima reflect the variability observed during specific seasons: the

$$\sigma(v')$$

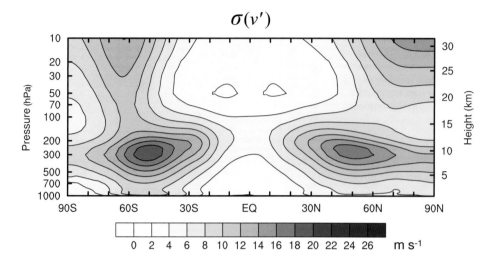

Figure 1.32 Zonally averaged temporal standard deviation of meridional wind component v based on daily data, referred to as *the transients*, for all calendar months, computed at each grid point and level. It is argued in the text that the temporal standard deviations about the seasonally varying climatological means $\sigma(v^*)$ and $\sigma(v'^*)$ are virtually indistinguishable, from which it follows that these patterns can be interpreted as representative of the transient eddies. The maxima in the variance at temperate latitudes are associated with the *storm tracks* in Fig. 1.34. Transient eddy kinetic energy (not shown) exhibits a qualitatively similar pattern.

Northern Hemisphere variance maximum in boreal winter (DJF) and the Southern Hemisphere maximum in the austral spring (SON). The other feature that stands out above the background continuum is the pair of maxima at $\sim 28°$N/S, just equatorward of the climatological mean tropospheric jet streams. We will show in Chapter 17 that this feature is the signature of the El Niño-Southern Oscillation (ENSO) phenomenon.

The only notable feature in the distribution of $\sigma([v]')$ is the pair of equatorial maxima, one in the boundary layer and the other just above the 200 hPa level. Anomalies in $[v]$ at these levels tend to occur out-of-phase with one another (not shown). These features are associated with nonseasonal meridional shifts in the Hadley cell that mirror the seasonal shifts shown in Fig. 1.27, but occur on a wide range of timescales.

Transient Eddies

After the variance associated with the stationary waves and the nonseasonal zonally symmetric variability is accounted for, there remains the variability associated with the transient eddies, which are neither anchored in place longitudinally nor zonally symmetric. It is the only kind of eddy-related variability that appears on a rotating planet with zonally symmetric, temporally invariant boundary conditions and external forcing. The distinction between (all) *transients* and *transient eddies* is subtle and has often been ignored in the literature. For example, in extratropical latitudes, where $[v] < 1$ m s^{-1} the *transient* and *transient eddy* components of the v field are virtually indistinguishable. Accordingly, we will denote the *transient* eddy component of v simply as v' rather than v'^*. Notable examples are extratropical cyclones and the so-called *baroclinic waves* in which they are embedded.

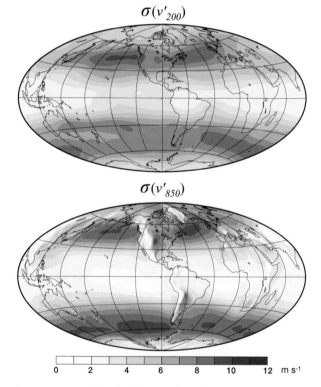

Figure 1.33 Temporal standard deviation of v' at the 200 hPa (top) and 850 hPa (bottom) levels based on year-round data. Note the signature of the mountain ranges in the bottom panel.

The meridional cross section of $\sigma(v')$ shown in Fig. 1.32, is dominated by broad maxima centered at $\sim 50°$N/S, just above the 300 hPa level. The transient eddies are stronger than the stationary waves (Fig. 1.30) by about a factor of two. At higher levels in the stratosphere, the variability

$$\sigma\,(\omega'_{700})$$

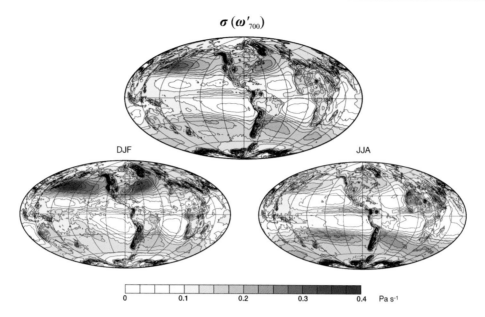

Figure 1.34 Temporal standard deviation of ω'_{700}. The maxima in temperate latitudes correspond to the oceanic storm tracks.

peaks at higher latitudes, around the periphery of the polar cap regions, where it is roughly comparable to that associated with the stationary waves at the same level. The corresponding distributions of zonally varying $\sigma(v')$ at the 200 hPa level (Fig. 1.33, top panel) confirm that the upper tropospheric, midlatitude maximum in transient eddy amplitude is a robust feature that is present at almost all longitudes. We will refer to this feature as the midlatitude *"storm tracks."* At the 850 hPa level (Fig. 1.33, bottom panel), the variability is markedly suppressed to the west of the Rocky and Andes Mountains and enhanced to the east of them.

In contrast to the rather amorphous patterns in $\sigma(v')$, the corresponding pattern for $\sigma(\omega'_{700})$, shown in Fig. 1.34, is more structured, with zonally elongated maxima over the western Pacific and Atlantic in the midlatitude storm tracks, as evidenced by their close correspondence with the preferred tracks of migrating cyclones and anticyclones.[6] It will be shown in Section 12.2 that these features are associated with baroclinic waves, which dominate the day-to-day variability of v, T, and Z at tropospheric levels, but in order to see their signature in fields of those variables it is necessary to highpass filter the data to remove the signatures of more slowly evolving phenomena. The elongated bands of enhanced precipitation over the extratropical North Pacific and North Atlantic and parts of the Southern Ocean in Figs. 1.14 and 1.15 are rectified signatures of the episodes heavy rain rate associated with the passage of the rainbands in extratropical cyclones in the storm tracks.

Exercises

1.2 Why is it necessary to insert extra contours in the tropics in Figs. 1.3, 1.4, and 1.5?

1.3 Explore the resources available to the community in the online *Visualization and Analysis Tools* on the companion web page. View instantaneous maps for a selection of variables and explore how the fields evolve over the course of a few days. Compare the instantaneous maps with the corresponding annual or seasonal mean climatologies.

1.4 The meridional extent of the meanders in the flow in the stationary waves embedded in a background zonal flow is inversely proportional to the background wind speed. Explain.

1.5 Are any of the features in the climatology of the Earth's skin temperature shown in Figs. 1.6–1.7 explainable without reference to the general circulation?

1.6 Contrast the annual mean sea level pressure and surface wind fields shown in Figs. 1.12 and 1.13 with their counterparts on an ocean-covered planet.

1.7 In most of the pole-to-pole cross sections of general circulation statistics presented in this book, the horizontal and vertical coordinates are latitude on a linear scale and log pressure, respectively. (a) For what vertical coordinate is the vertical spacing between levels proportional to the fraction of the mass of the atmosphere contained within the layer? What would be the disadvantage of using this as a vertical coordinate? (b) For what meridional coordinate do equal increments of distance along the x axis correspond to equal areas on the Earth's

[6] Klein (1957); Hoskins and Hodges (2002).

surface? In this coordinate system, what latitude lies halfway between equator and pole?

1.8   Make a rough (order of magnitude) comparison between the kinetic energies associated with the zonally symmetric zonal and mean meridional motions in the general circulation.

1.9   Compare the atmospheric mass transport in the annual mean Hadley cell with the transport of ocean water in the Gulf Stream, expressing both in units of sverdrups (Sv).

1.10   Compare the mass transport in the annual mean Hadley cell with that in the solstitial seasons DJF and JJA making use of the streamfunction fields in Fig. 1.27.

1.11   Why is the zonally averaged cross-equatorial overturning circulation stronger in JJA than in DJF?

1.12   Why can't the mean meridional circulation shown in Fig. 1.28 be represented in terms of a mass streamfunction, as in Fig. 1.27?

1.13   (a) Making use of the accompanying schematic of an idealized wave, explain why the zonal average, $\sigma(v^*)$ exhibits a single maximum, midway between equator and pole, flanked by a pair of maxima in $\sigma(u^*)$. (b) Relate your explanation to the extratropical stationary wave statistics shown in Fig. 1.30.

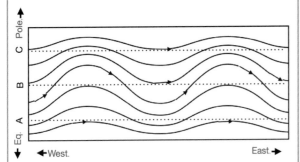

Figure for Exercise 1.13.

1.14   (a) If one had access to only irregularly spaced station data for v, why would it be easier to compute $\overline{[v'^2]}$ than $\overline{[v^{*2}]}$? (b) Making use of the formalism in Appendix A, show that the true transient eddy contribution $\overline{[v^{*'2}]}$ is equal to $\overline{[v'^2]} - \overline{[v]'^2}$. Based on the figures shown in this section, how might one justify neglecting $\overline{[v]'^2}$ in a study of the extratropical general circulation?

1.15   Based on the plots in the text and on the online mini-atlas, compare the amplitudes of the stationary waves in the upper tropospheric circulation in the winter and summer hemispheres.

1.16   Based on the data shown in Figs. 1.30 and 1.32, compare the transient eddy and stationary wave contributions to the kinetic energy (a) in the equatorial upper troposphere and (b) in the extratropical upper troposphere.

1.5 NUMERICAL SIMULATION OF THE ATMOSPHERIC CIRCULATION

The models used in atmospheric dynamics are based on the Navier–Stokes equations, the foundation of all of fluid dynamics; six equations that embody the conservation of momentum, energy, and mass, and the equation of state for an ideal gas, involving six dependent variables: the three velocity components, temperature, density, and pressure. The five conservation equations are said to be *prognostic equations* because they involve time derivatives. In a rotating coordinate system, the Navier–Stokes equations are capable of simulating three different families of waves:

- *acoustic waves* that exist by virtue of the compressibility of air,
- *gravity waves* and *inertio-gravity waves* that exist by virtue of the ambient static stability (the qualifier "inertio" applies to gravity waves with frequencies low enough so that oscillating air parcels feel the influence of the Earth's rotation),
- *Rossby waves* that exist by virtue of the meridional gradient of the Coriolis parameter $f = 2\Omega \cos \phi$, where Ω is the rotation rate of the Earth,

plus two "mavericks": the equatorial *Kelvin wave*, a gravity wave whose zonal wind component is in geostrophic balance with the pressure field, and the *mixed Rossby–gravity* (MRG) wave, which behaves as one or the other depending upon whether it is propagating westward or eastward. The three families of waves are clearly distinguishable in terms of their phase speeds, acoustic waves being the fastest and Rossby waves the slowest, but they exhibit a wide and overlapping range of frequencies and horizontal wavelengths.

Another way of characterizing the motions is by horizontal scale, *planetary scale* on the order of tens of thousands of kilometers, *synoptic scale* on the order of several hundreds of kilometers, and *mesoscale* 100 km or less. Rossby waves, which dominate the variability on planetary and synoptic scales, include synoptic scale baroclinic waves, planetary-scale extratropical stationary waves forced by orography and land–sea thermal contrasts, and a number of other phenomena that will be discussed in subsequent chapters. Planetary-scale Kelvin, mixed Rossby–gravity waves, and inertio-gravity waves play an important role in the tropical general circulation. Inertio-gravity waves are detectable at all latitudes, the fastest of them propagating at the rate of the speed of sound.

1.5.1 Filtering of the Navier–Stokes Equations

To simplify the equations and reduce the computations required in numerical simulations of the global circulation, the Navier–Stokes equations are usually simplified by treating certain of the dependent variables as if they were in a state of balance, thereby converting one of the prognostic equations to a diagnostic equation. The first level of simplification is to assume that the atmosphere is in *hydrostatic balance* (i.e., that the vertical pressure gradient is balanced by the acceleration due to gravity)

$$\frac{\partial p}{\partial z} = -\rho g,$$ (1.7)

where p is pressure, z is height, ρ is density, and g is the acceleration of gravity. Invoking hydrostatic balance is equivalent to assuming that the pressure at any level of the atmosphere is determined entirely by the weight of the overlying air, in effect, ignoring the role of buoyancy-induced vertical accelerations. Writing the equations in isobaric coordinates and using the ideal gas law $p = \rho RT$, Eq. (1.7) becomes

$$\frac{\partial \Phi}{\partial \ln p} = -R T,$$ (1.8)

where Φ is geopotential on a constant pressure surface, T is temperature, and R is the ideal gas constant for dry air. Equation (1.8) is referred to as the *hypsometric equation*. The resulting simplified set of equations, referred to as the *primitive equations*, is capable of simulating Rossby waves and inertio-gravity waves, but not the more rapidly propagating acoustic waves.

Another of the Navier–Stokes prognostic equations can be replaced by a diagnostic equation if it is assumed that the horizontal wind field **V** on geopotential surfaces is to first order in *geostrophic balance* whereby the horizontal component of the pressure gradient force is balanced by the Coriolis force:

$$-\frac{1}{\rho_o}\nabla p = f\,\mathbf{k} \times \mathbf{V}_g,$$ (1.9)

where ∇ is the horizontal gradient operator, ρ_o the mean state (reference) density, f is the Coriolis parameter, **k** is the unit vector in the vertical, and $\mathbf{V}_g$ is defined as the geostrophic wind. Invoking geostrophic balance is equivalent to assuming that the horizontal wind field at any level of the atmosphere is determined entirely by the horizontal pressure gradient in the direction transverse to the velocity vector, in effect, ignoring the role of pressure gradient force in accelerating the flow as it does in gravity waves. In isobaric coordinates, Eq. (1.9) assumes the form

$$-\nabla\Phi = f\,\mathbf{k} \times \mathbf{V}_g,$$ (1.10)

When combined together with the prognostic equation for the geostrophic vorticity $\zeta_g \equiv \mathbf{k}\cdot\nabla \times \mathbf{V}_g$, Eqs. (1.7) and (1.9) (or Eqs. (1.8) and (1.10)) comprise a simplified set of equations, referred to as the *quasi-geostrophic formulation* of the primitive equations, which is capable of simulating Rossby waves, but not the more rapidly propagating gravity waves and acoustic waves. Geostrophic balance is always

invoked in combination with hydrostatic balance. The two in combination are referred to as *thermal wind balance*

$$\frac{\partial \mathbf{V}_g}{\partial \ln p} = -\frac{R}{f}\mathbf{k} \times \nabla T.$$ (1.11)

This so-called *thermal wind equation* relates the vertical gradient (or *shear*) of the horizontal wind field at any given level to the transverse horizontal temperature gradient at that level. The distinctive quadrupole patterns (colored shading) centered on the westerly jet streams (contours) in Fig. 1.29 serve to illustrate this relationship. It follows that the zonally symmetric zonal wind and temperature fields cannot be fully understood by considering them independently: they are different expressions of the same balanced circulation. In subsequent chapters, we will show that the mean meridional circulations are instrumental in maintaining thermal wind balance.

The hydrostatic approximation inherent in the primitive equations eliminates the very high frequency, fast propagating acoustic waves and the further approximation that the horizontal wind field is nearly in geostrophic balance eliminates gravity and inertio-gravity waves. The "filtered" wind field retains the more slowly evolving Rossby waves, which account for most of the kinetic energy. Including the necessary reduction in the time-stepping, 16 times the computational resources are needed to run the same climate simulation or weather forecast with a model that has twice the resolution in all three spatial dimensions. Hence removing the acoustic waves through the use of a primitive equation model greatly reduces the computational resources required, and removing the inertio-gravity waves through the use of a quasi-geostrophic model reduces them further. Without these simplifications, the development of numerical weather prediction and the models used to make extended numerical simulation of the global general circulation could not have started nearly as early and proceeded as rapidly as it did.

1.5.2 The Types of Models

The types of models used in studies of the atmospheric general circulation can be categorized as one-, two-, or three-dimensional.

- 1-D models are not really dynamical but they are sometimes helpful in understanding the observed vertical profiles of temperature and moisture, which exert a strong influence upon the dynamics. An example will be discussed in Section 2.1.
- 2-D models are of two types: those based on the governing equations that determine the evolution of zonally averaged fields of zonal wind, temperature, and other variables, and those in which the variations are treated as separable functions of the horizontal coordinates (x, y) and the vertical coordinate. It is possible to model the latter using a system of equations developed for studying shallow water waves. Included in this category are barotropic structures

with no density perturbations and a more general class of structures, referred to as *equivalent barotropic*, in which the temperature field has the same horizontal spatial structure as the pressure or geopotential height field.

- 3-D models are generally referred to as *baroclinic models*. Included in this category are primitive equation and quasi-geostrophic models.

Any of the types of models mentioned above may be *linearized* to study the evolution of infinitely small perturbations about a prescribed basic state; for example, to determine whether they will evolve into a modal form that will amplify or decay with time.

1.5.3 General Circulation Models

Atmospheric general circulation models (GCMs) simulate both dynamical and physical processes. The various fields are represented by their values on a three-dimensional grid spanning the entire globe. The latent heat released within the atmosphere by the condensation of water vapor, the fluxes of heat, water vapor, and momentum at the bottom boundary, and other subgrid scale processes are "parameterized" (i.e., represented as functions of the gridded variables based on an understanding of the processes). Model parameterizations of subgrid scale processes play a more important role in GCMs than they do in models used in weather prediction, because the gridded fields are not constrained by observations and thus will drift toward an unrealistic climate unless the parameterizations are realistic. The quest for more realistic parametrization schemes has driven much of the process-oriented research in the atmospheric sciences, including large field experiments.

The first GCM was based on the quasi-geostrophic formulation of the primitive equations.[7] It consisted of only two layers and it did not include moisture. Nevertheless, it was able to simulate many of the features of the extratropical circulation documented in Section 1.4, such as the tropospheric jet stream, the surface westerlies, and baroclinic waves. About 10 years later, scientists at the Geophysical Fluid Dynamics Laboratory (GFDL) published a series of papers presenting results based on a nine-layer primitive equation model with a rudimentary hydrologic cycle that was capable of simulating many aspects of the observed general circulation with a surprising degree of fidelity, including the broad outlines of the tropical circulation. The models included a very simple parameterization scheme, referred to as *convective adjustment*, for representing the upward heat and moisture transfer by convection and the release of the latent heat of condensation of water vapor. The related notion of *radiative–convective equilibrium* provides a simple explanation of why Earth-like planets exhibit relatively well-mixed tropospheres capped by well-defined tropopauses, as documented in Fig. 1.20 and discussed further in Section 2.1.

The performance of GCMs has improved dramatically since the 1960s, paralleling the advances in state-of-the-art NWP models. The model development has involved a progression

- from lower toward higher resolution models,
- from quasi-geostrophic to primitive equation models,
- from dry models and, recently, to non-hydrostatic models that incorporate the hydrologic cycle,
- toward more realistic treatment of radiative transfer and parameterizations of subgrid scale physical processes,
- toward models that extend higher into the stratosphere and mesosphere,
- toward models that include prognostic equations for the concentrations of selected trace gas species and serials, taking into account chemical transformations and aerosol physics.

During the 1990s, when the understanding of human-induced climate change emerged as a prominent, if not the predominant justification for general circulation research, GCMs morphed into climate models or Earth System models in which the atmosphere is coupled to the ocean, the cryosphere, and the marine and terrestrial biospheres.

The Atmospheric Model Intercomparison Project (AMIP) and the Climate Model Intercomparison Project (CMIP) are currently providing gridded datasets generated in coordinated sets of simulations run with a suite of models to study the sensitivity of the general circulation (and climate) to factors such as seasonally varying insolation, concentrations of radiatively active trace gas species and aerosols, distributions of sea surface temperature, sea ice cover, and terrestrial vegetation, and the various parametrization schemes used in the models. Datasets generated by GCMs are also increasingly being used as surrogates for observational data in studies of climate variations on timescales of a decade or longer, for which the number of statistical degrees of freedom in the historical record is limited.

Exercises

1.17 Note that westerly jet streams in Fig. 1.29 are attended by quadrupole patterns in $[\widehat{T}]$. Show that this is generally true of westerly (and easterly) jet streams, provided that the zonally averaged wind and temperature fields are in thermal wind balance.

1.18 Show that the quadupoles centered around the westerly jets in the temperature departure field in Fig. 1.29 are consistent with the pinching together of the isentropes poleward of the westerly jet streams in Fig. 1.24.

1.19 On Earth, the strongest baroclinicity (near-surface temperature gradient) is located about 10 degrees of latitude poleward of the strongest vertical wind shear. Reconcile this poleward displacement with the thermal wind equation.

[7] Phillips (1956).

2 Heuristic Models of the General Circulation

This chapter introduces some of the fundamental concepts that underlie our understanding of the general circulation of planetary atmospheres:

- radiative–convective equilibrium,
- a mechanical energy cycle,
- a thermodynamic heat engine,
- stratification – how it develops and why it matters,
- the dynamical response to horizontal and vertical heating gradients,
- the influence of rotation,
- the far-reaching effects of frictional drag.

The discussion is framed in terms of a set of experiments, most of which have been or conceivably could be performed in a laboratory or with a numerical model.

2.1 RADIATIVE–CONVECTIVE EQUILIBRIUM

In an atmosphere in pure *radiative equilibrium*, at each level the net emission of infrared radiation equals the absorbed solar radiation so that energy is neither gained nor lost. Where processes other than radiative transfer come into play, it is still possible to define a *radiative equilibrium temperature*. If the observed temperature departs from radiative equilibrium, the radiative heating or cooling will be in the sense as to bring it into balance.

The Earth's atmosphere is relatively transparent to incoming solar radiation, so most of the incoming solar energy is absorbed at the surface. If it were in pure radiative equilibrium, the lapse rate $\Gamma = -dT/dz$ would have to be very steep just above the surface. The solution of Exercise (2.1) shows that in the limit of weak emissivity, the surface temperature would be the same as the equivalent blackbody temperature $T_E = 255$ K and the entire atmosphere would be isothermal at a temperature of $T_E(1/2)^{1/4} = 214$ K. Owing to the presence of greenhouse gases, the Earth's radiative equilibrium temperature decreases monotonically from $\sim$333 K at the surface to $\sim$190 K at the 10 km level, as shown in Fig. 2.1. Below 10 km the implied lapse rate exceeds the critical value for convective instability. Pure radiative equilibrium is thus not an achievable state.

In simple one-dimensional models the effect of convection can be taken into account by limiting the lapse rate to a prescribed maximum value, which is equivalent to cooling the surface and the air just above it and warming the air in the upper part of the convectively unstable layer. Reducing the lapse rate is thus a simple way of prescribing an upward "convective" transfer of energy. In this so-called *convective adjustment scheme*, at each level the combined upward transfer of energy by radiation and convection in the adjusted profile is equivalent to the transfer by radiation alone in the unadjusted profile (apart from a small correction due to the redistribution of water vapor). It is applied to the layer extending upward from the surface to the top of the convectively unstable layer in the radiative equilibrium temperature profile. The adjusted temperature profile is referred to as being in *radiative–convective equilibrium*. Two variants of a convectively adjusted temperature profile are shown in Fig. 2.1: one based on the dry adiabatic lapse rate 9.8 K km^{-1} and the other designed to mimic the observed global-mean tropospheric lapse rate of 6.5 K km^{-1}. The latter closely resembles the observed global mean temperature profile shown in the left panel of Fig. 1.20.

Such simple radiative–convective equilibrium calculations, pioneered by Manabe and Strickler (1964), provide deep insight into the temperature profiles of planetary and stellar atmospheres and they are fundamental to understanding the sensitivity of the Earth's atmosphere to the changes in greenhouse gas concentrations in the distant past and in the immediate future. However, they fall short of providing a complete understanding because they do not take into account the role of large-scale fluid motions, which affect not only the regional expressions of climate and climate variability, but even global mean temperature itself.

Exercise

2.1 Consider an isothermal atmosphere with emissivity ϵ. Prove that in the limit as $\epsilon \to 0$, the temperature of the surface of a planet approaches the planet's equivalent blackbody temperature T_E and the temperature of its atmosphere approaches $T_E(1/2)^{1/4}$. The atmospheric temperature is sometimes referred to as the "skin temperature",[1] not to be confused with the quantity relating to the satellite-derived temperature of the land and ocean surface in Figs. 1.6–1.7.

---

[1] Goody and Walker (1972).

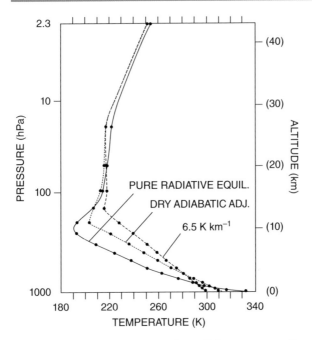

Figure 2.1 Radiative and radiative–convective equilibrium temperature profiles for an Earth-like planet. See the text for further explanation. From Manabe and Strickler (1964). Published (1964) by the American Meteorological Society.

2.2 THE CYCLING OF MECHANICAL ENERGY

To understand how large-scale fluid motions arise in the first place, it is useful to consider a simple thought experiment that could easily be performed in the laboratory. The left panel of Fig. 2.2 shows a tank filled with equal volumes of two homogeneous, immiscible liquids of differing densities, placed side by side and separated by a movable partition. The shaded liquid on the left is more dense than the unshaded liquid. In the middle panel, the partition has been removed and fluid motion has developed. We will not be concerned about the details of the motion, but only the final state after friction has brought the liquids to rest. In the new equilibrium configuration, shown in the right panel, the heavy fluid occupies the bottom half of the tank.

Now let us consider this sequence of events from the point of view of the conservation of *mechanical energy* – the sum of the potential energy and the kinetic energy. In the initial configuration, the center of mass of the fluid, denoted by the dot, is exactly halfway between the top and bottom of the tank and slightly displaced toward the heavier fluid. After the partition is removed, the center of mass drops as the denser liquid slides under the lighter one. Through the sinking of denser liquid and the rising of lighter liquid, potential energy is converted to the kinetic energy of fluid motions and the system as a whole becomes stably stratified. Frictional dissipation eventually converts all the fluid motions to random molecular motions so that, in the final state, the only evidence of the conversion that took place is the drop in the center of mass of the system and a very slight increase in the temperatures (or internal energy) of the liquids.

The energy transformations in this experiment are summarized schematically in Fig. 2.3. Note that only a small fraction of the potential energy inherent in the initial state is available for conversion to kinetic energy, since no matter what kind of motions develop, the center of mass cannot possibly come into equilibrium below the level indicated in the right panel of Fig. 2.2.

Atmospheric motions on virtually all scales contribute to the rising of warmer, more buoyant fluid and the sinking of colder, denser air that characterizes the release of potential energy and, in this sense, drives the general circulation. In the absence of rotation, the strengths of the horizontal and vertical motions would scale in accordance with the aspect ratio of the overturning circulations. Owing to the high static stability of the stratosphere, strong ascent and descent are confined to the troposphere, which is of order 10 km deep. It follows that on horizontal scales in excess of $\sim$100 km, virtually all the kinetic energy resides in the horizontal component of the wind field. The remainder of the release of potential energy is released in convective cells on scales ranging from microscale turbulence just above the Earth's surface to deep cumulus convection cells extending through the entire depth of the troposphere. Such nonhydrostatic motions are not resolved explicitly in climate global models: they are parameterized in terms of the resolved fields.

Exercises

2.2 In the experiment described in this section, suppose that the depth of the tank is 1 m, the two fluids have specific gravities of 1.1 and 0.9, and the specific heat of both fluids is 4×10^3 J kg^{-1} K^{-1}. What is the maximum possible root mean square (r.m.s.) velocity of fluid motion that can be realized in the experiment? How much does the mean temperature of the fluid increase as a result of frictional dissipation? [Assume that the r.m.s. velocity and the temperature rise are the same for the two fluids.]

2.3 By how much would the center of mass of the atmosphere have to drop in order to release enough potential energy to account for the observed kinetic energy of the general circulation? As a rough estimate of the kinetic energy, assume a root mean squared wind speed of 17 m s^{-1}.

2.4 What processes contribute to the irreversible mixing of denser and lighter air in the atmospheric boundary layer?

2.5 Consider an experiment starting with a tank of water at rest with saline water at the bottom and fresh water on the top. The tank is mechanically stirred until the salt is uniformly mixed. Describe the energy source and the energy transformations that take place during this experiment.

2.6 Suppose that the liquids in the experiment described in this section are miscible. How does the kinetic energy that can be realized when the

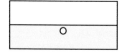

Figure 2.2 (Left) Heavier (shaded) and lighter (unshaded) fluids separated by a movable partition. The small circle represents the center of mass of the two-fluid system. (Middle) Fluids in motion following the removal of the partition. The center of mass is dropping. (Right) Equilibrium configuration of the fluids after the motion has been dissipated by friction.

Figure 2.3 Flow chart for the mechanical energy cycle relevant to the experiment in Fig. 2.2.

partition between the fluids is removed depend upon the degree to which the two fluids mix?

2.7 Suppose that the liquids in Fig. 2.2 are fresh water and salt water at the same temperature and that the partition is replaced by a porous partition that allows fluid molecules to pass through it in random molecular motions but not fluid motions. How will the outcome be different?

2.3 STEADY MOTIONS DRIVEN BY HEATING GRADIENTS

In the real atmosphere, potential energy is constantly being replenished by diabatic heating so that there is continuous cycling of energy through the reservoirs indicated in Fig. 2.3. Here we will consider a hypothetical "steady-state" laboratory analog of the general circulation that mimics this behavior. Such an experiment would not be as straightforward to design or execute as the one considered in the previous section. The right panel of Fig. 2.4 shows a tank full of liquid, which has internal heat sources along the bottom and along the left wall and a distributed heat sink in the interior, which mimic the dominant features in the observed distribution of diabatic heating rate. This configuration is inspired by the observations shown in the left panel, which were extracted from Fig. 1.27. The liquid expands with increasing temperature. The gradient of heating drives a slow, clockwise circulation cell ABC that mimics the observed Hadley cell. An important distinction that should be kept in mind, however, is that in this experiment, the distribution of heating and cooling is prescribed, whereas in the atmosphere it depends on the circulation.

As a parcel of fluid is carried around the tank in this cell, it warms as it passes along the bottom of the tank AB and as it ascends along the left wall BC. Then it cools slowly as it moves toward the right and sinks to complete the circuit. Consistent with these temperature changes, the isotherms must slope from lower left to upper right, as indicated by

the colored shading in the right panel. Since temperature increases and density decreases with height, the fluid in the tank is stably stratified.

At any given level in the tank, less dense fluid is rising along the left-hand side and an equal volume of heavier fluid is sinking in the interior. This exchange of equal volumes of fluid with different densities produces a net downward flux of mass that, if unopposed, would lower the center of mass of the fluid, just as in the experiment described in the previous section, thereby converting potential energy to kinetic energy.

Since the circulation is envisioned as being in steady state, the conversion of potential to kinetic energy proceeds at exactly the same rate as the kinetic energy of the fluid motions is being depleted by frictional dissipation. The lowering of the center of mass of the fluid by the overturning circulation is opposed by diabatic heating, which is always warming and expanding the fluid near the bottom of the tank, and cooling and compressing the fluid in the interior. In order to accommodate this expansion and compression, a very small mean ascent is required at intermediate levels. The associated upward mass flux exactly cancels the downward mass flux due to the circulation cell so that the center of mass of the fluid remains at the same level.

The maintenance of large-scale thermally driven circulations in the atmosphere requires both horizontal and vertical gradients of diabatic heating rate. In the absence of horizontal heating contrasts, the heat source/sink at the bottom and top of the tank would destroy the stable stratification and initiate convectively driven motions on a scale much smaller than the dimensions of the tank. In these convection cells, plumes of warm, light fluid rise, and are replaced by cooler, denser fluid from above. In convective equilibrium, the downward mass flux in the convection cells is just strong enough to maintain the center of mass of the fluid at a constant level. In the absence of a vertical gradient of heating, the thermally driven circulation cell would strengthen the stable stratification, lowering the center of mass. Under these conditions the static stability would increase and the strength of the thermally driven circulation would decrease indefinitely. In the real world, the downward diffusion of heat would intervene. The existence of meridional and vertical heating gradients in the same sense as indicated in Fig. 2.4 is clearly evident in the diagrams shown in Section 1.4.2.

In relating the experiment described in this section to the atmospheric general circulation, an important distinction needs to be kept in mind. In experiments of laboratory-size

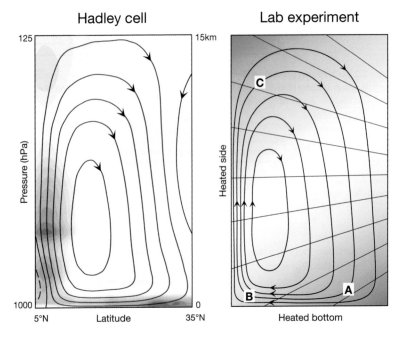

Figure 2.4 (Left) As in the top panel of Fig. 1.27 but limited to the Northern Hemisphere annual mean Hadley cell extending from 5°N to 35°N. The colored shading indicates the distribution of diabatic heating. (Right) A hypothetical laboratory experiment in which a steady state circulation cell (arrows) is driven by heating along the bottom and left walls and a distributed heat sink in the interior analogous to radiative cooling in the atmosphere. The colored shading is for specific volume, the inverse of density (blue represents lower and warm colors higher values). Specific volume in an incompressible fluid is analogous to potential temperature in a compressible fluid. Hence, specific volume, a proxy for potential temperature, increases upward and toward the left and is highest near point C. The straight contour lines represent constant pressure surfaces as inferred from the hydrostatic relationship. The slopes are exaggerated.

proportions, the horizontal and vertical components of the motion are likely to be of the same order of magnitude so that an appreciable fraction of the kinetic energy associated with the fluid motions resides in the vertical component. This energy is converted from potential energy when imbalances between the forces in the vertical propel buoyant parcels of fluid upward and denser parcels downward. In effect, gravity does work on the fluid, just as it does work on a falling object. In atmospheric motions with horizontal scales longer than 100 km, the kinetic energy associated with the vertical motion component is negligible. Kinetic energy is realized when the horizontal component of the pressure gradient force does work on parcels of fluid as they move across the isobars down the horizontal pressure gradient. The relation between this cross-isobar flow and the release of potential energy will become fully apparent when we consider the governing equations. However, one can get an intuitive feel for it by considering the slopes of the pressure surfaces in the right panel of Fig. 2.4. In agreement with the hypsometric equation, the vertical spacing between pressure surfaces (i.e., the "thickness") is greater on the warm side of the tank than on the cold side. It follows that the horizontal flow in the clockwise circulation cell must be primarily directed down the horizontal pressure gradient, as required for the generation of kinetic energy. Hence, the rising of warmer fluid and sinking of colder fluid implies cross-isobar horizontal flow toward lower pressure and vice versa.

Exercises

2.8 How would the overturning circulation depicted in Fig. 2.4 be different in a laboratory experiment in which the heat sink is a cold top and right side boundary, rather than in the interior of the fluid?

2.9 Explain why, in a heated room with a window, the heating element is ordinarily placed on or alongside the wall on the same side as the window rather than opposite sides of the room.

2.10 For a homogeneous (constant density) fluid in a tank with a flat bottom, vertical walls, and no top:
(a) Show that the potential energy per unit area, averaged over the area of the tank, is given by $\rho g <z^2> /2$ where ρ is density, z is the height of the free surface of the fluid and the brackets $<>$ denote an average over the area of the tank.
(b) Show that the potential energy per unit area averaged over the area of the tank that is available for conversion into kinetic energy is given by $\rho g <\widehat{z}^2> /2$, where $\widehat{z} = z - <z>$.
(c) Show that the average rate of conversion from potential to kinetic energy is given by $-\rho g <wz>$, where w is the vertical velocity of the free surface. Give a physical interpretation of this result in terms of the horizontal flow across the isobars.

2.11 Prove that on average over the Hadley cell, the horizontal flow is down the pressure gradient.

Figure 2.5 Examples from one of the Tropic World simulations showing SST and surface wind vectors for months when the SST *contrast* between warm and cold patches is increasing, maximum, decreasing, and minimum in that chronological order. Month numbers refer to months elapsed since the start of the numerical integration. The maximum wind vector is 10 m s^{-1}.

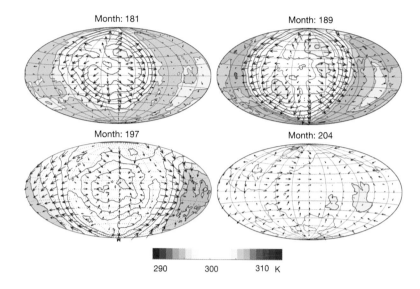

2.4 TROPIC WORLD: CONVECTION ON A PLANETARY SCALE

In this section, we will consider the circulation on a spherical planet consisting of an ensemble of convective cells that develop spontaneously in an atmosphere that is otherwise in radiative–convective equilibrium. The planet called "Tropic World" is Earthlike in many respects, but it is not rotating, the incoming solar radiation is evenly distributed over its entire surface, not just the side facing the sun, and the planet is covered by a "slab ocean" 50 m deep. Hence, there is no distinction between the climate at any location on the planet and any other location.

A slab ocean is a layer immediately beneath a model atmosphere that serves as a reservoir of heat. At each grid point, its temperature varies in accordance with the net energy flux at the air–sea interface. By storing heat when the overlying atmosphere is warm and subsequently releasing it when the atmosphere is cooler, it acts to reduce the damping at low frequencies and thereby redden the frequency spectrum.[2] The larger the prescribed heat capacity (i.e., the depth of the slab ocean), the lower the frequencies to which the reddening extends. Heat is not transported horizontally within the slab, and its global mean temperature never departs very far from radiative–convective equilibrium.

The circulation on Tropic World organizes itself into regions of relatively high ocean temperature (SST), where deep cumulus convection occurs and the mean vertical motion averaged over updrafts and their surroundings is upward, and regions in which the SST is lower and the air is subsiding. The warm and cold regions have no preferred location, but they tend to be of very large scale and slowly evolving. The SST contrast arises from an instability in which (1) the warm regions attract convection, (2) convection moistens the atmosphere above the boundary layer, (3) the moistening warms the surface further through the greenhouse effect. In contrast, cooler regions are

drier in the free atmosphere so a greater fraction of the infrared radiation emitted from the atmosphere is lost to space. Subsidence in these cool regions inhibits the entrainment of dry air into the boundary layer, which favors the development of low clouds. Consequently, less solar radiation reaches the surface of the planet, resulting in further cooling. These processes all serve to amplify the temperature contrasts between the warmer and cooler regions. The strength of the temperature contrasts and the lifetime of the individual warm and cold patches is limited by the overturning circulations, which transport energy from the warm regions to the cool regions, thereby damping the temperature contrasts. In the presence of these competing positive and negative feedbacks, the simulated circulation drifts back and forth irregularly between high- and low-contrast SST regimes with a period of 2–3 years.

Figure 2.5 shows the SST and surface wind fields at four times during one of these cycles in SST contrast. Surface winds are strongest where the SST contrast is strongest. During the growth phase when the SST contrasts are increasing, the strongest surface winds surround the newly formed, relatively small, but expanding cold patch, while the winds over the warm patch are weak, whereas during the decay phase the SST contrasts are weakening, and the patterns are becoming disorganized. The global mean evaporation and precipitation rates are largest at the time in the cycle when the SST contrasts are strongest. The enhanced evaporative cooling and sensible heat fluxes cool the slab ocean during the phase with strong SST contrasts and winds, resulting in a ~0.5 K drop in global mean surface air temperature. Hence, the general circulation in Tropic World mediates both regional temperature and global mean temperature.

The aggregate properties of the overturning cells are summarized in terms of the cross sections shown in Fig. 2.6. In lieu of horizontal distance, the ranking in the frequency distribution of SST between the coolest and warmest grid points is used as the x coordinate. A plot of the average

[2] Barsugli and Battisti (1998).

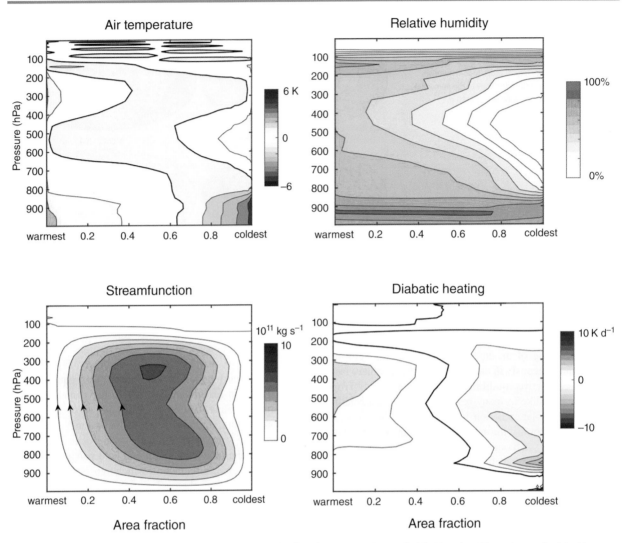

Figure 2.6 Vertical cross sections from the Tropic world simulation extending from the warmest regions on the left side to the coldest regions on the right side, as determined from the frequency distribution of monthly mean SST. (Top left) temperature, the departure from the "global" mean at each level, in K; (top right) relative humidity in percent; (bottom left) mass streamfunction in units of 10^{11} kg s^{-1}; and (bottom right) diabatic heating.

of a large number of conventional cross sections extending from a warm region on the left to a cold region on the right (not shown) exhibits similar patterns. Large departures from the height-dependent, domain averaged temperature are confined to the boundary layer (top left). As on Earth, relative humidity (RH, top right panel) is uniformly high in the boundary layer and just below the tropical cold point tropopause, and in the mid-troposphere it tends to be higher over the warm regions than over the cool regions, rendering them more opaque to OLR. The diabatic heating rate (bottom right panel) is also SST dependent: positive in the moist, ascending air and negative in the dry, subsiding air.

The mass streamfunction (Fig. 2.6, bottom left) assumes the form of a circulation cell analogous to the one in the previous section, with warm air rising and cold air sinking, releasing potential energy, and low-level flow from the colder to the warmer region and oppositely directed high-level flow from the warmer to colder region. The double peak in the

streamfunction contours and the indentation on the cool side are related to the vertical structure of the radiative cooling. The lower cell is driven in part by radiative cooling resulting from the high emissivity of the moist boundary layer air in combination with the dryness of the mid-tropospheric air, which renders it more transparent to the radiation emitted from below. The resulting low-level subsidence over the cool regions, where the mid-troposphere is driest, is balanced by ascent (mainly in the form of shallow convection) in the warm regions, in which the release of latent heat contributes to the driving of the cell. In a similar manner, the upper cell is driven in part by radiative cooling of descending air by emission of infrared radiation from water vapor molecules, which is balanced by deep convection above the region of warm SST. This "push–pull" relationship also shapes the vertical structure of convection in the Earth's atmosphere, but the processes that shape its horizontal structure and evolution are different.

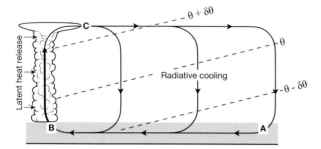

Figure 2.7 Atmospheric analog of the steady state, thermally driven circulation in Fig. 2.4. The gray shading represents the boundary layer in which air parcels acquire latent and sensible heat from the fluxes at the Earth's surface. The letters A, B, and C are for reference in the next figure. It is broadly applicable to the Lagrangian mean circulation of tagged air parcels and in particular to tropical motion systems in which the diabatic heating is dominated by the release of latent heat of condensation in deep cumulus convection. The atmosphere is stably stratified, with potential temperature increasing toward the left.

The steady overturning circulations in the hypothetical laboratory experiment considered in the previous section and the spontaneously developing cells in the "Tropic World" numerical simulation embody some of the characteristics of overturning circulations in the atmospheric general circulation, as shown in Fig. 2.7. Energy is added along the bottom AB by the fluxes of latent and sensible heat from the Earth's surface, which is distributed throughout the boundary layer by shallow convection. The heat source along the left side BC is the release of latent heat in deep cumulus convection, much of which is concentrated within the inner tropics. The water vapor condensed in deep convective clouds enters the atmosphere at the bottom boundary and is distributed through the boundary layer by the same shallow convection. The cooling along the path from C to A is accomplished mainly by the emission of infrared radiation to space, which tends to be strongest in regions of subsidence, where the warming due to adiabatic compression maintains temperature at a level above radiative equilibrium.

In contrast to the circulation cells in the laboratory experiment described in the previous section and in the "Tropic World" experiment, mean (i.e., zonally averaged) meridional motions on a rotating planet are constrained by the conservation of angular momentum, which inhibits meridional displacements of air parcels in much the same way that the static stability inhibits vertical displacements. Hence, for example, the meridional extent of the Hadley cell cannot be explained on the basis of thermodynamic considerations alone. As will be explained in Chapters 3 and 7, transports of angular momentum by the eddies play an important role in determining the configuration and strength of the mean meridional circulations.

Exercises

2.12 On Tropic World, more rain falls over the warm regions than over the cool regions. Where does the extra moisture come from?

2.13 In Tropic World, is there a gradient of net radiation between the warm and cool region at the top of the atmosphere? If so, in which direction is it and what balances it?

2.5 THE GENERAL CIRCULATION AS A HEAT ENGINE

Averaged over the mass of the atmosphere, the general circulation behaves as a heat engine, receiving energy at a higher temperature and losing it at a lower temperature, and doing work. Energy is added mainly by the latent heat fluxes at the lower boundary. The latent heat is transformed into sensible heat when the water vapor condenses in regions of deep cumulus convection and in regions of stratiform ascent in waves and cyclones. Energy is lost mainly by the emission of infrared radiation to space. The work done by the engine is equivalent to the generation of kinetic energy by the horizontal flow across the isobars from higher toward lower pressure. The thermodynamic processes in idealized heat engines can be represented in *thermodynamic diagrams* depicting the path of a typical air parcel (the "working fluid") in a two-dimensional phase space defined by the state variables as it undergoes a series of reversible processes which, in combination, constitute a cycle. Two widely used thermodynamic charts are

- the p–α *diagram*, with pressure as the ordinate and specific volume α as the abscissa, and
- the T–s *diagram* with temperature as the ordinate and entropy $s = c_p \ln \theta$ as the abscissa.

In both kinds of diagrams the trajectory of an air parcel assumes the form of a clockwise closed loop. On the p–α diagram, the rate at which expansion work is being done per unit mass is $dp\,\alpha$ and the output of work W done in a single cycle is proportional to the area enclosed by the loop. On the T–s diagram, the sensible heat added to the air per unit mass is $c_p\,T\,d\ln\theta$, and the difference between that added Q_A and the heat lost Q_L, is proportional to the area enclosed by the same loop. Since the conservation of energy requires that $W = Q_A - Q_L$, it follows that the output of the engine can be inferred from either diagram. In estimating the throughput of energy in the atmospheric general circulation we will make use of the $(T - s)$ diagram.

Temperature versus entropy plots, inspired by the experiments discussed in Sections 2.3 and 2.4, are shown in Fig. 2.8. In constructing both panels of this diagram, we have assumed that the working fluid is a perfect gas, and that the ratio of adiabatic temperature changes due to the expansion and compression of rising and sinking air parcels is roughly comparable to that experienced by air parcels moving about in tropospheric circulations like the Hadley cell or baroclinic waves. In the plot on the left, the heat sources and sinks are all along the boundaries and, in the one on the right, the heat sink is in the interior, mimicking radiative cooling in the real

Figure 2.8 (Top panels) Idealized laboratory experiments. The right panel is the same as the right panel in Fig. 2.4 with additional labeling to indicate where the cooling is occurring. The left panel is similar except that all the heating and cooling are applied along the boundaries. (Bottom panels) Temperature entropy diagrams for heat engines in which the working fluid undergoes the reversible processes indicated in the respective top panels. The heat added to the parcels is proportional to the areas under the curves AB, BC, and the heat lost by them is represented by the area under CD and DA (left) and CA (right), all represented here as straight lines. The rate at which work is done by the heat engine, which is equal to the heat added minus the heat rejected, is thus given by the shaded area ABCD (left) and ABC (right), divided by the time to complete the circuit AB, BC . . . .

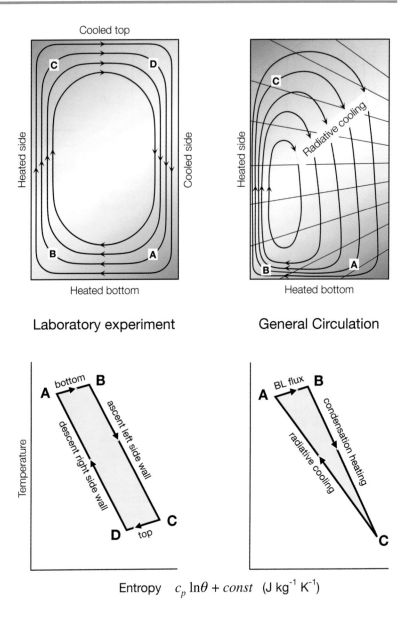

atmosphere. The former involves four steps, like the Carnot cycle. AB involves the addition of heat and a temperature rise as the air parcel moves along the bottom of its cyclic path, BC involves substantial additional diabatic heating as the parcel ascends along the side wall while cooling due to adiabatic expansion. CD involves the loss of heat while residing at a low temperature, but at a high potential temperature while near the top of its (geometric) path, and DA involves an additional heat loss as the parcel sinks along the other side wall, while warming due to adiabatic compression. If the cooling due to adiabatic expansion as air parcels flow down the horizontal pressure gradients in steps AB and CD were just strong enough to cancel the diabatic heating and cooling, then the top and bottom of the parallelogram would consist of horizontal lines as in the Carnot cycle. The work is proportional to the area of the parallelogram ABCD.

In the more realistic representation of the atmospheric general circulation depicted in the right panel of Fig. 2.4, the heat source at the bottom is specifically related to the sensible heat flux from the underlying surface and the heating in the rising branch is due to the release of the latent heat of condensation of water vapor. Heat is lost not in an isothermal process at the temperature of a "cold sink", as in the Carnot cycle, but continuously (by radiative cooling) beginning at the time when the air is detrained from the cloud tops C and continuing until the subsiding air parcels are entrained into the boundary layer A, as depicted in Fig. 2.7. Represented on a temperature–entropy diagram (Fig. 2.8, bottom right panel), the cycle assumes the form of a triangle ABC rather than a parallelogram ABCD. Despite the differences in shape, the insights about the output and thermal efficiency derived from the Carnot cycle are applicable to the atmospheric heat engine.

We will show in Section 6.4 that within certain limited volumes of the atmosphere, the flow of energy is in the reverse sense: rather than acting as a heat engine, these limited volumes act as *refrigerators*, receiving energy at a lower temperature and losing it at a higher temperature. In order to perform this feat, they must import potential (and/or) kinetic energy from neighboring volumes.

Exercises

2.14 Depict the thermodynamic processes in the classic Carnot cycle and in the Hadley cell on a thermodynamic chart with specific volume α as the x axis and pressure p on a linear scale but inverted as the y axis.

2.15 Suppose that the lighter and denser fluids in the experiment described in Section 2.2 are warmer and colder water. Explain how the transformations in this experiment result in an increase in the entropy.

2.6 THE INFLUENCE OF PLANETARY ROTATION

Rotation matters for several different reasons. The Earth's atmosphere is heated by radiation (insolation) from a distant sun, which subtends a very small arc of solid angle in the sky. Hence, the heating is zero on the night side and proportional to the cosine of solar zenith angle on the day side. In the absence of rotation, the response to the differential heating would consist of a circulation cell analogous to the one described in Section 2.3. The heating and ascent would occur on the day side and cooling and descent on the dark side. On a rotating planet the response to the day side/night side heating contrast assumes the form of a thermally driven *atmospheric tide* ("thermally driven" to distinguish it from the gravitationally driven tide). For a very thin planetary atmosphere, in which the radiative relaxation time is on the order of a day or less, tidal motions dominate the general circulation. Owing to its appreciable heat capacity (mass times specific heat), the radiative relaxation time of the Earth's atmosphere (on the order of 30 days) is long enough so that the tidal response to the differential heating is of secondary importance except in the upper atmosphere, as discussed in greater detail in Chapter 21.

The influence of rotation on the general circulation goes beyond its role in mediating horizontal gradients in diabatic heating rate. The Coriolis force induced by the planetary rotation profoundly influences the character of the motions, giving rise to nearly geostrophically balanced flows with horizontal temperature gradients much larger than could be sustained in the absence of rotation, as illustrated by the numerical experiments described in this section.

2.6.1 A "Spin Up" Experiment

Here we describe the events that transpire in the "spin up" of the atmospheric circulation of an idealized Earth-like planet, starting from a state of rest (i.e., no motion, with flat pressure surfaces, isothermal on each pressure level, and stably stratified). This experiment is performed using an *aquaplanet model* simulation of the circulation on a water-covered, but otherwise Earth-like planet. In experiments of this kind, SST may be prescribed to mimic the observed equator-to-pole gradient, or the ocean may be interactive, typically modeled as a motionless slab whose heat capacity is prescribed by assigning it a depth, typically on the order of tens of meters, to match the so-called *mixed layer depth* of the Earth's ocean. The schematics shown in Fig. 2.9 can be viewed as representing a composite of many such numerical "spin up" experiments, each with its own prescribed SST distribution or mixed layer depth ocean. Spin-up times vary, depending on the specifics of how the ocean is modeled, but the sequence of events depicted in Fig. 2.9 is typical when the planet's rotation rate is similar to that of Earth.

The first thing that happens is that the tropics begin to warm and the polar regions begin to cool in response to an imposed distribution of diabatic heating, which mimics the equator-to-pole heating gradient in the real atmosphere that derives from the meridional gradient of solar zenith angle. As the tropical atmosphere warms, thermal expansion causes the pressure surfaces in the upper troposphere to bulge upward, while the cooling of the polar atmosphere draws the pressure surfaces downward, in accordance with the hypsometric equation. These tendencies are evident even in the very first time step of the numerical integration. The sloping of the pressure surfaces gives rise to an equator-to-pole pressure gradient at the upper levels. The pressure gradient force, in turn, drives the poleward flow depicted in Fig. 2.9a, which comes into play in the second time step.

The poleward mass flux results in a latitudinal redistribution of mass, causing surface pressure to drop in low latitudes and rise in high latitudes. The resulting low-level pole-to-equator pressure gradient drives a compensating equatorward low-level flow. Hence, the initial response (the first few hours) to the heating gradient is the development of an equator-to-pole circulation cell, as shown in Fig. 2.9b. The Coriolis force in the horizontal equation of motion imparts a westward component to the equatorward flow in the lower branch of the cell, as shown in panel (c), and an eastward component to the poleward flow in the upper branch.

During the first few simulated days of the integration the flow becomes progressively more zonal, as the meridional component of the Coriolis force comes into geostrophic balance with the pressure gradient force. As required by thermal wind balance, the vertical wind shear between the low-level easterlies and the upper-level westerlies increases in proportion to the strengthening equator-to-pole temperature gradient forced by the meridional gradient of the diabatic heating. The Coriolis force acting on the relatively weak meridional cross-isobar flow is instrumental in building

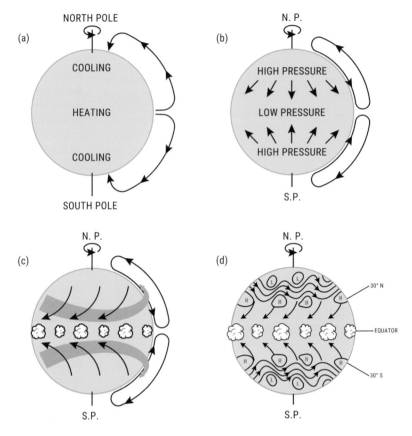

Figure 2.9 Schematic depiction of the numerically simulated general circulation as it develops from a state of rest for equinox conditions on an idealized aquaplanet with no land–sea contrasts. The arrows in the globes depict the low-level circulation. The clouds denote the equatorial rain belt and the cells denote the meridional overturning circulation. The contours in (d) represent sea level pressure and highs and lows are marked by H and L, respectively. Adapted from Wallace and Hobbs (2006). © Elsevier Inc. All rights reserved.

up the vertical shear of the zonal wind $\partial[u]/\partial z$, leading to a low-level westward acceleration, and an upper-level eastward acceleration. Frictional drag limits the strength of the surface easterlies, but the westerlies aloft become stronger with each successive time step of the numerical integration.

When the meridional temperature gradient reaches a critical value, the simulated circulation undergoes a fundamental reorganization: baroclinic instability spontaneously breaks out in midlatitudes, imparting a wave-like character to the flow. The fastest growing waves are eastward propagating, with a typical zonal wavelength of ~3500 km. Figure 2.10 shows a longitude–height section through an exponentially amplifying baroclinic wave in the Northern Hemisphere based on linear theory. The perturbations are prescribed to be sinusoidal in longitude. The perturbations in the meridional wind component v and temperature T are seen to be strongest at the Earth's surface, where they vary in phase with one another so that warmer air is flowing poleward and colder air is flowing equatorward, resulting in a poleward heat transport. The temperature perturbations decay exponentially with height with an e-folding depth much smaller than the scale height of the atmosphere. Hence, the poleward heat transport is largely confined to the lower troposphere. The geopotential height and meridional wind perturbations decay with height more slowly and they tilt westward with height, so as to bring the geopotential height perturbations into closer alignment with the temperature perturbations beneath

them, as required for hydrostatic balance (see Fig. 13.6 of Gill, 1980).

Within a few days of simulated time, the baroclinic waves amplify to the point where their poleward heat transports arrest the buildup of the equator-to-pole temperature gradient. As they amplify, their structure begins to depart significantly from the modal structure in Fig. 2.10. Horizontal wind speeds are no longer highest at the Earth's surface: most of the kinetic energy in the wave resides at the jet stream level, where the waves produce a poleward transport of westerly momentum from the tropics into middle latitudes. These transports of heat and momentum establish and maintain a statistically steady circulation in the configuration represented by the snapshot shown in Fig. 2.9d. From here on it gets more complicated. The remainder of this subsection is intended as an abstract for the more detailed development that will follow in the next two sections of the book.

Successive generations of baroclinic waves develop and evolve through their life cycles, emanating from their source region in the midlatitude lower troposphere, first dispersing upward toward the jet stream (10 km) level and thence equatorward into the tropics. (Here and throughout the book we use the term *disperse* to indicate the flow of wave-related mechanical energy, i.e., the group velocity, in the meridional plane, as distinguished from the direction of phase propagation of the waves.) The equatorward dispersion of the waves in the upper troposphere causes the axes of their

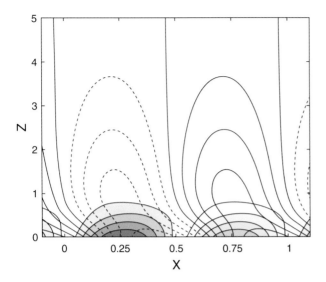

Figure 2.10 Structure of baroclinic waves in the longitude–height plane in the early stages of their development when they are amplifying exponentially with time in accordance with the normal mode solution derived from linear theory. The waves are propagating from west to east in a westerly background flow with temperature decreasing with latitude. This representation is based on the theoretical study of Charney (1947). See the *Solutions Manual* for discussion of an alternative representation based on Eady (1949), in which the phase relationships between geopotential height, wind, temperature and their respective vertical tilts in the lower troposphere are similar. The colored shading indicates the temperature perturbations (warm colors positive) and the contours indicate the perturbations in meridional wind (solid contours poleward, dashed contours equatorward, zero contour solid). Based on the theoretical analysis of Charney (1947). Courtesy of Ian James and Brian Hoskins.

ridges and troughs to tilt in the manner shown in panel (d). As a consequence of this tilt, poleward-moving air to the east of the troughs exhibits a stronger westerly wind component than the equatorward moving air to the west of the troughs. The greater angular momentum per unit mass of the poleward moving air parcels gives rise to a net poleward eddy transport of westerly momentum from the subtropics into middle latitudes. In response to the import of westerly momentum from the subtropics, the surface winds in midlatitudes shift from easterly in panel (c) to westerly in (d), consistent with the direction of the observed surface winds in midlatitudes.

Baroclinic waves drive their own weak mean meridional circulation cells in the midlatitudes, referred to as *Ferrel cells*. These cells are characterized by poleward, frictionally induced *Ekman drift* at the latitude of the storm tracks (45°N/S), with ascent on the poleward flank and descent on the equatorward flank of the storm track. Hence, with the spontaneous development of the baroclinic waves, the Hadley cells withdraw into the tropics, and a region of subsidence develops at subtropical (~30°N/S) latitudes. These regions of subsidence coincide with the subtropical anticyclones, which mark the boundary between the tropical trade winds and the extratropical westerlies. Most of the world's major desert regions lie within this latitude belt.

2.6.2 Varying the Rotation Rate Ω

Here we examine the effect of planetary rotation rate upon the character of the general circulation by comparing the simulated general circulations on four aquaplanets after they are fully spun-up from a state of rest as in the previous subsection, each with a different rotation rate. The model atmosphere is coupled to a passive, slab ocean, with which it exchanges energy and water. The depth of the slab is set to 2.5 m, so that when the model is run with an Earth-like rotation rate, the model's surface temperature resembles surface temperature on Earth. The experiments are all run for perpetual equinox conditions (i.e., with the Sun directly overhead at noon on the equator year-round) and daily averaged insolation: the only difference between them is the rotation rate. The solar constant and greenhouse gas concentrations are prescribed in all four experiments to be those in the year 2000. We examine conditions toward the end of the integrations, by which time the general circulations can be considered to be fully "spun up" and in statistical equilibrium.

Figures 2.11–2.14 show selected fields derived from the four experiments. In the absence of rotation the general circulation is dominated by equator-to-pole Hadley cells. The zonal wind field is virtually nonexistent, and the equator-to-pole temperature gradient is very weak. The presence of rotation radically transforms the circulation, giving rise to eddies and jet streams and confining the Northern and Southern Hemisphere Hadley cells to lower latitudes. The SLP patterns (Fig. 2.11) exhibit belts of high pressure, separating a tropical trade-wind belt from an extratropical westerly wind belt. As the rotation rate increases, the equator-to-pole temperature gradient increases, the cells become increasingly confined to low latitudes and the primary jet streams also shift equatorward (Fig. 2.13). In the run with the highest rotation rate (2Ω, twice the Earth's rotation rate), there is a suggestion of secondary jet streams at ~35°N/S, poleward of the primary jets, which exhibit a more barotropic structure. Simulations with even higher rotation rates (not shown) exhibit multiple jets. It is evident from Fig. 2.14 that the horizontal scale of the eddies decreases with increasing rotation rate.

In both hemispheres, the axes of the waves in the T_s field in Fig. 2.14 tilt eastward with increasing latitude. As in the observations, in each of the three runs with rotation, the jet streams are positioned directly above the subtropical anticyclones, and in the run with the rotation rate that matches the Earth's, the latitudes of the jets and other zonally averaged features in the SLP field are quite realistic. These features of the general circulation evidently derive directly from the zonally averaged heating distribution and the rotation rate, without reference to land–sea geometry or orography.[3]

---

[3] For a more in depth exploration of the effects of varying the planetary rotation rate, see Williams (1988) and Liu et al. (2017) and references therein.

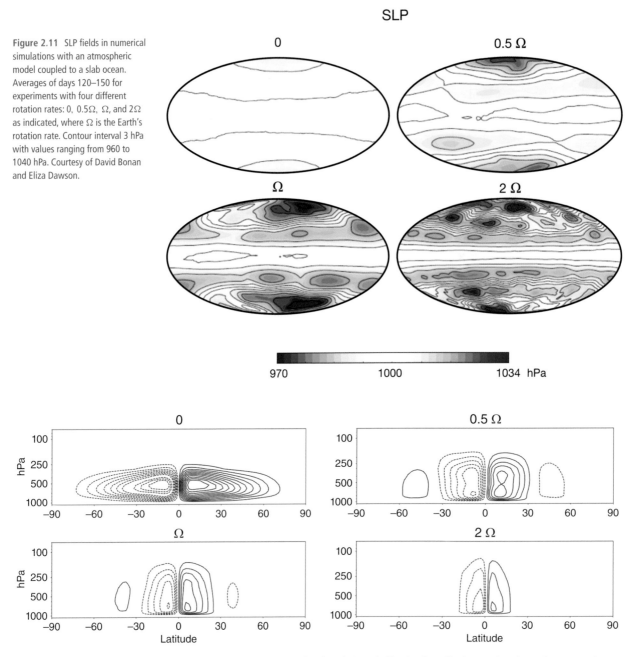

Figure 2.11 SLP fields in numerical simulations with an atmospheric model coupled to a slab ocean. Averages of days 120–150 for experiments with four different rotation rates: 0, 0.5Ω, Ω, and 2Ω as indicated, where Ω is the Earth's rotation rate. Contour interval 3 hPa with values ranging from 960 to 1040 hPa. Courtesy of David Bonan and Eliza Dawson.

Figure 2.12 As in Fig. 2.11, but for the mass streamfunction for the mean meridional circulations. The flow is poleward in the upper branches and equatorward in the lower branches. Contour interval 2×10^{10} kg s^{-1} (20 Sv), except in the case $\Omega = 0$, where it is 50 Sv. The zero contour is not shown. Courtesy of David Bonan and Eliza Dawson.

The sequence of experiments gives the impression that the general circulation can be divided into tropical and extratropical regimes, the former dominated by the thermally driven Hadley cells and the latter, largely "eddy-driven," dominated by baroclinic waves in the "storm tracks." We will show in later chapters how the waves interact nonlinearly with each other and with the zonally symmetric flow to produce a spectrum of eddies, dominated by Rossby waves ranging from planetary scales to the smallest resolvable scales. These interactions give rise to multiple jet streams,

temporary blocking of the zonal flow by intense quasi-stationary anticyclonic gyres, and a host of other phenomena. We will show in Section 7.6 that in the absence of eddies the tropical regime would comprise the entire general circulation on a rotating planet: the Hadley cells would be much weaker than – and poleward of – those observed, the atmosphere would be at rest, in thermal equilibrium with the heating.

The influence of planetary rotation rate has also been investigated using a laboratory analog of the general circulation – a circular or toroidal-shaped tank filled with a

$[\overline{u}]$

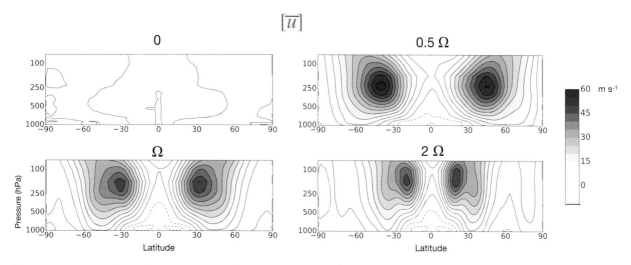

Figure 2.13 As in Fig. 2.11 but zonally averaged zonal wind $[\overline{u}]$. Contour interval 5 m s^{-1}. Courtesy of David Bonan and Eliza Dawson.

$\overline{T}_\mathrm{s}$

Figure 2.14 As in Fig. 2.11, but surface temperature T_s (in K) on Day 132 of the integrations. Courtesy of David Bonan and Eliza Dawson.

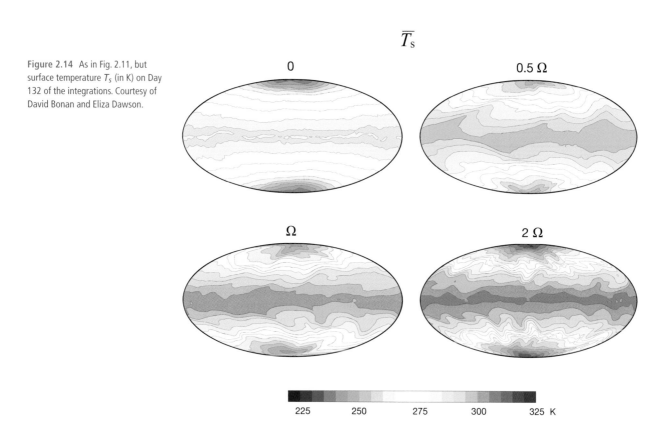

stratified fluid (typically, water whose salinity increases with depth) centered on the axis of rotation of a rotating turntable. To mimic the equator-to-pole gradient of diabatic heating, the tank is heated along its outer wall and cooled along its inner wall (or, in the case of a circular tank, along the bottom, close to the axis of rotation). Baroclinic waves develop and transport heat radially inward, driving a mean toroidal circulation in the same sense as the Ferrel cell, with radial inflow along the bottom of the tank. In a toroidal tank, whether the flow is steady, consisting of a wave with a single,

discrete zonal wavenumber, periodic (i.e., vacillating back and forth between two different wavenumbers), or aperiodic depends on the combination of differential heating rate and rotation rate. The remarkable intricacy of the behavior of this seemingly simple system has been the subject of many theoretical investigations.[4]

---

[4] See Fultz et al. (1959) for summary of foundational experiments of thermal convection in rotating cylinders.

2.16 (a) Summarize the salient features of the structure of exponentially amplifying baroclinic waves, as represented by the theoretical solution shown in Fig. 2.10. (b*) Have baroclinic waves been detected in the atmospheres of planets other than Earth's?

2.17 (a) Estimate the radiative relaxation time for the Earth's atmosphere. (b) On the basis of this estimate, explain why the thermally driven *atmospheric tide* with a period of a day is much more prominent in the Martian atmosphere than in the Earth's atmosphere.

2.18 Why are ozone and water vapor of particular importance in the thermal driving of atmospheric tides on Earth?

2.19* Read the spirited exchange of letters between Businger and Holton and Libby in the 29 March 1968 issue of *Science* and judge the merits of their arguments. Bear in mind that this exchange took place about a decade before general circulation modeling simulations of the kind described in this section were first performed.

2.7 INFLUENCE OF ORBITAL GEOMETRY

The seasons on Earth are almost exclusively due to obliquity: the 23.5° tilt of the axis of rotation about the perpendicular to the Earth–Sun plane. In aquaplanet simulations, the Northern and Southern Hemisphere dynamical responses to an obliquity-only forcing are identical but lagged relative to one another by half a year. On Earth, the dynamical response is complicated by the differing Northern and Southern Hemisphere land–sea configurations and the higher and much more extensive Northern Hemisphere mountain ranges. Through the mechanism of wave–mean flow interaction, these equatorial and zonal asymmetries in the boundary conditions profoundly affect the zonally symmetric wind and temperature fields, as will be explained in Chapters 8 and 11. For example, the wintertime stratospheric polar vortex is much stronger in the Southern Hemisphere than its Northern Hemisphere counterpart and the tropical cold point tropopause is more pronounced in DJF than in JJA. On other planets and at other times in the history of the Earth, orbital ellipticity comes into play as a secondary, and in some cases even the primary, forcing for the annual cycle, as illustrated in Fig. 2.20.

The Earth is presently closest to the Sun during the boreal midwinter. Hence, summer-to-winter contrasts in insolation over the large Northern Hemisphere continents are not quite as large as they would be if the Earth's orbit were perfectly circular. Five thousand years ago, the Earth was closest to the Sun around the time of the spring equinox, and summer insolation over the Northern Hemisphere continents was about

5% percent stronger than it is today. Numerical experiments with general circulation models suggest that in response to the stronger insolation, the monsoon circulations penetrated deeper into the Northern Hemisphere subtropics bringing rains during the growing season and thereby rendering some of today's arid and semiarid regions of North Africa and central Asia more capable of supporting agriculture. The juxtaposition of this *precession cycle* and the cycles in the *obliquity* and *eccentricity* of the Earth's orbit modulate the latitudinally and seasonally varying insolation and the atmospheric general circulation on time-scales ranging from thousands to hundreds of thousands of years.

A rich literature on orbitally induced variations in paleoclimate with numerical experiments exists dating back to around 1980.[5] But in this book, which focuses on the Earth's present climate, orbital influences other than seasonality are not considered further. Nor is there space to consider the remarkable diversity of planetary atmospheres that exists within our own solar system and beyond, to which orbital influences contribute. Other factors that contribute to the diversity include contrasts in (i) atmospheric mass, (ii) radiative properties of the various gaseous constituents, (iii) the conditions under which the various constituents undergo phase changes, (iv) the extent and type of cloud cover, (v) whether the underlying surface is solid or liquid or part solid and part liquid as on Earth, and (vi) the height of the planet's orography.[6]

2.20 Mars experiences seasonal variations in insolation large enough to cause a substantial fraction of the mass of its atmosphere, mainly carbon dioxide, to condense into a cap of dry ice surrounding wintertime pole. (a) Describe the seasonality of global mean surface pressure on Mars, taking into account the large eccentricity of its orbit. (b) Estimate the strength of the mass-weighted, vertically averaged meridional wind component at the times when the ice cap is expanding and contracting most rapidly.

2.21 The accompanying figure shows the insolation incident on a unit horizontal surface at the top of the atmosphere, expressed in units of 10^6 J m^{-2} integrated over the 24-h day, as a function of latitude and calendar month. Within the shaded regions the insolation is zero. Solar declination refers to the latitude at which the Sun is overhead at noon. (a) Why aren't the DJF and JJA global insolation profiles exact mirror images of one another? (b) How can it be that the insolation in the summer hemisphere is almost the same at all latitudes despite the fact that the solar zenith angle

[5] An introduction to Ice Age cycles can be found in Imbrie and Imbrie (1986) and Kump et al. (2004).
[6] For an introduction to the physics and dynamics of planetary atmospheres in a wide range of different settings, see Pierrehumbert (2010).

Figure for Exercise 2.22. The black curves show standardized time series of the strength of the monsoon across South and East Asia, as inferred from the oxygen isotopic composition of calcite (δ^{18}O, in ‰) in cave stalagmites. Negative values (positive upward) indicate a more intense summer monsoon. Superposed on each speleothem record is the summer (JJA) insolation at 30°N. The long-term mean has been removed from each cave record and, for ease of viewing, the insolation, indicated by the gray curves, has been scaled so that the standard deviation of insolation is identical to the standard deviation of the respective cave record. Adapted from Battisti et al. (2014). © American Geophysical Unio

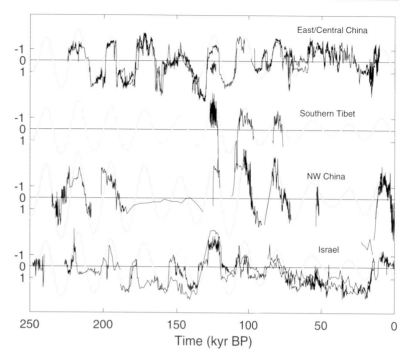

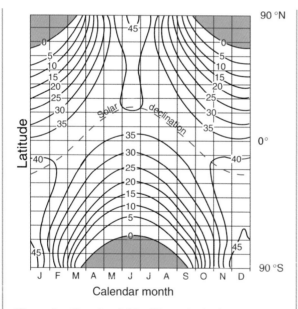

Figure for Exercise 2.21. Climatological mean top of atmosphere insolation. Adapted from Meteorological Tables (R. J. List, Ed.), 6th Ed., Smithsonian Institute (1951), p. 417. © Elsevier Inc. All rights reserved.

decreases from the subtropics to the poles? (c) In the absence of a meridional gradient of insolation in the summer hemisphere, why does the zonally averaged temperature decrease with latitude?

2.22 The proxy records of rainfall in South and East Asia shown in the accompanying figure exhibit large variations with a period on the order of 20 000 years over a number of arid and semiarid regions that were especially well marked around 200 000 years before present. Interpret these features in terms of orbital cycles.

2.8 THE LONG ARM OF FRICTIONAL DRAG

The atmosphere does not depart very far from a state of solid body rotation (i.e., being at rest with respect to the bottom boundary in a rotating frame of reference), as evidenced by the meridional profile of zonally averaged zonal wind $[u]$ at the jet stream level, shown in Fig. 2.15. It is not surprising that frictional drag limits wind speeds in the boundary layer but it is less clear how the winds at the jet stream level feel the influence of the lower boundary. In this section, we describe a laboratory experiment that shows how frictional drag confined to a very thin boundary layer exerts a strong influence on the laminar flow regime in the interior of the domain.

Consider a cylindrical tank filled with fluid that is sitting on a rotating table, in solid body rotation about its axis of symmetry (Fig. 2.16). At some instant in time the turntable motor is turned off so that the walls of the tank abruptly become stationary. Estimating how long it takes for the rotation of the fluid to slow down is referred to as the "spin-down problem." If the boundary effects were transmitted to the interior of the fluid solely by molecular diffusion, the spin-down time should be on the order of hours for a laboratory-size tank and minutes for a teacup-size demonstration model. By comparison, the observed spin-down time is remarkably short; minutes for a laboratory-size apparatus

Angular speed at 250 hPa

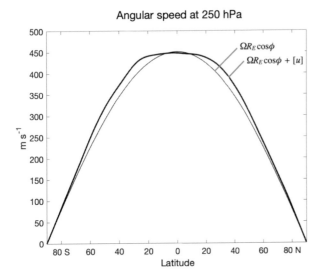

$\Omega R_E \cos\phi$

$\Omega R_E \cos\phi + [u]$

Figure 2.15 Meridional profile of angular velocity of the annual mean zonally averaged zonal wind at the jet stream level (250 hPa). The thin line is a reference profile for an atmosphere at rest with respect to the rotating Earth.

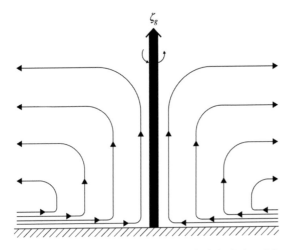

Figure 2.16 Streamlines depicting the overturning circulation in the radial plane in the laboratory "spin down" experiment described in the text. The bracket (bottom left) indicates the thickness of the frictional boundary layer. From Holton and Hakim (2012). © Elsevier Inc. All rights reserved.

and much less than a minute for a teacup. Evidently some other, much more efficient process is at work. The tendency for tea leaves to pile up in the middle of the cup during such an experiment provides a clue as to the nature of this process.

During the spin-down process, the tangential flow in the interior remains close to a state of balance between the inward-directed horizontal pressure gradient force associated with the parabolic shape of the free surface (not shown) and the outward-directed centrifugal force associated with the curvature of the circular trajectories. However, within the very thin molecular boundary layer adjacent to the outer wall and bottom of the tank, this balance is continually being upset by the frictional drag, which is slowing down

the tangential flow so that the centrifugal force is unable to balance the radially inward-directed pressure gradient force. As a result of the imbalance, fluid drifts inward toward the axis of rotation along the bottom of the tank (hence the pile of tea leaves in the middle of the teacup). This frictionally driven radial inflow is accompanied by a very slow radially outward return flow above the boundary layer, as shown in Fig. 2.16. The character of the return flow depends upon whether the fluid in the tank is homogeneous (constant density) or stably stratified. If it is homogeneous, then the flow must be barotropic (independent of height): there is no way to generate horizontal density gradients, so there can be no vertical shear of the tangential flow. Hence, the radial outflow must be the same at all levels. As an element of fluid drifts inward along the bottom of the tank, it loses most of its angular momentum, so that by the time it completes a full circuit of the radial circulation cell, the spin-down of the tank is almost complete. The stably stratified (baroclinic) case is more complicated.[7] The radial circulation cell in the tank is the analog of mean meridional cells in the general circulation.

2.9 GRAVITY WAVES, THE HIDDEN MESSENGERS

Inertio-gravity (IG) waves play an essential role in enforcing geostrophic balance and spreading the heat released in the mesoscale updrafts in deep convective clouds throughout the tropics. In this sense, they are the messengers that make the large-scale motion field aware of subgrid scale processes. They are also instrumental in extending the long arm of frictional drag in the boundary layer into the upper atmosphere. IG waves do most of their work in secret. They can be represented in numerical models and their presence is discernible in general circulation statistics, but they do not show up clearly on synoptic charts or in conventional climatological mean fields like the ones shown in the previous chapter. Most of the visual representations of them that appear on the web are limited to vertical cross sections through geographically localized, mesoscale, quasi-stationary mountain lee waves and animations of mesoscale downdrafts in short-lived convective storms (see, e.g., Fig. 3.2). They are much more ubiquitous than they might appear based on these examples.

Part of the scientific legacy of the catastrophic Krakatoa eruption in 1883 was the set of barograph traces from the global network of weather stations that recorded a distinctive wave signature that propagated back and forth several times between its point of origin in the Maritime Continent and its antipodal point in the Americas, traveling at roughly the speed of sound. The behavior of the pressure wave is concisely summarized in a set of maps compiled by the Royal Society (Fig. 2.17) showing its time of arrival at various stations. The arrival of the wave at the antipodal point about 16–17 hours after the explosion and its subsequent

---

[7] Holton (1965).

Figure 2.17 The first two of a series of figures included in the Royal Society report on Krakatoa (Symons, 1888) illustrating the progression of the pressure wave from the explosion as it propagated around the world and back several times. The innermost red contour surrounding Krakatoa labeled "4" in the top right map corresponds to the position of the wave at 0400 UTC, minutes after the explosion. Successive contours radiating outward from Krakatoa in the top right panel and inward toward the antipodal point in the top left panel are at two-hour intervals 0600, 0800, . . . , 2000 UTC. Hence, the signal traveled halfway around the world in slightly over 16 hours. Subsequent blue-green contours 22, 24, . . . , 38 in the bottom panels show the wave returning to its point of origin. From Strachey (1988).

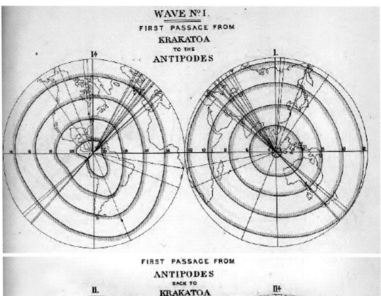

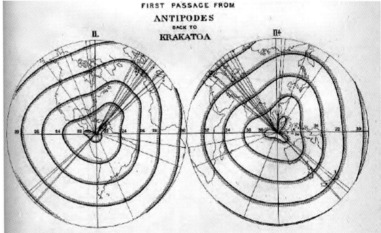

return to its source 16–17 hours after that are clearly revealed by the pressure traces. The phase speed inferred from the travel times is about 337 m s^{-1}, which is comparable to the speed of sound. Noting that the observed phase speed of this wave pulse from the Krakatoa explosion corresponded to that of a gravity wave in a homogeneous (constant density) ocean of roughly 10 km depth, Taylor (1929) argued that it could equally well be interpreted as a long, horizontally propagating IG wave. His interpretation has been elaborated on and refined in numerous subsequent studies, but it has withstood the test of time.

The wave from the January 15, 2022 Tonga eruption is shown in Fig. 2.18. The right panel shows the simulated surface pressure signature, which was marked by a pressure surge followed, about half an hour later, by a dip, as confirmed by barometric readings at thousands of stations around the world. The observed satellite imagery in the left panel is based on infrared radiation at wavelengths ~7.3 μm, in which most of the radiance is emitted by water vapor molecules in clear air (rather than by the Earth's surface and cloud tops). Features on the time-scale of the pressure wave are well resolved by the sequence of images from the geostationary satellites, in which successive frames over the

Pacific sector are 10 minutes apart. The pressure surge was preceded by adiabatic compression, which warmed the air, increasing the radiance in this channel, and the adiabatic expansion that accompanied the subsequent pressure drop created a cool ring immediately inside the warm ring. These adiabatic pressure changes were due, not to vertical motions, but to compression and rarefaction, as in an acoustic wave. To make this fast moving wave visible in the presence of the much stronger radiance contrasts associated with more slowly evolving synoptic and planetary-scale waves, it is necessary to highpass filter the data in the time domain. Both panels of Fig. 2.18 are snapshots from animations on the companion web page that show the wave propagating from its point of origin to the antipodal point and back over an interval of about a day and a half.

Volcanic eruptions are exceptional events, but mesoscale updrafts in deep cumulus convection are an everyday occurrence. Figure 2.19 shows the waves in sea-level pressure excited by a composite of thousands of such "mini Krakatoas" along the 6°S latitude circle in 10 years of hourly ERA5 gridded data. The wave can be seen emanating from the reference grid point and propagating along great circles in all directions, reaching the antipodal point about 20 hours

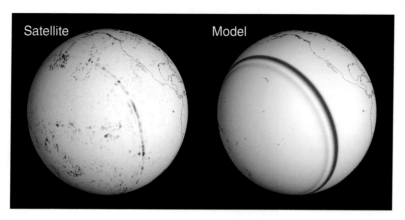

Figure 2.18 The atmospheric acoustic-gravity wave excited by the explosive Tonga eruption January 15, 2022, shown 6 hours after the time of the excitation of the wave. (Left) Infrared imagery in the water vapor channel, highpass filtered in the frequency domain to isolate the fast-moving waves. At these high frequencies the variations in the infrared radiances reveal the temperature perturbations induced by adiabatic compression and expansion as the wave passes. (Right) The corresponding pressure perturbations (the same phase at all levels, but amplifying exponentially with height) in a numerical simulation based on the shallow water wave equations, which will be introduced in Chapter 10. The pressure and temperature perturbations are in phase. The wave is propagating toward the antipodal point with a phase speed of 318 m s^{-1}. Courtesy of Mathew Barlow, simulations provided by Nedjeljka Žagar. The corresponding animations can be viewed on the companion web page.

later. (That it assumes the form of a wave train, as opposed to a single wave, is an artifact of the filtering used in producing the composite.) Given that the model used in producing the ERA5 Reanalyses is hydrostatic, it follows that this must have the properties of an IG wave.

Much more slowly propagating gravity waves, with phase speeds on the order of 50 m s^{-1}, are the important messengers in the Earth's atmosphere. Compelling observational evidence of their existence is afforded by satellite imagery of low cloud decks during volcanic eruptions, like the one shown in Fig. 2.20. The expanding ring of clear sky in this image is the signature of adiabatically induced subsidence warming in a downwelling gravity wave. As this ring of subsiding air passes a fixed point, the air warms adiabatically, while adiabatic expansion cools the air parcels rising in the volcanic plume. In this manner, the sensible heat injected into the atmosphere in the mesoscale volcanic plume is rapidly spread into the large-scale environment.

2.10 CONCLUDING REMARKS

The "experiments" considered in this chapter do not constitute a theory of the general circulation or even an outline thereof. They are intended more as a primer in how to frame and address questions relating to the global distributions of wind, temperature, and moisture – how do they come about? How are they maintained? How does energy (mechanical energy in particular) cycle through them?

2.10.1 Thermodynamic Insights

The concept of radiative–convective equilibrium, described in Section 2.1, explains the existence of a troposphere and

stratosphere on Earth and other planets, and it accounts for the distinction between the photosphere and the chromosphere on the Sun. That the observed tropospheric lapse rate tends to be less steep than the moist adiabatic lapse rate, especially in the extratropics of the winter hemisphere, is a consequence of vertical heat flux in large-scale motion systems such as baroclinic waves, in which warmer air rises and colder air sinks, even in an environment that is stable with respect to moist convection. The lapse rate in the extratropics is appreciably more stable during winter than during summer (Fig. 1.20, right panel) because of the stronger wintertime baroclinic wave activity.

In the hypothetical laboratory experiment described in Section 2.3, the circulation is intuited and shown to be consistent with energy balance considerations, rather than being derived from first principles. This thought exercise nonetheless provides insight into how prescribed horizontal gradients in diabatic heating drive atmospheric circulations and how the vertical temperature gradient (i.e., the static stability) and the horizontal temperature gradient are both determined by the combined effects of the heating and the induced circulation. Depending on the relative strengths of the horizontal and vertical heating gradients, the system may reside either in a convective regime or a stably stratified regime in which the overturning circulation is forced by the horizontal heating contrasts. When scaled to atmospheric dimensions, in which the horizontal length scale is two or three orders of magnitude larger than the depth scale, almost all of the kinetic energy of the forced motions in the stably stratified regime resides in the horizontal wind component. Potential energy is released and kinetic energy is generated when warm fluid rises and cold fluid sinks and fluid parcels flow down the horizontal pressure gradient.

The Tropic World simulation considered in Section 2.4 demonstrates how slowly evolving, planetary-scale

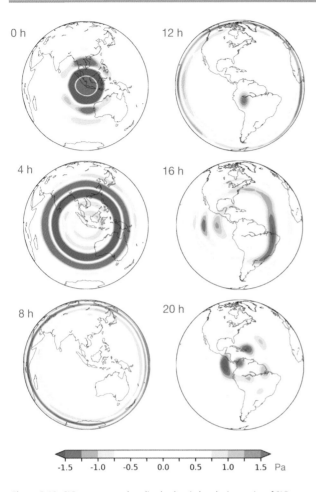

Figure 2.19 SLP response to short-lived pulses in hourly time series of SLP at 180 equally spaced, reference grid points along the 6°S latitude circle, which corresponds to the latitude of Krakatoa. To make these high frequency, rapidly propagating waves stand out in the presence of higher amplitude day-to-day variability and the diurnal tide and its higher harmonics, the SLP data are highpass filtered and the tidal signal is removed. One-point lagged-regression patterns are computed for each of the reference grid points and each one is shifted longitudinally to center it on Krakatoa. The panels in the left column show the waves radiating out from the reference grid points at 6°S in the first 8 hours following the pulses and those in the right column show them approaching their respective antipodal points at 6°N about 20 hours after the pulses. (The orange spot over the upper Amazon in the panel for 12 h is at 6°S, not the antipodal point. It is an artifact of the incomplete removal of the diurnal cycle). For further specifics on the selection criteria for the pulses in the vertical velocity time series, the high pass filtering of the SLP field, and the removal of the diurnal cycle and its higher harmonics, see the extended caption in Appendix F. Courtesy of Hamid A. Pahlavan.

circulations can develop spontaneously in an atmosphere that is initially at rest in radiative-convective equilibrium, even in the absence of horizontal heating gradients. The planetary-scale horizontal transports of energy and moisture and the moisture fluxes at the air sea interface play active roles in the evolution of these circulations. We will discuss how deep convection on Earth is organized on large scales in Chapter 15.

The overturning circulations considered in Sections 2.3 and 2.4 involve sequences of thermodynamic processes that result in the generation of kinetic energy. On average, air parcels in these circulations are gaining energy while they are warm and losing it when they are cold, thereby functioning as a heat engine, doing work while increasing the entropy of the universe. In Section 2.5, we followed an idealized air parcel in the overturning circulations as it picks up latent and sensible heat during its time in the boundary layer, realizes the latent heat as the moisture condenses in the updrafts of deep convective clouds, and cools by emission of infrared radiation while slowly sinking, until it is entrained into the boundary layer again. This sequence of processes resembles the idealized Carnot cycle in some respects, but there is not a one-to-one correspondence between the two. We will revisit thermodynamic diagrams and heat engines in the context of tropical cyclones in Chapter 20.

An important concept that recurs in several places in Part I of this book is *potential temperature* θ. It appears first in Section 1.4.2, where it is defined in Eq. (1.1) and is used in lieu of temperature for documenting the atmosphere's zonally averaged thermal structure (Fig. 1.24). In the hypothetical laboratory experiment described in Section 2.3, potential temperature in a compressible fluid plays a role analogous to specific volume in an incompressible fluid. Potential temperature appears again in Section 2.5 as a measure of entropy, a coordinate on thermodynamic diagrams. In an isentropic (constant entropy) process, no heat is added: in this sense, *isentropic* is synonymous with *adiabatic*. In *isentropic coordinates*, horizontal motions are adiabatic and vertical velocity is expressed in terms of diabatic heating rate Q.

We have seen how the vertical derivative of potential temperature determines the *static stability*, a measure of the *restoring force* that air parcels experience when they are displaced vertically, as defined quantitatively in Eq. (1.2). A *stably stratified* fluid supports the horizontal and vertical propagation of gravity waves, whose *buoyancy frequency* is proportional to the magnitude of the restoring force per unit vertical displacement. Static stability was also shown in Eq. (1.3) to be proportional to the difference between the dry adiabatic lapse rate Γ_d and the local environmental lapse rate Γ. It is much larger in the stratosphere than in the troposphere. The effective static stability of the tropical troposphere is smaller that it would appear to be by these metrics because of the pervasive effects of the release of latent heat in deep convection, as discussed in Chapter 15.

2.10.2 Dynamical Insights

Atmospheric motions are profoundly influenced by the Earth's rotation. In the subtropics and extratropics the horizontal wind field is nearly in geostrophic balance with the pressure field; so much so that the structure

Figure 2.20 Photo from the International Space Station June 12, 2009, showing the eruption of the Sarychev volcano in the Kuril Islands in the Russian Far East. The expanding "hole" in the low cloud deck is due to subsidence in an internal gravity wave radiating outward from the hot plume of gases. Earth Science and Remote Sensing Unit, NASA Johnson Space Center, eol.jsc.nasa.gov

of the wind field can be inferred from distributions of pressure on geopotential surfaces or geopotential height on pressure surfaces. Just as the vertical derivative of potential temperature (with sign reversed) determines the magnitude of the restoring force per unit mass in response to vertical displacements of air parcels, to first order, the Coriolis parameter $f = 2\Omega sin\theta$ determines the restoring force per unit mass in response to meridional displacements. f thus plays a role in determining the *inertial stability*, so named because the Coriolis force is an inertial force. Whereas N, as defined in Eq. (1.2), is the frequency of a *buoyancy oscillation*, f is the frequency of an *inertial oscillation* in which an air parcel executes a circular loop in response to the sideways Coriolis force.

These basic concepts are helpful in interpreting the results of the numerical experiments with the different rotation rates presented in Section 2.6.2. In the absence of rotation, the Hadley cells extend all the way to the poles: there are no westerly jet streams and the equator-to-pole temperature gradient is much weaker than observed. As the rotation rate increases, baroclinic waves develop. They are initially of planetary scale but they shrink in response to the increasing inertial stability. The mean meridional circulations weaken and retreat into lower latitudes, where the inertial stability

is not as strong. The eddy-driven jets and associated higher latitude features in the zonally averaged meridional cross sections become progressively narrower as the meridional displacements become more and more constrained by the higher rotation rates.

Gravity waves are closely related to buoyancy oscillations. They can propagate horizontally as well as vertically, in which case their frequency may be much lower than the buoyancy frequency. In fact, the gravity waves shown in Fig. 2.19 propagate purely horizontally, as explained further in Section 10.2. Rossby waves, on the other hand, are distinct from gravity and inertio-gravity waves. They exist not because of the inertial stability itself, but because on a spherical planet, the Coriolis parameter f increases with latitude. Unlike gravity and inertio-gravity waves in which the horizontal acceleration term is of first order importance, in Rossby waves it is nearly an order of magnitude smaller than the pressure gradient and Coriolis force terms, which are nearly in geostrophic balance.

The eddy statistics shown in Section 1.4.3 are dominated by Rossby waves that exist by virtue of the Earth's rotation. Rossby waves account for more that 90% of the mechanical energy of the global general circulation. To clearly discern the signature of inertio-gravity waves requires examining spectra or high-pass filtering the data in the time domain to emphasize the higher frequencies. The examples shown in Section 2.9 relate to what are referred to as *external IG waves*, which propagate at the speed of sound. These waves are clearly evident in the sea-level pressure field (Fig. 2.19) and they are equally prominent in data for the geopotential height field at stratospheric levels (not shown). More slowly propagating *internal gravity waves*, discussed in Sections 10.2, 15.2, and 21.4 play an important role in the general circulation, imposing geostrophic balance, extending the reach of friction, and spreading the heat released in the updrafts in mesoscale convection.

Part II   Balance Requirements for the General Circulation

Some of the most influential general circulation papers in the late 1940s, 1950s, and 1960s involved the formulation and diagnosis of *balance requirements* that can be applied to any scalar, conserved quantity. Averaged over a sufficiently long time interval, the conserved quantity can be regarded as being "maintained" in steady state such that the sources, sinks, imports, and exports of the conserved quantity averaged over any prescribed region must sum to zero (hence, the moniker "balance requirements"). This conservation principle was first applied to the atmospheric angular momentum budget[1] and subsequently to the energy budget,[2] and to the sources and sinks of water vapor.[3] Lorenz showed how conservation principles can be applied to the atmosphere's mechanical energy cycle, taking into account the generation of what he referred to as "available potential energy" by diabatic heating gradients, the conversion of potential energy to kinetic energy, and frictional dissipation.[4]

Priestly showed how the wind field and the fields of the conserved quantities can be partitioned into contributions from the (time and zonal) mean meridional circulations, the *transients*, and the so-called climatological mean *standing eddies* or stationary waves (see Appendix A).[5] Back at that time, when the analyses were based on irregularly spaced station data, computing these contributions individually was more informative and much less labor intensive than computing the total transports directly.

These balance requirements – angular momentum, mass, total energy, and mechanical energy – are the principal focus of Chapters 3, 4, 5, and 6, respectively. The material is similar to that which Starr presented in his graduate level course at MIT, but the treatment is updated to reflect the important new measurements and atmospheric datasets that have become available since the 1960s. It also incorporates new ways of thinking about the balance requirements themselves and reflects upon what can and cannot be learned from considering them.

[1] Starr (1948), Widger (1949), Starr and White (1951).
[2] White (1951).
[3] (Starr and White, 1955), Starr et al. (1958).
[4] Lorenz (1955).
[5] Priestly (1949).

3 The Angular Momentum Balance

In this chapter, we revisit one of the classical topics of atmospheric dynamics: the maintenance of the zonal mean zonal flow $[u]$ relative to the rotating Earth. The literature on this topic, discussed at length in the monograph of Lorenz,[1] dates back over 300 years and includes works of Edmund Halley (of Halley's comet), George Hadley (of the Hadley cell), and William Ferrel (of the Ferrel cell). Most of the early work on this topic was restricted to consideration of the role of mean meridional circulations in the transport of angular momentum within the atmosphere, a line of inquiry that proved inconclusive. Jeffreys[2] recognized that lateral mixing by large scale eddies must play an essential role in the general circulation, as lucidly explained in Eady's essay on the general circulation.[3] Starr framed the balance requirements in a more explicit and formal way, anticipating the numerous observational studies that would be conducted in the 1950s to show how they are satisfied.[4] By the late 1940s, a sufficiently large archive of upper air wind data was available for making quantitative estimates of the eddy transports.

In the years that followed, most published studies relating to atmospheric angular momentum focused on fundamental questions such as: what are the processes that cause the atmosphere to gain (or lose) angular momentum at the expense of the other components of the Earth system, and how is the observed distribution of surface winds, with tropical trade winds coexisting with midlatitude westerlies, maintained in the presence of frictional drag. Back at that time, researchers were aware of two processes by which the atmosphere gains and loses angular momentum: frictional torque and mountain-related torque. The mountain torque has subsequently come to be recognized as resulting from two somewhat different processes: the torque due to the climatological mean pressure difference across the mountain ranges and the breaking of orographically induced inertio-gravity (IG) waves on all scales. In terms of global and long-term averages, the frictional torque has proven to be the more important.

In the late 1940s it was confirmed that the eddies are indeed instrumental in providing the poleward transport of

angular momentum required to maintain the trade winds and the westerly wind belts, as postulated by Jeffreys. Both transient eddies and stationary waves contribute. The finding that the poleward eddy transport of angular momentum tends to be concentrated at the jet stream level rather than uniformly distributed in the vertical raised the new question of how it is transported upward in the deep tropics and downward in the midlatitudes. The eddy-driven mean meridional circulation cells (i.e., the Hadley and Ferrel cells) were found to be instrumental in producing the required vertical transports. By-products of these investigations were proofs of the existence of the Hadley and Ferrel cells and the first reliable estimates of their amplitudes.

Variations in globally integrated atmospheric angular momentum (AAM) can now be monitored making use of zonal wind data derived from today's global reanalysis datasets in combination with precise measurements of length of day (LOD). Comparing AAM and LOD time series provides a satisfying demonstration of the consistency of precise measurements relating to different components of the Earth system. The availability of reliable AAM measurements obviates the need for many of the qualifications and justifications provided in early general circulation courses.

3.1 ANGULAR MOMENTUM CONSERVATION FOR THE "EARTH SYSTEM" AS A WHOLE

The angular momentum of a zonally symmetric ring of air centered at latitude y and height z and with geometric thickness δy and δz is

$$M = (2\pi R_E \cos\phi \, \delta y \, \delta z) \, [\rho m], \qquad (3.1)$$

where the quantity in parentheses is the volume of the ring and

$$m = R_E \cos\phi \, (\Omega R_E \cos\phi + u) \qquad (3.2)$$

is the angular momentum per unit mass. Here R_E is the radius of the Earth 6.37×10^6 m and Ω its angular velocity 7.29×10^{-5} s^{-1}. The first term in Eq. (3.2) is the angular momentum associated with the Earth's solid body rotation and the second is that associated with the zonal motions relative to the rotating Earth. In regions of westerlies ($u > 0$) the atmosphere is rotating slightly faster than the Earth and

---

[1] Lorenz (1967).
[2] Jeffreys (1926).
[3] Eady (1950).
[4] Starr (1948).

in regions of easterlies ($u < 0$) it isn't rotating quite as fast. Since $\Omega R_E = 465$ m s^{-1}, it is clear from Fig. 2.15 that the solid body rotation term is dominant, except at very high latitudes.

Neglecting longitudinal variations in density so that $[\rho m] = [\rho][m]$, Eq. (3.1) can be written as

$$M = (2\pi R_E \cos\phi \; \delta y \; \delta z)[\rho][m], \qquad (3.3)$$

in which the mass and angular momentum per unit mass appear explicitly, or as

$$M = (2\pi R_E^2 \cos^2\phi \; \delta y \; \delta z[\rho]) \, (\Omega\cos\phi + [u]), \qquad (3.4)$$

which may be recognized as the product of the moment of inertia times the angular velocity in absolute coordinates.

The total atmospheric angular momentum AAM is obtained by integrating m over the mass of the atmosphere (i.e., from the top $p = 0$ to bottom $p = p_s$) in three steps: a vertical integration

$$\frac{1}{g} \int_0^{p_s} m \, dp,$$

followed by a zonal integration

$$\frac{R_E \cos\phi}{g} \int_0^{2\pi} \int_0^{p_s} m \, dp \, d\lambda,$$

which, ignoring the small contributions from the covariance between p_s and m along the latitude circle, reduces to

$$\frac{2\pi R_E \cos\phi}{g} \int_0^{[p_s]} [m] \, dp,$$

where [ ] denotes the zonal average, and g is the acceleration due to gravity. Finally, we perform a pole-to-pole integration, which yields

$$\text{AAM} = \frac{2\pi R_E^2}{g} \int_{-\pi/2}^{\pi/2} \int_0^{[p_s]} [m] \, \cos\phi \, dp \, d\phi. \qquad (3.5)$$

Substituting for m from Eq. (3.2), we obtain

$$\text{AAM} = \frac{2\pi R_E^4 \Omega}{g} \int_{-\pi/2}^{\pi/2} [p_s] \, \cos^3\phi \, d\phi$$
$$+ \frac{2\pi R_E^3}{g} \int_{-\pi/2}^{\pi/2} \int_0^{[p_s]} [u] \, \cos^2\phi \, dp \, d\phi. \qquad (3.6)$$

The first term is by far the larger of the two terms, but its variability is very small on time-scales of years or less, with which this book is concerned. Temporal variations in total AAM on the scales of interest are determined almost exclusively by the second term, which involves the zonal component of the motions relative to the rotating Earth.

For the atmosphere, oceans, and solid Earth as a single system, total angular momentum is conserved, except for tidal interactions with the moon, which are small. Subtle changes in the angular momentum of the solid Earth can be inferred from measurements of LOD. If the angular momentum of the molten core of the Earth is assumed to be constant on timescales of a few years or less, and if the exchange of angular momentum with the oceans is

neglected, changes Δ in LOD and angular momentum of the solid Earth should obey the linear relation

$$\Delta\text{LOD} = 1.68 \times 10^{-29} \, \Delta\text{AAM}, \qquad (3.7)$$

where ΔLOD is expressed in units of milliseconds and AAM is in kg m^2 s^{-1}.[5] The mean length of the day has varied by several milliseconds on the decadal timescale over the course of the past century which, according to the above relation, implies gains and losses in angular momentum an order of magnitude larger than the atmosphere could have possibly experienced on these timescales. These gradual changes are believed to be due to core-mantle coupling.[6]

The conservation of angular momentum for the atmosphere alone can be expressed in the form

$$\frac{d}{dt}\text{AAM} = \sum_i T_i, \qquad (3.8)$$

where d/dt is the time tendency and T_i are the vertically integrated torques resulting from the transfer of angular momentum between the atmosphere and the solid Earth. The torques T_i are the zonal forces F_i weighted by the moment arm $R_E \cos\phi$. Integrated over the surface of the Earth, they are

$$2\pi R_E^3 \int_{-\pi/2}^{\pi/2} [F_i] \, \cos^2\phi \, d\phi. \qquad (3.9)$$

Angular momentum is transferred between the atmosphere and the solid Earth by means of pressure torques and frictional torques.

The *pressure torque*, analogous to the form drag on an airfoil, derives from the pressure differences across mountain ranges. If the pressure is higher on the western side of a range than on the eastern side, angular momentum is transferred from the atmosphere to the solid Earth; that is, the air pushes the mountain range eastward, thereby increasing the angular momentum of the solid Earth at the expense of its own angular momentum, and vice versa. The globally integrated pressure torque is given by

$$-2\pi R_E^3 \int_{-\pi/2}^{\pi/2} \left[p_s \frac{dH}{dx} \right] \cos^2\phi \, d\phi, \qquad (3.10)$$

where p_s is the surface pressure and H is the terrain height. The pressure torque is restricted to the layer below the highest resolved terrain.

Even in models with horizontal resolutions on the order of tens of kilometers, the pressure torque, as inferred from Eq. (3.10), is seriously underestimated, because much of it is produced by terrain features that are not resolved. We will show in Section 11.2 that simply by enhancing the surface frictional drag (roughness) over the mountains in the frictional parameterization scheme, it is possible to obtain quite realistic simulations of the general circulation using models with horizontal resolutions comparable to those in

---

[5] Rosen and Salstein (1983).
[6] Lambeck (2005).

Figure 3.1 MODIS image for January 27, 2004 showing patterns of lee-wave clouds downstream of the Sandwich Islands in the South Atlantic that resemble the wakes of ships moving through the water. The flow is from left to right and the islands are situated at the vertices of the V-shaped cloud formations. The individual cloud bands in the pattern correspond to ridges of gravity waves like the one depicted schematically in the next figure. Flow over mountain ranges generates more complex wave patterns, but the individual waves are qualitatively similar to the ones shown here. Image courtesy Jacques Descloitres, MODIS Land Rapid Response Team at NASA GSFC. NASA Earth Observatory.

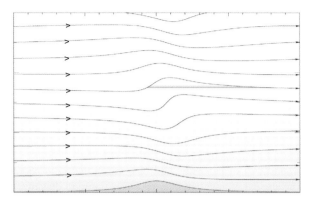

Figure 3.2 Streamlines in a vertically propagating gravity wave in flow over a mountain range with a wave cloud embedded in it, indicated by the lens-shaped unshaded feature. In this sketch, the flow is transverse to the mountain range from left to right and the elongated wave cloud parallels the mountain crest. If the flow is adiabatic, the shape of the streamlines can be inferred from the undulations of the potential temperature surfaces, which has been documented in numerous research aircraft missions over the years. That the westerly momentum of the descending air is greater than that of the ascending air is a consequence of the upstream tilt of wave crests with increasing height. The waves are dispersing upward into the stratosphere, where they will eventually break, extracting westerly momentum from the flow at the level where they break that will be transmitted to the solid Earth. Courtesy of Dale Durran.

the reanalyses. In these models, the missing pressure torque associated with the subgrid scale orography is, in effect, incorporated into the frictional torque term.

The atmosphere exerts a *frictional torque* on the ocean or land below it in the same sense as the surface wind; it is acting to slow down the rotation of the solid Earth equatorward of 30° latitude, where easterlies prevail, and to increase it in the midlatitude westerly wind belt. From the point of view of the atmosphere, the frictional torque acts as a source of angular momentum in the trade-wind belt and a sink in the westerlies. The globally integrated frictional torque on the atmosphere is given by

$$-2\pi R_E^3 \int_{-\pi/2}^{\pi/2} [\tau_x] \cos^2 \phi \, d\phi, \qquad (3.11)$$

where the zonal surface stress τ_x exerted by the atmosphere upon the underlying surface is determined from the surface wind using the bulk aerodynamic formula

$$\tau_x = \rho \, C_D \, (Vu), \qquad (3.12)$$

where V is the scalar wind speed and u the zonal wind component at the anemometer level and C_D is an empirically determined, dimensionless drag coefficient which is on the order of 1.2×10^{-3} over the sea,[7] and it increases slowly with wind speed. For a time average such as a climatological mean, τ_x involves both mean and transient terms. If we ignore the contribution of the meridional wind to the wind speed, the scalar value of the stress is given by

$$\overline{\tau_x} = \rho C_D (\overline{u}^2 + \overline{u'^2}).$$

Because of the nonlinearity inherent in these relationships, the stress tends to be concentrated in regions of high surface wind speeds and it also tends to be larger in regions of large temporal variability (such as the storm tracks) than over regions in which the winds vary less from one day to the next, such as the trade-wind belts.

Some of the exchange of angular momentum between the atmosphere and the solid Earth by the pressure torques is mediated by unresolved, orographically induced, vertically propagating gravity and IG waves like the ones shown in Fig. 3.1. The cloud bands correspond to the crests of mountain lee waves like the ones depicted in the schematic shown in Fig. 3.2. They can be oriented in any direction, depending upon the synoptic situation. Because the wave axes tilt upstream with increasing height, the subsiding air downstream of the crests possesses more momentum than the ascending air upstream of the crests. Hence, at any given level the waves are transporting momentum downward. Theory tells us the momentum that they are transporting downward is extracted from the air in the layer in which they break

[7] Smith (1980); Large and Pond (1981).

(i.e., where the undulating flow within them degenerates into smaller scale turbulence). Rather than being diffused downward, this momentum is transported directly by the pressure and velocity perturbations in waves, without any exchange with the air in the intervening layer. The level at which the wave-breaking occurs is determined by the vertical wind shear and the static stability that they encounter along the way. Many of the waves break in the vicinity of the jet stream level. Those that remain intact may continue to disperse upward, their displacement amplitudes increasing in inverse proportion to air density, until they reach the mesosphere. The angular momentum that the waves transport downward is deposited in the layer in which they are generated by pressure differences between the upwind and lee sides of mountain ranges, whereupon it is immediately transferred to the solid Earth, either by the resolved pressure torques or by the unresolved torques that are represented in the models as an enhanced surface roughness. Sections 9.3 and 10.3 offer a more quantitative treatment of the role of wave dispersion in the meridional and vertical transport of momentum in the general circulation.

The zonally symmetric flow in the oceans is much weaker than the zonal winds. It is strongest at ∼55°S, the only latitude belt in which the zonal flow is not interrupted by the presence of a continent. Hence, despite the much greater mass of the ocean, changes in the storage of angular momentum in the oceans play only a minor role in the Earth's angular momentum budget. The ocean serves as a conduit for the transfer of angular momentum from the atmosphere to the Earth's crust. The zonal mean wind stress on the ocean surface gives rise to zonal gradients in sea level that balance it, as documented in Fig. 3.3. For example, in the trade-wind belts sea level is higher on the western side of the oceans than on the eastern side. As a hydrostatic response to the gradients in sea level, the ocean pushes the continental shelves westward, as depicted schematically in Fig. 3.4, extracting angular momentum from the Earth's crust that replaces what was lost to the atmosphere due to frictional drag. In the westerly wind belts the angular momentum extracted from the atmosphere by the frictional torque is imparted to the Earth's crust by the same mechanism.

3.1.1 Temporal Variations in Atmospheric Angular Momentum

Now that we have defined AAM and identified the torques that serve as its sources or sinks, we are in a position to consider how the conservation of angular momentum plays out in the Earth system. Extended time series of AAM are shown in Fig. 3.5, together with LOD in milliseconds (ms) scaled in accordance with Eq. (3.7). Two AAM time series are shown: one for total AAM and the other for *relative* AAM (i.e., the second term on the right-hand side of Eq. (3.6)) alone. [Note that the relative angular momentum is only a small fraction of the total, but it dominates the variability.] Scaled in this manner, the three time series are virtually identical. When seasonal variations are included, instantaneous values

of relative AAM have ranged over almost a factor of two in recent decades. Prominent features of the time series are the pronounced dips that occur during the boreal summer of each year, when the Tibetan anticyclone is present at the jet stream level and the Northern Hemisphere westerlies are shifted far to the north of it. The r.m.s. amplitude of the AAM anomalies about the seasonally varying climatology is not as large as that of the seasonal cycle variations.

A daily time series of the time rate of the change of AAM is shown in Fig. 3.6 together with the sum of the pressure torque and the frictional torque terms, as defined above. Given (1) the fact that the mountains are not fully resolved in the 1° × 1° gridded dataset used in the calculation of the pressure torque, (2) the crudeness of the frictional parameterization scheme, and (3) the crudeness of the datasets that were available before the reanalysis products became available in the 1990s, the similarity between the two curves (the correlation coefficient based on 4 years of data is 0.86) is quite remarkable.

Now let us consider the roles of the pressure and frictional torques in generating and damping AAM anomalies. Figure 3.7 shows SLP (departures from the seasonally varying climatology) regressed upon the instantaneous AAM tendency, estimated using a 2-day centered difference. The pattern exhibits a distinctive pressure torque signature, with positive SLP anomalies to the east of the major mountain ranges, increasing the AAM by pushing backward against the Earth's crust. The prominent center of action in the Pacific to the west of Chile is the signature of one of the Southern Hemisphere's preferred wave patterns: when pressure is anomalously low over this region, it tends to be anomalously high to the east of the Andes, and vice versa.

Frequency spectra of the time series of the AAM tendency and the pressure and frictional torques are shown in Fig. 3.8. For fluctuations with periods less than ∼10 days, the AAM tendency and pressure torque terms are of comparable amplitude and the frictional torque is about an order of magnitude smaller. In this range of frequencies, variations in AAM are stochastically forced by variations in the pressure torque; that is, $d(\text{AAM})/dt$ varies in phase with the pressure torque, so that variations in AAM lag those in the pressure torque by about 1/4 cycle, consistent with Fig. 3.7.[8] In contrast, in variations with periods of seasons or longer, the frictional torque is among the leading terms. In the absence of an AAM tendency, the sum of the torques must be equal to zero.

Exercises

3.1   In the annual mean, the zonally averaged zonal wind $[u]$ at the tropospheric jet stream level is close to zero at 10°N/S. If a zonally symmetric ring of air at this latitude moved poleward while conserving angular momentum, what would its velocity be when it reached 30°N/S?

---

[8] Madden and Speth (1995).

Sea level, $\overline{\tau}$

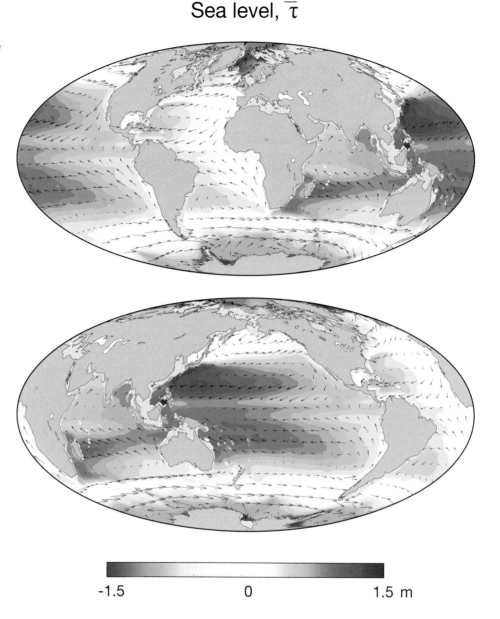

Figure 3.3 Annual mean steric sea level relative to an arbitrary reference level (colored shading) and wind stress on the ocean surface (vectors). The longest vectors correspond to 0.25 N m^{-2}. Courtesy of Xianyao Chen.

3.2 It is shown in Fig. 3.5 that the relative atmospheric angular momentum is roughly 15×10^{25} kg m^2 s^{-1}. How large a superrotation relative to the rotating Earth is this equivalent to?

3.3 Estimate the value of the solid body rotation term in Eq. (3.6), ignoring meridional variations in surface pressure. Rather than performing the indicated integration, a solution can be obtained by making use of the formula for moment of inertia of a spherical shell.

3.4 In units of 10^{29} kg m^2 s^{-1}, the angular momentum of the solid Earth, the atmosphere's solid body rotation $\Omega R_E^2 \cos^2 \phi$ integrated over the mass of the atmosphere, and the atmosphere's zonal winds relative to the rotating Earth are 58 600, 0.101, and 0.0013, respectively. If the atmosphere were to come to a state of rest under the conservation of angular momentum, by how much would the length of the day change?

3.5 Using the estimate of atmospheric angular momentum in Exercise 3.4 and estimates of the eastward mass transport in the Antarctic Circumpolar Current on the internet, make a rough, order of magnitude comparison of the relative angular momentum stored in the atmosphere versus the oceans.

3.6 Show that for a zonally symmetric ring of air, angular momentum per unit mass is equal to the circulation (as in Kelvin's circulation theorem) divided by 2π, provided that longitudinal variations in density can be ignored.

3.7 Prove that on any pressure surface that doesn't intersect the ground the zonally averaged zonal

Figure 3.4 Schematic of the zonal mean momentum balance along a latitude circle in an ocean basin. The westward wind stress τ_x on the ocean due to the trade winds causes the sea level to slope upward slightly toward the west, producing a net westward pressure difference δP_x on the continental shelves that acts as a torque on the solid Earth. The force on the shelves is equal to the surface integral of τ_x.

pressure gradient force $-[\partial\Phi/\partial x]$ is identically equal to zero.

3.8 The accompanying figure shows the tendency in $[u]$ observed in association with positive SLP anomalies at grid points to the east of the Rocky and Andes mountain ranges. Explain why the tendencies are positive. How will the positive tendencies affect the frictional torque?

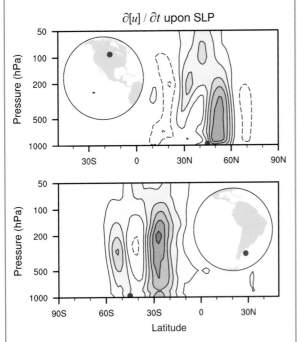

$\partial[u] / \partial t$ upon SLP

Figure for Exercise 3.8. $d[u]/dt$ regressed upon time series of standardized SLP anomalies at the grid points indicated by the red circles. Contour interval 0.01 m s^{-1}d^{-1}. Dashed contours indicate negative regression coefficients. The zero contour is omitted. Courtesy of Ying Li.

3.2 POLEWARD TRANSPORT OF ATMOSPHERIC ANGULAR MOMENTUM

Figures 1.12 and 1.13 show the climatological-mean distribution of near-surface winds. Bearing in mind that frictional drag in the boundary layer is always acting to reduce the surface wind speed, it is evident that the portion of the atmosphere that lies within 30° of the equator, where the trade winds prevail, is gaining angular momentum from the underlying surface, while the part that lies in the westerly wind belt poleward of 30°N/S is losing angular momentum to it. It follows that there exists a balance requirement for a poleward transport of angular momentum across 30°N/S. This requirement is fundamental to Earth-like planets and is not an accident of the configuration of the Earth's major mountain ranges: it exists in observations as well as in aquaplanet model simulations such as the one shown in Fig. 2.11. The nature of the balance is illustrated schematically in Fig. 3.9. It can be expressed quantitatively as

$$-2\pi R_E^3 \int_{0°}^{30°} [\overline{\tau_x}] \, \cos\phi \, d\phi = \frac{2\pi R_E^2 \cos\phi}{g} \int_0^{p_s} [\overline{mv}] \, dp \bigg|_{30°},$$
(3.13)

where $[\overline{\tau_x}]$ is the stress exerted by the atmosphere upon the underlying surface and the overbar refers to time averages over a long time interval such as a season. In order to keep the discussion focused on the most essential elements of the dynamics, we are emphasizing the frictional torque term, which is larger than the pressure torque in the long-term average and is present even in aquaplanet simulations.

The poleward transport of atmospheric angular momentum at 30°N/S is given by the integral

$$\frac{2\pi R_E \cos\phi}{g} \int_0^{p_s} [\overline{mv}] \, dp \bigg|_{30°} = \frac{2\pi \Omega R_E^3 \cos^3\phi}{g} \int_0^{p_s} [\overline{v}] \, dp \bigg|_{30°}$$
$$+ \frac{2\pi R_E^2 \cos^2\phi}{g} \int_0^{p_s} [\overline{uv}] \, dp \bigg|_{30°}.$$
(3.14)

The integration is performed over an imaginary wall extending all the way around a latitude circle at 30°N latitude, and from the Earth's surface to the "top" of the atmosphere.

The first term on the right-hand side represents the transport of the component of the angular momentum associated with the Earth's solid body rotation. It is proportional to the net poleward flux of mass across the latitude circle: if the atmosphere, as a whole, were to draw closer to its axis of rotation by moving poleward, it would rotate faster in the same sense as the Earth's rotation (from west to east), just as ice skaters spin more rapidly by drawing in their arms. The center of mass of the Earth's atmosphere undergoes a small latitudinal shift with the seasons, but it is not anywhere near large enough to account for the observed seasonal changes in zonal wind speed. When we examine the mass balance in Chapter 4, we will see that in the statistical average, water vapor is always being transported

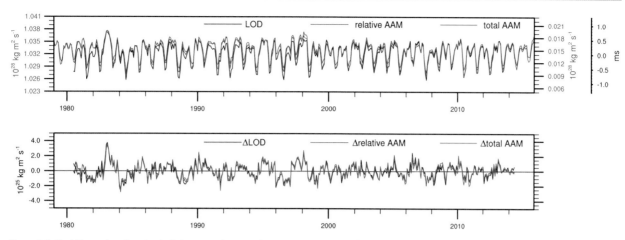

Figure 3.5 (Top) Time series of total and relative atmospheric angular momentum AAM as defined in Eq. (3.6) indicated by the red and blue curves, respectively, together with length of day LOD indicated by the black curve. The interdecadal variability in LOD attributable to motions in the Earth's mantle has been removed and the curve has been scaled in accordance with Eq. (3.7) to make it compatible with the AAM curves. The scales at the left and right are different because the total AAM contains the almost constant solid body rotation component. The LOD scale in units of milliseconds is also included at the far right. (Bottom) The corresponding anomalies with respect to the seasonally varying climatological means. Courtesy of Ying Li.

Figure 3.6 Daily time series of the atmospheric angular momentum (AAM) tendency (dotted) and the sum of the frictional and pressure torques (solid). From Iskenderian and Salstein (1998). © American Meteorological Society. Used with permission.

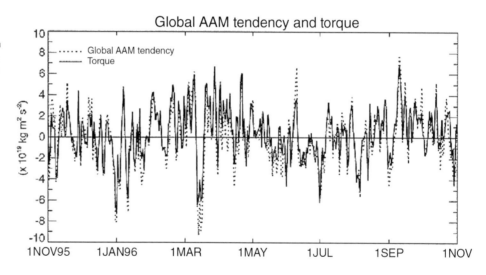

Figure 3.7 SLP regressed upon standardized AAM tendency based on a 2-day centered difference. Courtesy of Ying Li.

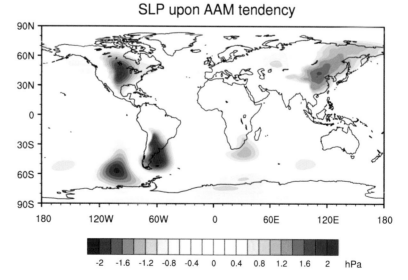

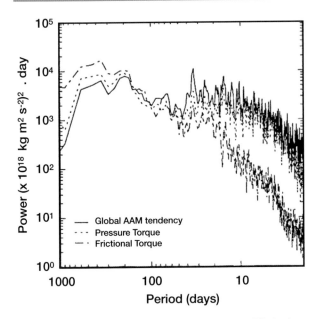

Figure 3.8 Power spectra of AAM tendency, pressure torque, and frictional torque on a log–log scale based upon data for the period February 1992 to November 1996. The spectra are smoothed using a five-point moving average in frequency. From Iskenderian and Salstein (1998). © American Meteorological Society. Used with permission.

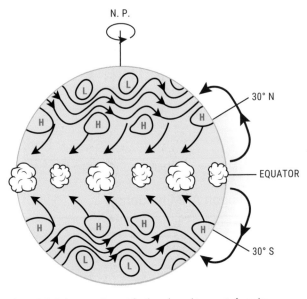

Figure 3.9 Balance requirement for the poleward transport of angular momentum across 30°N/S showing the subtropical trade winds, which serve as a source of AAM by virtue of the frictional torque, and the westerlies at temperate latitudes, which serve as a sink. The bold arrows indicate the flow of angular momentum upward to the tropospheric jet stream level, poleward, and downward, as discussed in the text.

poleward through the midlatitudes. However, it constitutes such a small fraction ($\simeq 1\%$) of the atmospheric mass that its contribution to the poleward angular momentum transport is negligible. In contrast, on Mars, where much of the atmospheric mass condenses into the ice cap at the winter

pole, the mass transport is of first order importance in the angular momentum balance.

The second term is the transport of angular momentum associated with the zonal motion of air parcels relative to the rotating Earth. It involves what is referred to as an *exchange process* (i.e., the exchange of equal amounts of mass containing differing amounts of relative angular momentum). The integrand [uv] can be evaluated by calculating the product uv at each grid point and averaging around latitude circles and over time, in either order. Here we will partition it into the various types of exchange processes, as explained in Appendix A:

$$[\overline{uv}] = [\overline{u}][\overline{v}] + \overline{[u]'[v]'} + [\overline{u^* \, v^*}] + \left[\overline{u^{*\prime} v^{*\prime}}\right]. \qquad (3.15)$$

The first term in this expansion is identified with the climatological and zonally averaged mean meridional motions. Since the net vertically integrated meridional mass transport is vanishingly small, it follows that this term is associated with mean meridional circulation cells with poleward motion at some levels and equatorward motions at others. The observed distribution of mean meridional motions is shown in Fig. 1.27. The $\sim 30°$N/S latitude circles, where the integration is performed, lie in the transition zones between the tropical Hadley cells and the midlatitude Ferrel cells. Hence, it is clear that the mean meridional cell term cannot account for the required transport of angular momentum from the tropics into the middle latitude westerly wind belt.

Throughout most of the troposphere, zonally averaged temperature $[\overline{T}]$ decreases from equator to pole, and zonally averaged zonal wind $[u]$ increases with height, as required by the thermal wind equation (Fig. 1.29). Hence, the transport of angular momentum is in the same sense as $\overline{v}$ in the upper branch of the cell; that is to say, the Hadley cell transports westerly angular momentum poleward and the Ferrel cell transports it equatorward. For example, in the upper branch of the Hadley cell (say, at 15–20° latitude), $[\overline{u}]$ and $[\overline{v}]$ are both positive in the Northern Hemisphere and hence their product $[\overline{u}]\,[\overline{v}]$ is positive there. In the lower branch, $[\overline{v}]$ is negative in the equatorward flow and $[\overline{u}]$ is negative in association with the trade winds. Hence, in the product $[\overline{u}][\overline{v}]$ is positive at low levels as well, and in the vertical integral $\overline{[\overline{u}][\overline{v}]}$ is certainly positive in the Northern Hemisphere and negative in the Southern Hemisphere.

The second term $\overline{[u]'[v]'}$ in Eq. (3.15) involves temporal correlations between [u] and [v] in the transient component of the mean meridional circulations. This term has generally been assumed to be small and has not been included in conventional general circulation atlases. It is shown in Exercise 3.12 that this assumption is justified for an integration performed at 30°N and 30°S.

The eddy transports are represented by the third and fourth terms in Eq. (3.15). Both terms are associated with the meridional tilt of the "eddies" or "waves" in the horizontal plane, as depicted in Fig. 3.10. In waves that tilt eastward

$$\psi = -[u]y + A\sin(kx - my)$$

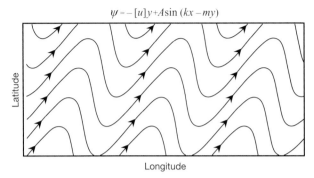

Figure 3.10 Streamlines in waves or eddies that produce a northward transport of westerly momentum superimposed on a uniform zonal mean flow showing the northeast–southwest tilt of the wave axes. The equation used to produce the streamline field is shown in the title. ψ is the streamfunction, A is the wave amplitude, k is the zonal wavenumber, m is a measure of the meridional tilt of the wave axes, and $[u]$ is the zonal mean flow.

with increasing latitude, poleward moving air carries with it more angular momentum (i.e., it has a more positive westerly wind component) than equatorward moving air. At any given level, the exchange of equal masses of air containing differing amounts of angular momentum per unit mass results in a net meridional transport of angular momentum. The third term in Eq. (3.15) represents the transport by standing eddies or stationary waves, longitudinally dependent features of the flow that appear on the climatological maps such as those in Figs. 1.4 and 1.5, while the fourth term represents the transport by the transient eddies.

Figure 3.11 shows the total northward transport of westerly momentum by the eddies $[\overline{u^*v^*}]$: the sum of the transient eddy $[\overline{u'v'}]$ and stationary wave $[\overline{u^*}\overline{v^*}]$ contributions. In the annual mean, the transports are largest and in the appropriate directions precisely where they are needed: that is, they are poleward across 30°N/S, the boundary between the trades and the westerlies. In both hemispheres the transports are strongest during winter, the time of year when the extratropical surface westerlies are strongest, and the strongest transport of westerly momentum across the 30° latitude circle is required to maintain them against frictional drag. The transports are equatorward across 65°N/S, the boundary between the midlatitude westerlies and zones of weak easterlies over the polar cap regions. Figures 3.12 and 3.13 show the contributions of the transient eddies and stationary waves to the total transport. Though the total poleward eddy transports in the Northern and Southern Hemispheres are quite similar, the stationary waves account for a larger fraction of the transport in the Northern Hemisphere owing to the much larger high latitude continents and the more extensive mountain ranges. It will be shown in Chapter 8 that baroclinic waves dominate the poleward transport by the transient eddies. From a careful inspection of Figs. 1.3 and 1.4 it is evident that at latitudes around 30°N, the major features in the climatological mean 200 hPa wintertime geopotential field in the Northern Hemisphere exhibit a

southwest–northeast tilt, consistent with the large poleward transport of angular momentum across this latitude by the stationary waves in Fig. 3.13.

Exercises

3.9 The poleward flux of angular momentum across 30° latitude is on the order of 30×10^{18} kg m^2 s^{-2}.[9] How does that compare with the angular momentum transfer from the tropical oceans to the atmosphere? Assuming that the source of the angular momentum that is transported poleward across 30° is the frictional stress on the zonal component of the trade winds at the Earth's surface, and that the trade winds cover roughly one-fifth of the surface area of the Earth, make a rough estimate of the root mean squared zonal wind speed in the trade-wind belt. [Hint: Apply the bulk aerodynamic formula, using your own rough estimates of the spatially averaged zonal wind and scalar wind speeds in the trade-wind belt at the Earth's surface and ignoring nonlinearities.]

3.10 Consider the balance requirement for the poleward transport of angular momentum across 15° N/S rather than 30° N/S. Compare the magnitudes of the mean meridional circulation and the eddy transport terms on the right-hand side of Eq. (3.14).

3.11 At 30°N at the jet stream level during DJF, $[\overline{u'^2}] \sim [\overline{v'^2}] \sim 120$ m^2 s^{-2}, and $\overline{u'v'} \sim 40$ m^2 s^{-2}. Use the product moment formula in Appendix A, Eqs. (A.17) and (A.18), to estimate the correlation coefficient between u and v in the transients.

3.12 Based on the data shown in Fig. 1.31, estimate the maximum possible contribution of nonseasonal transient variations in $[u]$ and $[v]$ to the annual mean poleward transport of westerly momentum $[uv]$ at 30°N at the jet stream level. Compare this contribution to the transport by transient eddies $\overline{[u'v']}$ in Fig. 3.12.

3.13 Suppose that the eddies consist of waves on an instantaneous map in which the perturbations in u and v both have r.m.s. amplitudes of 10 m s^{-1} and a 60° phase difference. Calculate $[\overline{u^*v^*}]$ and the correlation coefficient between u^* and v^* in the eddies.

3.14 How would the trade wind regimes be different if the Earth were rotating more or less rapidly? [Hint: Consider the experiment described in Section 2.6.2.]

3.15 Analyze the balance requirements for the conservation of angular momentum within an intense and rapidly deepening extratropical cyclone. Consider the transport across a circular "wall" concentric with the center of the cyclone at the Earth's surface

---

[9] For example, see Obasi (1963) and the references therein.

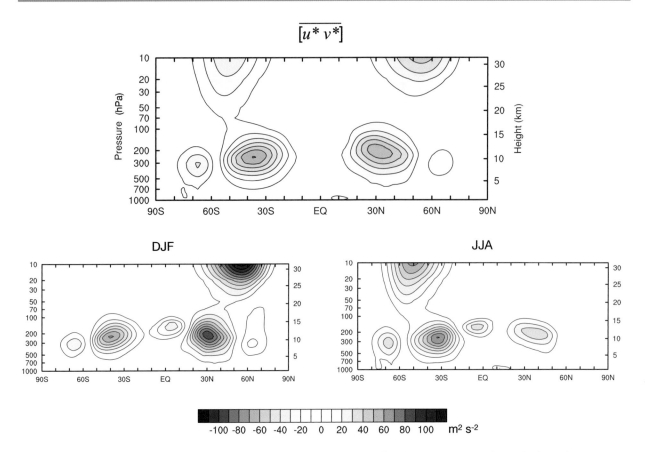

Figure 3.11 Northward transport of westerly momentum by the eddies, the sum of the transient eddy and stationary wave contributions. (Top) annual mean. Contour interval 10 m² s⁻²; zero contour omitted. The domain in this and the next two figures extends upward into the middle stratosphere, so as to include the wintertime polar night jet.

and enclosing the strong and intensifying low-level circulation. Assume that the associated trough at the jet stream level lies about one quarter of a wavelength to the west of the deepening surface cyclone. Take into account the vorticity transports associated with the nondivergent component of the wind at upper levels.

3.16 By means of sketches analogous to Fig. 3.10, show that eddies with axes that tilt eastward with latitude transport westerly momentum poleward regardless of whether a background flow is present or whether it is westerly or easterly.

3.3 THE VERTICAL TRANSPORT OF ANGULAR MOMENTUM

In the previous section, we showed that most of the poleward transport of angular momentum within the Earth's atmosphere takes place near the jet stream level. This begs the question of how the angular momentum gets transferred upward from the boundary layer to the jet stream level equatorward of 30°N/S and back down to the Earth's surface poleward of 30°N/S.

Starr was aware of the requirement for a downward transport of westerly momentum in midlatitudes and he speculated that the eddies might be responsible for that as well.[10] However, it is now well established that, on average, the ascending, poleward flowing air in advance of a typical trough of a wave at the jet stream level carries with it more westerly momentum than the subsiding, equatorward flowing air to the west of the cyclones. For documentation, see the cross sections showing the vertical transport of westerly momentum on the companion web page. Consistent with scaling considerations, the convergence ($\partial/\partial p$) of the vertical momentum fluxes $[u^*\omega^*]$ is almost an order of magnitude smaller than the convergence ($\partial/\partial y$) of the poleward momentum flux $[u^*v^*]$. Hence, even if the eddies transported momentum downward instead of upward as observed, the resulting zonal accelerations would not be large enough to carry the westerly momentum transported poleward by the eddies down to the Earth's surface. It follows that the mean meridional circulations (i.e., the Hadley and Ferrel cells) must be primarily responsible for the vertical transport of angular momentum in the atmospheric general circulation.

[10] Starr (1948).

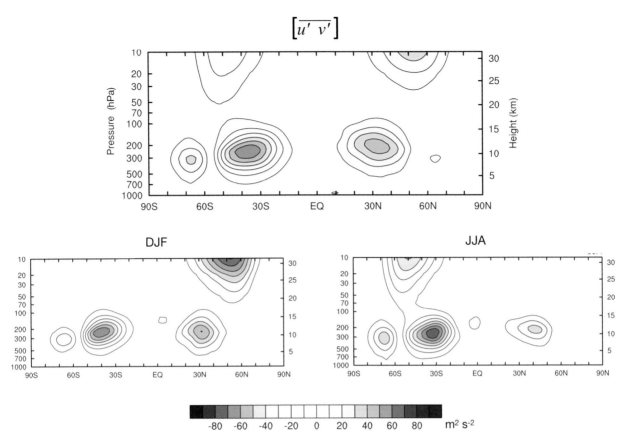

Figure 3.12 As in Fig. 3.11, but for the transient eddy contribution only. Contour interval 10 m² s⁻²; zero contour omitted.

The radial circulation cell depicted in the "spin down" experiment in Fig. 2.16 is the analog of mean meridional cells in the general circulation. Let us examine how such cells transport angular momentum vertically in the mean meridional circulations. The time-averaged upward flux of angular momentum by the zonally averaged mean meridional circulation is given by

$$-\int [\overline{\omega}][\overline{m}]dy = -R_E^2 \int [\overline{\omega}] \left(\Omega R_E \cos^2 \phi + [u]\right) \cos^2 \phi \, d\phi.$$

$$(3.16)$$

If the meridional integration extends over a range of latitudes comparable to the Hadley or Ferrel cells, the first term in the brackets in the above expression exhibits a much wider range of variation with latitude than the second term. For example, for the latitude range of the Hadley cell (0° to 30° latitude), $\Omega R_E \cos \phi$ varies from ~ 465 m s⁻¹ to 402 m s⁻¹ and for the latitude range of the Ferrel cell (30° to 60° latitude) it varies from 402 to 232 m s⁻¹, as shown in Fig. 2.15. Thus the air in the rising branch of the Hadley cell contains much more angular momentum per unit mass than the air in the sinking branch. It follows that the Hadley cell transports angular momentum upward from the planetary boundary layer into the upper troposphere. The eddies at the jet stream level transport this angular momentum poleward into middle latitudes. The Ferrel cell then transports the angular momentum downward into the boundary layer, where it serves to maintain the surface westerlies in the presence of frictional drag.

An alternative way of portraying the role of the mean meridional circulation in the vertical transport of angular momentum is to focus on the associated *Coriolis torques* (i.e., the Coriolis force times moment arm) in their upper and lower branches. Since the meridional mass fluxes crossing through any latitude circle in the upper and lower branches are equal and opposite, it follows that the associated Coriolis torques are equal and opposite so that in effect, the mean meridional circulations extract angular momentum from the flow at one level and deposit it in the flow at the other level. For example, meridionally integrated over the lower branch of the Hadley cell the equatorward flow induces a westward Coriolis torque that serves to maintain the easterly trade winds against friction, while meridionally integrated over the upper branch, the poleward return flow induces an eastward Coriolis torque that accelerates the upper level westerlies. The Hadley cell thus serves to increase the vertical shear of the geostrophic (zonal) flow and the Ferrel cell serves to decrease it. Meridionally integrated over the width of the Hadley and Ferrel cells, the vertical transport of angular momentum in their ascending and descending branches is equal to the transfer of momentum between the upper and lower branches of the cells by the Coriolis torques.

$$\left[\overline{u}^*\overline{v}^*\right]$$

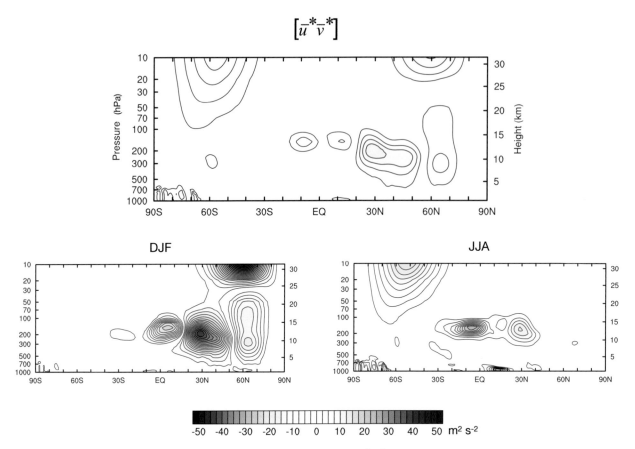

Figure 3.13 As in Fig. 3.11, but for the stationary wave contribution only. Contour interval 2.5 m² s⁻²; zero contour omitted.

It should be emphasized that the arguments presented in this section are not intended to explain the existence of the Hadley and Ferrel cells, but only to describe their role in the angular momentum budget. We will defer consideration of causality to Part III.

3.4 THE LOCAL, ZONALLY AVERAGED ZONAL MOMENTUM BALANCE

In Appendix B (Eq. (B.6)), it is shown that the time rate of change of zonally averaged zonal wind is given by

$$\frac{\partial [u]}{\partial t} = [v]\,[\eta] - [\omega]\frac{\partial [u]}{\partial p} - \frac{1}{\cos^2\phi}\frac{\partial}{\partial y}[u^*v^*]\,\cos^2\phi$$
$$- \frac{\partial}{\partial p}[u^*\omega^*] + [\mathcal{F}_x], \qquad (3.17)$$

where $x = R_E\,\lambda\,\cos\phi$ is the zonal coordinate, $y = R_E\,\cos\phi$ is the meridional coordinate, λ is longitude, ϕ is latitude, $\mathcal{F}_x$ is the zonal component of the frictional drag force, and η is the vertical component of absolute vorticity, the sum of the planetary vorticity f in the horizontal plane that exists by virtue of the Earth's solid body rotation and the *relative vorticity* ζ of the horizontal flow:

$$\eta \equiv f + \zeta = f + \frac{1}{R_E}\frac{\partial v}{\partial \lambda} - \frac{1}{\cos\phi}\frac{\partial}{\partial y}(u\cos\phi). \qquad (3.18)$$

It is readily verified that $[\zeta] = -\partial[u]/\partial y$.

In this section, we will use Eq. (3.17) to infer the direction and strength of the tropospheric mean meridional circulation cells. More specifically, we will evaluate $\overline{[v]}$ at the latitudes and levels in the meridional plane at which the eddy forcing and the frictional drag terms in Eq. (3.17) are strongest; that is, at the jet stream level and in the boundary layer. It is clear from Fig. 1.25 that these are the levels at which $\overline{[v]}$ is strongest, and from Fig. 1.27 that they correspond to the upper and lower branches of the Hadley and Ferrel cells. At these levels $\overline{[\omega]}$ is much smaller than in the mid-troposphere, so that the vertical advection terms in Eq. (3.17) can be neglected.

Neglecting the terms involving the vertical advection, Eq. (3.17) can be rewritten as

$$\frac{\partial [u]}{\partial t} = [v][\eta] + \mathcal{G} + [\mathcal{F}_x], \qquad (3.19)$$

where $\mathcal{G}$ represents the zonal average meridional convergence of the northward transport of westerly momentum by the eddies,

$$\mathcal{G} = -\cos^{-2}\varphi\,\partial/\partial y([u^*v^*]\cos^2\varphi). \qquad (3.20)$$

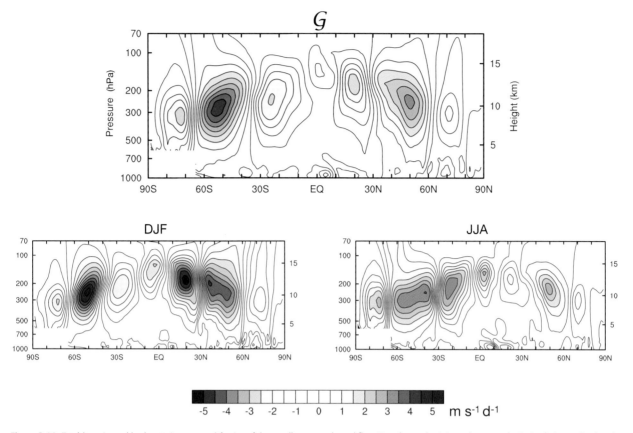

Figure 3.14 Total (transient eddy plus stationary wave) forcing of the zonally averaged zonal flow. Here the emphasis is on the tropospheric circulation so the domain extends upward only to 70 hPa. Contour interval 1 m s^{-1} d^{-1}; zero contour omitted.

Averaged over a season or longer, the time derivative term is much smaller than the individual terms on the right-hand side, so we can write

$$[v] \cong \frac{-\mathcal{G} - [\mathcal{F}_x]}{[\eta]}. \tag{3.21}$$

The numerator in Eq. (3.21) can be interpreted as the forcing of the mean meridional circulations. The absolute vorticity of the zonally symmetric flow $[\eta]$ is $(R_E cos\phi)^{-1}$ times the meridional gradient of angular momentum per unit mass, as shown in Exercise (3.19). It may be interpreted as the forcing required to produce a unit response in $[v]$. $[\eta]$ is thus a measure of the stability of the zonally symmetric flow with respect to perturbations involving mean meridional circulations. It is, in fact, a more exact form of the *inertial stability* introduced in Section 2.10.2, where it was approximated by the Coriolis parameter f. Including the zonally averaged relative vorticity as well as the planetary vorticity in the definition slightly increases the inertial stability on the poleward flank of the zonally averaged tropospheric jet stream and reduces it on the equatorward side. If $f[\eta]$ is less than zero anywhere in the domain or equivalently, if the angular momentum does not decrease monotonically going from equator to pole (see Exercise 3.20), the flow is

said to be *inertially unstable* and unforced mean meridional circulation cells may develop spontaneously and amplify, feeding upon the instability. Inertial instability is sometimes observed close to the equator in the upper atmosphere, where it gives rise to flat cells with aspect ratios reminiscent of a cross section through a stack of pancakes.[11]

As noted previously, the frictional drag term $[F_x]$ is important only in the boundary layer, where it is positive in the trade-wind belt equatorward of 30°N/S and negative in the westerly belt extending from 30 to 60°N/S. The eddy forcing term $\mathcal{G}$, whose distribution is shown in Fig. 3.14, is dominated by a tripole pattern, centered at the jet stream level, with easterly forcing in the subtropics, where the poleward transport of westerly momentum is increasing with latitude, westerly forcing in midlatitudes, where the transport is most rapidly decreasing with latitude, and weak easterly forcing at subpolar latitudes, on the poleward flanks of belts of equatorward transport.

Now let us consider the role of the mean meridional motions in balancing the forcing terms in Eq. (3.21), summarized schematically in the equator-to-pole meridional cross section shown in Fig. 3.15. It is particularly instructive

[11] Hitchman et al. (1987).

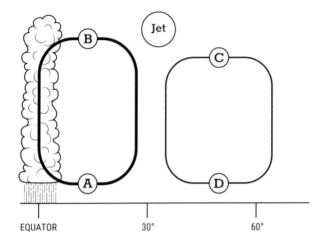

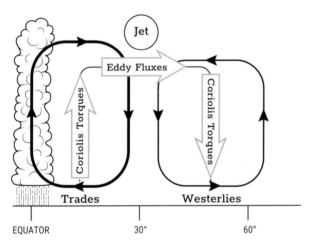

Figure 3.15 Schematic showing mean meridional circulation cells, eddy transports of westerly momentum, the tropospheric jet stream, and the trade-wind and surface westerly wind belts in the winter hemisphere. Top panel shows reference points A, B, C, and D used in the discussion in the text.

to evaluate the terms at points A, B, C, and D where the balances are between just two of the terms. Poleward eddy transports are dominant in the upper troposphere (points B and C), and frictional drag is important only within the boundary layer (points A and D). For convenience we will write $[\eta]$ as f^* (see Eq. (3.18)).

At point A in the figure there must be a balance between $f^*[v]$, which is producing a westward acceleration and the frictional drag $-[\overline{\tau_x}]$ on the surface easterlies. Thus, the easterly trade winds are maintained against frictional drag by the equatorward flow in the lower branch of the Hadley cell. In a similar manner, at D the surface westerlies in the middle latitudes are maintained against frictional dissipation by the poleward flow in the lower branch of the Ferrel cell. This is why the transition between easterly and westerly surface winds, near 30°N/S, coincides exactly with the transition between Hadley and Ferrel cells.

At point B, the poleward eddy transport of zonal momentum is increasing rapidly with latitude. Hence, $\mathcal{G} \ll 0$; i.e, there is a strong divergence of eddy transport out of this latitude belt. Thus at B, the $f^*[v]$ term in the upper branch of the Hadley cell supplies westerly momentum at the same rate that the eddies extract and export it poleward. At point C the picture is just the reverse. The westerly momentum transported poleward by the eddies is converging into this latitude belt so that $\mathcal{G} \gg 0$. The Coriolis term in the upper branch of the Ferrel cell removes westerly momentum as fast as the eddies bring it in, thus maintaining a steady state. We have purposely estimated $[v]$ at latitudes near 15° and 45° degrees, where the signs of the terms are unambiguous.

Exercises

3.17 The accompanying figure shows the distribution of the zonally averaged absolute vorticity $[\eta]$ (see Eq. (3.18)) for the Northern Hemisphere winter season DJF. Explain (with reference to the $[u]$ field) why the profile has the shape that it does, with relatively well-mixed extratropical regions and strong meridional gradients around 30°N.

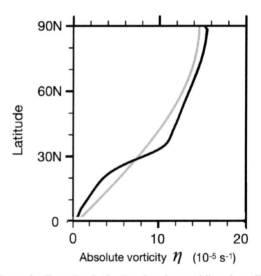

Figure for Exercise 3.17, showing the meridional gradient of absolute vorticity at the 200 hPa level during DJF (black) and the component due to the Earth's solid body rotation f (gray).

3.18 Derive Eq. (3.17) or (B.6), making more explicit use of the conservation of angular momentum. [Hint: follow the protocol in Appendix B using $[m]$ in place of $[u]$ until the final step.]

3.19 Show that, on planetary and synoptic scales, to a close approximation $R_E \cos \phi \, [\eta] = -\partial[m]/\partial y$.

3.20 For the zonal mean flow to be inertially stable, $f[\eta]$ must be positive everywhere in the domain. Show that this is equivalent to the condition that the absolute angular momentum $[m]$ must decrease monotonically going from equator to pole; that is,

$$f\frac{\partial[m]}{\partial y} = -f[\eta]R_E\cos\phi \leq 0.$$

3.21 In the annual mean in both hemispheres, the Hadley cell is much stronger than the Ferrel cell. Reconcile this with the angular momentum balance.

3.22 Consider a zonally symmetric flow with a jet directly over the equator. Is it an easterly jet or a westerly jet for which the flow is inertially unstable?

3.23 Show how the zonal momentum balance is satisfied at point D in Fig. 3.15.

3.24 Using Cartesian geometry for simplicity, prove that if u^* and v^* are nondivergent, $\mathcal{G} \equiv -\partial[u^*v^*]/\partial y = [\zeta^*v^*]$ where ζ^* is the eddy component of the relative vorticity $\partial v^*/\partial x - \partial u^*/\partial y$.

3.25 Prove that $[v] = [v_a]$, where the subscript a denotes the ageostropic component $[v] - [v_g]$.

symmetric rings of air in a manner analogous to the way in which *static stability* inhibits their vertical motion. In effect, the Coriolis force acts as a restoring force in the meridional direction in the same manner as gravitational stratification acts as a restoring force in the vertical. Given the prevalence of positive inertial stability, it follows that significant meridional transports of air cannot occur in the absence of eddy transports of angular momentum or frictional drag. Just as the static stability is proportional to $\partial\theta/\partial p$, the inertial stability is proportional to $f\eta$ (or equivalently to $-f\partial m/\partial y$).

It should be acknowledged that the balance requirement for the conservation of angular momentum considered in this chapter does not explain important features of the zonal wind distribution such as the positions and strengths of the jet streams. That the existence of a poleward transport of angular momentum across 30°N/S can be inferred from the balance requirement does not mean that it follows from the balance requirement. Indeed, the existence of the tropical trade wind and midlatitude westerly wind belts is a consequence of the poleward angular momentum transport, not the cause of it. And the poleward momentum transport is a reflection of the equatorward dispersion of waves originating in the extratropics, as will be explained in Chapter 8.

3.5 CONCLUDING REMARKS

One of the main motivations for the study of the balance requirement for the poleward transport of angular momentum is the centuries-long quest for an explanation for the coexistence of the trade-wind belts equatorward of ~30°N/S and the belts of surface westerlies poleward of that latitude.[12] The upper air observations documented in this chapter confirm Jeffreys' (1926) conclusion that the poleward eddy transport of westerly momentum across 30°N/S is instrumental in maintaining the trade winds and the westerlies against frictional drag.

An important concept introduced in this chapter is *inertial stability*, which inhibits the meridional motions of zonally

Exercises

3.26 (a) Prove that for the vertically integrated horizontal flow, a westerly jet stream on the equator cannot be sustained in the absence of equatorward eddy transports of westerly momentum. (b) Why doesn't this constraint apply level-by-level?

3.27 Are there conditions under which the atmosphere of a planet rotating at the same rate as the Earth might exhibit superrotation?

3.28* Compare the zonally symmetric circulations in the observations and the aquaplanet simulation described in Exercise 1.6 and discuss the implications for the angular momentum balance.

[12] Lorenz (1967).

4 Mass Balance of Atmospheric Trace Constituents

The first studies of the mass balance of atmospheric trace constituents were focused on water vapor. The earliest of these studies were motivated by the fact that the release of latent heat of condensation in precipitation is an important heat source in the global energy budget, the subject of Chapter 5. These early studies also provided new insights into the hydrologic cycle, particularly over land, and were helpful in explaining the observed salinity distribution in the ocean. In addition to considering zonal averages, as in the previous chapter, these studies also examined zonally varying balances for seasonal means, which shed light on features such as the ITCZ and the monsoons. Time-mean circulations were found to play an important role in the water vapor transport, especially in the tropics and subtropics: the transient eddies play an important role only at the higher latitudes. The sparse network of radiosonde observations that was available 50 years ago was sufficient to establish that the transports inferred from the observed distribution of climatological mean evaporation, the source for atmospheric water vapor, and precipitation, the sink, were broadly consistent with the observed transports. Using today's state-of-the-art global reanalysis products, it is possible to estimate the transports more accurately than the precipitation and evaporation. A mass balance approach can also be used to trace the dispersal of other trace species such as CH_4, CO, and CO_2 from point sources of combustion and, in the case of CO_2 to document its exchanges with the biosphere.

Measurements of a single tracer considered in isolation offer only fleeting glimpses of the time-dependent structure of the general circulation, and even collectively, tracer measurements provide only insights, not a complete picture. Yet, these measurements are valuable because they relate to the three-dimensional Lagrangian velocity field, which is different in some respects from the Eulerian velocity field revealed by the gridded data that have been used to create the diagrams in previous chapters. The tracers documented in this chapter are also of interest in their own right: water vapor because of its relationship to precipitation and latent heat release, and the carbon species because of their exchanges with the biosphere and their role in greenhouse warming. Distributions of ozone and other stratospheric tracers are considered in Chapter 9.

For water vapor and other tropospheric tracers x for which the horizontal transports are of primary interest, we consider the mass per unit area X, given by the vertical integral

$$X = \{x\} \equiv \frac{1}{g} \int_0^{p_s} x \, dp. \tag{4.1}$$

For which we can write the conservation equation

$$\frac{\partial X}{\partial t} \simeq \{S^+\} - \{S^-\} - \{Tr(x)\}, \tag{4.2}$$

where $\{S^+\}$ and $\{S^-\}$ are vertically integrated sources and sinks, respectively, and $\{Tr(x)\}$ is rate of change due to the vertically integrated horizontal transport. The sign convention is such that positive values denote *export*; that is,

$$\{Tr(x)\} \equiv \{\nabla \cdot x\mathbf{V}\} \tag{4.3}$$

where $\mathbf{V}$ is the horizontal velocity vector and $x\mathbf{V}$ is the mass transport vector for that constituent. Eq. (4.2) would be exact if the mass flux at the bottom boundary were identically equal to zero, but in pressure coordinates there exists a small apparent mass flux through the boundary due to the combined effects of the terrain slope and the cross-isobar flow at the Earth's surface. The neglected boundary layer term in this expression is small except in regions of steep terrain.

If the distribution of X is not changing with time, the term on the left-hand side of Eq. (4.2) vanishes and we can write

$$\{S^+\} - \{S^-\} = \{Tr(x)\}. \tag{4.4}$$

Horizontally integrated over a reservoir like the stratosphere or the entire atmosphere, the transport term vanishes so that

$$\frac{d}{dt}\mathcal{X} \geq \mathcal{S}^+ - \mathcal{S}^-, \tag{4.5}$$

where the scripted symbols refer to quantities integrated over the entire reservoir. The so-called residence time

$$\tau = \frac{\mathcal{X}}{\mathcal{S}^-} \tag{4.6}$$

is the time required for the sinks to completely drain the reservoir if the sources were turned off while the sinks remain unchanged. For averaging times much longer than the residence time of the constituent, the sink term must be very nearly balanced by the other two terms on the right-hand side of Eq. (4.2) and so the balance requirement expressed in Eq. (4.4) is applicable.

Table 4.1 *Atmospheric concentrations and residence times of the carbon species considered in this chapter. Data circa 2021 from www.esrl.noaa.gov/gmd/ccgg/trends.*

CO_2	415 ppmv	
CH_4	1.9 ppmv	9 years
CO	0.04–0.2 ppmv	2 months

4.1 CARBON SPECIES

The carbon trace species carbon monoxide, CO, methane, CH_4, and carbon dioxide, CO_2, are of particular interest because they are important greenhouse gases whose concentrations vary in space and time and are increasing in response to human activities. Table 4.1 shows present atmospheric concentrations of these species and the residence times for CO and CH_4. CO and CH_4 are both injected into the atmosphere by incomplete combustion: biomass burning is an important source for CO and the sink for both is oxidation within the atmosphere. Wetlands are also an important source for methane. CO_2 cycles back and forth between the atmosphere and the biosphere in synchrony with the annual growth and decay cycle of plants. In the annual average, there is strong cancellation between photosynthesis, which is concentrated during the growing season, and decay, which occurs throughout the year. This annual cycling back and forth between the atmosphere and the biosphere involves only a small fraction of the atmospheric mass of CO_2, but the rates of transfer are orders of magnitude larger than those between the atmosphere and the much more massive inorganic reservoirs in the oceans and the Earth's crust. Hence, it is difficult to define an unambiguous residence time for CO_2.

Figure 4.1 contrasts monthly mean distributions of CO and CH_4 at the 12 km (400 hPa) level for a single month. CO levels are elevated by about a factor of 2 (relative to the global mean background concentration) over southern Africa, the Amazon, and parts of the Maritime Continent in response to biomass burning, much of which is agricultural residue from the previous growing season. In contrast, CH_4 is more uniformly distributed, with a difference of only ~10% between localized peaks and background levels in the extratropics of the Northern Hemisphere. The dominant feature in the CH_4 pattern is the pole-to-pole gradient, which reflects the more widespread burning in the Northern Hemisphere. By virtue of its much longer residence time, the distribution of CH_4 in the atmosphere is not as concentrated within its source regions as that of CO. The smoother time-averaged concentrations shown in Fig. 4.2 more clearly reveal the interhemispheric differences. The long-term mean distribution of CO is much smoother than the mean distribution for a single month because agricultural fires are set in different regions in different months in accordance with the local planting and harvesting seasons.

CO_2 is relatively well mixed in the atmosphere, but it exhibits some interesting structure related to biospheric sources and sinks and atmospheric transports, as shown

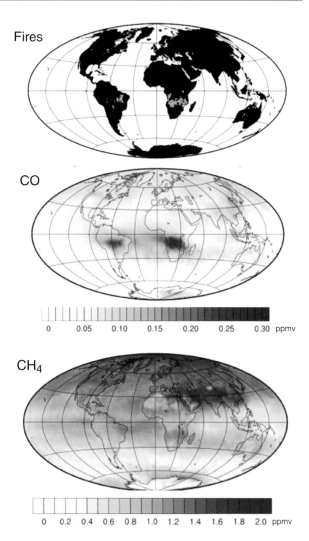

Figure 4.1 Means for September 2002. (Top) Fires, (middle) CO and (bottom) CH_4; CO and CH_4 at 12 km.

in Fig. 4.3. Averaged over the year, CO_2 concentration is highest in the Northern Hemisphere extratropics, where most of the burning of fossil fuels is occurring and lowest over Antarctica, which is surrounded by oceans that serve as a sink for atmospheric CO_2. Air parcels that transit from the Northern Hemisphere into the Southern Hemisphere on trajectories passing over the oceans tend to lose CO_2 along the way. The strength of this meridional gradient is determined by the strength of the difference between the Northern Hemisphere and Southern Hemisphere carbon sources, the timescale required for interhemispheric mixing of air parcels, and the strength of the sink at the ocean surface. It is notable that throughout much of the year, the meridional gradient in CO_2 mixing ratio tends to be concentrated in the equatorial belt, reflecting the slowness of the interhemispheric mixing. CO_2 concentrations in the Southern Hemisphere tropics rise from May to July, as air that has ascended in deep convection in the Asian monsoon flows southward across the equator and descends in the Southern Hemisphere subtropics.

CO

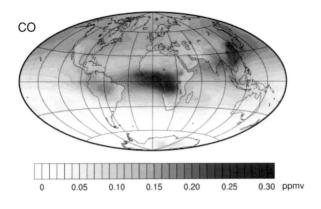

| | | | | | | |
0   0.05   0.10   0.15   0.20   0.25   0.30   ppmv

CH4

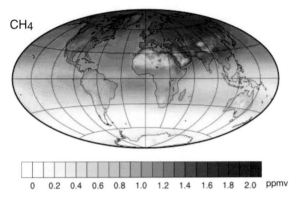

0   0.2   0.4   0.6   0.8   1.0   1.2   1.4   1.6   1.8   2.0   ppmv

Figure 4.2 As in the middle and bottom panels of Fig. 4.1, but averages for 2002–2017.

CO_2 concentrations in the lower troposphere are perturbed by the annual cycle in the uptake of carbon in the boreal forests during their summer growing season, dropping by ~ 10 ppm from May to July over the boreal polar cap region.[1] By September, mixing ratios over the polar cap region have already started to rise, but the deficit created by the biospheric uptake spreads southward as air parcels that resided over the boreal forests during summer mix with air parcels farther to the south.

That CO_2 concentrations decrease with height in the upper atmosphere is due to the general upward trend in CO_2 concentrations due to the burning of fossil fuels. Stratospheric air parcels "remember" what their concentrations were back at the time when they last resided in the troposphere. The strength of the vertical gradient is an indicator of the mixing time: the slower the ventilation of the upper atmosphere with tropospheric air, the "older" the air in the upper atmosphere and hence, the lower its CO_2 concentration. That the vertical gradient tends to be concentrated along the tropopause, even to the point of having a nearly vertical segment along the axis of the tropospheric jet stream during Northern Hemisphere winter, with carbon-rich tropospheric air on its equatorward flank, suggests that the exchange of air between the troposphere and the stratosphere is relatively slow. The strong interhemispheric differences in annual mean CH_4 and CO_2 are suggestive of rather slow mixing of trace species across the tropics.

Exercises

4.1 (a) Prove that Eq. (4.4) is applicable on time scales much longer than the residence time. (b) Show that the residence time of atmospheric CO_2 is much longer if it is estimated on the basis of annual mean data than on the basis of monthly mean data. (c) Why is it difficult to define a meaningful residence time for atmospheric CO_2?

4.2 (a) Consider an idealized tracer with a uniform, steady source of magnitude S in $kg\,s^{-1}$ that covers the entire Northern Hemisphere and a uniform, steady sink of the same magnitude that covers the entire Earth. Assume that the concentrations of the tracer are in steady state. Calculate the mass fluxes across the equator and across 30°N and 30°S, noting that half the surface area of the Earth lies poleward of 30°. (b) Infer the sign of the meridional gradient of tracer concentration. (c) Suppose that in addition to the steady source/sink, there exists a seasonally varying source/sink poleward of 30°N that mimics the drawdown in the boreal forests. How would the atmosphere respond to it?

4.3 How does the presence or absence of an anthropogenic source of CO_2 affect (a) the meridional concentration gradient in the troposphere, and (b) the vertical profile of global mean concentration in the upper atmosphere?

4.2 WATER VAPOR

As applied to atmospheric water vapor, Eq. (4.2) can be written as

$$\frac{\partial W}{\partial t} = E - P - \{Tr(q)\}, \tag{4.7}$$

where W is column-integrated water vapor $\{q\}$, also known as *precipitable water*, E is the evaporation rate at the Earth's surface, including transpiration through the stomata of plants, in units of $kg\,m^{-2}\,s^{-1}$, P is the precipitation rate at the Earth's surface, and $\{Tr(q)\}$ is the vertically integrated horizontal divergence of water vapor transport as given by Eq. (4.3). We ignore the storage of water in the form of cloud droplets, rain drops, and ice particles because cloud liquid (and solid) water content is generally much less than the mass of water in vapor except in deep convective clouds, which cover only a small fraction of the area of the Earth. The horizontal distribution of W is largely determined by concentration of water vapor within and just above the boundary layer, as explained more fully in the next

---

[1] Mixing ratio of a substance is the mass of the substance divided by the mass of dry air.

CO$_2$ 2010

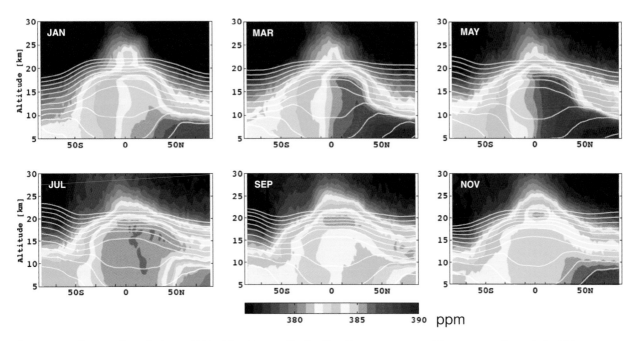

Figure 4.3 Zonally averaged CO$_2$ for the year 2010 as inferred from satellite data. The white contours are potential temperature surfaces as in Fig. 1.24. Adapted from Diallo et al. (2017). Courtesy of Mohamadou Diallo and Eric Ray.

subsection. Owing to the strong temperature dependence of saturation vapor pressure in accordance with the Clausius–Clapeyron equation, climatological mean W decreases with latitude.

The evolving distribution of column-integrated water vapor is intimately related to the atmospheric motion field. Equation (4.1) for the conservation of mass, from which Eq. (4.7) is derived, is similar in form to the equation for the conservation of relative vorticity. Synoptic meteorologists have long observed that front-like features in the temperature field are marked by narrow bands of strong cyclonic vorticity and high values of W. The similarity between the vorticity and the (directly and continuously observed) column-integrated water vapor field is exploited in synoptic meteorology and operational weather forecasting. Figure 4.4 shows a series of snapshots from a high resolution aquaplanet simulation in which the model is "turned on," starting from a state of rest, as in the experiment described in Section 2.6.1. In the top panel, the first generation of baroclinic waves is just beginning to develop. The W field reveals an amplifying wave train with a nearly normal mode-like structure when it first appears, and subsequently develops front-like features, tight cyclonic gyres, and cusps as the flow becomes progressively more nonlinear. The tongues of moist air eventually get drawn out into thin filaments, which lose their identity as they mix with the surrounding, drier air. In this manner, moist air originating in the tropics is irreversibly mixed with drier air at higher latitudes. The mixing of drier extratropical air into the tropics is less obvious, but upon close inspection, it is evident that the dry, subsiding air masses behind the

cold frontal zones gradually penetrate deeper and deeper into the tropics.

The bottom panel of Fig. 4.4 shows the observed W field on a day when these kinds of structures were present over the Pacific sector in association with two developing baroclinic waves. A particularly sharp warm, moist, front-like feature, referred to by synoptic meteorologists as an "atmospheric river," is impinging on the mountain ranges of central California, producing an extended interval of heavy rain rate. These systems entrain tropical moisture.

4.2.1 Mean Distribution of Water Vapor, Sources, and Sinks

Averaged over the Earth's surface, the column-integrated water vapor is ~25 mm and the average daily precipitation P is ~2.5 mm per day. Hence, the mean residence time of water vapor in the atmosphere is on the order of 10 days. Within the troposphere, water vapor concentration decreases exponentially with elevation with a scale height ~2 km. The main reason for this sharp dropoff is the decrease of temperature with height (6.5°C per km for the standard atmosphere) in combination with the strong temperature dependence of the saturation vapor pressure (~7%/°C; a doubling per 10°C). Another contributing factor is the fact that the relative humidity tends to be higher in the boundary layer than in the free atmosphere. In the vicinity of the tropical cold point tropopause (at ~17 km), the mixing ratio of water vapor is almost four orders of magnitude smaller than in the tropical boundary layer.

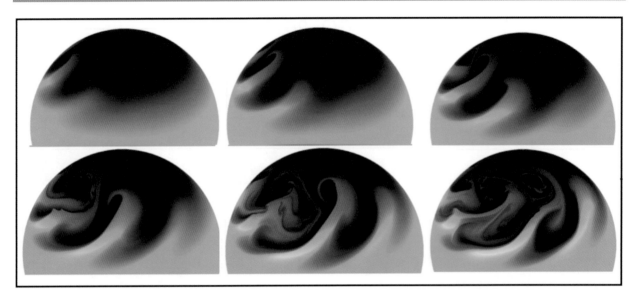

Figure 4.4 (Top) A sequence of six snapshots from a numerical simulation with a high resolution aquaplanet model showing the field of column-integrated water vapor W (brighter colors indicate higher values) provided by the Energy Exascale Earth System Model (E3SM) project supported by the Office of Science/ US Department of Energy. (Bottom) the observed W field on February 16, 2004. Adapted from Ralph et al. (2006). Courtesy of F.M. Ralph. The brightly colored bands in both panels correspond to low-level jets in which warm, moist tropical air flows poleward in the frontal rain bands of extratropical cyclones that develop in association with baroclinic waves.

Figure 4.5 shows a meridional cross section of annual mean specific humidity q extending upward into the stratosphere. The color bar is designed to bring out the subtle contrasts near and just above the tropopause level. At tropospheric levels the contours bulge upward over the tropics, paralleling the isotherms. Like temperature, specific humidity exhibits minimum values over the equator just above the equatorial cold point tropopause. That it increases with height toward the top of the section is a consequence of the oxidation of methane in the upper atmosphere.

The top panel of Fig. 4.6 shows the global distribution of climatological annual mean column water vapor or precipitable water W in units of mm of liquid water. The highest values are over the Indo-Pacific warm pool, the ITCZs, and the continental monsoon regions. The middle and bottom panels show the contributions of the boundary layer (1000–850 hPa) and the free atmosphere (850–500 hPa) to the total. Since the boundary layer relative humidity tends to be spatially rather uniform with values around 80%, the structure in the bottom panel mainly reflects the underlying distribution of skin temperature shown in Fig. 1.9, which determines the saturation vapor pressure at the Earth's surface. A similar structure is apparent in the free atmosphere, but the spatial contrasts are stronger because relative humidity is substantially higher over the moist regions than over the dry regions. The dry zones over the subtropical eastern oceans are much more prominent in the free atmosphere than in the boundary layer.

The observed distribution of annual mean precipitation P is repeated in the top panel of Fig. 4.7. The prominent features in this pattern were described in Chapter 1. In the corresponding zonally averaged distribution shown in

Fig. 4.8, the narrow peak along $7°$N corresponds to the Pacific and Atlantic ITCZs, and the more diffuse secondary peak a few degrees south of the equator to the South Pacific Convergence Zone (SPCZ), the band along the Indian Ocean, and several of the continental rain belts. The schematic circulation cell depicted in Fig. 2.7 can be viewed as a composite of zonally localized Hadley cells with low-level trade winds converging into the Atlantic and Pacific ITCZs and the continental equatorial rain belts. The subtropical dry zones centered near $20°$N/S are clearly evident in zonal averages shown in the top panel of Fig. 4.8 while the broad peaks at $\sim40°$N/S correspond to the extratropical storm tracks, which will be discussed in Chapter 12.

The distribution of evapotranspiration E, shown in the bottom panel of Fig. 4.7 can be inferred by applying the bulk aerodynamic formula[2] to fields of SST, surface humidity, temperature and wind speed derived from a global reanalysis. That E decreases with latitude is a consequence of the meridional gradient in downward radiation at the Earth's surface, which determines how much energy is available to convert water from its liquid state to its gaseous state. That E is largest in the trade-wind belts reflects the influence of surface wind speed, which appears in quadratic form in the bulk aerodynamic formulae.

4.2.2 Water Vapor Transport

For time averages over intervals much longer than the atmospheric residence time of water vapor, Eq. (4.7) reduces to

$$E - P = \{Tr(q)\} = \{\nabla \cdot q\mathbf{V}\}. \qquad (4.8)$$

---

[2] See Chapter 9 of Wallace and Hobbs (2006).

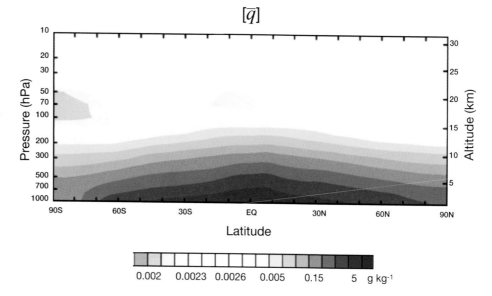

Figure 4.5 Zonally averaged climatological, annual mean specific humidity $[\overline{q}]$. The contour interval is logarithmic down to 0.0026 g kg^{-1} spanning four orders of magnitude, and it is linear below that value, revealing the much weaker gradients at stratospheric levels.

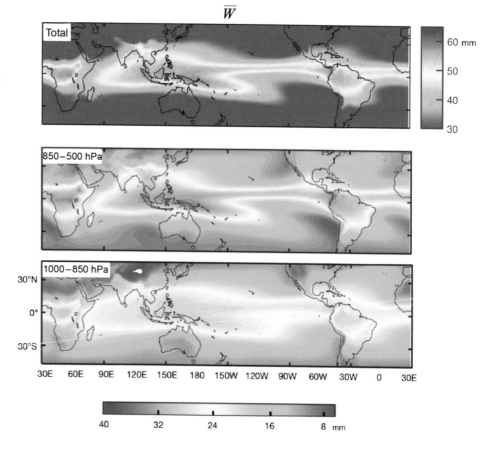

Figure 4.6 The climatological annual mean distribution of column water vapor or precipitable water W in units of millimeters of liquid water. (Top) Total; (middle) the free atmosphere (850–500 hPa) contribution; and (bottom) the boundary layer (1000–850 hPa) contribution. Courtesy of Ángel Adames.

The balance requirement for mass continuity thus requires that water vapor be exported out of regions of excess E and into regions of excess P. The distribution of zonally averaged evaporation minus precipitation $E - P$ is shown in Fig. 4.8 (bottom) and its geographical distribution is shown in Fig. 4.9 (top). Over the major rain belts, $P \gg E$. The patterns of P and $(E - P)$ look rather similar (but for the sign reversal), despite the fact E is far from spatially uniform. The corresponding vertically integrated water vapor transport $\{q\mathbf{V}\}$ indicated as vectors in the top panel of Fig. 4.9 diverges out of the dry zones and converges into the rain belts, as required for mass continuity.

Lest the reader be inclined to think that the water vapor transports *determine* the pattern of $E - P$, it should be

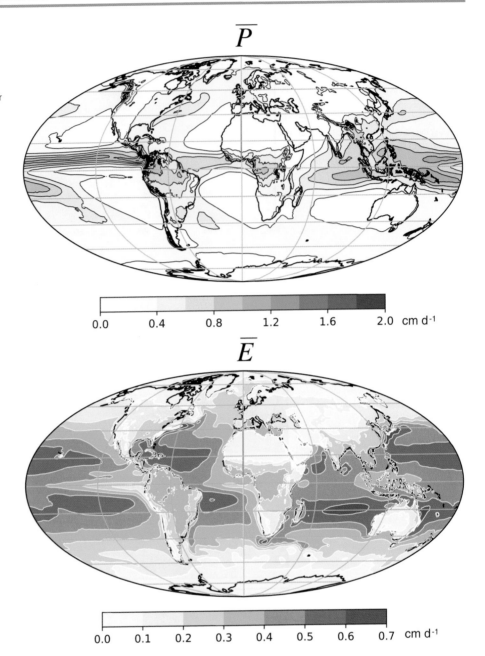

Figure 4.7 Distributions of the climatological, annual mean (top) precipitation and (bottom) evaporation over oceans and evapotranspiration over land; the differences between them are attributable to the transport.

emphasized that the fields of precipitation and water vapor transport are coupled, not only by the conservation of mass as expressed in Eq. (4.8), but also by way of the momentum and thermodynamic energy equations. Hence, the moisture transports can equally well be viewed as the planetary-scale *response* to the release of latent heat in regions of heavy precipitation. The manner in which individual requirements – not only the one for the mass of water vapor, but others as well – are satisfied does not tell the whole story.

Most of the vertically integrated water vapor transport occurs within the boundary layer; that is, $\{q\mathbf{V}\} \sim \{q\mathbf{V}\}_{BL}$,

where

$$\{q\mathbf{V}\}_{BL} \equiv \frac{1}{g} \int_{850}^{1000} q\mathbf{V} dp. \qquad (4.9)$$

Over this limited depth, the horizontal variations in $q\mathbf{V}$ are dominated by gradients in $\mathbf{V}$. Hence, the vertically integrated transport can be approximated as $\{\mathbf{V}\}_{BL}$, scaled by the climatological mean value of q spatially averaged over the tropical oceans. It follows that the vertically integrated moisture transport $\{Tr(q)\} = \{\nabla \cdot q\mathbf{V}\}$ is largely determined by the pattern of horizontal divergence of the boundary layer wind field.

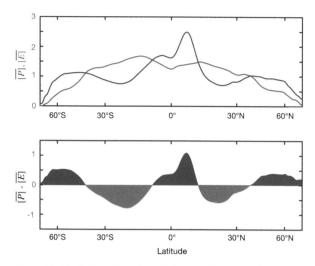

Figure 4.8 Distributions of the climatological, annual mean zonally averaged (top) precipitation $[\overline{P}]$ (blue), and evaporation plus evapotranspiration $[\overline{E}]$ (red). (Bottom) the vertically integrated net moisture sink $[\overline{P}] - [\overline{E}]$ in units of meters per year. Distance on the abscissa is proportional to sine of latitude.

The water vapor transports are revealed more clearly when the vertically integrated moisture transport $\{q\mathbf{V}\}$ in Eq. (4.8) is decomposed into irrotational (also referred to as divergent) and nondivergent (rotational) components:

$$\{q\mathbf{V}\} = \{q\mathbf{V}\}_\chi + \{q\mathbf{V}\}_\psi. \tag{4.10}$$

A similar decomposition can be applied to any vectorial field and we will be making use of it on several occasions later in the book. The transport from source to sink in Eq. (4.8) is determined by the irrotational component,

$$\{Tr(q)\} = \nabla \cdot \{q\mathbf{V}\} = \nabla \cdot \{q\mathbf{V}\}_\chi, \tag{4.11}$$

where

$$\{q\mathbf{V}\}_\chi = -\nabla \chi \tag{4.12}$$

and χ is the moisture weighted velocity potential. Hence,

$$\nabla^2 \chi = -(E - P). \tag{4.13}$$

The irrotational component $\{q\mathbf{V}\}_\chi$ is thus directed from source regions where $E > P$ toward sink regions where $P > E$, as shown in the bottom panel of Fig. 4.9, while the nondivergent component $\{q\mathbf{V}\}_\psi$ tends to circle around extrema of $E - P$.

Now let us consider how the zonally averaged water vapor balance is satisfied. Figure 4.10 shows the zonally averaged divergence of the water vapor transport partitioned into the contributions from the time-mean circulations (the sum of the zonally averaged mean meridional circulations and the stationary waves) and the transients. The former is dominated by the Hadley cell, which transports moist, boundary layer air equatorward and only trace amounts of moisture poleward in its upper branch. The transport of water vapor in the 20–35°N/S belts is poleward (see also Fig. 4.9)

and dominated by the climatological mean stationary waves. The poleward flow of warm, humid, boundary layer air on the western flanks of the climatological mean subtropical anticyclones, juxtaposed against the equatorward flow of cooler, drier air on their eastern flanks, as indicated by the vector field in the top panel of Fig. 4.9, contributes to the poleward transport of moisture in the 20-35° latitude belt, especially during summer. Much of the poleward flow of warm, humid air tends to be concentrated in *low-level jets* along the eastern slopes of mountain ranges, especially during summer.[3] Poleward of 40°, the transient eddies are of first order importance in the poleward transport of water vapor by virtue of the poleward transports in baroclinic waves. Poleward moving air in advance of developing extratropical cyclones tends to be warmer and more humid than the equatorward moving air behind them, so the waves produce a net poleward transport of water vapor. The observed and simulated front-like features in Fig. 4.4 exemplify the poleward transport of water vapor by the transient eddies.

In the frontal bands of vigorous extratropical cyclones, water vapor is transported hundreds and sometimes even thousands of km in the atmospheric branch of the hydrologic cycle before it condenses, whereas in summer "airmass thunderstorms", the source of the water vapor may be evaporation from the Earth's surface within a few hundred kilometers of the precipitation. Knowing where the water molecules entered the atmosphere is essential for interpreting time series of paleoclimate proxies such as ice cores, speleothems, and coral and lake carbonates, which record the isotopic composition of precipitation.

4.2.3 The Land Branch of the Hydrologic Cycle

Mass continuity for the land branch of the hydrologic cycle requires that

$$\frac{\partial \mathcal{W}}{\partial t} + R = (P - E) = -\nabla \cdot \{q\mathbf{V}\}, \tag{4.14}$$

where $\mathcal{W}$ is the vertically integrated water stored in the soil, lakes, and rivers (analogous to, but far greater than the atmospheric column water vapor W) and R is the runoff in rivers and subsurface aquifers. R is analogous to the transport term $\{Tr(q)\}$ in the atmospheric balance Eqn. (4.3), except that the liquid water transport on the land surface is channelized. Whereas the residence time of water vapor in the atmosphere is much less than a month, liquid water and ice may reside in lakes, aquifers, glaciers, and ice sheets over intervals much longer than a year. It follows that the storage term in Eq. (4.14) cannot necessarily be assumed to be in steady state. Variations in $\mathcal{W}$ are reflected in lake levels and water tables indicating the level below which the ground is saturated with water. Apart from the

---

[3] Rasmusson (1968); Stensrud (1996); Marengo et al. (2004).

$$\overline{E\text{-}P}, \{\overline{q\mathbf{V}}\}$$

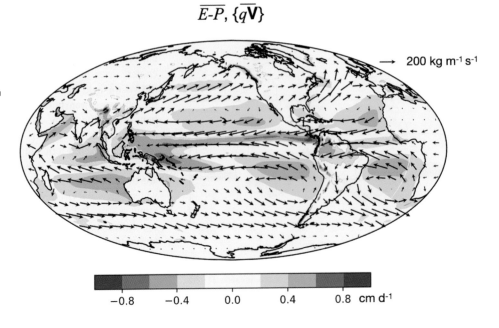

Figure 4.9 Distributions of the climatological, annual mean (top) vertically integrated moisture transport $\{\overline{q\mathbf{V}}\}$ and (bottom) the irrotational component of $\{\overline{q\mathbf{V}}\}$, both superimposed on the field of $\overline{E\text{-}P}$.

$$\overline{E\text{-}P}, \{\overline{q\mathbf{V}}\} \text{ irrotational}$$

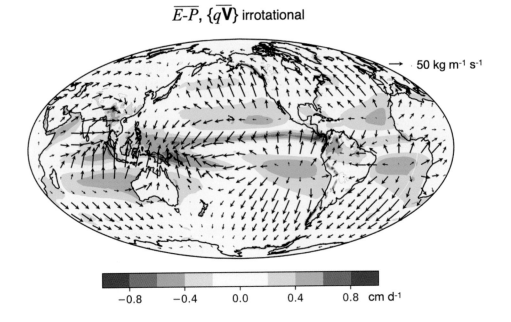

continental ice sheets, most of the storage of water on land is in the form of groundwater, only a few percent of which is recharged on a time-scale shorter than 50 years. The reservoir of "modern" groundwater based on that definition is estimated to be sufficient to cover the continents to a depth of only ∼3 m.[4]

The land branch of the hydrologic cycle is strongly coupled to the terrestrial biosphere. The viability of various plant species depends upon the sufficiency of precipitation during the growing season and on the potential

evaporation, which increases with increasing insolation and daily mean temperature. The vegetation coverage and type strongly influence the actual evaporation. Generally speaking, the more lush the vegetation, the higher the rate of photosynthesis, the higher the rate of evapotranspiration (evaporation plus transpiration from the leaves of plants). Where the agricultural productivity is water limited, growing more food by resorting to the use of fertilizers or relying on more productive plant species requires increasing amounts of irrigation, drawing upon the flow in rivers and/or groundwater. Over many regions dependent on groundwater, the present rate of extraction is much larger than the recharge

[4] Gleeson et al. (2016)

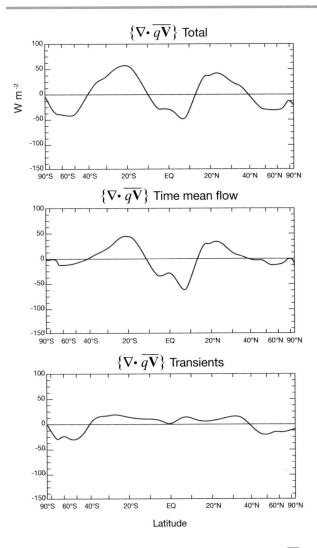

$\{\nabla \cdot \overline{q\mathbf{V}}\}$ Total

$\{\nabla \cdot \overline{q\mathbf{V}}\}$ Time mean flow

$\{\nabla \cdot \overline{q\mathbf{V}}\}$ Transients

Latitude

Figure 4.10 Annual mean zonally averaged moisture flux divergence $\nabla \cdot \{\overline{q\mathbf{V}}\}$ expressed in W m^{-2}. (Top) The total transport. (Middle) The transport by the time-mean circulation, the sum of the contributions from the mean meridional circulations and stationary waves. (Bottom) The contribution from the transient eddies. Adapted from Fig. 2 of Trenberth and Stepaniak (2003). © American Meteorological Society. Used with permission.

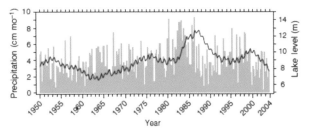

Figure 4.11 The black curve shows variations in the depth of the Great Salt Lake relative to a reference level. Depth scale at right. Blue bars indicate seasonal-mean precipitation at nearby Logan, Utah, scale to the left. Courtesy of John D. Horel and Todd P. Mitchell.

In the interior of large land domains there may be considerable local recycling of water vapor and liquid water, mediated by evapotranspiration from plants. That (1) the health of the plants depends upon the availability of sufficient moisture, (2) the local evaporation depends upon the health of the plants, and (3) the rainfall depends in part upon the local evaporation gives rise to a positive "soil moisture feedback" that may act to prolong droughts and increase the sensitivity of the climate to human-induced alterations in the ground hydrology. Modern reanalysis-based estimates of water vapor transports arguably yield more reliable estimates of $E - P$ than direct measurements. Atmospheric datasets can thus be useful in diagnosing natural and human-induced year-to-year and decade-to-decade variations in stream flow and the extraction of groundwater for irrigation.

The water balance over land exhibits substantial departures from steady state, even in averages over years or decades. Consider, for example, the case of a land-locked watershed such as the Great Basin over the states of Utah and Nevada. Integrated over such a region, $P - E$ must be equal to the time rate of change of storage. Figure 4.11 shows how the level of the Great Salt Lake has varied in response to variations in $P - E$ over its watershed. From the time of its historic low in 1963 to the time of its high in 1987, the level of the lake rose by more than 6 m, its area increased by a factor of 3.5, and its volume by a factor of 4. The average precipitation at a nearby station during this 14-year interval was substantially greater than in the long-term average. The lake level time series is much smoother than the precipitation time series because it reflects the time integral of the precipitation surplus or deficit $P - E$. In this respect, the behavior of lake level is somewhat analogous to a "random walk" process with a weak negative feedback due to the increase in evaporation as a lake expands.

4.2.4 Water Vapor Transport and Ocean Surface Salinity

The annual-mean pattern of surface salinity over the world ocean, shown in Fig. 4.12, bears the imprint of the atmospheric water vapor transports. The most saline regions coincide with the subtropical anticyclones, where $E > P$, resulting in a net loss of fresh water at the sea surface and an increase in salinity of the residual sea water. Another oceanic

rate. Current extraction is increasingly drawing upon the much larger but not inexhaustible reservoir of "fossil groundwater."

For a land domain with well-defined boundaries such as a continent, a drainage basin defined by a river and its tributaries, or a land-locked basin, there must be a long-term balance between the net import of water from the atmosphere, expressed as the surface integral of $P - E$, and the export in rivers and subsurface aquifers. The import can also be expressed as the line integral of the water vapor transport into the domain. As an illustration of this balance requirement, the water vapor transport from sea to land must be equal to the total transport of liquid water from the continent to the surrounding oceans by rivers and subsurface aquifers. It follows that over the continents as a whole, $P > E$.

Surface salinity

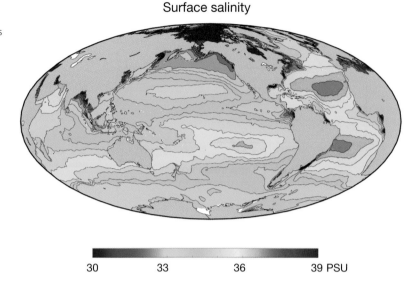

Figure 4.12 Annual mean ocean surface salinity. Values in the Arctic range as low as 18 PSU (g kg^{-1}). From the World Ocean Atlas (2018). Courtesy of Xianyao Chen.

30   33   36   39 PSU

region of high salinity and high surface evaporation is the Mediterranean Sea. In contrast, the water beneath the ITCZs in the rain belts around the Maritime Continent is relatively fresh. Outflow from the delta that encompasses the mouths of the Ganges and Brahmaputra Rivers results in a patch of very fresh water in the Bay of Bengal.

The surface waters of the Atlantic are more saline than those of the Pacific. Contributing factors are (1) the trade-wind flow from the Caribbean across the Isthmus of Panama, which carries with it large amounts of water vapor, as shown in Fig. 4.9 and (2) the presence of the Rocky Mountains, which cause much of the water vapor evaporated from the North Pacific to return in westward flowing rivers, rather than crossing the continent and reaching the Atlantic basin.[5] Over the long term, the transport of atmospheric water vapor from the Atlantic to the Pacific is balanced by a freshening of the ocean water that flows across the Arctic from the Pacific to the Atlantic side and by the influx of water from northward-flowing rivers. Much of the flow of water from the Arctic into the Atlantic is in the form of a virtual river of southward flowing sea ice along the east coast of Greenland.[6]

Exercises

4.4   Estimate the scale height for atmospheric water vapor assuming constant relative humidity and a lapse rate of 6.5°C per km.

4.5   Show that the numerical values of zonally averaged evaporation and precipitation shown in Fig. 4.8 are consistent with the values of water vapor transport shown in Fig. 4.10.

4.6   On the basis of the meridional profile in the top panel of Fig. 4.10, estimate the latitudes of (a) the largest equatorward transport of water vapor; (b) the latitudes at which the poleward transport of water vapor changes sign from negative to positive; and (c) the latitudes at which the poleward transport of water vapor is strongest.

4.7   Based on the water vapor transport data shown in the bottom panel of Fig. 4.9, show that the vertically integrated poleward transport of atmospheric mass associated with the poleward transport of water vapor in the hydrologic cycle does not make an appreciable contribution to maintaining the surface westerlies in midlatitudes.

4.8   Precipitation tends to be much more concentrated in space and time than evaporation. For example, in the annual mean climatologies shown in Fig. 4.7, P ranges up to about twice as large as E locally, and the disparity is even larger in instantaneous distributions. Why is this the case?

4.9   (a) Show that the transport term in Eq. (4.11) can be decomposed into a moisture advection term and a term involving $\nabla \cdot \mathbf{V}$. (b) Show that the term involving the $\nabla \cdot V$ is more important in the boundary layer. (c) Show that much of the structure of the E-P field can be inferred from satellite data alone.

4.10   The relative vorticity field tends to get stretched out into long streamers like the ones in column water vapor (W) in Fig. 4.4, and the bands of high W and strong cyclonic vorticity tend to be collocated. What is it about the governing equations that makes them behave like this?

4.11   How have civilizations structured their agricultural practices to adapt to the constraints inherent in the water vapor budget, as expressed in Eq. (4.14)?

[5] Broecker (1991).
[6] Wijffels et al. (1992).

Figure 4.13 Climatological mean DJF potential vorticity, PV (colored shading), potential temperature, θ (red contours) and angular momentum m (blue contours). The dynamical tropopause, which corresponds to the 2 PVU surface, is indicated by the dotted black line and the tropospheric jet stream is indicated by the **W**. Potential temperature increases with height: contour interval 5 K in the troposphere and 10 K in the stratosphere. The bolded contour corresponds to $\theta = 320$ K. Angular momentum decreases with latitude: contour interval 10^8 m^2 s^{-1}. An increment of distance along the x axis is linearly proportional to sine of latitude rather than latitude, so that it is linearly representative of the area of the corresponding annulus on a sphere. With the x axis defined in this manner and the y axis linearly proportional to pressure, the area of (m, θ) grid cells is inversely proportional to PV.

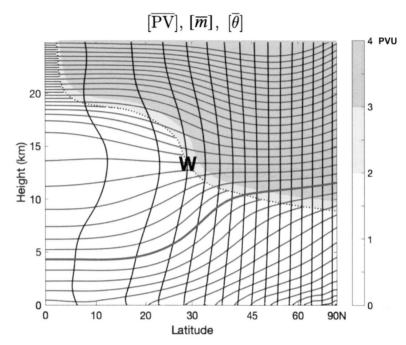

$[\overline{\text{PV}}], [\overline{m}], [\overline{\theta}]$

4.3 POTENTIAL VORTICITY

In addition to the chemical tracers described earlier in this chapter, we will have occasion to refer to a dynamical tracer known as *potential vorticity*, approximated in pressure coordinates as:

$$\text{PV} = \frac{1}{\rho} \, \eta \, \frac{\partial \theta}{\partial z} = -g \, \eta \, \frac{\partial \theta}{\partial p}, \qquad (4.15)$$

where $\eta = \zeta + f$ is the vertical component of absolute vorticity, a measure of the dynamic stability, equal to the $-(R_E \cos \phi^{-1}) \partial [m]/\partial y$ as shown in Exercises 3.19 and 3.20. The vertical gradients of temperature $-\partial \theta/\partial p$ and $\partial \theta/\partial z$ are measures of the static stability. PV has units of K m^2 s^{-1} kg^{-1}, commonly referred to as potential vorticity units PVU. [7]

Figure 4.13 shows the climatological mean PV distribution for the boreal winter season DJF, together with the potential temperature and surfaces of constant angular momentum per unit mass. The reversal of the meridional temperature gradient at the level of the tropospheric jet stream is in accordance with the thermal wind equation, which requires that the θ surfaces be spread apart equatorward of the jet and pinched together poleward of it: that is, that static stability be higher on the poleward flank of a westerly jet than on its equatorward flank. The angular momentum surfaces bulge poleward in the vicinity of the jet in proportion to the zonal wind speed, enhancing $-\partial m/\partial y$ and the absolute vorticity η on the poleward flank of the jet and reducing them on its equatorward flank. These distortions of the contours shrink the (m, θ) grid cells on the poleward flank of the jet and enlarge them on its equatorward flank. PV is thus seen to be large where grid cells are small and vice versa. It is shown in Exercise 4.13 that PV is inversely proportional to the area of grid cells in plots like Fig. 4.13, but with pressure on a linear scale as the vertical coordinate. It follows that a westerly jet corresponds to a dipole in the PV field relative to the background distribution, with high values on its poleward flank and low values on its equatorward flank.

The distribution of climatological mean PV in Fig. 4.13 may be compared with the left panel of Fig. 1.24, which shows the corresponding section for the zonal wind and temperature fields. The pronounced tropopause break at the latitude of the tropospheric jet stream is clearly evident in the PV field, with much higher values, indicative of stratospheric air, poleward of the jet by virtue of the cyclonic shear. The 2 PVU contour, indicated by the black line in Fig. 4.13, closely parallels the tropopause along most of its length.

The instantaneous distribution of PV on potential temperature surfaces tends to be deformed by the horizontal wind field in baroclinic waves, forming extended streamers of high PV air oriented along the frontal zones in developing extratropical cyclones analogous to the streamers of moist air in Fig. 4.4.

---

[7] For further specifics relating to PV, including how it resembles and differs from the mixing ratio of a chemical tracer, see Haynes and McIntyre (1990); see also section 4.5 of Vallis (2017).

Exercises

4.12 Estimate the potential vorticity per unit mass in the core of the annual mean zonal mean tropospheric jet stream in potential vorticity units (1 PVU = 10^{-6} K m^2 kg^{-1} s^{-1}).

4.13 Show that in a meridional cross section with pressure as the ordinate plotted on a linear scale and sine of latitude as the abscissa, potential vorticity PV is inversely proportional to the area enclosed by adjacent, intersecting m and θ surfaces spaced at fixed intervals δm and $\delta\theta$.

4.4 CONCLUDING REMARKS

The narrative in this chapter is largely focused on atmospheric dynamics as it relates to the hydrologic cycle, and to the transport of trace substances in general. It does not do justice to the body of literature in which the emphasis is on the sources, sinks, and transports of specific chemical trace species, motivated by basic research on geochemical cycles. Examples include compounds of nitrogen and sulfur, halogens including CFCs, metals such as mercury, and a variety of stable and radioactive isotopes. Other transport studies, more in the realm of applied research, are motivated by concerns about the adverse impacts of specific chemical agents such as herbicides, pesticides, and various toxic waste products upon human health or environmental quality. Many of these studies make use of so-called *chemical transport models* to trace the trajectories of tagged clouds of the tracer in question as the air moves downstream from the sources and disperses, simulating the chemical transformations that take place along the way. In contrast to GCMs, in which the wind, temperature, and moisture fields evolve in accordance with the governing equations from a prescribed initial state, chemical transport models rely on wind, temperature and moisture fields based on assimilated data derived from operational models or reanalyses. In some of these studies observations relating to the tracers are also assimilated, making use of the same model.

The treatment of the balance requirement for the conservation of mass in this chapter has focused mainly on the vertically integrated transport of trace species, which for the carbon species and water vapor is dominated by the tropospheric transport. The distributions of these same trace species offer insights into the exchange of mass between the troposphere and the stratosphere and transport within the stratosphere. We will reconsider some of them, along with ozone, in Chapter 9, which introduces the reader to the general circulation of the stratosphere. In Section 9.2, we will show how a chemical transport model with sources and sinks can be used to track the spread of a hypothetical tracer released at the Earth's surface as it spreads into the upper atmosphere.

5 The Balance of Total Energy

Total energy connotes the sum of the internal and mechanical (i.e., internal plus potential plus kinetic) energy, where the kinetic energy is ordinarily neglected, as justified in Exercise 5.4. Observational studies of the long-term mean global energy balance dating back to the 1950s demonstrate the central role of the poleward eddy heat transports.[1] Using space-based measurements of radiative fluxes through the top of the atmosphere, it is now possible to partition the total poleward transport of energy between the atmosphere and the oceans[2] and to monitor seasonal and nonseasonal variations in energy storage in the oceans.[3]

This chapter describes the dynamical mechanisms that contribute to the atmospheric poleward energy transports. The transports are partitioned into both time-mean and transient contributions and into zonal mean and eddy contributions. Whether the mean meridional circulations transport energy poleward or equatorward across midlatitudes depends on whether the analysis is performed in pressure coordinates or in isentropic coordinates.

The first section summarizes the globally averaged energy balance in terms of (i) the radiative fluxes at the top-of-atmosphere (TOA) and at the Earth's surface; (ii) the exchanges of energy and water vapor between the atmosphere and the underlying surfaces; and (iii) the release of the latent heat of condensation of water vapor in clouds. The second section introduces a metric that has come to be widely used in atmospheric thermodynamics, which takes into account the work done by or on air parcels as they expand and contract, but not those due to phase changes of water that take place within the atmosphere. This easily calculated quantity, referred to as *moist static energy* (MSE) is conserved in the absence of radiative heating and sub-grid scale fluxes of latent and sensible heat. MSE, defined as the sum of the *enthalpy* $c_p T$, the latent heat of condensation Lq, and the geopotential Φ.

The meridional profile of TOA net radiation described in the third section is characterized by a surplus of incoming solar radiation, relative to the outgoing infrared radiation (OLR) emitted by the Earth system at low latitudes, and a deficit at high latitudes. The TOA radiative fluxes act to

steepen the equator-to-pole temperature gradient while the poleward, down-gradient transport of moist static energy acts to oppose this tendency. In equilibrium, the convergence of the transport of MSE by the "fluid Earth" consisting of atmosphere and oceans is balanced by the TOA net radiative flux.

The fourth section describes the partitioning of the poleward energy transport between atmosphere and oceans. The atmospheric contribution is dominant, but the contribution of the oceans is significant, especially at the lower latitudes where most of the poleward transport is associated with a frictionally driven overturning circulation, with a poleward flow of warmer surface water and an equatorward flow of colder water below the thermocline. Most of the heat carried poleward by the oceans is handed off to the atmosphere in the western boundary currents, where the upward latent and sensible heat fluxes through the air–sea interface tend to be concentrated.

The fifth section describes the partitioning of the atmospheric transport between the climatological (time) mean component and the transients, and between the mean meridional circulations and the eddies. Baroclinic waves dominate the transport at extratropical latitudes while the Hadley cell is dominant in the outer tropics and subtropics. The air in the upper branch of the Ferrel cell contains much more MSE per unit mass than the air in the poleward flowing lower branch, and consequently, the transport of energy by the Ferrel cell is equatorward. As in the angular momentum balance considered in Chapter 3, the configuration of mean meridional motions (i.e., the existence of the Hadley and Ferrel cells) can be deduced from the energy balance requirement. The consistency of the mean meridional circulations as inferred independently on the basis of the angular momentum and energy budgets will be explained in Chapter 7.

When the general circulation is viewed in isentropic coordinates rather than pressure coordinates, the Ferrel cell disappears and a single Hadley-like cell with separate low and high latitude centers extends from equator to pole.[4] Explaining these apparently contradictory results requires going beyond the treatment of the angular momentum balance in

---

[1] White (1951).
[2] Oort and Vonder Haar (1976).
[3] Hartmann et al. (1986).

[4] The idea of representing the mean meridional circulations in isentropic coordinates was proposed by Dutton and Johnson (1967) and implemented by Gallimore and Johnson (1981) and Townsend and Johnson (1985).

Chapter 3, considering the nature of the eddy heat transports and their relation to the mean meridional circulations.[5]

In the sixth section, it is shown that in the lower troposphere the poleward heat transport resembles down-gradient eddy diffusion. In the upper troposphere and in the winter stratosphere, where the zonal flow $[u]$ is much faster than the phase speed c of the waves, air parcels move through the waves so quickly that they barely feel the meridional gradient of diabatic heating. The "eddy" temperature perturbations T^* in the waves are mainly induced not by diabatic heating, but by adiabatic expansion and compression of air parcels as they ascend on the poleward side of the storm track and descend on the equatorward side. The poleward eddy heat transport across the storm tracks, exclusive of that associated with diabatic heating, can be inferred by considering the three dimensional trajectories of the air parcels passing through the waves.

In the penultimate section of this chapter, it is shown that the rectified component of these helical trajectories is analogous to the so-called Stokes drift in water waves. Adding it to the Eulerian mean meridional circulation yields a circulation that is arguably a close approximation to the Lagrangian mean meridional circulation; that is, the average circulation of a large number of tagged air parcels. A more rigorous and complete development of this argument will be offered in Section 8.2.

5.1 THE GLOBALLY AVERAGED ENERGY BALANCE

The annual mean globally averaged energy balance is summarized in Fig. 5.1. The flux of geothermal energy through the Earth's surface, which is estimated to be on the order of 0.1 W m^{-2}, is neglected, as is the imbalance of incoming over outgoing radiation associated with global warming, currently estimated to be on the order of 0.8 W m^{-2}. Averaging over the entire globe and over a full year eliminates the large seasonal imbalances in top-of-atmosphere net radiation associated with changes in energy storage of the ocean mixed layer. The incoming solar radiation per unit area of the Earth's surface is ~342 Wm^{-2}, represented in this diagram as 100 units, of which 70 units (~239 Wm^{-2}) are absorbed by the Earth system after the reflection associated with Earth's albedo is accounted for.

The atmosphere is heated by the absorption of solar radiation and by the absorption of infrared radiation emitted by the Earth's surface. It loses energy at a rate of 29 units by net radiative transfer, which is balanced by an upward flux of latent and sensible heat from the Earth's surface. The latent heat is converted into sensible heat when water vapor condenses in clouds, most of which occurs in the lower and middle troposphere. The net emission of infrared radiation from

---

[5] The material presented in the last two sections of this chapter lays the groundwork for the introduction of the transformed Eulerian mean (TEM) formalism presented in Section 8.2. It presumes a higher level of familiarity with atmospheric dynamics on the part of the reader than Sections 5.1-5.5 and most of the material in Chapter 6.

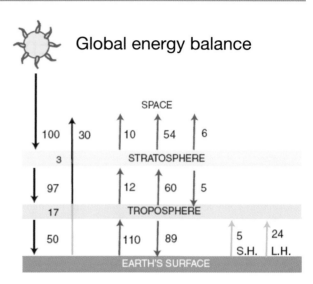

Figure 5.1 The climatological annual mean global energy balance for the Earth–atmosphere system expressed as a percentage of the insolation incident upon the top of the atmosphere (342 W m^{-2}). 50% of it passes through the atmosphere and is absorbed at the Earth's surface. Another 20% is absorbed in the atmosphere. The remaining 30% is reflected back to space from the Earth's surface, clouds, and air molecules. The leftmost column of red arrows represents the infrared radiation emitted from the Earth's surface (110%), of which only 10% passes through the atmosphere without being absorbed and reemitted. The middle column of red arrows represents the emission of infrared radiation from the troposphere (89% downward and 60% upward), of which 54% passes through the stratosphere without absorption. The rightmost column represents the emission of infrared radiation from the stratosphere. The blue arrows farther to the right represent the turbulent fluxes of latent heat (L.H.) and sensible heat (S.H.) from the Earth's surface. From Wallace and Hobbs (2006). © Elsevier Inc. All rights reserved.

the Earth's surface represents the difference between a large upward irradiance (110 units) from the surface, determined by the Stefan–Boltzmann Law assuming that the Earth's surface emits as a blackbody, and a smaller downward irradiance (89 units) from the atmosphere. Clouds play a pervasive role in the balance. They cool the Earth's surface by reducing the amount of solar radiation that reaches the ground, but they warm it by contributing to the greenhouse effect.

The net radiation at the top of the atmosphere is the difference between the net downward (i.e., incident minus reflected) solar radiation and the OLR emitted by the Earth–atmosphere system, both of which can be monitored to an accuracy of ~1 Wm^{-2} with sensors carried aboard satellites. The energy balance at the Earth's surface is more complicated, with upward and downward irradiances of both solar and terrestrial radiation, plus latent and sensible heat fluxes, none of which are easily monitored from space because of the intervening cloud cover.

Exercises

5.1 Using Fig. 5.1 in conjunction with the Stefan–Boltzmann Law, calculate the effective temperatures (i.e., the temperatures of black bodies that emit the equivalent amounts of radiation) of (a) the

Earth system, as viewed from space, and (b) the atmosphere as felt by the Earth's surface. (c) Why does the atmosphere emit more infrared radiation in the downward direction than it emits in the upward direction?

5.2 How much energy would it take to raise the mean temperature of a unit area (1 m^2) column of the atmosphere by 1 K?

5.3 The thermal capacity of the atmosphere is equivalent to that of an ocean how many meters deep? The specific heat of liquid water is 4218 J kg^{-1} K^{-1}.

5.2 MOIST STATIC ENERGY

The thermodynamic energy equation can be written as

$$c_p \frac{dT}{dt} = Q + \alpha \frac{dp}{dt}. \tag{5.1}$$

where Q is the diabatic heating rate per unit mass.[6] The second term on the right-hand side is the contribution of adiabatic expansion or compression. Using the chain rule, assuming that the flow is in hydrostatic balance, it can be expanded and written as

$$\alpha \frac{dp}{dt} = -g \frac{dz}{dt} = -\frac{d\Phi}{dt}. \tag{5.2}$$

Substituting in Eq. (5.1) and rearranging yields

$$\frac{d}{dt} \text{DSE} \equiv \frac{d}{dt} \left(c_p T + \Phi \right) = Q, \tag{5.3}$$

where the quantity in parentheses is referred to as *dry static energy* DSE. It is shown in Exercise 5.4 that kinetic energy is much smaller than the internal and potential energy provided that the wind speed is much smaller than the speed of sound. That temperature in the expression for DSE is multiplied by c_p rather than c_v reflects the fact that as an air parcel warms at constant pressure, it expands and thereby does work on its environment. Hence, a greater amount of diabatic heating is required to produce a given temperature increase than if its volume remained fixed. The term $c_p T$, referred to as sensible heat or enthapy, is thus the sum of the internal energy and the work that an air parcel would have to do in order to expand from zero volume to its present volume.[7]

Expanding the total derivative in Eq. (5.3) and vertically integrating yields

$$\frac{\partial}{\partial t} \{\text{DSE}\} = \{Q\} - \{Tr(\text{DSE})\}, \tag{5.4}$$

where $\{Tr\}$ is the rate of change due to the vertically integrated horizontal transport, as defined in Eqs. (4.1) and (4.3). This expression can also be obtained by substituting DSE for X in Eq. (4.2).

[6] Here as in Section 5.1, the heating rate Q is expressed in W kg^{-1} and evaluated per square meter.
[7] For a more complete derivation, see Holton and Hakim (2012), pp. 50–52 and Chapter 4 of Randall (2015).

The total vertically integrated diabatic heating rate $\{Q\}$ can be expanded into a radiative component Q_R, a component LP that represents the release of latent heat of condensation of water vapor and a component F_{SH} that represents the upward flux of sensible heat through the bottom boundary; that is,

$$\{Q\} = \{Q_R\} + LP + F_{SH}, \tag{5.5}$$

where the precipitation (or rain rate) P is expressed as a mass flux in units of kg m^{-2} s^{-1} and L is the latent heat of vaporization, 2.5×10^6 J kg^{-1} at 0°C. Substituting for $\{Q\}$ in Eq. (5.4) yields

$$\frac{\partial}{\partial t} \{\text{DSE}\} = \{Q_R\} + LP + F_{SH} - \{Tr(\text{DSE})\}. \tag{5.6}$$

The latent heat release in this expression is difficult to evaluate accurately because precipitation is the least well observed of the variables related to the energy balance. A more useful energy conservation equation can be obtained by including the latent heat of vaporization of water vapor as a part of the energy. Noting that $W = \{q\}$, multiplying Eq. (4.7) by L and adding it to Eq. (5.6) yields

$$\frac{\partial}{\partial t} \{\text{MSE}\} = \{Q_R\} + F_{SH} + F_{LH} - \{Tr(\text{MSE})\}, \tag{5.7}$$

where

$$\text{MSE} \equiv c_p T + \Phi + Lq. \tag{5.8}$$

and $F_{LH} = LE$. MSE is analogous in many respects to equivalent potential temperature θ_e (see Exercise 5.5).

Figure 5.2 shows a meridional cross section of zonally averaged MSE together with the contributions of the sensible heat, latent heat, and geopotential terms in Eq. (5.8), all plotted using the same contour interval so their gradients can easily be compared. The sensible and latent heat contributions are largest in the tropical lower troposphere. Both contributions decrease with height and with latitude. The geopotential term increases with height more rapidly than the sum of the sensible and latent heat terms decreases, so that MSE increases with height. The one notable exception is at the top of the tropical boundary layer, where the transition from moist air below to dry air above renders typical soundings conditionally unstable; that is, convection will occur provided that moist air parcels below the temperature inversion can be forcibly lifted to their so-called *lifting condensation level*, at which point they become saturated, and farther to the *level of free convection*, beyond which point they become warmer and therefore buoyant relative to the environmental air at the same level. Air parcels ascending in deep convective clouds remain warmer than the environmental air until they reach their so-called *equilibrium level*, at which they run out of buoyancy. By the time they reach the upper troposphere, virtually all the water vapor that they acquired while residing in the boundary layer has been condensed out. Hence, the equilibrium level for deep convective clouds corresponds to the level in the stably stratified upper troposphere at which the MSE is comparable

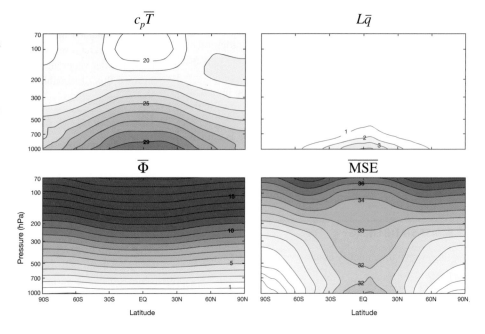

Figure 5.2 Meridional cross sections of climatological annual mean moist static energy MSE and its three components: sensible heat $c_p\overline{T}$, latent heat $L\overline{q}$, and geopotential $\overline{\Phi}$, as indicated. The contours are labeled in units of 10^4 J kg^{-1} in all four panels and darker shading implies higher values. Courtesy of Aaron Donohoe.

to that of boundary layer air. It is evident from Fig. 5.2 that in the tropics, the equilibrium level is $\sim$200 hPa.

Under steady state conditions

$$\{Tr(\mathrm{MSE})\} = [Q_R] + F_{SH} + F_{LH}, \tag{5.9}$$

where

$$\{Tr(\mathrm{MSE})\} = \{Tr(\mathrm{DSE})\} + \{Tr(Lq)\}. \tag{5.10}$$

It is instructive to revisit the energetics of the idealized circulation cell in Fig. 2.8 as viewed from the perspectives of dry and moist static energy. The reader may find it helpful to refer also to Fig. 5.3, which depicts the overturning circulation in terms of a more conventional vertical cross-section. An air parcel gains DSE due to the turbulent flux of sensible heat during the time that it resides in the boundary layer, moving along the path AB. Its DSE increases further due to the release of latent heat as it ascends in clouds along the path BC, cooling at the moist, rather than the dry adiabatic lapse rate. The DSE gained along the path ABC is lost by radiative cooling as the parcel descends through the free troposphere to the top of the boundary layer, along CA. Viewed from an MSE perspective, the cycle looks somewhat different. While residing in the boundary layer (AB), the parcel acquires MSE due to the fluxes of both latent and sensible heat in shallow convection. It conserves MSE while ascending in clouds (BC), and it loses all the MSE that it gained from the boundary layer fluxes while descending in the free troposphere (CA).

The characteristic time scales of the processes described in Figs. 2.8 and 5.3 are entirely different. A typical air parcel resides in the boundary layer (AB), for a few days; ascends in clouds (BC) within a few hours, and descends while undergoing radiative cooling (CA) for a week or longer. Since the upward and downward mass fluxes must nearly cancel when averaged over the tropics, it follows that deep

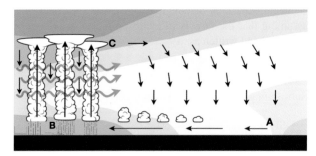

Figure 5.3 Idealized overturning circulation as in Fig. 2.7, also showing the distribution of moist static energy shaded (gold high, blue low) and a more realistic depiction of the clouds. The gray squiggly arrows represent small-scale gravity and IG waves that disperse the heat released in convective scale updrafts to the synoptic and planetary-scale environment. **A** represents the point at which air parcels that have been descending through the free troposphere are entrained into the boundary layer, **B** the point at which they ascend through cloud base, and **C** at which they are detrained at the cloud top level. Adapted from Wallace and Hobbs (2006). © Elsevier Inc. All rights reserved.

convection can occupy only a small fraction of the area of the tropics. Tropical rainfall is, in fact, highly concentrated: at any given time, half of it is falling in less than 1% of the area.[8] But although the release of latent heat is concentrated within the convective scale updrafts, the induced temperature tendency is manifested on the planetary scale due to the action of gravity and IG waves radiating outward from the convective cells. The rectified effect of the waves is to induce just enough subsidence and adiabatic compression outside the cells to neutralize the large temperature gradients that would exist on the mesoscale if the warming were confined to the cells, as described in greater detail in Section 15.2.

The moist static energy of a typical air parcel is lowest at A, at the end of its long, slow descent, when it arrives

---

[8] See Venugopal and Wallace (2016) and the references therein.

at the top of the boundary layer. After it is entrained into the boundary layer, it mixes with air that has been warmed and moistened by the fluxes of sensible and latent heat at the Earth's surface: hence the higher MSE of the boundary layer air. The boundary layer airflow is directed from the cooler regions of tropics, most notably the equatorial dry zones, toward the warmer rain belts.

The importance of atmospheric water vapor in these large-scale thermally driven circulations is worth emphasizing. Latent heat added to the atmosphere in the lower branch of the cell is converted into sensible heat when the water vapor condenses in the rising branch. This additional source of heating greatly enhances the horizontal heating contrasts relative to what would have existed in a dry atmosphere. As a result, circulations driven by horizontal heating gradients tend to be much stronger than they would be in an atmosphere without water vapor. This distinction is clearly brought out in the pioneering general circulation model experiments, comparing results from a dry model with no latent heat release term in the thermodynamic energy equation[9] with those from a moist model that included a hydrologic cycle.[10] The invigoration of the circulation due to the inclusion of moist thermodynamics is most pronounced in the tropics.

Exercises

5.4 Show that the kinetic energy is a negligible fraction of the total energy of an air parcel, provided that the wind speed is much smaller than the speed of sound.

5.5 Enumerate the similarities and differences between moist static energy and equivalent potential temperature.

5.3 MAINTENANCE OF THE OBSERVED STABLE STRATIFICATION

From the foregoing section, it is evident that radiative heating and the fluxes of latent and sensible heat at the Earth's surface are acting to reduce the static stability of the troposphere. The overturning circulation depicted in Figs. 2.8 and 5.3 plays an important role in maintaining the stable stratification, especially in the tropics. Deep convective clouds transport undiluted boundary layer air with high moist static energy to the upper troposphere, forming anvil-shaped clouds. Detrained air spreads out until it occupies almost the entire volume of the free atmosphere, subsiding slowly in cloud-free air until it reaches the top of the boundary layer. The slow cooling of these descending air parcels due to the emission of infrared radiation gives rise to the vertical

gradient of moist static energy and thus acts to maintain the stable stratification in the troposphere.

In a similar manner (but upside down), the thermohaline circulation in the oceans maintains the strong stratification in the thermocline. Cooling in the high latitudes generates negative buoyancy and the plumes of dense (cold and/or saline) water sink in the form of deep convection. In the much slower upward return flow through the thermocline, cold advection is balanced by the downward diffusion of heat from the ocean surface.

Another process that contributes to maintaining the stable stratification – implicit but not clearly represented in Fig. 5.3 – is the vertical flux of sensible heat (or equivalently, of dry static energy) in deep cumulus convection. This flux is not resolved in the numerical models that are used in producing the reanalyses, but it is implicitly represented in their parameterization schemes. The updrafts warm the air in the layer in which they detrain and downdrafts cool the air in the layer below cloud base. Shallow boundary layer convection acts in a similar manner to distribute the sensible heat received from the underlying surface by way of the boundary flux term F_{SH} through the depth of the boundary layer. In both cases, convective heating renders the stratification more stable than it would be in its absence, but it does so only locally in patches of deep convection, and in boundary layer turbulence. The high-level heating and low-level cooling are conveyed to synoptic scale and planetary-scale motion systems by gravity and IG waves, as depicted schematically in Fig. 5.3.

Even in the absence of convective heating by subgrid-scale overturning circulations, a stable stratification would be maintained by the upward flux of dry static energy by the large-scale circulations in which warm air rises and cooler air sinks, most notably baroclinic waves. As discussed in Section 2.3 (Fig. 2.4 and Exercise 2.9), the stronger the horizontal heating gradients, the stronger the overturning circulations, the larger the upward flux of static energy, and the stronger the resulting stable stratification. It is not surprising, then, that the most stable lapse rates are observed in middle and high latitudes of the winter hemisphere, where the meridional heating gradient is very strong.

5.4 BALANCE REQUIREMENTS AND ENERGY TRANSPORT IN THE EARTH SYSTEM

Averaged over many years and over the entire globe, the incoming and outgoing radiation at the top of the atmosphere (TOA) is nearly in balance. Hence, the horizontal distribution of net radiation (i.e., the difference between TOA incoming and outgoing) must be consistent with the divergence in the horizontal transport of energy within the Earth system as a whole. The top panel in Fig. 5.4 shows the annual mean TOA net (downward) shortwave radiation, which reflects the geographical variations in solar declination angle and local albedo. Values range from over 300 W m^{-2} in the tropics, where the Sun is nearly

---

[9] Smagorinsky et al. (1965).
[10] Manabe et al. (1965).

Figure 5.4 Global distributions of the climatological annual-mean radiation at the top of the atmosphere. (Top) Absorbed solar radiation (insolation multiplied by (1 minus the local albedo)). (Bottom) Outgoing longwave radiation (OLR).

Absorbed solar radiation

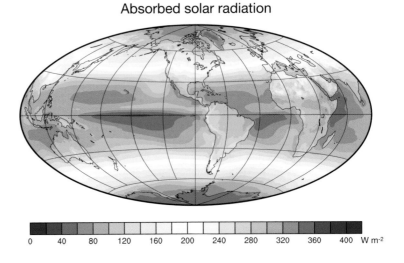

| 0 | 40 | 80 | 120 | 160 | 200 | 240 | 280 | 320 | 360 | 400 | W m-2 |

Outgoing longwave radiation

| 110 | 130 | 150 | 170 | 190 | 210 | 230 | 250 | 270 | 290 | 310 | 330 | W m-2 |

directly overhead at midday throughout the year, to less than 100 W m^{-2} in the polar regions, where winters are dark and the continuous summer daylight is offset by the high solar zenith angles, widespread cloudiness, and the high albedo of ice-covered surfaces.[11]

The corresponding distribution of TOA outgoing long wave radiation (OLR), shown in the bottom panel of Fig. 5.4, is linked, via the Stefan–Boltzmann Law, to the effective temperature T_E at each grid point. It exhibits a much gentler equator-to-pole gradient than the absorbed solar radiation, and more regional variability related to cloud coverage, especially within the tropics. As shown in Exercise 5.10, the observed equator-to-pole contrast in surface air temperature is sufficient to produce a 2:1 difference in outgoing OLR between the equatorial and polar regions, but this is partially offset by the fact that cloud tops and the top of the moist layer are higher in the tropics than over high latitudes. The regions of conspicuously low OLR over the Maritime Continent and

parts of the tropical continents reflect the prevalence of deep convective clouds with high, cold tops. The ITCZs are also regions of local OLR minima, but they are not as pronounced as those over the tropical continents and the extreme western Pacific and Maritime Continent because the cloud tops are not as high. Areas with the highest annual mean OLR are the deserts and the equatorial dry zones over the tropical Pacific and Atlantic, where the atmosphere is relatively dry and cloud free, allowing more of the radiation emitted at the Earth's surface to escape to space unimpeded.

The net downward radiation at the top of the atmosphere, obtained by taking the difference between the two panels of Fig. 5.4, is shown in the top panel of Fig. 5.5. The imbalances in the zonally averaged TOA net radiation are characterized by a low latitude surplus and a high latitude deficit of incoming radiation, with a crossover point $\sim38°$N/S, as shown in Fig. 5.6, from which it can be inferred that there exists a continuous energy transport from lower toward higher latitudes by the atmosphere and oceans.

The subsections that follow focus on the poleward transport of energy as required to fulfill the balance requirement

[11] Donohoe and Battisti (2011).

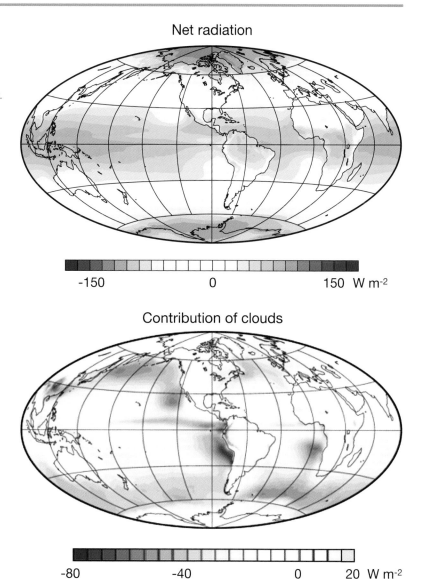

Figure 5.5 (Top) Global distribution of the difference between the climatological annual-mean absorbed solar radiation and outgoing longwave radiation in the previous figure. Positive values indicate a downward flux. (Bottom) The cloud radiative effect (CRE): the difference in the mean net downward radiation based on all pixels minus the mean net radiation based on cloud-free pixels only. Note that the color bars are different.

in Fig. 5.6 We will emphasize the annual-mean global energy balance, in which the energy stored in the important reservoirs (the atmosphere, oceans, and cryosphere) can be assumed to be constant, apart from the very small imbalances associated with climate change.

5.4.1 Partitioning of the Poleward Transports between Atmosphere and Ocean

In principle, the contributions of the atmosphere and oceans to the poleward energy transport, shown schematically in Fig. 5.7, can be estimated independently on the basis of observations, and checked for consistency. However, out of concerns about the quality and completeness of the ocean observations, a common practice has been (1) to compute the combined (atmosphere plus ocean) poleward transport by integrating the TOA net radiation over latitude, starting from one of the poles, to estimate the cumulative deficit

or surplus at each latitude, (2) to estimate the atmospheric contribution to the transport based on observations, and (3) to infer the oceanic contribution as a residual. This practice dates back to the mid-1970s, when ocean observations were sparse and sporadic.[12] More systematic observations became available starting in the late 1990s, with the deployment of ARGO floats that provide a *real-time observing system* for monitoring temperature and salinity in the uppermost 2000 m of the ocean. Data from this system are assimilated into ocean models to produce gridded datasets that provide a basis for estimating the oceanic heat transport independently of atmospheric measurements.

Estimates of the poleward transport of energy broken down into atmospheric and oceanic contributions based on the above methodology are shown in Fig. 5.8. The strongest poleward transports occur at ~38°N/S, consistent with the

---

[12] Following Oort and Vonder Haar (1976).

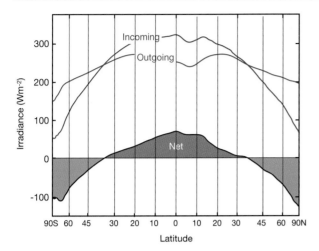

Figure 5.6 (Top) Meridional profiles of climatological annual mean, zonally averaged TOA absorbed (i.e., incident minus reflected) solar radiation and outgoing infrared radiation. (Bottom) TOA net radiation, the excess of incoming over outgoing. Distance from the equator is proportional to sine of latitude.

balance requirement shown in Fig. 5.6. The atmosphere accounts for most of the required transport but the ocean makes an appreciable contribution equatorward of ~40°N/S, most of which is associated with the transport of near-surface water from the deep tropics, where it gains energy, to the subtropics and extratropics, where it is transferred to the atmosphere by way of the latent and sensible heat boundary fluxes. The near-surface water gains heat in regions of equatorial upwelling in the Pacific and Atlantic, where water that is much colder than its radiative equilibrium temperature is continually being lifted to the surface to replace water that is moving poleward in the so-called *Ekman drift* induced by the trade winds. It would exist even on an aquaplanet, where the equatorward atmospheric mass flux in the frictionally induced trade winds would exactly balance the poleward oceanic mass flux in the Ekman drift. The existence of a narrow zone of upwelling along the equator is confirmed by color imagery (Fig. 5.9) and the Ekman drift is clearly evident in near surface currents in both the Pacific and Atlantic oceans shown in Fig. 5.10. The net sensible plus latent energy flux at the air–sea interface is shown in Fig. 5.11; the greatest net flux into the ocean is in the regions of equatorial upwelling. The Ekman drift extends poleward to ~30 °N/S, the latitude of the subtropical anticyclones, but in the Northern Hemisphere the energy gained in the equatorial belt is carried even farther poleward by the western boundary currents, the Gulf Stream and the Kuroshio. Air masses crossing over these warm currents often become unstably stratified, especially during winter, resulting in large, episodic fluxes of latent and sensible heat in the regions of the orange shading in Fig. 5.11. In these zones the energy carried poleward by the oceans is, in effect, handed off to the atmosphere, which carries it farther poleward.

The conventional zonally and time-averaged mass streamfunction in pressure coordinates ψ_p is defined as

$$\psi_p = \frac{1}{g} \int [\bar{v}] dp, \qquad (5.11)$$

where $\bar{v}$ is the climatological northward flow on pressure surfaces, and $[\cdot]$ denotes a zonal average. The boundary condition is $\psi_p = 0$ at the Earth's surface (approximated as $p = 1000$ hPa) on each latitude circle. The corresponding mass streamfunction in height coordinates ψ_z is

$$\psi_z = - \int [\overline{\rho v}] dz, \qquad (5.12)$$

where ρ is the density of the fluid. Rather than being binned by pressure level, as in Eq. (5.11), the ρv data are binned by height z before being averaged. The lower boundary condition is that $\psi_z = 0$ at sea level.

In analogy with Eq. (5.11), we can define a mass streamfunction ψ_h in isentropic (ϕ, h) coordinates where ϕ is latitude and h denotes moist static energy for the atmosphere and internal energy for the ocean:

$$\psi_h = - \int [\overline{\rho v}] dh. \qquad (5.13)$$

The bottom boundary condition is that $\psi_h = 0$ at $h = h_-$, where h_- is below the range of observed values.

The poleward energy transport at a specified latitude ϕ is given by $-\int h \, d\psi_h$, can be inferred from a contour plot of ψ in the meridional plane, as shown in Fig. 5.12. Such plots provide a global survey of the transports, taking both the atmospheric and the oceanic contributions into account. The atmospheric pattern is dominated by a pair of equator-to-pole cells in which air with higher values of h flows poleward in the upper branch, and air with lower values of h flows equatorward in the lower branch. The direction of the circulation in the cells is thus indicative of a poleward transport of h in both media, and in the atmosphere it extends all the way from equator to pole. That the cells slope downward (i.e., toward lower values of h) with increasing latitude is a reflection of the radiative cooling of the poleward flowing air in the upper branch and the moistening and heating of the equatorward-flowing boundary layer air by the latent and sensible heat fluxes through the underlying surface. An estimate of the oceanic transport of internal energy based on an ocean model is included in the figure. In agreement with Fig. 5.8, the heat transport in the oceanic cell is ~1/4 as strong as the transport in the atmospheric cell and it is restricted to lower latitudes.

5.4.2 The Poleward Transport of Atmospheric Moist Static Energy

The poleward transport of moist static energy MSE, across a latitude circle is

$$\{[\overline{v\,\text{MSE}}]\} = \frac{c_p}{g} \int_0^{p_s} [\overline{vT}] dp$$
$$+ \frac{L}{g} \int_0^{p_s} [\overline{vq}] dp + \frac{1}{g} \int_0^{p_s} [\overline{v\Phi}] dp . \qquad (5.14)$$

The transport of MSE by the mean meridional circulations depends mainly upon the difference between typical values

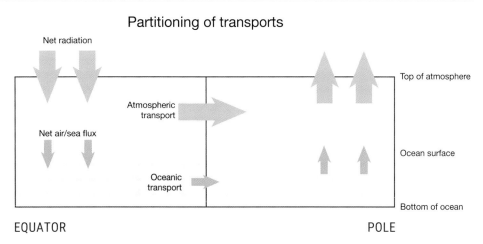

Figure 5.7 Partitioning of the energy transports between atmosphere (upper layer) and ocean (lower layer). The top row of vertical arrows represents the fluxes through the top of the atmosphere and the bottom row represents the fluxes through the atmosphere–ocean interface.

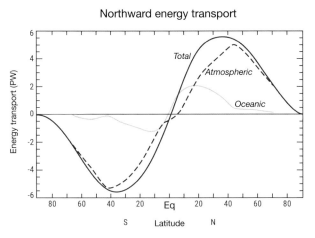

Figure 5.8 Climatological, annual mean northward transport of total energy (solid black curve) partitioned into atmospheric MSE (dashed black) and oceanic internal energy (gray-blue) contributions. The total is calculated by meridionally integrating the net radiation at the top of the atmosphere starting from either pole. The atmospheric transport is calculated from the reanalysis, while the ocean transport is inferred from the difference, total minus atmospheric transport. From Fig. 7 of Trenberth and Caron (2001). © American Meteorological Society. Used with permission.

Figure 5.9 Ocean color imagery showing phytoplankton blooms in the narrow band of equatorial upwelling. Mean conditions for the boreal winter. Dark blue and purple hues denote the clearest water; colors that range from deep blue through turquoise to yellow indicates successively higher concentrations of chlorophyll. Imagery courtesy of SeaWiFS Project, NASA/GSFC and ORBIMAGE, Inc.

of MSE in the upper and lower branches. As shown in Fig. 5.2, MSE is larger in the upper branch than in the lower branch because of the much larger values of geopotential. The mean difference is determined mainly by the mean tropospheric lapse rate and to a lesser degree by the moisture field, as illustrated by the following quantitative example. At $15°$N near the midpoint of the Hadley cell, temperatures at mid-depth in the boundary layer (925 hPa) are about $+20°$C, the specific humidities are on the order of 12 g kg^{-1} (80% relative humidity) and the height above sea level is about 0.8 km. The upper branch is centered near 160 hPa (12.8 km) where $T \sim -55°$C and the specific humidity is negligible. The difference in MSE (upper branch minus lower branch) is

$$(-55 - 20)c_p - 0.012L + 12,000\,g.$$

Substituting values of c_p, L, and g we obtain

$$-75,000 - 30,000 + 117,600 = +12,600 \text{ J kg}^{-1}.$$

It follows that the mean meridional circulations transport net MSE in the direction of the flow in the upper branch, where MSE is larger. Hence, the Hadley cell contributes to the required poleward transport of energy in the Earth system and the Ferrel cell works against it. In general, provided that the atmosphere is stably stratified, circulations in which warm air is rising and cooler air is sinking, such as the one in Fig. 2.4, transport energy horizontally from the heat source to the heat sink.

It is instructive to consider the separate contributions of the transports of DSE $= c_p T + \Phi$ and latent heat Lq to

Near surface currents

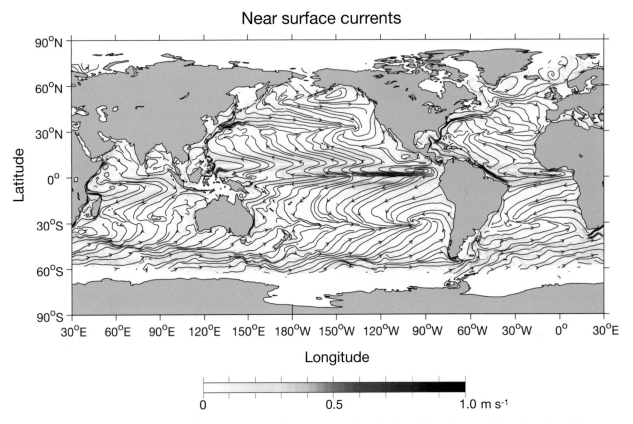

Figure 5.10 Mean current speeds (colors, in m s^{-1}) from near-surface drifter data with streamlines (black lines). Streamlines are calculated from the spatially smoothed velocity field to indicate flow direction and qualitatively illustrate large-scale circulation features, including surface divergence. Adapted from the climatology presented in Lumpkin and Johnson (2013). Courtesy of Gregory C. Johnson (NOAA/PMEL).

Net upward sea / air energy flux

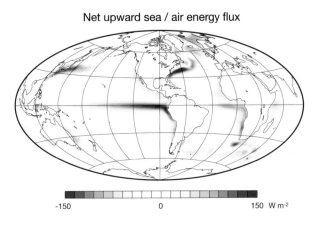

Figure 5.11 Climatological annual-mean net energy flux at the air–sea interface, including radiative fluxes. Blue shading denotes a net flux of energy into the ocean and the warm colors indicate a flux from the ocean to atmosphere.

Northward energy transport streamfunction

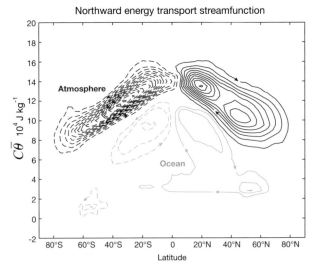

Figure 5.12 Climatological annual-mean mass streamfunction within constant energy layers, ψ_h. The contour interval is 10 Sv = 10^{10} kg s^{-1}, dashed when circulating counterclockwise. The y axis is an energy coordinate $C\theta$ in units of 10^4 J kg^{-1}, equivalent in the atmosphere to moist static energy and in the ocean to internal energy. Atmospheric calculations are performed on reanalysis data; ocean calculations are performed using output from an ocean model forced by the observed annual cycle of wind stress and surface buoyancy flux. From Czaja and Marshall (2006). © American Meteorological Society. Used with permission.

the transport of MSE. The contribution of latent heat can be inferred simply by multiplying the water vapor transports discussed in Section 4.2.2 by L. Figure 5.13 shows the meridional divergence $\partial/\partial y$ of the vertically integrated transports of DSE, the energy in the latent heat of water vapor Lq,

Figure 5.13 Meridional profiles of the divergence of the climatological annual-mean, zonally averaged, vertically integrated poleward transport of moist static energy. (Top) Total transport. (Middle) Transport by the time-mean flow. (Bottom) Transport by the transient eddies. From Trenberth and Stepaniak (2003). © American Meteorological Society. Used with permission.

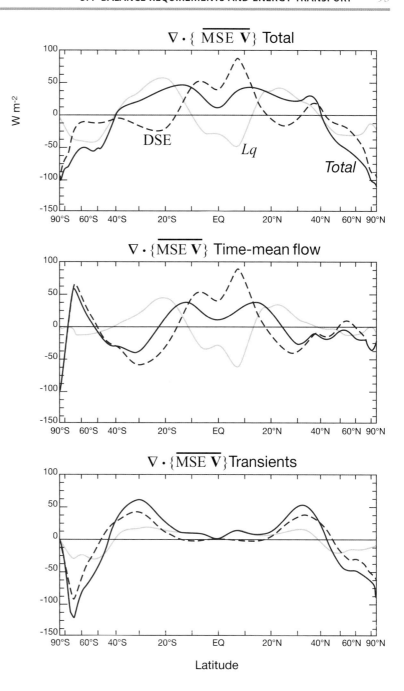

and their sum, MSE. In overturning circulations such as the Hadley cell, the transports of dry static energy and latent heat tend to be in opposite directions: the former in the direction of the upper-level flow and the latter in the direction of the low-level flow. Hence, in the top panel of Fig. 5.13, latent heat converges into latitude belts out of which dry static energy diverges and vice versa; for example, the equatorial rain belt imports latent heat and exports dry static energy. The total transport of MSE is seen to be divergent out to 40°N/S and convergent poleward of that latitude. Because of the ocean transports, this crossover latitude is a few degrees

poleward of its counterpart in TOA net radiation shown in Fig. 5.6. Other apparent influences of the ocean transports are the small dip in the divergence of atmospheric MSE transport over the band of equatorial upwelling, where the oceans take up heat, and the small bulge in the profile near 40°N, where the oceans impart latent and sensible heat to air masses passing over the Gulf Stream and the Kuroshio current.

At low latitudes, the transports of DSE and Lq are all dominated by the time-mean flow shown in the middle panel of Fig. 5.13, which includes both the Hadley cell contribution and the standing eddy or stationary wave contributions. In contrast, the transients dominate the transports at the higher latitudes. The zero crossings in the transient profile in the bottom panel of Fig. 5.13 near 45°N/S correspond to the "storm tracks," in which the poleward transports of dry static energy and latent heat by baroclinic waves exhibit their maximum amplitudes. Contrasting the curves for MSE and DSE in the top panel of Fig. 5.13 reveals the important role of water vapor in the global energy balance. In the subtropical belt (20–40°N/S) the divergence of MSE is dominated by the transports of latent heat Lq. Water vapor derived from the excess of evaporation over precipitation in the subtropics is carried equatorward in the lower branches of the Hadley cells and poleward in the eddies, mainly in the poleward low-level flow on the western sides of the climatological mean subtropical anticyclones (see Fig. 1.12). Trailing frontal rain bands of extratropical cyclones like the ones shown in Fig. 4.4 contribute to the poleward transport. Roughly half the poleward transport of moist static energy across lower midlatitudes is attributable to the poleward transport of latent heat.

The zonally averaged meridional transports of latent heat by the eddies are concentrated in the boundary layer (not shown), whereas the eddy transports of sensible heat exhibit a more interesting vertical structure, as shown in Figs. 5.15 (transient eddies), 5.16 (stationary waves), and 5.14 (total). For reasons that will be explained in Section 5.6, the poleward transport of sensible heat is indicative of a poleward transport of dry or moist static energy only in the lower troposphere. Higher in the atmosphere the poleward eddy transport of sensible heat is largely cancelled by the equatorward transport of geopotential. The transports in planetary waves in the winter stratosphere exhibit a similar cancellation.

The poleward transient eddy (mainly heat) transport in the troposphere and lower stratosphere is year-round, stronger in the winter hemisphere, with maxima at the Earth's surface and in the lower stratosphere ~200 hPa at 50°N/S. The total eddy transports in the Northern and Southern Hemispheres are quite comparable, while the transports by the stationary waves play a significant role only in the Northern Hemisphere and only during winter.

Exercises

5.6   The bottom panel of Fig. 5.5 shows contribution of clouds to the net radiation shown in the top panel, known as the cloud radiative effect (CRE). Why is the CRE overwhelmingly negative, with well-defined maxima at subtropical latitudes on the eastern sides of the oceans?

5.7   The text does not address the question of what sets the magnitude of the observed combined poleward energy transport by the atmosphere and ocean. How (qualitatively) would the balance requirement for the conservation of energy be satisfied in the absence of any transport?

5.8   The annual mean surface air temperature ranges from roughly +25°C in the tropics to −25°C in the polar cap regions. (a) On the basis of the Stefan–Boltzmann Law, estimate the ratio of the irradiance of the infrared radiation emitted from the Earth's surface in the tropics to that in the polar cap region. Start by making an estimate ignoring the atmosphere altogether, with the observed meridional profiles of TOA shortwave and longwave radiation shown in Fig. 5.6. (b) Making use of Fig. 4.5, show that the level of unit optical depth of water vapor is 2–3 km higher at the equator than at the poles. How much does this contribute to flattening of the meridional profile of infrared radiation?

5.9   In Fig. 5.5 the net (downward) radiation at the top of the atmosphere exhibits minima over all the subtropical deserts and negative values over some of them. The same is true in the seasonal (DJF and JJA) means. (a) List the factors that contribute to this feature. (b) How can the deserts be so hot if they exhibit a deficit of net radiation at the top of the atmosphere?

5.10   Describe the seasonal distributions of TOA absorbed solar, outgoing infrared, and net radiation and the net energy flux at the Earth's surface on the companion web page and explain why they are so much different from the annual mean distributions shown in the text.

5.11   (a) Why are the peaks in the poleward atmospheric energy transport in Fig. 5.8 about 6 degrees of latitude poleward of the nodes (zero crossings) in the TOA net radiation in Fig. 5.6? (b) Why is the near equatorial node in the atmospheric transport north of the node in the oceanic transport?

5.12   Why is the clearest ocean water in Fig. 5.9 located beneath the subtropical anticyclones? [Hint: these are also oceanic zones in which floating debris tends to collect.]

5.13   Based on the curve showing the oceanic contribution to the poleward energy transport in Fig. 5.8, qualitatively sketch the meridional profile of the zonally averaged energy flux at the air–sea interface and relate the features in it to equatorial upwelling and the strong fluxes of latent and sensible heat over the western boundary currents.

5.14   Reconcile the fact that the atmospheric mass flux in the Northern Hemisphere circulation cell in Fig. 5.12 is substantially stronger at 20°N than at 40°N, where the observed poleward energy transport is strongest.

5.15   In the plot analogous to Fig. 5.12 but for dry static energy, the circulation is much more concentrated in the latitude range of the Hadley cell (see, e.g., Fig. 5 of Czaja and Marshall (2006)). Explain why this is the case.

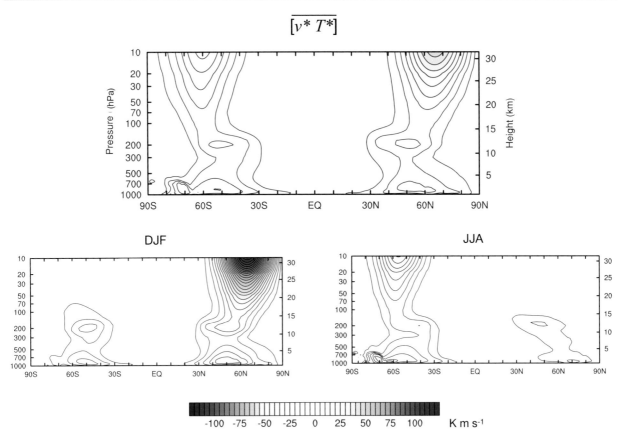

$$\overline{[v^* \, T^*]}$$

DJF   JJA

-100 -75 -50 -25   0   25   50   75   100   K m s⁻¹

Figure 5.14 Climatological northward transport of temperature by the eddies, the sum of the transient and standing eddy (stationary wave) contributions. Contour interval 5 K m s⁻¹; zero contour not shown. The top panel shows the annual mean. Although the stratospheric contribution to the poleward energy transport is small, the domain of this and the next two diagrams extends upward into the middle stratosphere in order to document the eddy transports in the vicinity of the polar night jet.

5.16 Vertical profiles of temperature over the polar cap regions are marked by strong inversions just above the surface, especially during wintertime. Surface temperatures rise abruptly when warmer cloud layers pass overhead, emitting downward infrared radiation, and when warmer air from above the inversion mixes down to the surface during storms. How is the relative warmth of the air above the inversion maintained?

5.17 At 15° N, the mean meridional motions in the Northern Hemisphere Hadley cell are ∼2 m s⁻¹ during DJF and they extend through a layer of depth ∼150 hPa (1.5×10^4 Pa) in the upper troposphere and in the boundary layer. Using the estimates of the moist static energy in the upper and lower branches given in Fig. 5.2, estimate the vertically integrated transport of MSE across 15° N.

5.18 (a) Review the update and extension of Fig. 5.8 in Figure 1 of Armour et al. (2019) and discuss the implications for the interpretation of the global energy transport. (b) How would the partitioning of the poleward energy transports by the various dynamical mechanisms be different on an aquaplanet?

5.19 A robust feature of global warming is "polar amplification," the tendency for the equator-to-pole temperature gradient to weaken as the Earth warms. A contributing factor is that the poleward energy transport increases in response to global warming, and vice versa, as evidenced by experiments with aquaplanet models. Why is this the case?

5.20 (a) Explain why, at the Earth's surface, cold advection is dominant over most of the globe. (b) How is the cooling balanced?

5.5 THE ZONALLY AVERAGED HEAT BALANCE

To derive an expression for the time rate of change of zonally averaged temperature $[T]$ we expand the terms in the conventional thermodynamic energy equation[13]

$$\frac{\partial T}{\partial t} = \left(\kappa \frac{T}{p} - \frac{\partial T}{\partial p} \right) \omega - \mathbf{V} \cdot \nabla T + Q \tag{5.15}$$

[13] From here onward, the heating rate Q is expressed in units of K d⁻¹ unless otherwise noted.

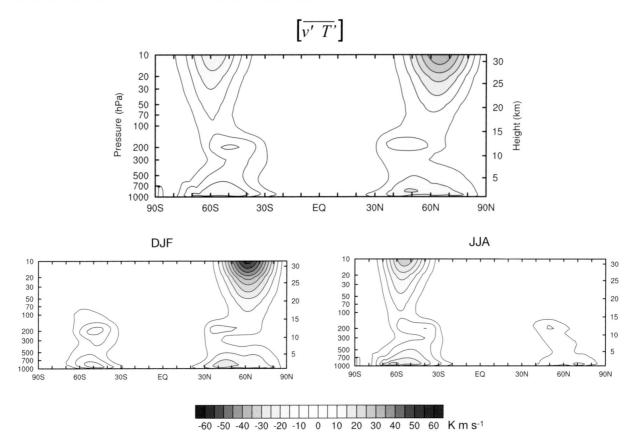

Figure 5.15 As in Fig. 5.14, but for the transient eddies.

into zonal mean $[\cdot]$ and eddy $(\ )^*$ components and proceed following a similar set of steps as in Appendix B to obtain

$$\frac{\partial [T]}{\partial t} = s[\omega] + \left(\kappa \frac{[\,\omega^* T^*\,]}{p} - \frac{\partial}{\partial p}[\,\omega^* T^*\,] \right) - \frac{[v]}{\cos \phi} \frac{\partial}{\partial y}[T] \cos \phi$$

$$- \frac{1}{\cos \phi} \frac{\partial}{\partial y}[v^* T^*] \cos \phi + [Q], \qquad (5.16)$$

where

$$s = \kappa \frac{[T]}{p} - \frac{\partial [T]}{\partial p} \qquad (5.17)$$

is a measure of the static stability, and ω is the vertical velocity in pressure coordinates ($\omega = dp/dt$). The horizontal temperature advection by the mean meridional motions is demonstrably an order of magnitude smaller than the $s[\omega]$ term and consequently will be neglected. The vertical eddy heat flux term is negligible in the stratosphere, but in baroclinic waves in the troposphere it ranges up to half as large as the corresponding meridional eddy heat transport term. We are going to neglect it anyway here and in deriving the TEM formalism in Chapter 8. In defense of this cavalier treatment, it can be argued that the neglected terms are analogous to the terms associated with the vertical flux of sensible heat by the parameterized subgrid scale motions. The unresolved fluxes are, in effect, incorporated into the diabatic heating

that contributes to maintaining the stable stratification of the troposphere in the presence of the upward flux of sensible heat at the Earth's surface. The neglected resolved upward heat fluxes could have been incorporated into the heating in a similar manner, but it would not have materially changed the conclusions drawn from the analysis that follows.

With these simplifications Eq. (5.16) reduces to

$$\frac{\partial [T]}{\partial t} = [\omega] s + H + [Q], \qquad (5.18)$$

where H is the zonal average of the convergence of the meridional eddy heat transport

$$H = -\frac{1}{\cos \phi} \frac{\partial}{\partial y}[v^* T^*] \cos \phi. \qquad (5.19)$$

Note the structural similarity between Eqs. (3.19) and (5.18). Just as, under steady state conditions, $[v]$ can be inferred from Eq. (3.21), $[\omega]$ can be inferred from Eq. (5.18):

$$[\omega] = -\frac{H + [Q]}{s}. \qquad (5.20)$$

In analogy with Eq. (3.21), s can be interpreted as the forcing required to produce a unit vertical velocity.

The time derivative term on the left-hand side of Eq. (5.18) can be related to the time rate of change in thickness $-\partial \Phi / \partial p$ (the spacing of zonally averaged pressure surfaces) just as the time derivative term in Eq. (3.17) can be related

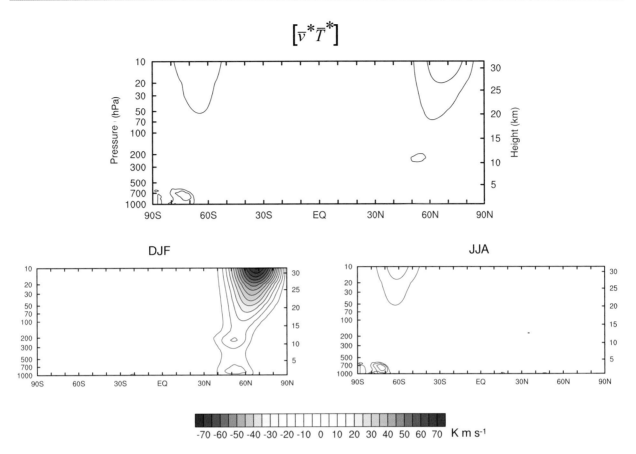

Figure 5.16 As in Fig. 5.14, but for the stationary waves.

to $-\partial\Phi/\partial y$, the meridional spacing between the pressure surfaces. In place of the inertial stability $[\eta]$ times $[v]$, we have the static stability s times $[\omega]$. Diabatic heating takes the place of friction and the eddy heat transport convergence term H is analogous in form to the expression for the eddy momentum transport $\mathcal{G}$ in the zonal momentum equation. Averaged over a sufficiently long interval, the time derivative term on the left-hand side of Eq. (5.18) is much smaller than the three terms on the right-hand side. Since the distributions of H and $[Q]$ are known, $[\omega]$ can be determined as a residual in the same way that $[v]$ is determined as a residual in Eq. (3.19). Before the advent of modern data assimilation schemes, vertical velocity and the divergent wind component were sometimes estimated as residuals in these equations.

The zonally averaged heat balance is thus, to first order, a three way balance between H, $[Q]$, and $s[\omega]$. The distribution of H can be inferred by differentiating the northward eddy heat transports shown in Fig. 5.14 with respect to latitude in spherical coordinates. The distribution for DJF, shown in Fig. 5.17, consists of a dipole pattern in the lower troposphere with cooling just equatorward of the strongest eddy heat transports (i.e., the zonally averaged storm track), where the eddies are extracting sensible heat, and heating just poleward of the storm track, where they are depositing it. The associated heating and cooling rates range up to 1 K per day.

The corresponding distribution of diabatic heating rate $[Q]$, shown in Fig. 5.18 is dominated by tropical heating that shifts back and forth across the equator with the seasons, flanked by belts of cooling at subtropical latitudes. The longitudinally dependent features that contribute to this zonally averaged pattern, shown in Fig. 1.27, are dominated by belts of heavy climatological mean rainfall in which air parcels ascend moist adiabatically (while conserving MSE) in deep convective scale updrafts. Air outside these "hot towers" undergoes a much more gradual forced descent as shown schematically in Fig. 5.3. Adiabatic compression maintains the temperature of descending air parcels above radiative equilibrium, inducing infrared cooling to space. The cooling is strongest in the upper troposphere near the level of (wavelength-integrated) unit optical depth of water vapor. The release of latent heat in baroclinic waves in the storm tracks contributes to the lack of cooling in the midlatitude lower and mid-troposphere.

To first order, the Eulerian mean meridional circulations inferred from the balance requirement for the conservation of energy can be viewed as being eddy forced, but

$$H$$

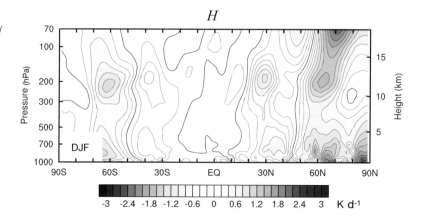

Figure 5.17 Climatological distribution of the forcing by eddy heat transports H (Eq. 5.19) in the zonally average energy balance for DJF. Contour interval 0.2 K per day.

$$[\overline{Q}]$$

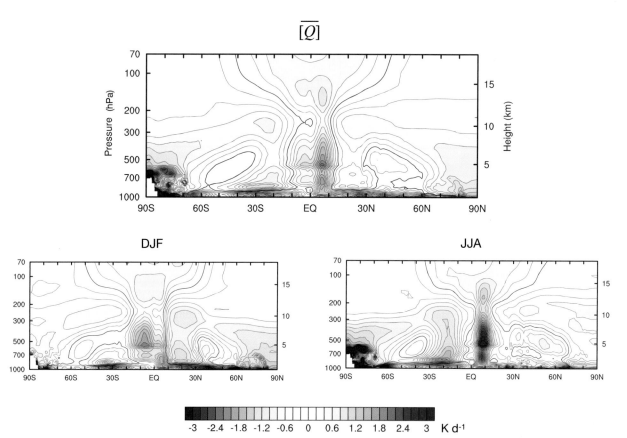

Figure 5.18 Climatological distribution of zonally averaged diabatic heating rate $[Q]$. (Top) annual mean; (bottom left) DJF and (bottom right) JJA. The zero contour is bolded.

with tropical Hadley cells strongly augmented by diabatic forcing, consistent with the schematic shown in Fig. 5.19. Figures 3.15 and 5.19 are analogs, the former based on the angular momentum balance, and the latter based on the global energy balance. That the same distribution of mean meridional circulations emerges from consideration of two different balance requirements is no accident. It follows from the fact that zonal wind and temperature fields are in thermal wind balance, as will be further explained in Chapter 7.

5.6 DYNAMICAL INTERPRETATION OF THE EDDY HEAT TRANSPORT

Just as the existence of a poleward eddy transport of westerly momentum implies that the wave axes tilt eastward with increasing latitude, the existence of a poleward eddy heat transport implies that the wave axes tilt westward with height. If the poleward moving air is warmer than the equatorward moving air, it follows from the hypsometric equation that the pressure surfaces must be more widely

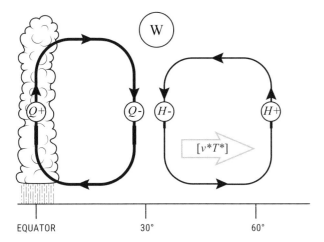

Figure 5.19 Inferred direction of the zonally averaged vertical velocity based on the thermodynamic energy equation Eq. (5.20) as inferred from the forcing terms Q and H. The meridional motions are inferred from the continuity of mass. See text for further explanation.

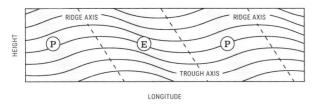

Figure 5.20 Idealized longitude–height section through a wave or eddy that produces a poleward transport of heat, showing the sloping pressure surfaces. Colored shading indicates the axes of the warmest and coldest air and P and E denote the areas of the strongest meridional wind v perturbations: P poleward and E equatorward. In accordance with the hypsometric equation, the vertical spacing between the pressure surfaces (i.e., the thickness) is proportional to the mean absolute (virtual) temperature of the layer. The warmest air in the wave is collocated with the poleward flow to the east of the trough in the pressure surfaces. These constraints require that the wave axes tilt westward with increasing height.

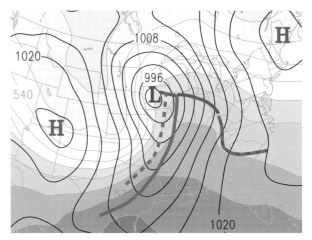

Figure 5.21 Synoptic chart depicting a deepening extratropical cyclone over the central United States associated with a developing baroclinic wave. The 1000–500 hPa thickness field, a surrogate for lower tropospheric temperature, is indicated by colored shading, labels in tens of meters and sea-level pressure is indicated by the black contours, contour interval 4 hPa. The low-level geostrophic meridional wind field can be inferred from the pressure field. That these waves produce a strong poleward heat transport $[v^*T^*]$ is evidenced by the coincidence between the highest temperatures and the strongest poleward flow. The vertical velocity (not shown) is upward in the poleward flow and downward in the equatorward flow. From Wallace and Hobbs (2006). © Elsevier Inc. All rights reserved. .

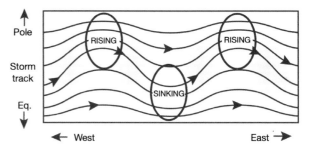

Figure 5.22 Schematic showing the fields of horizontal wind, temperature, and vertical velocity in an idealized synoptic or planetary-scale wave at the jet stream level, where air parcels are moving through the waves from west to east and the flow is nearly adiabatic. The graphic is centered on the storm track, where the amplitudes of the meridional wind component v^* and the vertical velocity w^* are strongest. The zonal wind u^* perturbations are strongest poleward and equatorward of the storm track. Air parcels cool due to adiabatic expansion as they pass through the ridges so that they are cold (blue) when they are moving equatorward and warm (gold) when they are moving poleward.

spaced in the vertical (i.e., the "thickness" must be larger) in the poleward flowing warm air than in the equatorward flowing cold air, as illustrated schematically in Fig. 5.20. It follows that the wave ridges and troughs must tilt westward with height so that they lie above the warmest and coldest air, respectively.

The transient eddy heat transport in the troposphere is primarily associated with baroclinic waves, which derive their energy from the meridional temperature gradient that develops in response to the strong equator-to-pole gradient of shortwave radiation absorbed at the Earth's surface. Much of the eddy heat transport occurs in developing baroclinic waves and their attendant extratropical cyclones in which warm, subtropical air flows poleward in the warm sector to the east of the surface low while a cold "polar" air mass flows equatorward to the west of it, as shown in Fig. 5.21. In agreement with the theory of baroclinic instability, the poleward eddy heat transports in Fig. 5.15 are largest just above the Earth's surface and they decrease with height up to the tropopause level.

The shallow secondary maximum in the poleward eddy heat transport just above the tropopause in Fig. 5.15 requires further explanation. At this level, the ambient zonal average zonal wind speed $[u]$ is much stronger than the roughly 10 m s^{-1} phase speed of baroclinic waves, such that air parcels are flowing through the waves from west to east, as shown in Fig. 5.22. At these levels the diabatic heating is weak, so the warming and cooling of air parcels passing through the waves is mainly due to adiabatic compression and expansion, rather than to diabatic processes. It follows that air parcels must be rising and cooling as they pass through the ridges and sinking and warming as they pass

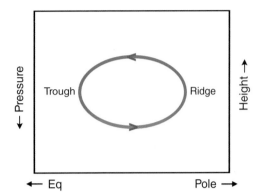

Figure 5.23 Schematic showing the projection of the helical trajectory of air parcels in Fig. 5.22 onto the meridional plane, to first order accuracy. The orange/blue shading of the trajectory indicates whether the temperature of the parcel is warmer/colder than the zonally averaged temperature at that latitude and level. Air parcels ascend at high latitudes while passing through the ridges of the waves and descend at low latitudes, while passing through the troughs, and thus move equatorward on higher geopotential surfaces than they move poleward.

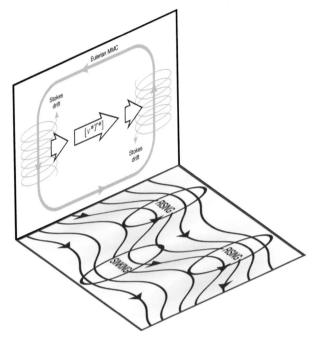

Figure 5.24 Isometric depicting the three-dimensional trajectories of air parcels moving from west to east through waves confined to a limited range of latitudes (i.e., a "storm track.") The projection of the trajectories upon the horizontal plane, is the same as in Fig. 5.22. The projection on the latitude–height plane is also shown but this time to second order accuracy. The big arrows depict the poleward eddy heat transport. The thick, counterclockwise circulating arrows represent the Eulerian mean meridional circulations driven by the eddy forcing. The helical arrows poleward and equatorward of the storm track indicate the flow through the eddies, whose rectified vertical component is referred to as the "Stokes drift." If the flow in the eddies is purely adiabatic and in steady state, the Eulerian MMC and the Stokes drift cancel so that the Lagrangian mean meridional circulation is zero.

through the troughs, as shown. Alternatively, we can think of the potential temperature surfaces as being elevated in the colder, equatorward flowing air to the east of the ridge and depressed in the warmer poleward flowing air to the west of it. As air parcels move eastward through the ridge while conserving potential temperature, they must be rising. When the helical paths of these air parcels are projected onto the meridional plane, as shown in Fig. 5.23, to first order (in a mathematical sense) the two-dimensional "orbit" is elliptical, with ascent in the high latitude ridges of the waves and descent in the lower latitude troughs (i.e., in the same sense as in the Ferrel cell). Because the air parcels flow equatorward at a higher level in the atmosphere than they flow poleward, they transport potential energy Φ equatorward. If the flow is purely adiabatic, the meridional transports of sensible heat and geopotential exactly cancel and there is no net transport of dry static energy.

To second order, the paths of the air parcels projected onto the meridional plane exhibit an additional "rectified" motion in the meridional plane, as shown schematically in Fig. 5.24. At the latitude of the storm track (i.e., the latitude at which the amplitude of the waves is greatest), the trajectories of air parcels are exactly as depicted in the previous figure, but poleward and equatorward of that latitude they are somewhat different. Because of the longitudinal variations in the wave-related zonal wind perturbations u^*, a parcel located poleward of the storm track spends more time descending as it passes through the troughs of the waves than it spends ascending as it passes through the ridges. Hence, in passing through an integral number of wavelengths it undergoes a small net downward displacement, as indicated by the looping high latitude trajectories depicted in Fig. 5.24. The meridional gradient of wave amplitude also contributes to the rectified descent: poleward of the storm track, a descending parcel in a wave trough is closer to the storm track than an ascending parcel passing through the ridge, so the rate of

descent is slightly stronger than the rate of ascent. These same arguments lead to the conclusion that orbits of air parcels located equatorward of the storm track must exhibit a rectified ascent. These rectified parcel displacements are in the opposite sense as the Eulerian mean meridional motions in the vicinity of the storm track.

The poleward heat transports at the 50 hPa level and above in Figs. 5.15 and 5.16 are nearly adiabatic and primarily associated with planetary-scale perturbations in the stratospheric polar night jet, which encircles the winter pole at a latitude of around 60°N/S (Fig. 1.19). The boreal wintertime stationary waves also contribute to the poleward heat transport at the higher latitudes. The orbits of air parcels passing through these waves also exhibit a rectified downward displacement at high latitudes, which has important implications for the Lagrangian mean meridional circulation.

Exercises

5.21 Figure 5.20 is analogous to Fig. 3.10 except that it involves the vertical tilt of the wave axes rather than the meridional tilt. Explain the similarities and differences between these two figures and the rationales used in constructing them.

5.22 Prove that in a dry atmosphere the poleward energy transport is proportional to the area of the elliptical projection of the parcel trajectory on the $(v, c_p T + \phi)$ plane divided by the time required for an air parcel to execute a circuit around it.

5.23 Describe the idealized trajectories of air parcels in Fig. 5.23 as they would appear in isentropic coordinates (i.e., with θ or DSE, rather than pressure as the vertical coordinate).

5.7 EULERIAN VERSUS LAGRANGIAN MEAN MERIDIONAL CIRCULATIONS

To reinforce the concept of a wave-induced, rectified Lagrangian motion, it is useful to consider the two-dimensional oscillatory flow in water waves, as first elucidated by Stokes (1847). Tagged water parcels in water waves propagating from left to right from the perspective of the observer exhibit clockwise orbits. As the parcel circulates, it spends more time moving toward the right, in the upper half of its orbit, where it is moving in the same direction as the wave is propagating, than in the bottom half, where it is moving in the opposite direction. This inequality induces what is commonly referred to as the *Stokes drift*,[14] a mass transport in the same direction as the wave propagation. In deep water waves there is an additional contribution from the decay of wave amplitude with depth.

In analogy with the Lagrangian parcel displacements in water waves, the term "Stokes drift" is also applicable to the Lagrangian displacements in Fig. 5.22 relative to the Eulerian background mean meridional circulation. The total Lagrangian mean meridional circulation can thus be interpreted as the sum of the Eulerian circulation and the Stokes drift.

In the absence of diabatic heating, the isentropes are "frozen into the flow" (i.e., they are twisted by Lagrangian mean meridional circulation cells): sloping isentropes flatten in cells like the one in Fig. 2.4, in which warm air is rising and cold air is sinking, and they steepen in cells in which it is the colder air that is rising. It follows that if the flow is adiabatic and in steady state, the Lagrangian mean meridional circulation must be zero. If that were not the case, the zonally averaged distributions of passive tracers, including potential temperature, would be changing in response to advection by the mean meridional circulation, and that would imply changes in $[T]$. And if $[T]$ is not changing, then the temperature tendency induced by the poleward eddy heat transport must be cancelled by the eddy-induced Eulerian mean meridional circulation; that is, $s[\omega_E] = -H$, where $[\omega_E]$ is the Eulerian-mean, zonally averaged vertical velocity. Under these special conditions, the Stokes drift $[\omega_S]$

must be equal and opposite to the Eulerian mean meridional circulation.

The Lagrangian mean vertical velocity – estimated on the basis of Eulerian mean statistics and denoted by the symbol $[\omega]^*$ – is the sum of the Eulerian mean vertical velocity and the Stokes drift $[\omega_S]$,

$$[\omega]^* = [\omega] + [\omega_S]. \tag{5.21}$$

Noting that the Stokes drift associated with the adiabatic eddy motions is $[\omega_S] = H/s$, we can combine Eqs. (5.20) and (5.21) to obtain

$$[\omega]^* = [\omega] + \frac{H}{s} = -\frac{[Q]}{s}. \tag{5.22}$$

To distinguish this estimated Lagrangian mean meridional vertical velocity inferred from Eulerian statistics $[\omega]^*$ from the true Lagrangian mean meridional motion $[\omega_L]$, $[\omega]^*$ is referred to as the *transformed Eulerian mean* (TEM) vertical velocity. The corresponding TEM meridional motion $[v]^*$ can be inferred from the distribution of $[\omega]^*$ making use of the zonally averaged continuity equation

$$\frac{1}{a\cos\phi}\frac{\partial}{\partial\phi}([v]^* \cos\phi) + \frac{\partial[\omega]^*}{\partial p} = 0. \tag{5.23}$$

The TEM formalism also offers a more incisive way of thinking about the eddy transport of westerly momentum, as will be explained in Chapter 8.

The observed TEM climatological mean circulation, shown in Fig. 1.29 and repeated in Fig. 5.25 superimposed upon the diabatic heating field, is dominated by equator-to-pole circulation cells with rising branches in regions of strong diabatic heating in the tropics, sinking branches in regions of cooling extending over much of the extratropics, and poleward flow in the upper troposphere and stratosphere. The cells and the associated meridional gradients of diabatic heating rate are stronger in the winter hemisphere than in the summer hemisphere. Compared with their Eulerian counterparts shown in Fig. 1.27, the low latitude "Hadley cell" segments are similar, but there is no Ferrel cell and the poleward mass transport at the higher levels extends all the way to the polar cap regions. The differences at the higher latitudes reflect the strong Stokes drift associated with the adiabatic eddy heat transports across the storm tracks.

To conclude this section, we show in Fig. 5.26 a schematic of the Lagrangian mean meridional circulation as simulated in a general circulation model whose Eulerian mean circulation exhibits a realistic midlatitude Ferrel cell. To reveal the Lagrangian mean circulation, circular rings of air parcels are inserted into the three-dimensional flow at designated latitudes and levels. The positions of the parcels in each ring, projected into the meridional plane, quickly disperse into clouds of points, which become progressively more diffuse as the parcels in the eddies disperse in all directions. The centroids of the "clouds" originating in different latitude/height rings are tracked as they are advected by the Lagrangian mean meridional circulation. The motions of a selected set of centroids (thick arrows) indicate the existence

---

[14] After G.G. Stokes (1847).

Figure 5.25 The climatological TEM circulation superimposed upon the distribution of diabatic heating rate in K day^{-1}. Top: Annual mean; middle: DJF; bottom: JJA. The velocities are multiplied by the cosine of latitude, and the reference arrows denote 0.001 m s^{-1} (vertical) and 0.4 m s^{-1} (meridional) in the stratosphere, 0.01 m s^{-1} (vertical) and 4 m s^{-1} (meridional) in the troposphere. Courtesy of Ying Li.

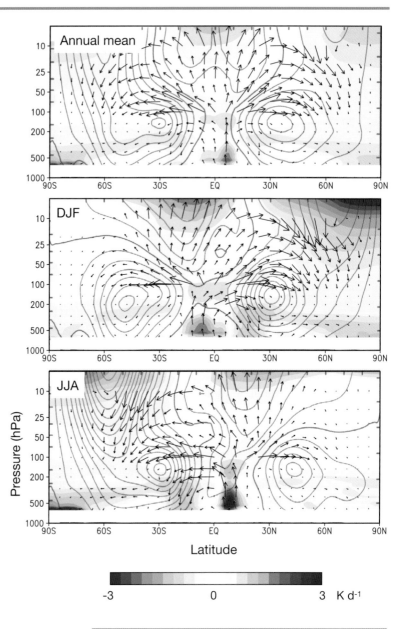

of an equator-to-pole cell extending upward into the stratosphere. The thinner arrows indicate the shapes of the trajectories of individual air parcels. At levels above 500 hPa, the trajectories resemble the idealized helical trajectories in Fig. 5.24 in which the parcels exhibit a rectified downward displacement as they circulate around counterclockwise loops. In the lower troposphere, the trajectories are suggestive of a diffusive regime in which the clouds of tracked parcels disperse meridionally. That the trajectories are looping at the upper levels derives from the fact that above 500 hPa the zonal wind speed $[u]$ is much larger than the phase speed of the waves, so the air parcels pass through the waves from west to east nearly adiabatically, as in Fig. 5.22. As in the observed TEM circulation shown in Fig. 5.25, Eulerian and Lagrangian mean meridional motions in low latitudes are similar, but in the extratropics they are entirely different because of the contribution of the Stokes drift.

Exercises

5.24 Describe the Stokes drift associated with two-dimensional deep water waves, taking account of the exponential dropoff of wave amplitude with depth.

5.25 Show that the mass streamfunction for the Stokes drift (exclusive of the boundary-forced component) in the meridional plane is $[v^*T^*]/s$.

5.8 CONCLUDING REMARKS

The treatment of the total energy balance in this chapter parallels the treatment of the angular momentum balance in Chapter 3. Both begin with consideration of global sources

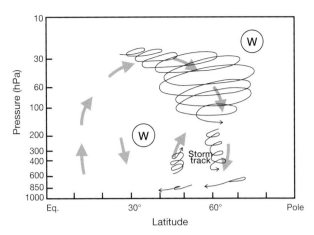

Figure 5.26 Schematic illustration of the Lagrangian mean meridional motion as inferred from the motions of the tracers inserted into the flow in a simple general circulation model. Looping arrows show tracks of individual air parcels and heavy arrows show tracks of centroids of "clouds" of parcels emanating from points in the meridional plane. W denotes westerly jets. Modified from Kida (1977) to highlight the process discussed in this chapter.

and sinks and then proceed to examine balance requirements relating to meridional and vertical atmospheric transports. Eddy transports of westerly momentum are shown to be related to the meridional tilt of the eddies and heat transports to the vertical tilt. In both chapters, the balance requirements are used as a basis for inferring the distribution of Eulerian mean meridional circulations and the picture that emerges

is essentially the same, not fortuitously as will presently be shown.

The energy balance is more complicated, and for some readers more interesting than the angular momentum balance because it requires consideration of thermodynamics as well as dynamics. Specifically, (i) it involves radiative sources and sinks at the top of the atmosphere, (ii) the ocean participates in its meridional transport, (iii) energy comes in different, but interchangeable forms, (iv) energy transport is inevitably accompanied by the performance of mechanical work and, last but not least, (v) it requires consideration of phase changes between water vapor and liquid water in the hydrologic cycle, as discussed in the previous chapter.

The last three sections of this chapter diverge from the energy balance theme and consider in greater detail the role of the eddy heat fluxes in the general circulation, not just in moving energy around, but also in the transport of mass in the Lagrangian mean meridional circulation. As such, these sections lay the groundwork for a more complete and rigorous treatment of the Lagrangian circulation and wave–mean flow interaction in Part III.

The treatment of the balance requirement for total energy considered in this chapter does not do justice to kinetic energy. The storage of kinetic energy is so small that its contribution to the total energy transports and conversions has been entirely neglected in this chapter and relegated to the final chapter of Part II, which comes next.

6 The Mechanical Energy Cycle

The *total energy* per unit mass of an air parcel is the sum of its internal, potential, and kinetic energy. It can be shown (see Exercise 6.1) that integrated over a column of unit area, the sum of the potential P plus internal I energy is given by $P + I = \int c_p T \, dp$. The integrand in this expression may be written as $(c_v + R)T$, in which the first term is the contribution to I and the second to P. It can also be shown that $c_p T$ is equal to $R/c_v \sim 2/5$ of the square of the speed of sound c_s. From these relationships it follows that $K/(P+I) \sim (V/c_s)^2$, where V is the root mean squared velocity in atmospheric motion systems. In the Earth's atmosphere V and c_s are on the order of 10 and 300 m s^{-1}, respectively. Hence, less than a thousandth of the total energy resides in the form of kinetic energy. Furthermore, it will soon become apparent that only a small fraction of the potential energy P is available for conversion into kinetic energy K. Because of this large disparity between the sizes of the available and total potential energy reservoirs, the cycling of *mechanical energy* – here defined as *the sum of the available potential energy and the kinetic energy* – needs to be considered separately from that of the total energy.

In Sections 2.2 and 2.3, the reader was introduced to the notion of a mechanical energy cycle in which potential energy, generated by differential heating, is converted to kinetic energy which, in turn, is subject to frictional dissipation. The examples provided there involved incompressible fluids. This chapter provides a quantitative framework for diagnosing the energy generation, conversion, and dissipation terms in planetary atmospheres, taking into account the adiabatic temperature tendencies associated with compression and expansion as well as those induced by diabatic heating.

The quantification of available potential energy by Lorenz (1955) provided a framework for the diagnosis of the atmosphere's mechanical energy cycle. By making a few key approximations that include treating static stability as a function of height only (i.e., $s = s(p)$), it is possible to represent the potential energy that is available for conversion to kinetic energy in terms of the spatial variance of temperature on pressure surfaces. This so-called *available potential energy* (A) is generated by the horizontal gradients in diabatic heating and released when warm air rises and cold air sinks, flattening the isentropes. Kinetic energy is generated when the air parcels flow down the horizontal pressure gradient and it is dissipated by frictional drag. The

release of available potential energy can be shown to be identical to the generation of kinetic energy. It is notable that, even though the kinetic energy generation is exclusively due to the cross-isobar flow, most of the kinetic energy resides in Rossby waves, in which the flow is parallel to the isobars.

Lorenz's formulation is framed in terms of the atmosphere as a whole, but with the inclusion of boundary terms it can be applied to layers of the atmosphere (e.g., troposphere versus stratosphere) or to latitude belts (e.g., tropics versus extratropics). It can also be partitioned into different classes of motions (e.g., zonally symmetric versus eddies, time mean vs. transients, or planetary scale versus smaller scales). Lorenz's treatment includes a partitioning of the energy cycle into zonal mean and eddy components, which is the basis for Section 8.1.1. Chapter 6 concludes with a section showing and interpreting the mechanical energy spectra for the general circulation as a function of zonal wavenumber.

Exercises

6.1 Show that the sum of the potential plus internal energy of a unit column of the atmosphere is given by[1]

$$P + I = \frac{c_p}{g} \int_0^{p_s} T \, dp. \qquad (6.1)$$

Hint: To derive this expression, make use of the identity

$$\int_0^{\infty} p \, dz = \int_0^{p_s} z \, dp.$$

6.1 QUANTIFICATION OF AVAILABLE POTENTIAL ENERGY

Lorenz proposed that the available potential energy of the whole atmosphere be defined as the difference between the globally integrated potential energy P and that which would exist if the mass of the atmosphere were redistributed under the conservation of potential temperature such that the isentropes became horizontal.[2] From this definition it

---

[1] Haurwitz (1941).
[2] Lorenz (1955).

follows that the available potential energy A and the kinetic energy K have the following properties:

(1) $A + K$ is conserved under adiabatic, frictionless flow;
(2) A is completely determined by the distribution of mass;
(3) A is zero if the potential temperature surfaces are flat and the flow is stably stratified;
(4) If A is nonzero, it must be positive definite.

If all the existing A were converted into K, it would result in an unrealistic, purely barotropic circulation. In atmospheric motion systems such as baroclinic waves, temporal variations in kinetic and available potential energy tend to be positively correlated in time, which limits the rate at which the conversion can proceed. The factors that determine the ratio of kinetic to available potential K/A locally are considered in Exercise 6.8.

6.1.1 A Simplified Expression for Available Potential Energy

Consider an atmosphere at rest, but with undulations in the potential temperature surfaces (isentropes) as sketched in Fig. 6.1 so that locally

$$T(x, y, z) = \langle T(z) \rangle + \widehat{T}(x, y, z), \tag{6.2}$$

where $\langle T \rangle$ with angle brackets denotes the global average of T at height z, and $\widehat{T}$ the departure from the global average. If the isentropes are allowed to flatten out by means of an adiabatic readjustment of mass, potential energy will be released as colder, heavier air sinks and warmer, lighter air rises in the process, so that the center of mass of the atmosphere drops by a small amount.

We will now derive an expression for A. The downward directed acceleration that exists on any air parcel, of unit mass, by virtue of the undulations in the isentropes can be expressed as

$$-\frac{\widehat{T}}{\langle T \rangle} g = \frac{g(\Gamma_d - \Gamma)}{\langle T \rangle} \widehat{z}, \tag{6.3}$$

where $\widehat{z}$ is the vertical displacement of the air parcel from the mean level of its isentrope, Γ is the lapse rate $-\,dT/dz$, and

$\Gamma_d = g/c_p$ is the dry adiabatic lapse rate.[3] The work done by gravity (or the potential energy released) when the air parcel ascends or descends is given by

$$W = \frac{g(\Gamma_d - \Gamma)}{\langle T \rangle} \int_0^{\widehat{z}} \zeta \, d\zeta = \frac{g(\Gamma_d - \Gamma)}{\langle T \rangle} \frac{\widehat{z}^2}{2}, \tag{6.4}$$

where ζ is a dummy variable. We assume that the undulations are small enough so that $\langle T \rangle$ and $(\Gamma_d - \Gamma)$ can be taken outside the integral sign. Since $\widehat{z} = \widehat{T}/(\Gamma_d - \Gamma)$, we can write

$$W = \frac{g}{2\langle T \rangle} \frac{\widehat{T}^2}{(\Gamma_d - \Gamma)}. \tag{6.5}$$

To obtain an expression for available potential energy, we integrate W over the entire mass of the atmosphere and divide it by the area of the Earth. These operations are equivalent to integrating Eq. (6.5) over pressure divided by g, and averaging it over the area of the Earth. Hence, the globally averaged available potential energy A (per unit area) is

$$A = \frac{1}{2} \int_0^{p_s} \frac{1}{\langle T \rangle} \frac{\langle \widehat{T}^2 \rangle}{(\Gamma_d - \langle \Gamma \rangle)} dp, \tag{6.6}$$

where p_s is the globally averaged value of pressure at the Earth's surface, and $\langle T \rangle$ and $\langle \Gamma \rangle$, refer to globally averaged values of the environmental temperature and lapse rate at pressure level p.

In the above derivation we have made three kinds of approximations:

(1) We have done some linearizing in getting to equations Eqs. (6.3) and (6.4). This approximation is based on the assumption that the isentropes in the Earth's atmosphere are relatively flat.
(2) In the final step we have treated the lapse-rate as if it were a function of pressure only. Strictly speaking, we should have written

$$A = \frac{1}{2} \int_0^{p_s} \frac{1}{\langle T \rangle} \left\langle \left[\frac{\widehat{T}^2}{(\Gamma_d - \Gamma)} \right] \right\rangle dp. \tag{6.7}$$

In effect, the approximation in Eq. (6.6) amounts to a further linearization. This approximation should be valid in the free atmosphere, where $\widehat{T}^2$ and $(\Gamma_d - \Gamma)$ are both positive definite, but it could be problematical if the boundary layer is included in the integration.
(3) By terminating the vertical integration at the pressure surface p_s rather than at a potential temperature surface, we have ignored the fact that isentropes intersect the Earth's surface because of orography and, more importantly, because of the substantial horizontal temperature gradients at the Earth's surface. This complication is dealt with more explicitly in Lorenz's more complete derivation.

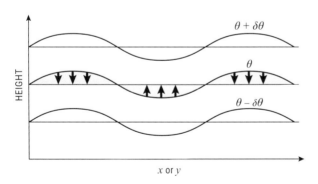

Figure 6.1 Schematic showing undulating isentropic surfaces relaxing back to a flat reference state, releasing available potential energy. At any given level, warmer (higher potential temperature) air is rising and cold (lower potential temperature) air is sinking.

---

[3] For a more thorough derivation, see Wallace and Hobbs 2nd ed., Exercise 3.11, p. 88.

6.2 Roughly estimate the globally averaged, column integrated K, A, and P in units of J m^{-2}.

6.3 If convective scale motions were treated as an integral part of the general circulation itself, would the mechanical energy cycle look different?

6.4 Describe how the cooling of a limited part of the atmosphere can result in an increase of A.

6.5 Suppose that the atmosphere is divided into parts by inserting imaginary walls, for example, at 30°N/S. Show that A of the whole atmosphere is equal to or greater than the sum of the available potential energy of its parts. Under what condition does the equality hold?

6.6 Suppose that the atmosphere is divided into layers by inserting one or more imaginary dividers coinciding with isentropes. Show that A of the whole atmosphere is equal to the sum of the available potential energy of its layers.

6.7 Are the assumptions that underlie the definition of available potential energy more valid in the troposphere or in the stratosphere?

6.8 How does the ratio of kinetic to potential energy per unit mass differ with respect to the horizontal and vertical scales of atmospheric disturbances?

6.9 How does the ratio of kinetic to potential energy per unit mass differ between troposphere and stratosphere? How does it vary with latitude?

6.10 Derive an expression for the available potential energy of a homogeneous (constant density) fluid with a free surface enclosed within a container with a flat bottom and vertical walls. (See Exercise 2.10).

6.2 SOURCES AND SINKS OF AVAILABLE POTENTIAL ENERGY

It is evident from Eq. (6.6) that the contribution of a layer of the atmosphere to A can be changed by any process that is capable of changing the variance of temperature on pressure surfaces, its mean static stability, or its mean temperature. Changes in the global mean temperature are unimportant for the purposes of this discussion. The spatial gradients in zonally averaged diabatic heating (e.g., as shown in Fig. 1.27) are in the sense as to increase the available potential energy, both by increasing the equator-to-pole temperature contrast and by destabilizing the lapse rate.

In a similar manner, atmospheric motions draw on the reservoir of available potential energy in two different ways: by flattening the potential temperature surfaces, thereby decreasing the spatial variance of temperature on pressure surfaces, and by stabilizing the lapse rate and thereby lowering the center of mass of the atmosphere. Hence, to perform a really comprehensive diagnosis of the budget of available potential energy requires consideration of changes

in the variance $\widehat{T}^2$ and in static stability ($\Gamma^d - \Gamma$) in Eq. (6.7).[4]

In most general circulation studies the static stability is treated as if it does not vary with time; variations of static stability are assumed to be in a state of balance. With this simplifying assumption, the task of keeping track of its budget is essentially reduced to accounting for changes in the variance of temperature on pressure surfaces. When we consider the general circulation from a global perspective, this simplification is justifiable because globally averaged static stability varies very little with time.

With this simplification, the expression for global available potential energy Eq. (6.6) can be written in the form

$$A = \frac{1}{2}\frac{R}{g}\int_0^{p_s}\frac{\langle\widehat{T}^2\rangle}{s}\,d\,lnp, \qquad (6.8)$$

where

$$s(p) \equiv \frac{R\langle T\rangle}{g\langle p\rangle}(\Gamma_d - \langle\Gamma\rangle), \qquad (6.9)$$

is a measure of the stratification in pressure coordinates.

6.3 CONVERSION FROM AVAILABLE POTENTIAL TO KINETIC ENERGY

In formulating the balance requirement for the global available potential energy, we consider only the departure $(\widehat{T})$ from the global mean temperature $\langle T\rangle$ on each pressure surface. Consistent with treating static stability as a function of pressure only, we can write the thermodynamic energy equation as

$$\frac{\partial\widehat{T}}{\partial t} = -\mathbf{V}\cdot\nabla\widehat{T} + s\widehat{\omega} + \widehat{Q}. \qquad (6.10)$$

Multiplying the terms in Eq. (6.10) by $\widehat{T}$ and globally averaging, we obtain

$$\frac{1}{2}\frac{d}{dt}\langle\widehat{T}^2\rangle = s\,\langle\widehat{\omega}\widehat{T}\rangle + \langle\widehat{Q}\widehat{T}\rangle. \qquad (6.11)$$

The global average of the horizontal temperature advection term in Eq. (6.10) is not identically equal to zero, but it is small, as argued in Exercise 6.12. Accordingly, we neglect it in the global budget Eq. (6.11), but we will show in Section 8.1.2 that it is a leading term in the exchange of available potential energy between the zonal average flow and the eddies.

The first term on the right-hand side of Eq. (6.11) is associated with the conversion from available potential energy to kinetic energy. On average in the Earth's atmosphere warm air rises and cold air sinks, so there is a negative correlation between ω and T on pressure surfaces, which tends to flatten the potential temperature surfaces (Fig. 6.2, top panel), thereby decreasing the variance of temperature and releasing available potential energy. The reservoir of available potential energy is maintained by the diabatic heating term

---
[4] See Dutton and Johnson (1967); Johnson (1970).

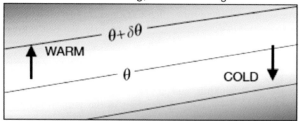

Warm air rising, cold air sinking

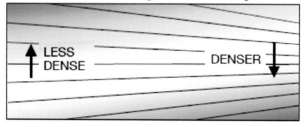

Less dense air rising, denser air sinking

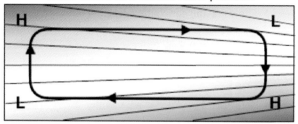

Cross-isobar flow toward lower pressure

Figure 6.2 Different ways of visualizing the conversion from available potential to kinetic energy ($C = A \rightarrow K$). The sloping lines in the top panel are isentropes; the lines in the middle and bottom panels are pressure surfaces. The arrows indicate the sense of the overturning circulation. The letters H and L denote highs and lows in geopotential height on pressure surfaces. Colored shading indicates potential temperature; warm colors denote higher values. See also Fig. 2.3.

$\langle \widehat{QT} \rangle$, which involves the spatial correlation between diabatic heating rate and temperature. At tropospheric levels Q and T tend to be positively correlated, since on average, the warmer tropical atmosphere receives more heat than it radiates to space and the colder polar atmosphere loses more than it receives, as documented in Fig. 5.6. There is an additional positive correlation between Q and T along latitude circles because of the tendency for precipitation and latent heat release to occur preferentially in rising warm air masses.

Combining Eq. (6.8) and Eq. (6.11) we can express the conversion from available potential energy to kinetic energy in the form

$$\boxed{\frac{dA}{dt} = -C + G,} \tag{6.12}$$

where

$$C = -\frac{R}{g} \int_0^{p_s} \langle \widehat{\omega T} \rangle \, d \ln p \tag{6.13}$$

and

$$G = \frac{R}{g} \int_0^{p_s} \frac{\langle \widehat{QT} \rangle}{s} \, d \ln p. \tag{6.14}$$

is the generation. We have chosen the sign convention for C such that it is positive when A is converted to K, which is prevalent in the Earth's atmosphere. Substituting for $\widehat{T}$ from the equation of state $p\widehat{\alpha} = R\widehat{T}$ we can rewrite the conversion as

$$C = -\frac{1}{g} \int_0^{p_s} \langle \widehat{\omega \alpha} \rangle \, dp, \tag{6.15}$$

which can be interpreted as the rising of lighter air, and the sinking of denser air (Fig. 6.2, middle panel) which results in a net downward flux of mass and a lowering of the center of mass of the atmosphere.

Substituting $-\partial \widehat{\Phi} / \partial p$ for $\widehat{\alpha}$ in Eq. (6.15) yields

$$C = \frac{1}{g} \int_0^{p_s} \left\langle \widehat{\omega} \frac{\partial \widehat{\Phi}}{\partial p} \right\rangle dp = \frac{1}{g} \int_0^{p_s} \frac{\partial \langle \widehat{\omega \Phi} \rangle}{\partial p} \, dp$$
$$- \frac{1}{g} \int_0^{p_s} \left\langle \widehat{\Phi} \frac{\partial \widehat{\omega}}{\partial p} \right\rangle dp. \tag{6.16}$$

The term $\langle \widehat{\omega \Phi} \rangle / g$ can be viewed as the mechanical stirring of one layer of the atmosphere by another layer. Note that $\langle \omega \Phi \rangle / g$ has units of (force × velocity)/area or, alternatively work/(area × time) and that $\langle \widehat{\omega \Phi} \rangle$ on pressure surfaces is almost identical to $\langle \widehat{pw} \rangle$ on surfaces of constant geopotential, which is perhaps more easily recognizable as a work term. If the air parcels on a pressure surface are being pushed up in high pressure regions and down in lows, then $\langle \widehat{\omega \Phi} \rangle < 0$ and the layer of air below the pressure surface is doing work on the layer of air above it. Now it is clear that $\langle \widehat{\omega \Phi} \rangle = 0$ at the top of the atmosphere, since the atmosphere cannot be doing mechanical work on empty space. The spatial correlation between ω and Φ does not vanish at the Earth's surface or on the 1000 hPa level. However, it is true that ω at the Earth's surface tends to be about an order of magnitude smaller than at mid-tropospheric levels. Were it not for the existence of large mountain ranges, the work done on the atmosphere by the Earth's surface or vice versa would play only a minor role in the atmosphere's kinetic energy budget. For the present, we will treat the surface of the Earth as if it were flat, so that the lower boundary corresponds to a surface of constant geopotential Φ and we will also assume that the 1000 hPa pressure surface is also rather flat and that ω is small there so that $\widehat{\omega \Phi}$ is vanishingly small at the bottom boundary. With this assumption

$$\frac{1}{g} \int_0^{p_s} \frac{\partial \langle \widehat{\omega \Phi} \rangle}{\partial p} \, dp = \langle \widehat{\omega \Phi} \rangle|_{p=p_s} - \langle \widehat{\omega \Phi} \rangle|_{p=0} = 0 \tag{6.17}$$

so that

$$C = -\frac{1}{g} \int_0^{p_s} \left\langle \widehat{\Phi} \frac{\partial \widehat{\omega}}{\partial p} \right\rangle dp. \tag{6.18}$$

Making use of the continuity equation, Eq. (6.18) can be rewritten as

$$C = \frac{1}{g} \int_0^{p_s} \langle \widehat{\Phi} \nabla \cdot \mathbf{V} \rangle \, dp. \tag{6.19}$$

Energy is being converted from A to K wherever $\widehat{\Phi}$ and $\nabla \cdot \mathbf{V}$ are positively correlated on pressure surfaces so that air diverges out of highs and converges into lows. We can express C explicitly in terms of the cross-isobar flow by integrating Eq. (6.19) by parts, which yields

$$C = \frac{1}{g} \int_0^{p_s} \nabla \cdot \langle \mathbf{V}\widehat{\Phi} \rangle \, dp - \frac{1}{g} \int_0^{p_s} \langle \mathbf{V} \cdot \nabla \widehat{\Phi} \rangle \, dp. \tag{6.20}$$

The term $\mathbf{V}\widehat{\Phi}$ is the horizontal counterpart of the $\widehat{\omega \Phi}$ term, but it involves the horizontal rather than the vertical flux of geopotential and, in (x, y, z) coordinates, has units of pressure $\times$ velocity or (force/area) $\times$ velocity; it can be written as $\widehat{\langle p \nabla \cdot \mathbf{V} \rangle}$ on geopotential surfaces. It also represents the horizontal transport of energy across a vertical "wall," such as a latitude circle, by mechanical stirring. Now since the whole atmosphere has no horizontal boundaries, there is no external region for it to do work on. Hence, Eq. (6.20) is equivalent to

$$C = -\frac{1}{g} \int_0^{p_s} \langle \mathbf{V} \cdot \nabla \widehat{\Phi} \rangle \, dp \tag{6.21}$$

and the conversion can be interpreted as the cross-isobar flow from higher toward lower pressure, as depicted in the bottom panel of Fig. 6.2.

To express the change in kinetic energy K in terms of C, consider the horizontal equation of motion, expressed in pressure coordinates:

$$\frac{\partial \mathbf{V}}{\partial t} = -\mathbf{V} \cdot \nabla \mathbf{V} - \nabla \widehat{\Phi} - f\mathbf{k} \times \mathbf{V} + \boldsymbol{\mathcal{F}}, \tag{6.22}$$

where $\boldsymbol{\mathcal{F}}$ is the frictional force per unit mass. Taking the dot product with $\mathbf{V}$, averaging over the area of the Earth's surface, noting that the advection term averages out to zero, and integrating over mass dp/g, we obtain

$$\frac{dK}{dt} = \frac{1}{2}\frac{d}{dt} \int_0^{p_s} \langle \mathbf{V} \cdot \mathbf{V} \rangle \, dp$$
$$= -\int_0^{p_s} \langle \mathbf{V} \cdot \nabla \widehat{\Phi} \rangle \, dp + \int_0^{p_s} \langle \boldsymbol{\mathcal{F}} \cdot \mathbf{V} \rangle \, dp, \tag{6.23}$$

where K is global averaged kinetic energy. Equation (6.23) can also be expressed in symbolic form

$$\boxed{\frac{dK}{dt} = C - D,} \tag{6.24}$$

where D is the globally average frictional dissipation, defined as

$$D = -\frac{1}{g} \int_0^{p_s} \langle \boldsymbol{\mathcal{F}} \cdot \mathbf{V} \rangle \, dp, \tag{6.25}$$

which is positive, since $\boldsymbol{\mathcal{F}}$ and $\mathbf{V}$ are in the opposite directions. Finally, we combine Eqs. (6.12) and (6.24) to arrive at the conservation of total mechanical energy

$$\boxed{\frac{d}{dt}(K + A) = G - D.} \tag{6.26}$$

Since the existing reservoirs of A and K are maintained over long periods of time, then C, G and D must all be equal to one another (see Eqs. (6.12) and (6.24)). We can express this equality in terms of the "flow chart" of the energy cycle shown in Fig. 2.3. *Thermally direct circulations* are those in which $A \to K$, so that warm, light air is rising, cold, dense air is sinking, and air is flowing across the isobars toward lower pressure, as depicted in the bottom panel of Fig. 6.2. *Thermally indirect circulations* are those in which the opposite conditions prevail. The mass-weighted global general circulation is thermally direct, as in a heat engine, but there are local exceptions in which the atmosphere behaves as a refrigerator. The work required to run these refrigerators is supplied by the $-\frac{\partial}{\partial p}\widehat{\omega \Phi}$ and/or $-\nabla \cdot \mathbf{V}\widehat{\Phi}$ terms, as discussed in more detail in Section 6.5.

Exercises

6.11 Prove that $\langle \widehat{\omega \Phi} \rangle$ on a pressure surface is almost, but not quite, equivalent to $\widehat{\langle pw \rangle}$ on a nearby geopotential surface.

6.12 Show that the horizontal advection term $-\widehat{T}\mathbf{V} \cdot \nabla \widehat{T}$ can be neglected in the global integral in Eq. (6.11).

6.13 (a) Why is it that $\langle \widehat{\omega} \rangle = 0$ on pressure surfaces? (b) In view of this identity, show that the $\langle \rangle$ can be dropped on covariance terms such as $\langle \widehat{\omega T} \rangle$ and $\langle \widehat{\omega \alpha} \rangle$.

6.14 Show that Eq. (6.15) can be written as

$$C = -\int_0^\infty \langle \omega \rangle \, dz,$$

which tells us that if $A \to K$ then in an average over the volume of the atmosphere, air parcels are undergoing a decrease in pressure. The bias toward flow down the pressure gradient is evident in the bottom panel of Fig. 6.2, in which air parcels are flowing across the isobars toward lower pressure in both upper and lower branches of the cell. The same is true of the working fluid in any heat engine, as implied by the clockwise circulation around the loop in a plot of pressure versus volume.

6.15 Bearing in mind the negative spatial correlation between geopotential height and relative vorticity in a rotating atmosphere, show that positive values of the conversion term C are conducive to increasing the r.m.s. amplitude of the perturbations in the vorticity field on pressure surfaces, making use of Eq. (6.19).

6.4 THE OBSERVED MECHANICAL ENERGY CYCLE

The middle and right panels of Fig. 6.3 show two different estimates of the size of the reservoirs of A and K and the rate of conversion from A to K, the first one based on an analysis of station data and the second, more recent

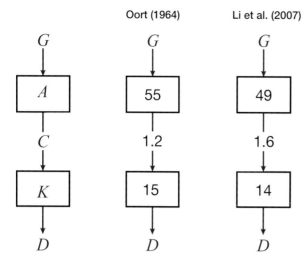

Oort (1964)   Li et al. (2007)

Figure 6.3 (Left) Symbolic representation of the mechanical energy cycle; (middle and right) estimates of the energy storage in the *A* and *K* reservoirs in units of J m^{-2} × 10^5 and the throughput in units of W m^{-2} based on two published studies, the earlier one based on radiosonde station data and the later one based on an early reanalysis product.

one, based on a reanalysis product. The overall similarity is reassuring. Both studies indicate that *K* and *A* are on the order of 15 and 50 × 10^5 J m^{-2}, respectively, which the reader is invited to show in Exercise 6.16 is consistent with an r.m.s. horizontal wind speed of 12 m s^{-1}, and an r.m.s. amplitude of the temperature perturbations on pressure surfaces on the order of 24°C, which corresponds to roughly half the equator-to-pole temperature contrast in the lower troposphere.

In both estimates shown in Fig. 6.3, the rate of conversion, *C* is ∼1.5 W m^{-2}, or slightly under 1.5 × 10^5 J kg^{-1} per day. At this rate, it would take about 10 days to replenish the *K* reservoir if it were empty to begin with. However, in view of the constraint of thermal wind balance, it is more appropriate to frame the question as, "At the current rate of conversion of 1.5 W m^{-2}, how long would it take to replenish both *A* and *K*, assuming that the current ratio $K/A \approx 15/50 = 0.30$ is maintained?" In terms of this metric, the replenishment timescale would be more like one month.

From the meridional cross sections of the Lagrangian mean meridional circulation and temperature shown in Fig. 1.29, it is evident that in the troposphere, the energy cycle is reminiscent of the steady-state laboratory experiment described in Section 2.3. Warm air is rising in the equatorial rain belt and cold air is sinking at higher latitudes, indicative of a thermally direct circulation in which kinetic energy is generated by the release of available potential energy that derives from the equator-to-pole contrast in diabatic heating; the release of latent heat in the tropics and radiative cooling in the polar cap regions where temperatures are maintained above radiative equilibrium by the convergence of the poleward eddy heat transports. A disproportionate fraction of the overturning is in the Hadley cells,

which are confined to within about 30° of the equator. Hence, the energy conversion is not as large as it would be if more of the descent occurred over the colder polar cap region.

In the stratosphere, the Lagrangian mean meridional circulation (i.e., the Brewer–Dobson circulation (BDC)) in Fig. 1.29 is in the same sense as in the troposphere, with tropical ascent and descent over the winter polar cap region. At these levels, it is the colder air that is rising and the warmer air that is sinking. It follows that from a thermodynamic perspective, the BDC in the lower stratosphere is thermally indirect; that is, it is maintained by the adiabatic temperature tendencies associated with dynamically induced expansion and compression. Kinetic energy is being converted into available potential energy: causality is in the opposite sense as in the laboratory experiment described in Section 2.3. In the seasonal sections in Fig. 1.29, it is evident that adiabatic compression of subsiding air contributes to the relative warmth of the midlatitudes in the winter hemisphere. To account for this behavior, it is necessary to invoke the existence of a process by which the tropospheric circulation can do work on the stratospheric circulation. We will consider how this is accomplished in the next section.

In the vicinity of the stratopause (1 hPa) level, where the absorption of solar radiation by ozone in photochemical reactions is a strong heat source, the circulation is thermally direct, as in the troposphere, with ascent over the strongly heated summer pole and descent over the cold winter pole. In contrast, in the upper mesosphere where photochemical reactions are relatively weak, the mean meridional circulation is thermally indirect, as evidenced by the remarkable coldness of the summer pole (not shown).

Adiabatic compression of sinking air in thermally indirect circulations is responsible for the exceptional warmth of some of the world's major desert regions, as evidenced by the observed deficit in net radiation over the Sahara and the Middle East in Fig. 5.5. Subsidence warming also contributes to the temperature extremes observed in transient "heat wave" events.

Exercise

6.16 Based on the data on the sizes of the *K* and *A* reservoirs presented in Fig. 6.3 and assuming a mean tropospheric lapse-rate of 6.5°C km^{-1} and mean temperature of 250 K, estimate the root-mean-squared wind speed and r.m.s. temperature departures from the global mean at the same level.

6.5 THE LOCAL MECHANICAL ENERGY CYCLE

The equality between the release of available potential energy, as manifested in the $-\widehat{\omega}\widehat{T}$ or $-\widehat{\omega}\widehat{\alpha}$ terms, and the generation of kinetic energy, as manifested in the $-\mathbf{V} \cdot \nabla \widehat{\Phi}$ term holds only for integrals over the entire mass of the atmosphere. Locally there may be large imbalances between

Figure 6.4 Idealized vertical profiles of (top left) the release of potential energy, (top right) the generation of kinetic energy by the cross-isobar flow, (bottom left) the difference between the top panels (see Eq. (6.27)), and (bottom right) the upward flux of geopotential. The heavy blue arrows represent the direction of the transport of geopotential and the mechanical work.

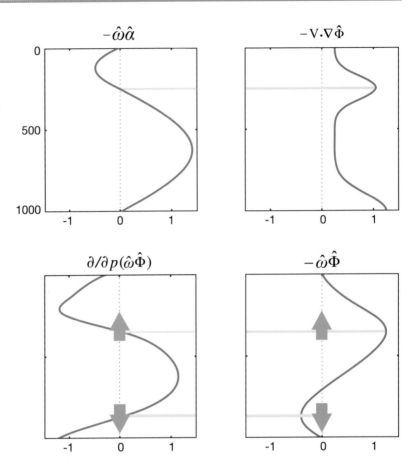

these terms that are compensated by imports or exports of energy through the $-\frac{\partial}{\partial p}\widehat{\omega\Phi}$ or $-\nabla \cdot \mathbf{V}\widehat{\Phi}$ terms. For example, most of the release of potential energy in the Earth's atmosphere takes place in the middle troposphere, where the vertical motions are largest, whereas most of the kinetic energy dissipation occurs in the planetary boundary layer and in the vicinity of the jet stream level. The middle troposphere is continually doing work on the upper and lower troposphere through the $\widehat{\omega\Phi}$ term in order to maintain the balance. Let us consider how this works in more detail.

Using the hypsometric and continuity equations, we can write

$$-\langle\widehat{\omega\alpha}\rangle = \frac{\partial}{\partial p}\langle\widehat{\omega\Phi}\rangle - \langle\mathbf{V}\cdot\nabla\widehat{\Phi}\rangle. \qquad (6.27)$$

Hence, all the available potential energy released at a particular level can be accounted for either in terms of export to the layers above or below, through the work term, or generation of kinetic energy at that same level. Figure 6.4 shows idealized vertical profiles of the three terms in Eq. (6.27) in the Earth's atmosphere. We know that $\langle\widehat{\omega\Phi}\rangle > 0$ at the top of the boundary layer because (1) high pressure regions at the Earth's surface tend to be clear and dry and low pressure regions, tend to be cloudy, often with precipitation; and (2) frictionally induced cross-isobar flow toward lower pressure implies a downward mass flux in the highs and an upward mass flux in the lows at the top of

the boundary layer (see Fig. 6.5). In contrast to conditions at the Earth's surface, at upper tropospheric levels regions of ascent and enhanced precipitation correspond to upper air ridges which is consistent with a reversal in the sign of $\langle\widehat{\omega\Phi}\rangle$, between the upper and lower troposphere. The shape of the $\langle-\mathbf{V}\cdot\nabla\widehat{\Phi}\rangle$ idealized profile in Fig. 6.4 is inspired by direct measurements[5] and is consistent with estimates of the vertical distribution of kinetic energy dissipation in turbulence in the boundary layer and in the free atmosphere. It exhibits a broad peak centered at the jet stream level, where the vertical wind shear is often strong enough to support the spontaneous development of Kelvin–Helmholtz waves. The $\langle\widehat{\omega\alpha}\rangle$ and $\langle\widehat{\omega\Phi}\rangle$ profiles are also well established on the basis of observations. There are some minor differences between profiles for the tropics and mid-latitudes, but the shapes of all the profiles are qualitatively similar to those shown in Fig. 6.4.

It is also evident from Fig. 6.4 that the troposphere does work on the stratosphere. As noted above, the lower stratosphere is one of those passive regions of the Earth's atmosphere, whose motions are driven by an influx of mechanical energy from elsewhere, which is mainly accomplished through the mechanical stirring brought about by the $\widehat{\omega\Phi}$ term. There also exist situations in which the $\mathbf{V}\widehat{\Phi}$ term acts as

---

[5] Kung (1966); Kung and Smith (1974).

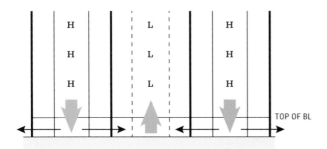

Figure 6.5 Vertical cross section through a wave with a barotropic vertical structure. Frictionally induced vertical motions and ageostrophic cross-isobar flow are indicated by the arrows. Vertical lines are perturbation isobars or geopotential height contours (dashed contours designate negative perturbations), lettering denotes Highs and Lows.

a mechanical stirring of a passive region of the atmosphere. For example, some of the transient disturbances in the tropics are driven by an influx of energy from higher latitudes as manifested in the observed negative correlation between $\mathbf{V}$ and $\widehat{\Phi}$ in the subtropical upper troposphere.[6] Evidently, some of the available potential energy released in middle latitudes is being used to generate kinetic energy in the tropics. A full treatment of the energy cycle for localized regions of the atmosphere requires consideration of additional terms involving boundary fluxes of kinetic energy and available potential energy. In this discussion we have considered only the leading terms.

Exercises

6.17 Consider the hypothetical situation in which the flow in the free atmosphere above the boundary layer is purely barotropic so that all the work done on the boundary layer by the $-\widehat{\mathbf{V}} \cdot \nabla \widehat{\Phi}$ term is at the expense of the kinetic energy of the free atmosphere. Complete the sketch of the pressure perturbations and the cross-isobar flow in an idealized vertical cross section in Fig. 6.5 (i.e., fill in the cross-isobar flow in the free atmosphere, assuming that the vertical motion and the cross-isobar flow satisfy continuity in the plane of the section and the vertically integrated cross-isobar flow is equal to zero). Make an analogous sketch for the contrasting situation in which the kinetic energy generated in the boundary layer is all at the expense of the available potential energy of the free atmosphere. [Hint: imagine a two-layer free atmosphere consisting of a baroclinic lower layer and an upper layer with no horizontal pressure gradients. Assume that all the cross-isobar flow required to satisfy continuity is concentrated in the upper layer.]

6.18 In the Brewer–Dobson circulation the poleward flow is directed down the horizontal pressure gradient, indicative of a thermally direct circulation, yet colder air is rising over the tropics and (relatively) warmer air is sinking at middle and high latitudes, indicative of a thermally indirect circulation. Explain this apparent paradox.

6.6 EDDY TRANSPORTS OF GEOPOTENTIAL

We have shown that one part of the atmosphere can dynamically force (i.e., export potential and kinetic energy to) another part through the horizontally averaged transports of geopotential through the interface between them, which might take the form of a pressure surface or a latitude circle.

We can derive a quantitative relationship between the poleward eddy heat transport by the geostrophic flow $[v_g{}^*T^*]$ and the vertical flux of geopotential $[\omega^*\Phi^*]$ in a linearized, adiabatic wave superimposed upon a background flow with zonal velocity $[u]$. Consider the adiabatic temperature tendency experienced by an air parcel as it passes through a wave from west to east in the background flow $[u]$, rising and cooling by adiabatic expansion in the ridges and sinking and warming by adiabatic compression in the troughs, as depicted in Figs. 5.22 and 5.24. For an adiabatic stationary wave, the thermodynamic energy equation assumes the form

$$\frac{DT^*}{Dt} = [u]\frac{\partial T^*}{\partial x} - s\,\omega^* = 0.$$

Multiplying by Φ^* and zonally averaging, we obtain

$$[u]\left[\Phi^*\frac{\partial T^*}{\partial x}\right] = s\left[\omega^*\Phi^*\right]. \tag{6.28}$$

Rewriting $[\Phi^*\partial T^*/\partial x]$ as

$$\frac{\partial[\Phi^*T^*]}{\partial x} - \left[T^*\frac{\partial\Phi^*}{\partial x}\right]$$

and noting that $[\partial/\partial x] \equiv 0$ and $\partial\Phi^*/\partial x = fv_g{}^*$, Eq. (6.28) can be rewritten as

$$f[u]\left[v_g{}^*T^*\right] = -s[\omega^*\Phi^*], \tag{6.29}$$

where v_g is the geostrophic component of v. The prevalence of poleward heat transports in temperate latitudes thus implies an upward flux of geopotential, which indicates that the atmosphere below any prescribed pressure surface is doing work on the layer above.

In a similar manner, we can write the equation for the time rate of change of zonal wind in a stationary wave

$$\frac{Du^*}{Dt} = [u]\frac{\partial u^*}{\partial x} - fv_a{}^* = 0, \tag{6.30}$$

---

[6] Mak (1969). See also Section 19.3 of this book.

Figure 6.6 Schematic of the mechanical energy cycle partitioned into (left) zonally symmetric (*Z*) and eddy (*E*) components and (right) time mean (*M*) and transient (*T*) components.

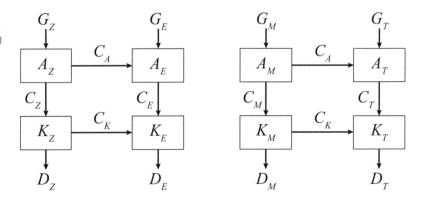

where v_a is the ageostrophic component of v. Multiplying by Φ^*, zonally averaging, and using the identity

$$\left[\Phi^* \frac{\partial u^*}{\partial x}\right] = -\left[u^* \frac{\partial \Phi^*}{\partial x}\right], \qquad (6.31)$$

we obtain

$$\left[v_a^* \Phi^*\right] = -[u]\left[u^* v_g^*\right]. \qquad (6.32)$$

Hence, if the zonal flow is eastward, the transport of geopotential and westerly momentum are in opposing directions; that is, the prevalence of poleward transports of westerly momentum in the 20–45° latitude band at the jet stream level implies an equatorward flux of geopotential, which may be interpreted as indicating that the eddies in the extratropics are doing work on the eddies in the tropics. These derivations can be extended to a propagating wave with eastward phase speed c simply by replacing $[u]$ by $[u] - c$ wherever it appears.

6.7 PARTITIONING OF THE MECHANICAL ENERGY

The Lorenz paper[7] is notable not only for its useful definition of available potential energy, but also for its subdivision of the energy cycle in terms of zonally averaged and eddy components, as illustrated in the left panel of Fig. 6.6. Here the zonal kinetic energy K_Z is the kinetic energy of the zonal wind component $[u]^2/2$, and the eddy kinetic energy K_E is $(u^{*2} + v^{*2})/2$. The kinetic energy associated with mean meridional motions is neglected because $[v]^2 \ll [u]^2$.

In order to partition A into zonal and eddy components, A_Z and A_E, we expand the temperature at each pressure level in terms of a global mean $\langle T \rangle$ and the departures from the global mean $\widehat{T}$. The latter are then decomposed into a zonally averaged component $[\widehat{T}]$ and an eddy component T^*

$$T = \langle T \rangle + [\widehat{T}] + T^* \equiv \langle T \rangle + \widehat{T}, \qquad (6.33)$$

from which it follows that the global A has contributions

$$\langle \widehat{T}^2 \rangle = \langle [\widehat{T}]^2 \rangle + \langle T^{*2} \rangle. \qquad (6.34)$$

Other variables $(u, v, \omega, \phi, \text{and } \alpha)$ are similarly partitioned. The various quadratic terms associated with the conversion terms C_Z and C_E (i.e., $-\omega T$, $-\omega \alpha$, $-\mathbf{V} \cdot \nabla \Phi$, etc) can be partitioned into C_Z and C_E in a similar manner, where the zonal component

$$C_Z = -\frac{1}{g}\int_0^{p_s}\langle [\widehat{\omega}][\widehat{\alpha}]\rangle \, dp = \frac{1}{g}\int_0^{p_s}\left\langle [\widehat{\omega}]\frac{\partial[\widehat{\Phi}]}{dp}\right\rangle dp \quad (6.35)$$

is associated entirely with mean meridional motions. The zonally averaged zonal flow $[u]$ is, by definition, nondivergent and parallel to the zonally averaged geopotential height contours. Hence the mean meridional motions play a crucial role in the kinetic cycle, even though they account for only a minute fraction of the kinetic energy reservoir. These simple two-dimensional circulations provide a convenient illustration of the energy conversion process: The Hadley circulation is obviously thermally direct, with $C_Z > 0$. It is characterized by the rising of warm, light air in equatorial latitudes and the sinking of colder, denser air in subtropical latitudes $(-\omega \alpha > 0)$; low-level equatorward flow out of the subtropical high pressure belt into a belt of low pressure along the equator $(-\mathbf{V} \cdot \nabla \widehat{\Phi} > 0)$; and upper tropospheric flow out of the equatorial belt and into subtropical latitudes in the presence of a westerly flow (again $-\mathbf{V} \cdot \nabla \widehat{\Phi} > 0$). The middle latitude Ferrel cell, which extends between approximately 30° and 60° latitude is, in some respects, a mirror image of the Hadley cell. Relatively warm, light air sinks in the subtropics while colder, denser air rises at higher latitudes. The cross-isobar flow is directed poleward, toward lower pressure at low levels, and equatorward, toward higher pressure at high levels. However, since $[\widehat{u}_g]$ increases with height at these latitudes, (i.e., $-\partial[\widehat{\Phi}]/\partial y$ is larger aloft) the up-gradient flow in the upper troposphere must dominate in the average over the cell. Hence, the Ferrel cell is thermally indirect with $C_Z < 0$. Observations indicate that the conversions in the Ferrel and Hadley cells nearly cancel one another, so that in the global average the numerical value of C_Z is a small difference between two large numbers, each of which has some uncertainty associated with it. Hence, even the sign of C_Z is uncertain in a global average.

Returning to Fig. 6.6, we note that C_E represents the conversion of eddy potential energy to eddy kinetic energy, $-\omega^* \alpha^*$ or $-\mathbf{V}^* \cdot \nabla \Phi^* > 0$. It is readily verified that

---

[7] Lorenz (1955).

$$\langle \widehat{\omega \alpha} \rangle = \langle [\widehat{\omega}][\widehat{\alpha}] \rangle + \langle \omega^* \alpha^* \rangle \qquad (6.36)$$

and

$$\langle \mathbf{V} \cdot \nabla \widehat{\Phi} \rangle = \left\langle [v] \frac{\partial [\widehat{\Phi}]}{\partial y} \right\rangle + \left\langle u^* \frac{\partial \widehat{\Phi}^*}{\partial x} \right\rangle + \left\langle v^* \frac{\partial \widehat{\Phi}^*}{\partial y} \right\rangle. \qquad (6.37)$$

The eddy conversion C_E is associated with phenomena such as monsoons and baroclinic waves, which are characterized by strong deviations from zonal symmetry with large and systematically interrelated fluctuations in T^*, α^*, ω^*, u^*, v^*, and Φ^*.

The conversion C_A of zonal potential energy A_Z to eddy potential energy, A_E (i.e., $[\widehat{T}] \to T^*$) involves the advection of zonal mean temperature. Similarly, the conversion C_K of zonal kinetic energy, K_Z to eddy kinetic energy K_E (i.e., $[u] \to u^*$) involves the advection of zonal mean momentum. For example, if horizontal temperature advection is tending to distort the isotherms from a zonally symmetric configuration into a more perturbed one, with larger values of T^*, then $A_Z \to A_E$. In a similar manner, if the meridional transport of westerly momentum by the eddies is into the latitude belt of the westerly jet stream, then $K_E \to K_Z$. We will consider these processes in further detail in Section 8.1.2.

The generation term $\langle \widehat{QT} \rangle$ in Eq. (6.11) can be partitioned in a manner entirely analogous to Eq. (6.36) and dissipation can be classified as D_Z or D_E. It is evident that G_Z constitutes a major input of energy into the energy cycle because of the large equator-to-pole heating gradient in the troposphere. Latent heat release in the troposphere makes strong positive contributions to both G_Z and G_E while Newtonian cooling at stratospheric levels represents a sink of both zonal and eddy available potential energy ($G < 0$). Most of the input of energy is through G_Z and energy flows from the A_Z reservoir to the other reservoirs and is destroyed mainly by frictional dissipation D_Z and D_E.

The zonal mean and eddy fields are spatially orthogonal so that spatial variance and covariance quantities such as $-\widehat{\omega T}$, $\widehat{QT}$, $-\widehat{\mathbf{V}} \cdot \nabla \widehat{\Phi}$, etc. can be partitioned as in Eq. (6.34), Eq. (6.36), and Eq. (6.37). A similar partitioning could be carried out among any set of spatially orthogonal fields. For example, the eddy energies and conversion processes in Fig. 6.6 could be expanded in terms of zonal wavenumber components so that A is partitioned into A_1, A_2, $A_3 \cdots A_k$, where k refers to zonal wavenumber, and similarly for G, C, and D, as illustrated in Fig. 6.7.

A and K are exchanged among the various harmonic components by means of advective processes analogous to those described above. The nonlinear wave–wave interactions as well as the interactions between each wave component and the zonally symmetric flow can be expressed in terms of formulae involving the coefficients of the various zonal harmonics[8] and they can be further partitioned in terms of temporal frequency.[9] With these formulations it is possible to

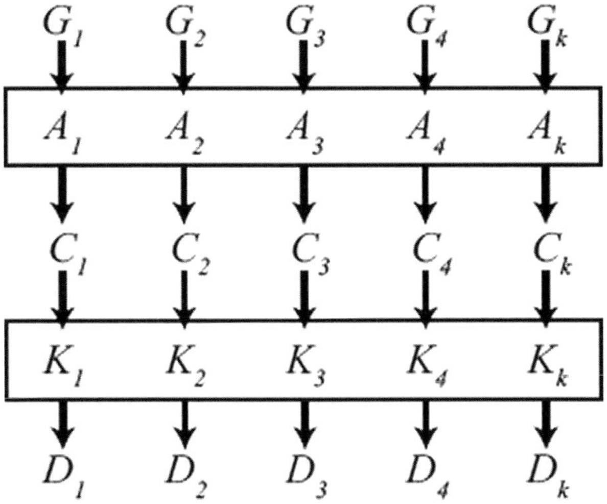

8 Saltzman (1970).
9 Kao (1968); Kao and Chi (1978).

Figure 6.7 Schematic of the eddy component of the mechanical energy cycle partitioned into individual zonal wavenumbers. Exchanges of A and K among the various zonal wavenumber components (not explicitly shown) would need to be represented by conversion tensor C_{ijk}.

gain some insight into the contributions of various dynamical entities to G and C. For example, the processes associated with the monsoon circulations are represented, for the most part, by $k = 1 - 3$, whereas baroclinic waves fall largely within the range $k = 6 - 10$. Because of the large number of possible interactions between the various wave components, the job of keeping track of the energy cycle depicted in Fig. 6.7 could become rather complex and tedious. A considerable reduction in complexity can be realized if the waves are grouped into a few categories such as planetary waves ($k = 1 - 3$), long waves ($k = 4 - 5$), synoptic scale waves ($k = 6 - 15$) etc..

Relationships analogous to Eq. (6.34) , Eq. (6.36), and Eq. (6.37) can also be derived from components of the general circulation which are orthogonal in the time domain. The simplest subdivision of this type is a partitioning between time-mean components, A_M, K_M, C_M, etc. and time varying or transient components A_T, K_T, C_T, etc., as illustrated in the right panel of Fig. 6.6. Such a breakdown illuminates the distinction between geographically fixed phenomena such as the monsoons and the stationary waves, versus propagating disturbances such as baroclinic waves and transient planetary waves. We will show in Section 10.2 how a field on a particular latitude circle at a particular level can be analyzed by means of harmonic analysis in the wavenumber domain in combination with power spectrum analysis in the frequency domain, which yields two-sided spectra, with eastward propagating waves on one side and westward propagating waves on the other. Other mixed space–time decompositions are possible and spherical harmonics or other spatially orthogonal functions can be used in place of zonal harmonics. It is also possible to subdivide motion systems in terms of vertical structure (e.g., barotropic versus baroclinic systems).

Figure 6.8 Height patterns for 500 hPa on Days 30, 40, and 50 of numerical simulations with an aquaplanet model, started from a state of rest, as described in Section 2.6.1. Courtesy of David Bonan.

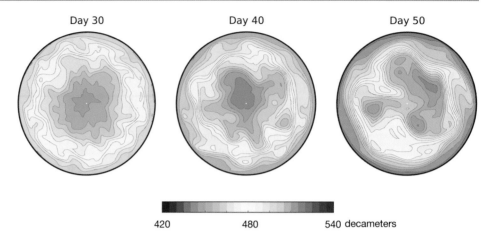

6.8 THE MECHANICAL ENERGY SPECTRUM

In the numerical experiment described in Section 2.6.1, the atmospheric general circulation is "spun up" starting from a state of rest. A critical transition point in this experiment is the time at which baroclinic waves first start to develop and interact with each other and with the zonally symmetric flow. In the Earth-like experiment described in Section 2.6.2, that transition occurs a few days before Day 30. At first (Fig. 6.8, left panel) the waves assume a nearly modal form consisting of zonally oriented wave trains that are strongest in midlatitudes and have zonal wavelengths on the order of 3500 km, which matches the scale of the most unstable baroclinic waves. By Day 40 (middle panel), disturbances with longer zonal wavelengths (i.e., lower zonal wavenumbers) are developing. By Day 50 (right panel), most of the variance of the geopotential height field resides in planetary waves with $k < 5$, where it remains throughout the remainder of the integration.

Another characteristic feature of these "spin up experiments" is revealed by Fig. 4.4, which shows a sequence of snapshots of the distribution of column-integrated water vapor using a higher resolution model. As in Fig. 6.8, the pattern that first develops exhibits the distinctive signature of the most rapidly amplifying baroclinic waves, but the evolution of the water vapor field is dominated by the development of front-like features that are stretched by the deformation field until they become so thin that they are obliterated by horizontal diffusion. The fields of column water vapor, potential vorticity (PV), and other tracers (not shown) exhibit a similar evolution.

The contrasting evolution of these two different fields illustrates how kinetic energy that resides initially in synoptic scale disturbances with zonal wavelengths of a few thousand kilometers is transferred toward larger, planetary-scale waves, which come to dominate the extratropical geopotential height field, while at the same time, the variance of fields of relatively passive tracers like water vapor is transferred toward smaller subsynoptic scale filaments that are subject to strong dissipation.

Quantitatively diagnosing the transfer of kinetic energy from synoptic scale baroclinic waves to disturbances with a much wider range of space scales requires consideration of the spatial *kinetic energy spectrum*, constructed by subdividing the atmospheric K reservoir into a large number of small reservoirs on the basis of horizontal wavenumber. Spectra can be computed for kinetic energy K or available potential energy A separately, or for mechanical energy $K + A$.

Figure 6.9 shows the global mechanical energy spectrum plotted on a log–log scale together with two theoretical reference spectra, both of which assume the form of power laws. Presenting spectra in this format is useful for comparing observations with power laws derived from theoretical considerations. In this representation, the contribution of a band of wavelengths to the total mechanical energy is not proportional to the area under that segment of the spectrum. Baroclinic waves occupy a range of zonal wavelengths centered around $k = 7$, which corresponds to a break in the spectrum between a gently sloping planetary-scale segment, in which mechanical energy is roughly proportional to $k = -1$ and a steeper synoptic-scale segment, in which it is more closely aligned with the $k = -3$ power law. We will defer further interpretation of the observed spectrum until Section 21.4, by which time the reader will have been introduced to some important distinctions

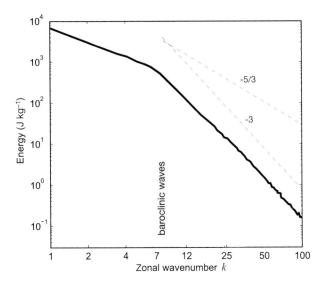

Figure 6.9 The globally averaged annual mean mechanical energy ($A + K$) spectrum based on the horizontal wind field from ERA Interim reanalysis plotted on a log–log scale. The horizontal arrows indicate the transfer of energy from baroclinic waves to larger and smaller scale perturbations. Adapted from Žagar et al. (2017). © American Meteorological Society. Used with permission.

between its Rossby wave and inertio-gravity wave components.

6.9 CONCLUDING REMARKS

Lorenz's formulation of the mechanical energy cycle expresses the concepts developed in Sections 2.2 and 2.3 relating to mechanical energy in a quantitative form that can be used to estimate the energy storages and rates of generation, conversion, and dissipation of mechanical energy (A and K) in the general circulation. It provides an indication of the "spin-up" or "spin-down" time of the general circulation and the processes that determine it. The notion of local thermally indirect circulations is helpful in explaining such phenomena as the coldness of the tropical cold point tropopause and to explain why the Sahara is extraordinarily hot, despite the fact that the top-of-atmosphere net radiation directly above it is negative. We will find it useful in Section 8.1.2 for diagnosing wave–mean flow interactions and in Sections 11.1.1, 11.3.2, and 14.5 for diagnosing the energetics of the stationary waves. It has proven less informative when applied to tropical tropospheric motion systems in which the generation of potential energy G is dominated by the release of latent heat in moist adiabatic ascent and the conversion of potential to kinetic energy C is occurring within the same overturning circulation and proceeding almost at the same rate as the generation term. In effect, the cycle is being short circuited: the A reservoir is being bypassed, as will be explained further in Chapter 15.

Part III Dynamics of the Zonal Mean Flow

The balance requirement approach covered in Part II provides a reasonable explanation of how the atmosphere satisfies the various budget constraints imposed by the conservation of mass, momentum, total energy, and mechanical energy, but it does not go very far in addressing such questions as:

- Why is there a single pair of tropospheric jet streams located around 30°N/S?
- Why do the eddies transport angular momentum poleward, across 30°N/S, maintaining the trade-wind and westerly wind belts?
- Why do the diagnoses based on the angular momentum balance in Chapter 3 and the energy balance in Chapter 5 yield the same configuration of mean meridional circulations?

Nor does the balance requirement approach provide any insight into the time dependence of the general circulation. Hand-waving arguments can only take one so far in diagnosing a system with as many feedbacks as the atmosphere.

The goal of Part III is to go beyond the various balance requirements to seek an understanding of why the mean circulation assumes the form that it does, and of how it evolves in response to various time-varying forcings. To this end, Chapter 7 goes back to basics (i.e., the governing equations), this time retaining the time derivatives of the zonally averaged zonal wind and temperature fields, while imposing the constraint that they remain in thermal wind balance with one another as the flow evolves.

7 Dynamics of the Zonal Mean Flow

On a rotating planet, the zonally symmetric zonal wind and temperature fields are in thermal wind balance. By applying this dynamical constraint, it is possible to go beyond the consistency arguments for steady state balances in Eqs. (3.21) and (5.20) and deduce how the flow will evolve in response to specified, time varying distributions of diabatic heating rate, frictional drag, and the eddy transports of zonal momentum and heat. In this zonally averaged version of the primitive equations, which dates back to Eliassen,[1] the mean meridional circulations play a critical role in enforcing the constraint that the zonal wind and temperature fields remain in thermal wind balance as the flow evolves.

The chapter begins with an example showing qualitatively how the mean meridional circulations keep the zonal flow in the mesosphere in thermal wind balance with the climatological mean temperature field as it evolves in response to the annual cycle in insolation. The second section assembles the governing equations, discussed in previous chapters, and describes how they behave as a time-dependent system. The section that follows expresses them in the more concise form of two-dimensional vector fields in the meridional plane. The third section presents the solution of the governing equations, and the fifth shows how these solutions are related to the "omega equation" and the "geopotential tendency equation," which are widely used in synoptic meteorology. The sixth section describes what the zonally symmetric flow on an otherwise Earth-like planet would look like in the absence of eddy transports. The final section consists of a chapter summary and concluding remarks.

7.1 AN EXAMPLE: THE ANNUAL CYCLE OF THE MESOSPHERIC CIRCULATION

The mesospheric circulation provides a simple example of how the zonally symmetric zonal wind and temperature fields and associated mean meridional circulations respond to the annual cycle in insolation. We choose this particular example because at these high levels in the atmosphere, the forcing in the zonal momentum equation is related to the breaking of gravity and inertio-gravity waves forced by flow over mountainous terrain, as discussed in Section 3.1. The breaking of these waves tends to relax the mean flow

toward the phase speed c of the waves. Since the waves are stationary ($c = 0$), it follows that the acceleration of the flow due to gravity and inertio-gravity wave-breaking can be represented in terms of a simple linear damping. To a close approximation, $[u]$ is in geostrophic balance and $[v] = kf[u]$, where k is the linear damping rate. But in addition to this balanced mean meridional circulation, there exists a small, seasonally varying, radiatively forced component that induces the observed seasonal reversals in $[u]$.

At the times of the solstices, the mesospheric zonal flow is characterized by a westerly jet stream in the winter hemisphere and an easterly jet stream in the summer hemisphere, as depicted schematically in Fig. 7.1. By mid-February, the Northern Hemisphere polar cap region is beginning to warm (panel b). The warming is reflected in the thickness field, so that geopotential height in polar latitudes at mesospheric levels must rise and as it does, the meridional pressure gradient weakens and the westerly jet in middle latitudes becomes *supergeostrophic*; that is, the poleward pressure gradient force is no longer quite strong enough to balance the equatorward Coriolis force. Air parcels in the vicinity of the jet accelerate in the direction of the stronger force, giving rise to an ageostrophic, equatorward mean meridional circulation as shown in the figure. This equatorward flow is accompanied by high latitude ascent at the stratopause level, in which the adiabatic cooling counteracts, but does not completely cancel, the radiative warming. Meanwhile, in the Southern Hemisphere, the easterly jets also become supergeostrophic as the polar regions begin to cool at the stratopause level and the elevated pressure surfaces over the polar cap region begin to drop. The unbalanced Coriolis force induces poleward flow in excess of what is needed to balance the frictional drag. The associated subsidence-induced adiabatic warming over high latitudes counteracts, but does not completely cancel, the radiative cooling.

This radiatively forced mean meridional circulation, which persists through late winter and spring, assumes the form of a single pole-to-pole cell, with rising over the colder polar cap and sinking over the warmer one and with cross-isobar flow toward higher pressure. It is in the sense as to maintain the existing meridional temperature gradient at the stratopause level, while acting as a brake on the jet streams. Kinetic energy is being converted into potential energy which, in turn, is being destroyed by radiative heating over the cold pole and radiative cooling over the warm pole.

[1] Eliassen (1951).

Figure 7.1 Schematic showing the seasonal evolution of the zonally averaged circulation in the mesosphere with snapshots at 45 day intervals, starting at the time of the Northern Hemisphere winter solstice, as indicated by the solar declination angle at the top of each panel. ⊙ and ⊗ indicate the positions of the centers of the westerly and easterly jet streams and the circles with arrows indicate the direction of the radiatively induced pole-to-pole mean meridional circulation when it is present. The colored shading indicates the temperature departures from the global mean at each level: blue denotes cold and gold warm. The squiggly arrows labeled Q provide an indication of the polarity and strength of the seasonally varying net diabatic heating (the departure from the global mean level by level) that drives the seasonal variations in the circulation. The arrows labeled C and P refer to the Coriolis and pressure gradient forces on zonally symmetric rings of air coincident with the jet streams. The small imbalances between them are exaggerated in order to make them clearly visible.

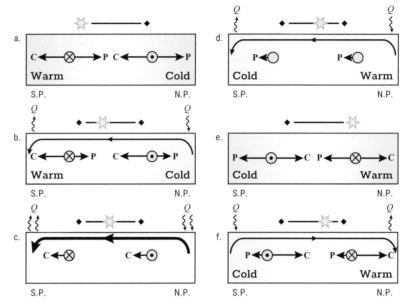

The mesospheric jet streams and the meridional temperature gradients at the stratopause level are both weakening at rates just sufficient to keep them in thermal wind balance with one another. If the decrease in the temperature gradient were to get a little bit ahead of the decrease in the strength of the jets, the winds would become more supergeostrophic and the radiatively driven mean meridional circulation would speed up a little until the near balance was restored.

Shortly after the time of the equinox, around April 1 (panel d), the mesospheric jets completely disappear at the same time that the pole-to-pole temperature gradient at the stratopause level reverses. The mean meridional circulation continues in the same sense as in late winter, but from this point onward it is characterized by rising over the warmer pole, sinking over the colder pole, etc. This continuing, pole-to-pole mean meridional circulation gives rise to a new pair of jets whose polarity is the opposite of those that existed previously, as indicated in the panels e and f.

Shortly after the solstice (July 1), temperatures at the stratopause level equilibrate with the thermal forcing and the net radiative heating and cooling stop. Afterward they resume with reversed polarity, leading to the development of a radiatively forced mean meridional circulation in the opposite sense, as indicated panel f.

In the above example, the mean meridional circulations develop in response to the imbalance between the Coriolis force $-f[u]$ and the meridional pressure gradient force $-\partial[\Phi]/\partial y$. In this explanation, we have invoked the zonally averaged prognostic equation for the meridional wind component in the system of primitive equations. In the system of equations for the zonally symmetric flow, presented in the next section, this prognostic equation is replaced by a diagnostic equation that follows directly from the constraint that the $[u]$ and $[T]$ fields are and must remain in thermal wind balance as the flow evolves.

Exercise

7.1   Show that in the upper stratosphere and meso-sphere, the energy conversion from available potential to kinetic energy C_Z exhibits a semiannual cycle.

7.2 THE GOVERNING EQUATIONS

In Part II of this book, we considered the angular momentum and heat balances separately. In this chapter, we will make use of a single set of equations to diagnose the behavior of the geostrophically balanced zonally symmetric flow and the associated temperature field. For simplicity, we write these equations in Cartesian coordinates, retaining only the leading terms.[2] When we include the zonally averaged thermal wind and continuity equations, the governing equations are:

$$\frac{\partial[u]}{\partial t} = f[v] - \frac{\partial[u^*v^*]}{\partial y} + \mathcal{F}_x \tag{7.1}$$

$$\frac{\partial[T]}{\partial t} = [s][\omega] - \frac{\partial[v^*T^*]}{\partial y} + Q \tag{7.2}$$

$$\frac{\partial[u]}{\partial p} = \frac{R}{fp}\frac{\partial[T]}{\partial y} \tag{7.3}$$

$$\frac{\partial[v]}{\partial y} + \frac{\partial[\omega]}{\partial p} = 0, \tag{7.4}$$

where $[s]$ is the static stability $\kappa[T]/p - \partial[T]/\partial p$, Q is the zonally averaged diabatic heating rate in deg K per unit time, and $\mathcal{F}_x$ is the zonally averaged source or sink

---

[2] See Andrews et al. (1987) for the equations in spherical coordinates.

of zonal momentum per unit mass due to friction. (Smaller terms neglected in Eqs. (7.1) and (7.2) are the accelerations due to the vertical eddy flux of zonal momentum $[u^*\omega^*]$, the vertical eddy flux of temperature, and the advection of temperature and zonal momentum by the zonally averaged meridional wind component.) The set of equations (7.1)–(7.4) involves four unknowns: $[u]$, $[T]$, $[v]$, and $[\omega]$. We can reduce it to three equations in three unknowns by expressing the mean meridional motions in terms of the gradient of a mass streamfunction ψ

$$[\omega] = -\frac{\partial \psi}{\partial y} \quad \text{and} \quad [v] = \frac{\partial \psi}{\partial p}, \qquad (7.5)$$

ensuring that the continuity equation (Eq. (7.4)) is satisfied. A local maximum in ψ corresponds to a clockwise rotation in the latitude–height plane, plotted with the North Pole to the right.

We can achieve some further simplification by making use of the equation of state to transform Eq. (7.2) into an expression for the time rate of change of specific volume α or geopotential thickness $-\partial \Phi / \partial p$. Since all the terms in Eq. (7.1)–(7.4) are zonally averaged, we can drop the brackets notation and write

$$\frac{\partial u}{\partial t} = f\frac{\partial \psi}{\partial p} + \mathcal{G} + \mathcal{F}_x \qquad (7.6)$$

$$\frac{\partial \alpha}{\partial t} = -\sigma\frac{\partial \psi}{\partial y} + \mathcal{H} + \mathcal{Q} \qquad (7.7)$$

and

$$\frac{\partial u}{\partial p} = \frac{1}{f}\frac{\partial \alpha}{\partial y}, \qquad (7.8)$$

where

$$\mathcal{G} = -\frac{\partial[u^*v^*]}{\partial y} = [v^*\zeta^*] \qquad (7.9)$$

$$\mathcal{H} = -\frac{R}{p}\frac{\partial[v^*T^*]}{\partial y} = -\frac{\partial[v^*\alpha^*]}{\partial y} \qquad (7.10)$$

$$\mathcal{Q} = \frac{R}{p}Q, \qquad (7.11)$$

and

$$\sigma = \frac{R}{p}[s] = \frac{R}{p}\left(\frac{\kappa[T]}{p} - \frac{\partial[T]}{\partial p}\right) = \frac{\kappa[\alpha]}{p} - \frac{\partial[\alpha]}{\partial p}. \qquad (7.12)$$

In the above, $\mathcal{G}$ is the zonally averaged source or sink of zonal momentum per unit mass due to meridional eddy transports,[3] $\mathcal{H}$ is the zonally averaged time rate of change of temperature due to meridional eddy heat transports. It is represented by a script $\mathcal{H}$ to distinguish it from its counterpart H defined in Eq. (5.19) because it relates to specific volume or thickness rather than to temperature. $v^*\zeta^*$ in Eq. (7.9) is the poleward transport of relative vorticity (strictly speaking, by the nondivergent component of the wind; see Exercise 3.24). The zonal mean zonal wind u and temperature T are assumed to be in thermal wind balance (Eq. (7.8)). Equations

(7.6)-(7.8) thus constitute a closed set of equations. Given prescribed time varying fields of $\mathcal{F}_x, \mathcal{Q}, \mathcal{G}$, and $\mathcal{H}$ and a suitable bottom boundary condition, they constitute a time-dependent, two-dimensional model of the evolving zonally symmetric flow.

At any instant in time, u is changing in response to $\mathcal{F}_x$ and $\mathcal{G}$, while α is changing in response to $\mathcal{H}$ and $\mathcal{Q}$. Yet despite the lack of any functional relation between $\mathcal{G}, \mathcal{F}_x$, $\mathcal{H}$ and $\mathcal{Q}$, α and u must be changing in a manner consistent with the thermal wind equation (7.8): the meridional temperature gradient cannot change unless the vertical wind shear changes, and vice versa. How does the temperature field know how the wind field is changing, and vice versa? Any change in one field without a compatible change in the other would immediately give rise to a small departure from geostrophic balance (that is to say, the zonal wind would become subgeostrophic or supergeostrophic). The resulting imbalance between the meridional components of the pressure gradient force and the Coriolis force induces an acceleration in the meridional direction and hence gives rise to mean meridional motions, together with the corresponding vertical velocities as required by the continuity equation. Hence, it is the terms involving mean meridional motions (the ψ terms) in Eqs. (7.6) and (7.7) that keep the zonal wind and temperature fields in thermal wind balance.

It is instructive to consider the instantaneous response to "turning on" an eastward eddy-induced body force $\mathcal{G}$ within a localized region of the meridional plane, shown in Fig. 7.2. The zonal flow accelerates in the vicinity of the forcing and thermal wind balance is maintained through the action of a mean meridional circulation. Equatorward flow in the vicinity of the forcing opposes the eastward body force $\mathcal{G}$ by inducing a westward Coriolis force, while the associated vertical motions induce meridional temperature gradients consistent with the induced vertical wind shears above and below the level of the forcing. The circulation cells spread the eastward acceleration vertically beyond the domain of the forcing and they induce weak westward tendencies (not shown) along the flanks of the westerly forcing.

When viewed in the context of a primitive equation model without the assumption of quasi-geostrophy, thermal wind balance between the $[u]$ and $[T]$ fields is being maintained by the gravity and inertio-gravity waves – the hidden messengers in Section 2.9 – which come into play whenever the fields get out of balance.

7.3 A VECTORIAL REPRESENTATION OF THE GOVERNING EQUATIONS[4]

We will now define a set of vectors that represent zonally averaged fields in the meridional plane in which $\mathbf{j}$ is the unit vector in the meridional direction (positive poleward

---

[3] See Exercise 3.24 and Appendix B.

[4] This section may help some visually oriented readers to understand the behavior of the governing Eqs. (7.6), (7.7), and (7.8). Little of what follows in subsequent sections is crucially dependent on it.

Figure 7.2 Instantaneous response to an eastward body force $\mathcal{G} > 0$, localized in the Northern Hemisphere middle latitudes at the jet stream level. Mean meridional motions and the induced temperature and net zonal wind tendencies are indicated.

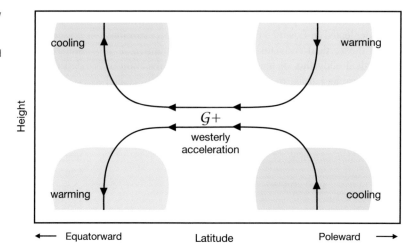

in the Northern Hemisphere) and $\mathbf{k}$ is the unit vector in the vertical (positive downward, toward increasing pressure). The zonally averaged pressure gradient force vector is

$$\Sigma \equiv -\nabla\widehat{\Phi} = -\partial\widehat{\Phi}/\partial y\,\mathbf{j} - \partial\widehat{\Phi}/\partial p\,\mathbf{k},$$

where the carat $^\wedge$ denotes the departure from a global average. Since the temperature field is in hydrostatic balance and the zonal mean zonal wind field is in geostrophic balance, we can write

$$\Sigma \equiv -\nabla\widehat{\Phi} = fu\,\mathbf{j} + \widehat{\alpha}\,\mathbf{k}. \tag{7.13}$$

Hence, in regions of zonally averaged westerlies, the poleward component of Σ is balanced by the Coriolis force; in regions of the meridional plane that are warmer (i.e., exhibit larger thickness) than the global mean temperature at that level, pressure tends to be high above and low below, relative to global means at the same levels, and so Σ exhibits a downward component. Note that the thermal wind equation Eq. (7.8) follows directly from the fact that Σ is the gradient of a potential function:

$$\nabla \times \Sigma = -\nabla \times \nabla\Phi = 0. \tag{7.14}$$

The top panel of Figure 7.3 shows the climatological mean pressure gradient force vector Σ for annual mean conditions superimposed upon the geopotential height field $\widehat{\Phi}$, the departure from the global mean at each level. The vectors diverge out of the highs in the $\widehat{\Phi}$ field and into the lows.

The *state vector*, defined as

$$\mathbf{S} \equiv \frac{u}{f}\,\mathbf{j} + \frac{\widehat{\alpha}}{\sigma}\,\mathbf{k} \tag{7.15}$$

may be interpreted as follows. Starting with a stably stratified atmosphere at rest relative to the rotating Earth, we displace zonally symmetric rings of air meridionally, letting $\partial u/\partial t = fv$ and vertically, letting $\partial\widehat{\alpha}/\partial t = \sigma\omega$ until u and $\widehat{\alpha}$ attain the observed values. It may be helpful to think of these displacements as relating to an imaginary stretched membrane in the meridional plane, which is of uniform thickness when the atmosphere is at rest, but becomes stretched and thinned in some places and compressed and thickened in others as

zonally symmetric rings of air are displaced meridionally and vertically in accordance with the state vector. The distribution of membrane thickness in the meridional plane is thus determined by the u and T fields.

The middle panel of Fig. 7.3 shows the state vector field for the annual mean circulation superimposed on the zonal wind field. The vectors point poleward in the westerlies, downward in latitude belts in which temperatures are higher than the global average and upward in latitude belts in which temperatures are lower than the global average. In order to satisfy the thermal wind equation, the vectors must diverge on the equatorward flanks of westerly jet streams and converge on the poleward flanks. The convergence of the state vector is given by

$$-\nabla \cdot \mathbf{S} = -\frac{\partial}{\partial y}\frac{u}{f} - \frac{\partial}{\partial p}\frac{\widehat{\alpha}}{\sigma} \equiv \frac{\widehat{q}}{f}, \tag{7.16}$$

where $\widehat{q}$ is the *pseudo-potential vorticity*, expressed per unit mass, an approximate form of the potential vorticity PV (Eq. (4.15)), as explained in Exercise 7.2. Convergence of $\mathbf{S}$ is indicative of positive $\widehat{q}$ and a thickening of the membrane relative to its resting state and vice versa. It is indicative of cyclonic relative vorticity and/or enhanced static stability relative to the global mean.[5] The bottom panel of Fig. 7.3 shows the observed distribution of $\nabla \cdot \mathbf{S} = -\widehat{q}/f$ for annual mean conditions. Jets correspond to regions of strong meridional gradients of pseudo-potential vorticity; $\widehat{q}$ is positive (cyclonic) on the poleward side of a westerly jet and negative (anticyclonic) on the equatorward side.

The pressure gradient force vector and the state vector are similar in form, and

$$\Sigma = \left(f^2, \sigma\right)\mathbf{S}. \tag{7.17}$$

The factor (f^2, σ) is like a two-dimensional modulus relating to the "stiffness": that is, the force required to maintain a unit displacement of a ring of air, represented here by an element within the membrane. The meridional stiffness increases

---

[5] Pseudo-potential vorticity is similar to, but not quite the same as, quasi-geostrophic potential vorticity defined in Section 7.5.

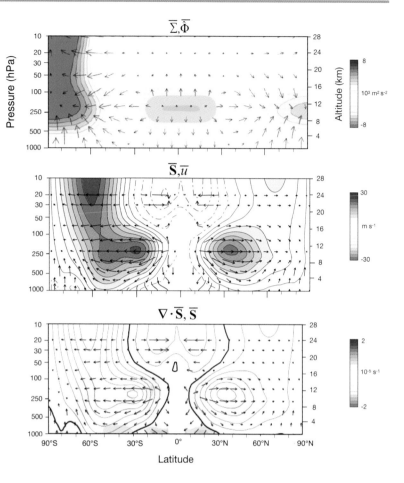

Figure 7.3 Representations of some of the fields defined in this section based on climatological annual mean conditions. (Top) The pressure gradient force vector $\overline{\Sigma}$ defined in Eq. (7.13) superimposed upon the distribution of geopotential, the departure from the global average at each level, $\overline{\widehat{\Phi}}$; (middle) the state vector $\overline{S}$ defined in Eq. (7.15) superimposed upon the distribution of zonally averaged zonal wind $[\overline{u}]$, indicated by the colored shading and contours; and (bottom) the divergence of the state vector (colored shading) in Cartesian coordinates defined in Eq. (7.16), proportional to $-[\widehat{\overline{q}}]/f$, the time-varying part of the zonally averaged pseudo-potential vorticity. The contours in the bottom panel indicate zonally averaged zonal wind $[u]$; the zero contour is bolded. The scaling of the vectors is arbitrary.

with latitude and the vertical stiffness is much larger in the stratosphere than in the troposphere. In height coordinates, the scaling factor can be written as (f^2, N^2), where N is the buoyancy or Brunt Väisälä frequency. An advantage of expressing the static stability parameter in height coordinates is that N tends to be relatively uniform in the extratropical troposphere, with a value on the order of 1.2×10^{-2} s^{-1} (equivalent to a uniform lapse rate $\Gamma = 6.5°$C km^{-1}). The relative stiffness of the membrane in the meridional and vertical directions determines the aspect ratio of the mean meridional cells; that is, the ratio of the meridional scale L to the depth scale D is equal to N/f. It can be shown that the same considerations apply to the aspect ratio of the flow in the eddies.[6]

The energy required to displace a zonal ring of air to its "observed position" in the distorted membrane is equal to $\int$ Force $\cdot\ d$(displacement) $= \int \Sigma \cdot d\mathbf{S}$, which for small displacements is equal to

$$\int \Sigma \cdot d\mathbf{S} = \frac{u^2}{2} + \frac{\widehat{\alpha}^2}{2\sigma}. \tag{7.18}$$

It is readily verified that when this expression is integrated over the mass of the atmosphere, the result is the mechanical energy

[6] Hoskins and James (2014).

$$\int_0^{p_o} \langle \Sigma \cdot d\mathbf{S} \rangle \, dp = K_Z + A_Z, \tag{7.19}$$

where the angle brackets in the integral represents the global average. In terms of the membrane analogy, K_Z is the energy associated with the meridional stretching and A_Z is the energy associated with the vertical stretching.

Let us define the zonal mean motion vector

$$\Psi \equiv \mathbf{i} \times \nabla\psi = v\mathbf{j} + \omega\mathbf{k}, \tag{7.20}$$

where $\mathbf{i}$ is the unit vector in the x direction, and the forcing vector

$$\Gamma \equiv \frac{(\mathcal{G} + \mathcal{F}_x)}{f}\mathbf{j} + \frac{(\mathcal{H} + \mathcal{Q})}{\sigma}\mathbf{k}, \tag{7.21}$$

which incorporates the eddy forcing together with diabatic heating and friction. Using the definitions in Eqs. (7.20) and (7.21), Eqs. (7.6) and (7.7) can be written in condensed vectorial form

$$\frac{\partial \mathbf{S}}{\partial t} = \Gamma + \Psi, \tag{7.22}$$

because the mean meridional motions are nondivergent,

$$\frac{\partial \widehat{q}}{\partial t} = -f\frac{\partial}{\partial t}\nabla \cdot \mathbf{S} = -f\nabla \cdot \Gamma. \tag{7.23}$$

The convergence (in the meridional plane) of the forcing vector Γ controls the thickening or thinning of the imaginary

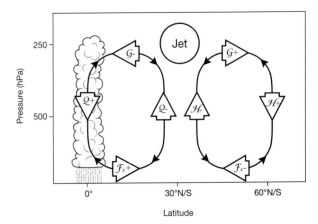

Figure 7.4 A vectorial representation of the climatological mean forcing vector Γ in the meridional plane.

membrane (i.e., the pseudo-potential vorticity tendency). The mean meridional circulations do not affect it.

For climatological mean (steady state) conditions, Eq. (7.22) reduces to

$$\Psi = -\Gamma. \tag{7.24}$$

Figure 7.4 illustrates the application of this identity to the observed climatological mean tropospheric flow. It is a vectorial representation of the information in Figs. 3.15 and 5.19. It follows from Eqs. (7.20) and (7.24) that for the climatological mean flow, the forcing vector Γ must be non-divergent in the meridional plane and it is subject to the same (oppositely signed) bottom and lateral boundary conditions as the mean meridional circulations. If, at some point in time, the forcing vector does not satisfy the boundary conditions, displacements (i.e., changes in zonal wind and temperature) must occur, and Γ must evolve in response to those displacements until the boundary conditions are satisfied.

The treatment of the governing equations in this section illustrates how the hydrostatically and geostrophically balanced $[u]$ and $[T]$ fields can be viewed as components of a single two-dimensional state vector in the meridional plane. We are accustomed to thinking about the mean meridional motions and the continuity of mass in this manner, but here the two-dimensional treatment is extended to the pressure field, and $[u]$ and $[T]$ are represented as meridional and vertical components of a two-dimensional state vector. The hydrostatic and geostrophic equations follow from the constraint that the pressure field be irrotational, and the conservation equations for zonal momentum and energy appear as components of a single vectorial equation involving eddy forcing, diabatic heating, and friction. If the system is in steady state, the net forcing in the meridional plane must be nondivergent and balanced by the mean meridional circulation vector. Pseudo-potential vorticity is of the convergence of the state vector, and the (global) zonal mechanical energy $A_Z + K_Z$ is related to the work done in deforming the state vector. It remains to be shown what happens when the zonally averaged flow is not in a steady state.

7.4 SOLUTION OF THE GOVERNING EQUATIONS

When the zonally averaged flow is not in steady state, the field of net forcing Γ exhibits an irrotational component that is not balanced by the mean meridional circulations, and the state vector $\mathbf{S}$ evolves in response to the imbalance. Under these conditions, the governing equations (7.6)–(7.8) involve both a diagnostic equation for the mean meridional circulation and a prognostic equation relating to the evolution of the distribution of the zonally averaged pseudo-potential vorticity field in the meridional plane in response to forcings. The eddy forcings in both equations involve gradients of $\mathcal{G}$ and $\mathcal{H}$ within the domain, as discussed in the first two subsections below. The prognostic equation also involves the nonzero heat transport at the Earth's surface, which induces a response in the interior, as described in the third subsection below. We will show in Section 7.5 that the governing equations presented in this section are analogous in many respects to the quasi-geostrophic "omega" and geopotential tendency equations and can be expressed in a similar form.

7.4.1 The Mean Meridional Circulation

To obtain a diagnostic equation for the mean meridional circulations, we must eliminate the time-dependent terms in Eqs. (7.6) and (7.7) and solve directly for ψ in terms of $\mathcal{G}$, $\mathcal{F}_x$, $\mathcal{H}$, and $\mathcal{Q}$. To this end we differentiate Eq. (7.6) with respect to pressure, and Eq. (7.7) with respect to latitude and substitute the right-hand sides of both equations into Eq. (7.8) to obtain, after some minor rearranging,

$$A(\psi) \equiv \sigma\frac{\partial^2\psi}{\partial y^2} + f^2\frac{\partial^2\psi}{\partial p^2} = -f\frac{\partial}{\partial p}(\mathcal{G}+\mathcal{F}_x) + \frac{\partial}{\partial y}(\mathcal{H}+\mathcal{Q}).$$

$$\tag{7.25}$$

Since $A(\psi)$ is defined as the (weighted) Laplacian of ψ, it follows that the distributions of $A(\psi)$ and ψ in the meridional plane should be qualitatively similar but of opposing sign, with the ψ distribution being smoother. Hence, distinct maxima and minima in ψ should correspond roughly to centers of circulation cells in the meridional plane. Based on this definition, the Northern Hemisphere Hadley cell should circulate around a maximum in the ψ field and the Ferrel cell around a minimum.

The forcing term on the right-hand side involves the curl of the forcing in the meridional plane (stretched by the modulus $-(f^2, \sigma)$). Distributions of $\mathcal{Q}$, $\mathcal{G}$, and $\mathcal{H}$ are shown in Figs. 1.27, 3.14, and 5.17, respectively, and the distribution of $\mathcal{F}_x$ can be inferred from the meridional profile of zonal wind at the Earth's surface. The curl of the forcing field, counterclockwise in the Northern Hemisphere tropics and clockwise in middle latitudes, is clearly evident in the distribution of all four forcings, as depicted schematically in Fig. 7.4. In accordance with Eq. (7.24), in the long term mean the combined forcing $\mathbf{\Gamma}$ must be cancelled by the accelerations induced by the mean meridional circulations.

Using the mean meridional circulations derived from Eq. (7.25), with the boundary condition that the vertical velocity vanish at the Earth's surface, the prognostic equations (7.6) and (7.7) can be solved to infer how the zonal wind and temperature fields are evolving without any further manipulation of the governing equations. However, the interpretation of the evolution of the zonally symmetric flow in terms of pseudo-potential vorticity in the next subsection is helpful for understanding how the system responds to the divergent component of the forcing and how the poleward heat transport at the bottom boundary influences the steady state wind and temperature fields in the interior.

A more exact equation analogous to Eq. (7.25) can be derived from the complete zonally averaged versions of the zonal momentum equation and the First Law of Thermodynamics in spherical coordinates. The resulting equation contains a $\partial^2 \psi / \partial y \partial p$ term and first order derivatives, but it is still elliptic, provided that angular momentum per unit mass m decreases with latitude on isentropes, which corresponds to the criterion for stability with respect to zonally symmetric disturbances.

7.4.2 Evolution of Pseudo-Potential Vorticity

To diagnose the evolution of the zonally symmetric flow, it is instructive to consider how the potential vorticity is changing in response to the divergence of the forcing vector $\mathbf{\Gamma}$. To this end we substitute Eq. (7.21) into Eq. (7.23) and obtain

$$\frac{1}{f}\frac{\partial \widehat{q}}{\partial t} = \frac{\partial}{\partial t}\left(-\frac{\partial}{\partial y}\frac{u}{f} - \frac{\partial}{\partial p}\frac{\widehat{\alpha}}{\sigma}\right)$$

$$= -\frac{\partial}{\partial y}\left(\frac{\mathcal{G} + \mathcal{F}_x}{f}\right) - \frac{\partial}{\partial p}\left(\frac{\mathcal{H} + \mathcal{Q}}{\sigma}\right), \qquad (7.26)$$

where $\widehat{q}$ is the pseudo-potential vorticity. The right-hand side of Eq. (7.26) may be recognized as the convergence (in the meridional plane) of the forcing vector $\mathbf{\Gamma}$, which controls the

thickening or thinning of the imaginary membrane (see Eqs. (7.16) and (7.23)). To get a sense of how the system responds to a forcing, consider the case in which diabatic heating is the only forcing. The nondivergent component is $\sigma^{-1}\partial \mathcal{Q}/\partial y$ and the irrotational component is $\sigma^{-1}\partial \mathcal{Q}/\partial p$. The former drives a compensating mean meridional circulation, while the latter changes the distribution of $\widehat{q}$, which involves changes of the distribution of u and $\widehat{\alpha}$. The forcing is compensated by the effects of the mean meridional circulations, but not completely. The zonal wind and temperature distribution change in a manner qualitatively consistent with the forcing, but the change is not as large as it would have been if the forcing had not been opposed by the mean meridional circulations.

7.4.3 Response to the Boundary Forcing

To complete the analysis of how the zonally symmetric circulation responds to the observed forcing, we need to take into account the fact that the poleward heat transport across the storm track shown in Fig. 5.14 extends all the way down to the Earth's surface. The low-level response to the heating and cooling by the eddies needs to be treated separately because close to the Earth's surface, the mean meridional circulation is ineffective in enforcing thermal wind balance. In this subsection, we consider this boundary-forced temperature tendency, which decays exponentially with height in the first few kilometers above the Earth's surface.

Poleward of the storm track, the convergence of the eddy heat flux induces positive temperature (and specific volume) tendencies, strongest at the Earth's surface and decaying with height. The isentropes, which slope upward toward the pole tend to be flattened. The expansion of the air poleward of the storm track and contraction of the air equatorward of it implies an equatorward vertically integrated mass flux, which serves to strengthen the SLP gradient across the storm tracks. The induced SLP tendencies are consistent with a strengthening of the surface westerlies, and thus a weakening of the (westerly) vertical wind shear, consistent with the weakening of the meridional temperature gradient across the storm track.

To understand how the potential vorticity is affected by the poleward heat transport at the bottom boundary, it is instructive to refer to the cross section of PV, as represented by a grid defined by intersecting potential temperature θ and angular momentum m contours shown in Fig. 4.13. The eddy-induced cooling equatorward of the storm track pulls θ surfaces away from the bottom boundary, pulling new (m, θ) grid cells into the domain while squeezing the existing squares above the boundary (i.e., increasing the static stability) and forcing them to spread poleward, across the storm track, consistent with the eddy-induced strengthening of the westerlies at and just above the surface. Poleward of the storm track, where the eddy heat transports are converging, the θ surfaces are pushed downward into the boundary. The grid cells just above the Earth's surface are vertically stretched and meridionally squeezed. In this

Maintaining thermal wind balance   Determining the tendency

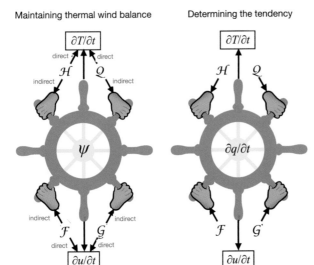

Figure 7.5 Schematic summarizing the processes that determine how the zonally averaged zonal wind u and temperature T fields evolve in response to prescribed time-varying forcings $\mathcal{Q}$, $\mathcal{H}$, $\mathcal{F}$, and $\mathcal{G}$. See text for explanation.

manner, the poleward eddy heat transport at the bottom boundary reshapes the distribution of m and θ within a layer of finite thickness immediately above the Earth's surface, irrespective of the PV changes that might be occurring in the interior. In effect, PV is being transported equatorward within a "reservoir" within the boundary, as θ contours emerge from it on the equatorward side of the storm track and are pushed down into it on the poleward side. Meanwhile, m contours are pulled poleward across the storm track, resulting in an eastward acceleration at and just above the Earth's surface.[7]

As will be explained in Section 11.4, the eddy-induced eastward acceleration of the surface winds is concentrated in the zonally varying storm tracks over the oceanic sectors. Most of the frictional drag in the westerly wind belts is likewise concentrated in these same regions of strong poleward heat transports at the bottom boundary. Were it not for this strong angular momentum sink, the observed equilibrium equator-to-pole mean meridional circulation in isentropic coordinates shown in Fig. 5.12 could not be maintained.[8]

7.4.4 Summary

The evolution of the geostrophically balanced $[u]$ and $[T]$ fields in response to the prescribed forcings $\mathcal{Q}$, $\mathcal{H}$, $\mathcal{F}$, and $\mathcal{G}$, as determined by the governing equations, is summarized in Fig. 7.5, a seafaring analog in which the forcings are represented by the hands on the helm.

In the left panel the emphasis is on keeping the $[u]$ and $[T]$ fields in thermal wind balance as the circulation evolves,

as described in Section 7.4.1. The forcings are depicted as exerting a direct influence on the evolution of the fields as well as an indirect influence by way of the mean meridional circulations, represented in the schematic as applying a torque on the wheel. Just as the wheel turns in accordance with the combined torque on it, the zonal wind tendency (and the temperature tendency) are influenced by both the mechanical forcings $\mathcal{F}$ and $\mathcal{G}$ and the thermal forcings $\mathcal{Q}$ and $\mathcal{H}$.

In the right panel of Fig. 7.5, the emphasis is on the evolution of the geostrophically balanced $[u]$ and $[T]$ fields, as described in Section 7.4.2. Here the torques exerted by the hands on the helm are represented as determining the pseudo-potential vorticity tendency, which implicitly determines the $[u]$ and $[T]$ tendencies, as indicated by the arrows. If the system is in steady state, the net torque must be zero and the direct and indirect tendencies in the left panel must sum to zero as well. It was shown in Section 7.4.3 that the poleward eddy heat transport at the Earth's surface makes an important contribution to the pseudo-potential vorticity tendency.

In the next section, we show that the conceptual way of thinking about the zonally symmetric circulation embodied in this schematic is equally applicable to the governing equations based on quasi-geostrophic scaling.

7.5 THE MID-LATITUDE QUASI-GEOSTROPHIC SYSTEM

An important advance in dynamical meteorology, concurrent with the discovery of baroclinic instability by Charney and Eady in the late 1940s, was the scale analysis for extratropical synoptic and planetary-scale motion systems. In framing this analysis, the horizontal flow was partitioned into a geostrophic wind component $\mathbf{V}_g$ that is in thermal wind balance with the temperature field, and a much smaller, residual, ageostrophic wind $\mathbf{V}_a$.[9] Owing to the variation of the Coriolis parameter f with latitude, the geostrophic wind field is not quite nondivergent. To circumvent this complication, f in the scaled equations has to be expanded into a constant reference value f_o plus a first order, latitudinally varying component. Subject to this scaling, the system of governing equations is quite analogous to the one considered in the previous section, with f replaced by f_o. Zonally averaged *quasi-geostrophic vorticity* is the Laplacian of zonally averaged geopotential height, from which it follows that the prognostic equation, elliptic in form like Eq. (7.26), can be inverted by solving Poisson's equation to predict how the zonally averaged geopotential height and geostrophic wind fields will change in response to a prescribed forcing. This so-called *invertibility principle* for potential vorticity is one of the cornerstones of atmospheric dynamics.

---

[7] Bretherton (1966); Vallis (2017).
[8] For further discussion of the response of the boundary forcing, see Section 8.2 and Schneider et al. (2003) and Pfeffer (1987).

[9] See James (1995) for a detailed description of the development of the Eady model of baroclinic instability.

For mid-latitude quasi-geostrophic scaling,[10] $f = f_o + \beta y$ and Eq. (7.23) reduces to

$$\frac{\partial \widehat{q}}{\partial t} \to \frac{\partial [q_g]}{\partial t} = -f_o \nabla \cdot \boldsymbol{\Gamma} = -\frac{\partial}{\partial y}(\mathcal{G} + \mathcal{F}_x) - f_o \frac{\partial}{\partial p}\left(\frac{\mathcal{H} + \mathcal{Q}}{\sigma}\right) \quad (7.27)$$

where $[q_g]$ is the zonally averaged quasi-geostrophic potential vorticity for the zonally averaged zonal flow:

$$[q_g] = f - \frac{\partial [u]}{\partial y} - f_o \frac{\partial}{\partial p}\left(\frac{[\widehat{\alpha}]}{\sigma}\right). \quad (7.28)$$

The zonally averaged transport of quasi-geostrophic potential vorticity by the geostrophic eddies is

$$[v^* q_g^*] = \left[v^*\left(\zeta^* - f_o \frac{\partial}{\partial p}\left(\frac{\alpha^*}{\sigma}\right)\right)\right] = \mathcal{G} - f_o \frac{\partial}{\partial p}\frac{[v^* \alpha^*]}{\sigma}. \quad (7.29)$$

$\mathcal{G}$ may be recognized as the meridional component of the forcing vector $\boldsymbol{\Gamma}$, which is related to the acceleration of the zonal wind component. Substituting Eq. (7.29) into Eq. (7.27) yields

$$\frac{\partial [q_g]}{\partial t} = -\frac{\partial}{\partial y}[v^* q_g^*] - \frac{\partial \mathcal{F}_x}{\partial y} - f_o \frac{\partial}{\partial p}\left(\frac{\mathcal{Q}}{\sigma}\right). \quad (7.30)$$

As explained more fully in Section 8.2, the meridional eddy transports of both heat and momentum force accelerations in $[u]$.

7.5.1 The Diagnostic Equation for Vertical Velocity

For mid-latitude, quasi-geostrophic scaling, the equation for the zonally averaged ageostrophic meridional circulation Eq. (7.25) reduces to

$$A(\psi) \equiv \sigma \frac{\partial^2 \psi}{\partial y^2} + f_o^2 \frac{\partial^2 \psi}{\partial p^2} = -f_o \frac{\partial}{\partial p}(\mathcal{G} + \mathcal{F}_x) + \frac{\partial}{\partial y}(\mathcal{H} + \mathcal{Q}). \quad (7.31)$$

The analogous equation for the vertical (ageostrophic) motion in the three-dimensional midlatitude quasi-geostrophic system, referred to as the "omega equation" is

$$A(\omega) = +f_o \frac{\partial}{\partial p}(\mathbf{V}_g \cdot \nabla \zeta_g) + \nabla^2(\mathbf{V}_g \cdot \nabla \alpha), \quad (7.32)$$

where $\mathbf{V}_g$ is the zonally varying geostrophic wind. Hence, the forcing term in Eq. (7.31) that involves the vertical gradient of the zonally averaged vorticity transport $\mathcal{G}$ by the eddies is quite analogous to the term in the conventional omega equation used by synopticians that involves the vertical derivative of the vorticity advection. Similarly, the meridional gradient of $\mathcal{H}$ may be recognized as being equivalent to the Laplacian of the zonally averaged heat transport by the eddies, which is analogous to the Laplacian of horizontal temperature advection in the omega equation. Hence, the role of the ageostrophic mean meridional motion

driven by the zonally averaged eddy transports of heat and vorticity is analogous to the role of the zonally varying ageostrophic flow in the three-dimensional midlatitude quasi-geostrophic system (e.g., in baroclinic waves). In both cases, the ageostrophic motions ensure that, to first order, the wind and temperature tendencies generated by the eddy transports are in thermal wind balance.[11]

7.5.2 The Geopotential Tendency Equation

The geopotential tendency equation is obtained by approximating f by f_o, multiplying Eq. (7.26) by f_o^2 and writing $[u]$ and $[\widehat{\alpha}]$ in terms of gradients of $[\Phi]$:

$$\frac{\partial}{\partial t}\left(\frac{\partial^2 \Phi}{\partial y^2} + \frac{\partial}{\partial p}\frac{f_o^2}{\sigma}\frac{\partial \Phi}{\partial p}\right)$$
$$= -f_o \frac{\partial}{\partial y}(\mathcal{G} + \mathcal{F}_x) - f_o^2 \frac{\partial}{\partial p}\left(\frac{\mathcal{H} + \mathcal{Q}}{\sigma}\right). \quad (7.33)$$

The elliptical operator on the left-hand side of Eq. (7.33) is helpful in understanding why membrane material is attracted toward the boundary poleward of the storm track, where potential vorticity storage is changing on the boundary due to the convergence of the poleward heat transport.

The analogous equation for the vertical (ageostrophic) motion in the zonally varying three-dimensional midlatitude quasi-geostrophic system of equations is the geopotential tendency equation

$$\frac{\partial}{\partial t}\left(\nabla^2 \Phi + \frac{\partial}{\partial p}\frac{f_o^2}{\sigma}\frac{\partial \Phi}{\partial p}\right)$$
$$= -f_o \mathbf{V}_g \cdot \nabla(\zeta_g + f) - f_o^2 \frac{\partial}{\partial p}\left(\frac{-\mathbf{V}_g \cdot \nabla \alpha}{\sigma}\right). \quad (7.34)$$

Hence, the zonally averaged vorticity and heat transports shape the evolution of the pseudo-potential vorticity of the zonally averaged flow in much the same way that the horizontal advection of vorticity and heat by the geostrophic wind shape the quasi-geostrophic potential vorticity in the zonally varying quasi-geostrophic system. We will have occasion to make use of this equation in diagnosing the forcing of the climatological mean stationary waves by the transients in Section 11.4.

Exercises

7.4 Relate the zonally averaged geopotential height field to the distribution of quasi-geostrophic potential vorticity $[q_g]$.

7.5 Explain and interpret the role of the bottom boundary condition in the invertibility principle.

7.6 Describe how the poleward eddy heat transport across the storm tracks at the Earth's surface affects the (a) heat, (b) mass, and (c) zonal momentum

---

[10] See Section 5.4 of Vallis (2017) for details.

[11] Before the advent of reanalyzed fields with reliable vertical velocities, the so-called "omega equation" Eq. (7.32) was used as a means of estimating the vertical velocity field in the extratropics.

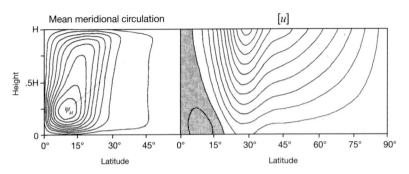

Figure 7.6 Streamfunction for the mean meridional circulations (left) and the zonal wind field (right) in a numerical simulation of the response to an idealized equator-to-pole heating gradient. The only forces in the zonal momentum equation are boundary layer friction and a weak viscosity ($\nu = 10$ m^2 s^{-1}) in the interior. The cell is restricted to the tropics, where the distribution of angular momentum is almost spatially uniform by virtue of the strong anticyclonic meridional shear of the zonal wind. The maximum in the streamfunction field ψ_M is 2680 m^2 s^{-1} and the contour interval is 0.1 ψ_M. The contour interval for $[u]$ is 5 m s^{-1} and easterlies are shaded. From Held and Hou (1980). © American Meteorological Society. Used with permission.

balances of the layer of air immediately above the surface.

7.7 (a) Describe the SLP, near surface wind, and temperature fields in a growing synoptic scale wave. (b) Sketch the corresponding wave in the PV field. (c) Under what conditions will the wave amplify?

7.8 Consider an idealized zonally symmetric circulation in the vicinity of a storm track in which the poleward heat transport increases from zero at the surface to a maximum at the top of the boundary layer and matches the observed profile above that level. The flow is in steady state, with the eddy-induced Stokes drift cancelling the Eulerian mean meridional circulation. Now suppose that the poleward eddy heat transport at the top of the boundary layer is instantaneously extended all the way down to the Earth's surface so that it more closely reflects the observed distribution of eddy heat transports. (a) Describe how the wind and temperature fields would adjust so as to eventually produce the steady state response described in Section 7.4.3. (b) Describe the field of specific volume $\hat{\alpha}$. (c) Describe how extending the heat transport all the way down to the Earth's surface affects the forcing vector Γ.

7.9 Under what conditions can a forcing ($\mathcal{Q}, \mathcal{H}, \mathcal{F}$, and $\mathcal{G}$) induce (a) changes in the zonal flow in the absence of mean meridional circulations and (b) mean meridional circulations in the absence of tendencies in zonal wind or temperature?

7.6 THE ZONALLY SYMMETRIC FLOW IN THE ABSENCE OF EDDY FORCING

To place wave–mean flow interaction in context and illustrate its importance in accounting for the observed features of the general circulation, let us consider what the general circulation would look like in the absence of eddy transports,

drawing upon results of numerical experiments in which the only forcings are the imposed equator-to-pole heating gradient and the frictional drag in the boundary layer. Under these circumstances, meridional motions cannot exist in the absence of momentum sources and sinks except in regions of very weak inertial stability, where zonally symmetric rings of air may circulate freely in the meridional plane while conserving angular momentum. Figure 7.6 shows that the mean meridional circulation in this eddy-less atmosphere is confined to the tropics and the tropospheric jet stream is unrealistically strong and displaced equatorward relative to the observed. Equatorward of the jet maximum, the anticyclonic zonal wind shear $\partial[u]/\partial y$ is so strong that the distribution of angular momentum is almost spatially uniform (in contrast to the observed distribution shown in Fig. 4.13). The poleward limit of the meridional circulation corresponds to the latitude at which the vertical shear of $[u]$ becomes so strong that it cannot be balanced by the meridional temperature gradient. For heating that is symmetric about the equator, the poleward limit of the cell is about 25°N/S. The mean meridional circulation in the angular momentum conserving regime consists of a single thermally direct cell in each hemisphere analogous to the annual mean Hadley cell (albeit much weaker than observed) and there exist subtropical jets with peak intensity along its outer edge. Poleward of the angular momentum conserving regime the atmosphere is in a thermal equilibrium regime in which there are no mean meridional motions and the $[u]$ field is in thermal wind balance with the prescribed radiative equilibrium temperature field. The coexistence of the angular momentum conserving and thermal equilibrium regimes is predicted by theory and has been verified in experiments with zonally symmetric general circulation models.[12]

In the absence of eddies, the general circulation would be largely restricted to the tropics. There would be no eddy-driven extratropical circulations like the ones in the numerical simulations described in Section 2.6.2. The tropospheric

---

[12] For further specifics, see Schneider and Lindzen (1977), and Schneider (1977).

jet streams would be about twice as strong as the observed, while the thermally driven Hadley cells would be only about 1/5 as strong. The trade winds and near-surface midlatitude westerlies would be very weak. The idealized modeling results presented in this section thus serve to illustrate how the response to zonally symmetric thermal forcing is mediated by the eddy transports.

7.6.1 The Solstitial Circulation

An angular momentum conserving regime also exists when the heating is asymmetric about the equator.[13] Moving the heating centroid from the equator into the summer hemisphere increases the strength of the winter (cross-equatorial) Hadley cell and the attendant jet at its poleward edge, and it decreases the strength of the summer Hadley cell. Displacing the heating from the equator also causes the total width of the flanking Hadley cells that define the width of the angular momentum conserving regime to increase slightly. These findings imply that the annual mean Northern and Southern Hemisphere Hadley cells are wider and stronger than they would be if the Sun were overhead on the equator year round. They also have implications for the stratospheric circulation, as discussed in Section 9.3.3.

Exercises

7.10 The vertical cross section of annual mean r.m.s. amplitude of the meridional wind component $[v]'$ exhibits a strong equatorial maximum in the upper troposphere (Fig. 1.31). Based on the meridional distribution of inertial stability, can you offer an explanation for the existence of this feature?

7.11 Revisit Exercise 3.28 in light of the theoretical concepts discussed in this section.

7.7 CONCLUDING REMARKS

In this chapter, we have considered the governing equations for the evolution of the zonally symmetric circulation; that is, the conservation equations for angular momentum and energy, subject to the constraints that the zonally averaged zonal wind $[u]$ and temperature $[T]$ fields remain in thermal wind balance and the mean meridional motions satisfy a two-dimensional continuity equation in the latitude–height plane. The equations can be solved to deduce the evolution of the balanced flow in response to specified, time dependent fields of diabatic heating, frictional drag, and eddy forcing, as summarized schematically in Fig. 7.5. If all the terms are included, this formalism is applicable to within a degree of the equator, but here the equations are simplified through the use of Cartesian geometry and the neglect of second order

terms. The solution can be expressed as a pair of prognostic equations for the state variables $[u]$ and $[T]$, and also as an equation for the evolution of the scalar *pseudo-potential vorticity* field, whose distribution determines the distributions of $[u]$ and $[T]$. A by-product of the calculation is the field of

Eulerian

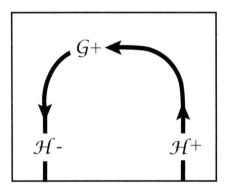

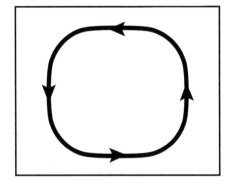

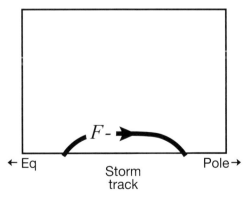

Figure 7.7 Idealized steady state eddy-induced mean meridional circulations in the vicinity of the storm track in the midlatitude troposphere (i.e., the Ferrel cells), ignoring the diabatic heating and the small increase in the poleward eddy heat transports with height in the boundary layer. (Top) Response to the eddy forcing terms $\mathcal{G}$ and $\mathcal{H}$; (bottom) response to frictional drag that balances the eddy-induced westerlies at the Earth's surface, and (middle) total eddy-induced Eulerian mean meridional circulation, the sum of circulations shown in the top and bottom panels. The small contribution due to the diabatic heating gradient across the storm track is not shown.

mean meridional circulations, which is determined through the use of a diagnostic equation.

The governing equations are expressed in a more compact vectorial form in Section 7.3, which provides additional insight into the way the system adapts to the various forcings imposed on it and it reveals the existence of the constraints on the forcing fields under steady state conditions. The various vectorial fields that define the system can be visualized in relation to a stretched membrane, whose thickness is proportional to the pseudo-potential vorticity.

An overview of the eddy-driven climatological mean meridional circulation in the vicinity of the storm track is presented in Fig. 7.7. In the interests of simplicity, the diabatic heating and the small increase of poleward eddy heat transport with height in the boundary layer are neglected: for justification, see Figs. 5.17 and 5.18. The middle panel shows the total mean meridional circulation field as inferred from Eqs. (7.25) or (7.31). It resembles the Ferrel cell in many respects. It is the sum of the eddy-forced component shown in the top panel and the frictionally forced component in the bottom panel. To satisfy the bottom boundary condition while ensuring that the total vectorial forcing field is nondivergent in the meridional plane, the low-level zonal winds must adjust so that at each latitude, the frictionally induced poleward mass flux in the bottom panel balances the equatorward mass flux induced by the vertically integrated forcing $\mathcal{G}$ in the top panel. In this sense, $\mathcal{F}_x$ should be viewed as an integral part of the eddy forcing.

Whenever the forcings drive the $[u]$ and $[T]$ fields out of thermal wind balance, the mean meridional circulations intervene to restore it. The $[u]$ and $[T]$ tendencies that they produce are generally in opposition to the forcing, but not strong enough to completely cancel it unless the vectorial forcing field is nondivergent in the meridional plane. In this sense, the role of the mean meridional circulation is analogous to the role of the ageostrophic motions in the three-dimensional midlatitude quasi-geostrophic system.

To achieve closure in the governing equations it is necessary to be able to parameterize the eddy forcing terms $\mathcal{G}$ and $\mathcal{H}$ as a function of the state variables $[u]$ and $[T]$, and this requires an understanding of the two-way interactions between the eddies and the mean flow, the subject of the next chapter.

8 Wave–Mean Flow Interaction

Wave–mean flow interaction has played a central role in studies of the general circulation, dating back to the foundational works of Rossby, Starr, and collaborators. In the early studies the waves were usually referred to as "eddies" (as in "turbulent eddies") without regard for the specific kind of instability or forcing mechanism that gave rise to them. Starr was particularly intrigued with the countergradient transports of angular momentum equatorward of the tropospheric jet stream.[1]

With the development of numerical models it became possible to diagnose the interactions between the zonally symmetric zonal mean flow and specific kinds of waves, taking into account the nonlinear processes that come into play when waves developing out of an infinitesimal initial perturbation reach finite amplitude and begin to feed back upon the zonal mean flow through the transports of heat and momentum. Notable examples are the relationship between the zonal mean flow and the life cycles of baroclinic waves,[2] discussed later in this chapter, stratospheric sudden warmings, and the QBO, which will be discussed in Part IV. Hence, rather than considering just the response of the zonally symmetric flow to observed or idealized configurations of eddy forcing, as we have done in previous chapters, we will show results of numerical simulations in which the time-varying eddy transports are simulated on the basis of the governing equations.

The first section introduces the topic of wave–mean flow interaction. It describes the interaction mechanisms, viewed in the context of the mechanical energy cycle partitioned into zonal mean and wave components. It demonstrates its relevance to an understanding of the general circulation, using baroclinic wave life cycles as an exemplar.

The second section recasts the zonal momentum and heat balances in terms of the transformed Eulerian mean (TEM) formulation, in which the eddy forcing of the zonally symmetric flow is expressed in terms of the divergence of the *Eliassen–Palm* (EP) flux, a vector field whose meridional component is the poleward momentum transport and whose vertical component is the poleward heat transport scaled by the static stability. The TEM formulation includes the Stokes drift – the rectified component of the helical trajectories of air parcels passing through the waves – and is thus similar to

the mean meridional circulation in isentropic coordinates.[3] This same EP flux vector traces the flux of *wave activity* in Rossby waves and gravity waves as they disperse in the meridional plane from the regions in which they are generated to the regions in which they are dissipated. Included in this section are subsections reinterpreting the baroclinic wave life cycle simulations and some of the climatological mean statistics in light of the TEM diagnostics.

The third section provides an integrated treatment of the balance requirements for heat and angular momentum, previously covered in Chapters 5 and 3, respectively, as well as for pseudo-potential vorticity. With the benefit of the TEM formulation it is possible to provide a more satisfying answer to the question of what maintains the tropical trade winds and the extratropical westerlies in the presence of frictional drag. The nonzero poleward heat transport at the bottom boundary is seen to play an important role in the pseudo-potential vorticity budget.

In the last two sections of this chapter, it is shown how interactions between baroclinic waves and the zonally symmetric flow give rise to transient fluctuations in hemispheric-scale structures in the troposphere with a timescale on the order of a few weeks: the *barotropic* and *baroclinic annular modes*. Examples of wave–mean flow interaction in the stratosphere are found in Chapter 10.

8.1 INTRODUCTION TO WAVE–MEAN FLOW INTERACTION

This section provides an introduction to the topic of wave–mean flow interaction. The first subsection describes the methodology for decomposing the mechanical energy cycle into zonally symmetric and eddy (or wave) components, with emphasis upon the dynamical interpretation of the barotropic and baroclinic conversion terms that involve exchanges of energy between the waves and the zonally symmetric flow. The second subsection describes the nonlinear life cycles of baroclinic waves, with emphasis on the two-way interactions between the waves and the mean flow. The final, very short subsection presents some background information on wave–mean flow interaction in the winter stratosphere, as viewed through the lens of the cycling of mechanical energy.

---

[1] Starr (1968).
[2] Simmons and Hoskins (1978).

[3] Andrews et al. (1987).

8.1.1 The Partitioned Mechanical Energy Cycle

Section 6.7 described how the mechanical cycle can be partitioned in various ways, including zonal mean versus eddy (or wave) categories. Figure 8.1 shows the partitioning in greater detail. To simplify the mathematical expressions, Cartesian geometry is used throughout this subsection. To explain how the various terms come about, we begin by deriving an expression for the time rate of change of zonal kinetic energy K_Z. This is done by multiplying Eq. (3.19) by $[u]$ and integrating it over the mass of the atmosphere to obtain

$$\underbrace{\frac{d}{dt}\frac{1}{2g}\int_0^{p_s}\langle [u]^2\rangle dp}_{\frac{\partial}{\partial t}K_Z} = \underbrace{\frac{1}{g}\int_0^{p_s}[\eta]\langle [u][v]\rangle dp}_{C_Z} + \underbrace{\frac{1}{g}\int_0^{p_s}\langle [u]\mathcal{G}\rangle\, dp}_{-C_K}$$

$$+ \underbrace{\frac{1}{g}\int_0^{p_s}\langle [u][\mathcal{F}_x]\rangle\, dp}_{-D_Z}, \qquad (8.1)$$

where $\mathcal{G}$ is the zonally averaged convergence of zonal momentum by the eddies (Eq. (7.9)), and $\langle\ \rangle$ denotes a global integral (i.e., the pole-to-pole integral) on a pressure surface.[4] Approximating $[\eta]$ by f, the C_Z term on the right-hand side may be rewritten as $-\frac{1}{g}\int_0^{p_s}\langle [v]\frac{\partial[\Phi]}{\partial y}\rangle dp$, in which form it may be recognized as representing the down-gradient cross-isobar flow. This term is related to the release of available potential energy in the zonal average overturning circulation $-[\omega][\alpha]$ in Eqs. (6.15) and (6.36). The sign convention in Eq. (8.1) is such that if C_Z is positive, the circulation is converting zonal available potential energy to zonal kinetic energy; that is, a positive sign denotes a thermally direct mean meridional circulation like the Hadley cell. The D_Z term represents frictional dissipation. Since the frictional drag is generally in the opposite direction of the wind itself, this term represents a drain on the zonal kinetic energy ($D_Z > 0$).

The barotropic conversion term $-C_K$ in Eq. (8.1) represents the conversion from eddy kinetic energy to zonal kinetic energy. Averaged over the atmosphere as a whole, $-C_K > 0$; that is, eddy kinetic energy is converted into the kinetic energy of the zonally symmetric flow. The tendency in globally averaged zonal kinetic energy K_Z will be positive if the convergence of westerly momentum transport by the eddies takes place in a region in which the wind is already westerly and the divergence of westerly momentum transport is out of a region in which the wind is already easterly. At the jet stream level this term exhibits a negative extremum near $20°$ latitude, where westerly momentum is diverging out of a region of relatively weak[5] westerlies on the equatorward flank of the jet stream, and a positive extremum near $45°$ latitude, where the same amount of zonal momentum is being deposited by the eddies in a region of somewhat stronger westerlies. Hence, in the global average this term represents a source of kinetic energy for the zonally symmetric flow.

To derive an analogous expression for eddy kinetic energy we start with the equations of motion for the zonal and meridional wind components u and v and subtract from each of the corresponding expressions for the zonally averaged components $[u]$ and $[v]$ to get equations for the local time rates of change of the eddy components, with the advective terms on the right-hand side. Multiplying the expression for $\partial u^*/\partial t$ by u^* and the expression for $\partial v^*/\partial t$ by v^*, combining them and integrating over the mass of the atmosphere yields, after some manipulation analogous to that in Appendix B,

$$\frac{dK_E}{dt} = -\frac{1}{g}\int_0^{p_s}\left\langle [u^*v^*]\frac{\partial[u]}{\partial y}\right\rangle dp - \frac{1}{g}\int_0^{p_s}\left\langle [u^*\omega^*]\frac{\partial[u]}{\partial p}\right\rangle dp$$

$$-\frac{1}{g}\int_0^{p_s}\left\langle [v^{*2}]\frac{\partial[v]}{\partial y}\right\rangle dp - \frac{1}{g}\int_0^{p_s}\left\langle [v^*\omega^*]\frac{\partial[v]}{\partial p}\right\rangle dp$$

$$-\frac{1}{g}\int_0^{p_s}\left\langle\left[u^*\frac{\partial\Phi^*}{\partial x}+v^*\frac{\partial\Phi^*}{\partial y}\right]\right\rangle dp$$

$$+\frac{1}{g}\int_0^{p_s}\left\langle\left[u^*\mathcal{F}_x{}^*+v^*\mathcal{F}_y{}^*\right]\right\rangle dp, \qquad (8.2)$$

where $K_E = \frac{1}{2g}\int_0^{p_s}\langle u^{*2}+v^{*2}\rangle dp$. The two terms involving vertical fluxes of momentum can be neglected on the basis of scaling considerations, though there is no reason that they could not be retained and evaluated in a more exact treatment. Comparing meridional profiles of $[u]$ with $[v]$, and $[u^*v^*]$ with $[v^{*2}]$, taking into account both the amplitudes and the shapes, it is left to the reader to verify in Exercise 8.2 that we are justified in neglecting the $[v^{*2}]\frac{\partial[v]}{\partial y}$ term, which involves the exchange of kinetic energy between the eddies and the mean meridional circulations. With these approximations, Eq. (8.2) reduces to

$$\frac{dK_E}{dt} = \underbrace{-\frac{1}{g}\int_0^{p_s}\left\langle [u^*v^*]\frac{\partial[u]}{\partial y}\right\rangle dp}_{C_K}$$

$$\underbrace{-\frac{1}{g}\int_0^{p_s}\left\langle\left[u^*\frac{\partial\Phi^*}{\partial x}+v^*\frac{\partial\Phi^*}{\partial y}\right]\right\rangle dp}_{C_E}$$

$$+\underbrace{\frac{1}{g}\int_0^{p_s}\left\langle\left[u^*\mathcal{F}_x{}^*+v^*\mathcal{F}_y{}^*\right]\right\rangle dp}_{-D_E}. \qquad (8.3)$$

Since the momentum transport across the poles is zero, it follows that $\frac{\partial\langle[u][u^*v^*]\rangle}{\partial y} = 0$. Hence, the second term in Eq. (8.1) is equal and opposite to the first term in Eq. (8.3). Since they both involve the conversion between K_Z and K_E, it is appropriate to label them as the conversion term C_K in the mechanical energy cycle. The sign of this term in Eq. (8.3) depends upon whether the eddy transport of westerly momentum is directed up or down the meridional gradient of zonal wind. If it is down-gradient, the eddies will be gaining kinetic energy from the mean zonal flow. This is the mechanism by which waves can amplify by barotropic instability. At the jet stream level the poleward momentum transports are strongest at the latitudes of the zonally averaged jets, near $30°$N/S. Equatorward of those

---

[4] In Lorenz's original paper, η in the C_z term is approximated by f.
[5] About 10 m s^{-1} in the annual mean, as shown in Fig. 1.19.

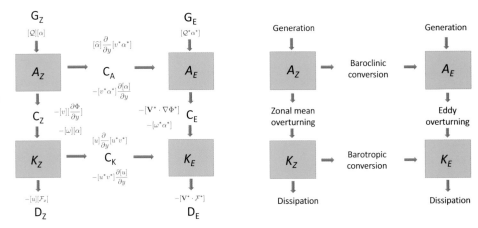

Figure 8.1 Reservoirs and conversion rates in the mechanical energy cycle discussed in Section 8.1.1 (see also Fig. 8.3). (Left) Mathematical summary, (right) verbal summary.

latitudes they are up (counter) gradient, while poleward of that latitude, up to about $60°$ N/S, where they undergo a sign reversal, they are down-gradient. It is evident from the top panel of Fig. 1.19 that the meridional gradient of westerly momentum is stronger on the equatorward side of the jet stream than on the poleward side, so the transport is predominantly countergradient. The equatorward transports observed at higher latitudes are also countergradient. Hence, consistent with the analysis of the C_K term in Eq. (8.1), we find that this conversion is primarily from the eddies to the zonal flow. Evidently, barotropic instability of the zonally symmetric flow is not the main energy source for the eddies.

The other terms in Eq. (8.3) may be recognized as the conversion due to the down-gradient cross-isobar eddy circulation labeled C_E (see Eq. (6.21) and Section 6.3) and the frictional dissipation labeled D_E. The conversion C_E must be large enough to balance dissipation plus the conversion from eddy to zonal kinetic energy. We already know that circulations in baroclinic waves are thermally direct, so it is reasonable to expect that C_E will be positive as labeled in Fig. 8.1 and quite large.

The corresponding equations for the time rates of change of A_Z and A_E are derived in an analogous manner. The resulting expressions are

$$\frac{d}{dt}A_Z = \underbrace{\frac{1}{g}\int_0^{p_s}\left\langle\frac{\mathcal{H}[\widehat{\alpha}]}{\sigma}\right\rangle dp}_{-C_A} + \underbrace{\frac{1}{g}\int_0^{p_s}\langle[\omega][\widehat{\alpha}]\rangle\,dp}_{-C_Z}$$
$$+ \underbrace{\frac{1}{g}\int_0^{p_s}\left\langle\frac{[\mathcal{Q}][\widehat{\alpha}]}{\sigma}\right\rangle dp}_{G_Z}, \qquad (8.4)$$

and

$$\frac{d}{dt}A_E = \underbrace{-\frac{1}{g}\int_0^{p_s}\left\langle\frac{[v^*\alpha^*]}{\sigma}\frac{\partial[\widehat{\alpha}]}{\partial y}\right\rangle dp}_{C_A} + \underbrace{\frac{1}{g}\int_0^{p_s}\langle[\omega^*\alpha^*]\rangle\,dp}_{-C_E}$$
$$+ \underbrace{\frac{1}{g}\int_0^{p_s}\left\langle\frac{[\mathcal{Q}^*\alpha^*]}{\sigma}\right\rangle dp}_{G_E}, \qquad (8.5)$$

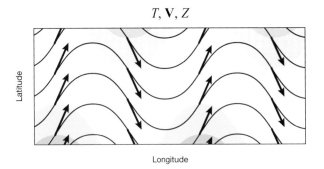

Figure 8.2 Temperature (shaded gold for warm, blue for cold), horizontal wind vectors, and streamlines (contours) in an idealized horizontal flow pattern in which eddy available potential energy is increasing at the expense of the available potential energy of the zonally symmetric background flow.

where $\mathcal{H}$ is the zonally averaged convergence of heat transport by the eddies, $\mathcal{Q}$ is diabatic heating, ($\widehat{}$) represents the departure from the global mean and σ is a measure of the static stability (Eq. (7.12)). The expressions for the global available potential energy expressed in terms of α are (c.f., Eq. (6.8))

$$A_Z = \frac{1}{2g}\int_0^{p_s}\left\langle\frac{[\widehat{\alpha}]^2}{\sigma}\right\rangle dp, \quad A_E = \frac{1}{2g}\int_0^{p_s}\left\langle\frac{[\alpha^{*2}]}{\sigma}\right\rangle dp. \tag{8.6}$$

The interpretations of Eqs. (8.4) and (8.5) are analogous to those of Eqs. (8.1) and (8.3). The first term on the right-hand-side, which involves the up- or down-gradient heat transport, represents the conversion C_A. It is clear that this term must be positive (i.e., in the sense of the arrow in Fig. 8.1), since the observed poleward eddy heat transport throughout the troposphere is down-gradient, from the warm tropics to the colder higher latitudes. This conversion can be viewed as distorting the shapes of the isotherms so as to make them less zonally symmetric and more wavelike, as depicted in Fig. 8.2. Note how temperature advection is increasing the amplitude of the wave and, at the same time, weakening the zonally averaged meridional temperature gradient. Total available potential energy is conserved, but the partitioning between the zonal and eddy reservoirs is altered.

The $\omega\alpha$ terms in Eqs. (8.4) and (8.5) are the energy conversions by the zonal average and eddy overturning circulations C_Z and C_E, respectively; these terms have already been discussed in the context of the kinetic energy reservoirs. The $Q\alpha$ terms represent the generation of available potential energy. It can be inferred from the distribution of net radiation at the top of the atmosphere (Figs. 5.5 and 5.6), together with the distribution of lower tropospheric temperature (Fig. 1.10), that the zonally averaged term G_Z must be very large, particularly in the winter hemisphere. Since C_Z is quite small, it follows that in the long-term average, the generation of zonal available potential energy must nearly balance the conversion to eddy available potential energy. Latent heat release in regions of precipitation is a source of eddy available potential energy (G_E). The mass-weighted generation term G_E, which is dominated by the tropospheric contribution, is also positive because precipitation and latent heat release in baroclinic waves tend to occur preferentially in warm air masses.

Two quantitative representations of the mechanical energy cycle are shown in Fig. 8.3. The first, produced at MIT in Starr's research group, is based on what would now be considered quite primitive operational analyses that were restricted to the domain poleward of 20°N. The second, published 43 years later, is based on global reanalysis products. The two analyses are similar in most respects: the generation of potential energy term G is positive for both the zonal mean flow and the eddies but substantially larger for the zonally symmetric component, whereas the dissipation D is much larger in the eddies; the baroclinic conversions C_A and eddy overturning conversion C_E are both large; the barotropic conversion C_K is much smaller and directed from the eddies to the zonally symmetric flow; and the conversion by the mean meridional circulations C_Z is so small as to be of uncertain sign.

The poleward, down-gradient heat transport near the lower boundary, as reflected in the conversion C_A, plays a central role in the linear theory of baroclinic waves. The most rapidly growing wave structures predicted by linear theory tend to be concentrated in the lower troposphere and their wave axes exhibit a strong westward tilt with height (see Fig. 2.10). The relationship between the wind and temperature fields is as shown in Fig. 8.2. On latitude circles near the "storm track," warm, poleward moving air rises, while cooler, equatorward moving air sinks, converting available potential energy to kinetic energy. This conversion is essential for the amplification of the waves. Without it the eddy kinetic energy embodied in the amplitude of the v-component of the wind could not increase to remain in thermal wind balance with the amplifying wave in the temperature field. At the same time, the conversion slows the growth of the temperature perturbations because poleward moving air cools as it rises and equatorward moving air warms as it sinks.

The relative importance of the meridional temperature advection vs. the temperature changes induced by the vertical motions is determined by the slope of the air trajectories

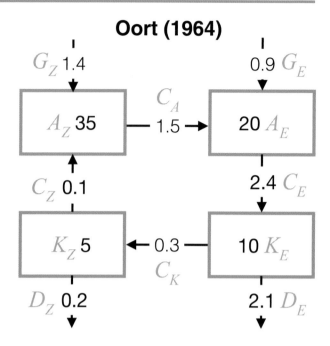

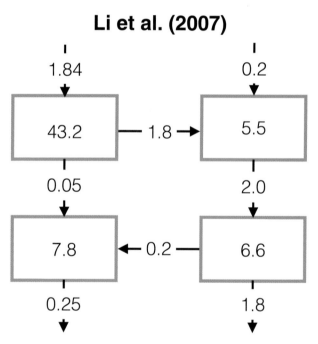

Figure 8.3 Two estimates of the terms in the mechanical energy cycle based on the decomposition described in Section 8.1 and summarized in Fig. 8.1. Energy in the various reservoirs expressed in units of J m^{-2} × 10^5 and the conversions in units of W m^{-2}.

in the meridional plane, as shown in Fig. 8.4. In constructing these trajectories it is assumed that the geopotential height and temperature fields in the waves are in quadrature as depicted in Fig. 8.2. In the case of purely horizontal trajectories (left), mixing by the eddies is acting to flatten the zonal

Figure 8.4 Idealized trajectories of air parcels in waves in which the streamlines and isotherms are sinusoidal and in quadrature with one another, projected onto the meridional plane. (Left) Pure horizontal flow; (right) adiabatic flow, and (middle) trajectories with an intermediate slope.

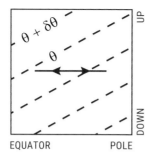
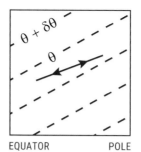
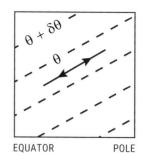

temperature gradient and thus convert zonal potential to eddy potential energy ($C_A > 0$), but $\omega^* = 0$ and so the conversion $C_E = 0$ and the wind perturbations do not amplify. In the case of adiabatic wave motions (right panel), K_E is increasing in time ($C_E > 0$) at the expense of the eddy potential energy, but since $\sigma\omega^*$ exactly cancels $-v^*\partial[T]/\partial y$, the wave in the temperature field does not amplify with time ($C_A = 0$). In the most rapidly amplifying waves, the trajectory slopes are about halfway between these values, as illustrated schematically in the middle panel.

The fastest growing baroclinic waves are not necessarily the waves that achieve the largest peak amplitude during their nonlinear life cycle. The other factor that needs to be considered in this regard is the vertical scale of the waves, which is directly proportional to their horizontal scale. Other things being equal, waves that tap into a deeper reservoir of zonal available potential energy achieve a larger amplitude. Hence, it can be argued that the waves that will eventually achieve the largest peak amplitudes are the ones whose vertical scale is comparable to the depth of the troposphere ($D \sim 10$ km). Bearing in mind that the trajectory slopes in the meridional plane w/v are on the order of $f/N \sim 10^{-4}/10^{-2} = 1/100$, it follows that the favored horizontal scale is on the order of 1000 km and a zonal wavelength of $\sim 2\pi \times 1000$ km, which corresponds to zonal wavenumber $k = 5$, which is slightly lower than the most unstable (fastest growing) waves, for which $k \sim 8$.

The generation, conversion, and dissipation of mechanical energy summarized in Fig. 8.3 applies to the atmosphere as a whole, and their numerical values are dominated by the tropospheric contribution. However, there are localized regions of the atmosphere in which the cycle looks much different. For example, in the lower stratosphere below the 50 hPa level, $[T]$ increases with latitude out to $\sim 55°$N/S (Fig. 1.21) in the presence of a predominantly poleward, (countergradient) heat transport $[v^*T^*]$. The zonally averaged diabatic heating rate $[Q]$ is in the sense as to weaken the existing temperature gradient. Hence, the local contributions to the mass-weighted integrals that determine the values of C_A and G_Z are negative, indicative of a (local) thermally indirect circulation.

8.1.2 Baroclinic Wave Life Cycles

To understand how baroclinic waves interact with the mean flow, it is necessary to consider them as finite amplitude dis-

turbances evolving through a characteristic "life cycle." Here we summarize results of a series of life cycle experiments with a high resolution primitive equation model with spherical geometry. Each integration was started from initial conditions consisting of a very weak wave of zonal wavenumber k, whose meridional and vertical structure matches that of the fastest growing normal mode, superimposed upon a zonally symmetric basic state flow. The two different zonal mean fields used as initial conditions are shown in the top panels of Fig. 8.5. The temperature fields are identical, with a deep baroclinic zone centered at $45°$ latitude. The zonal wind fields are both in thermal wind balance with the temperature field, but in the one shown in the left panel, $[u]$ is set equal to zero on the bottom boundary, whereas in the right panel a depth-invariant cyclonic shear is added such that at the Earth's surface, $[u] = +10$ m s^{-1} at $20°$ latitude and -10 m s^{-1} at $50°$ latitude.

During the linear growth phase, the waves that develop on the two zonal mean flows are virtually identical to the prescribed normal mode, but after they achieve finite amplitude and stop growing exponentially, they evolve quite differently. When the waves that develop in the zonal mean flow in the left panel break, the Lagrangian circulation at the jet stream level assumes the form of a series of anticyclonic gyres and hence the wave-breaking is said to be anticyclonic. In contrast, their counterparts in the right panel exhibit cyclonic wave-breaking, as illustrated below. The former (anticyclonic wave-breaking) life cycle (LC) is referred to as "LC1" and the latter (cyclonic wave-breaking) as "LC2." Figure 8.6 contrasts the life histories of the simulated waves as they evolve through their simulated life cycles, as reflected in their kinetic energy. The LC1 wave grows somewhat more rapidly until it peaks sharply on Day 7 and rapidly decays between Day 7 to Day 9. In contrast, the LC2 wave continues to grow until Day 10, by which time it is stronger than the LC1 wave when it was at its peak, and it decays gradually over the course of the next week.[6] From the time history of the energy conversions

---

[6] The material on baroclinic wave life cycle simulations presented in this chapter is drawn from Simmons and Hoskins (1978) and Thorncroft et al. (1993). As a result of minor differences in the experimental setup in these two studies, the LC1 wave grows and decays somewhat more rapidly in the latter study. The same is probably true of the LC2 wave, but it was not analyzed in the earlier study. Figures 8.5, 8.6, and 8.9 are based on the later study and 8.7 and 8.8 on the prior study.

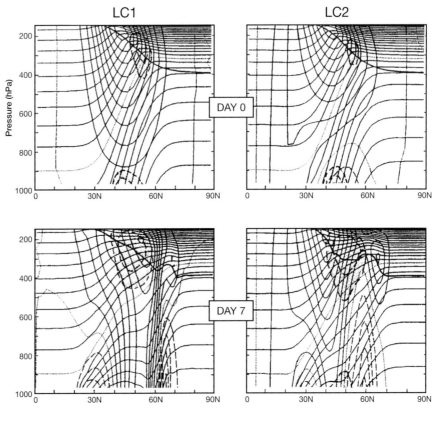

Figure 8.5 Zonally averaged zonal wind [u] and potential temperature [θ] fields in numerical simulations of LC1 and LC2 life cycles for zonal wavenumber $k = 6$. (Top panels) as specified in the initial conditions and (bottom panels) 7 days into the integration. Contour intervals 5 m s^{-1} for zonal wind; the zero contour is indicated by a dotted curve, and dashed contours denote easterlies. The dashed, stippled regions near the surface and near the jet core aloft denote regions of maximum eddy kinetic energy. The solid line that slopes upward towards the tropics is the 2 PVU surface. The contour interval for potential temperature is 5 K; dotted contour is 300 K. From Thorncroft et al. (1993). © Royal Meteorological Society.

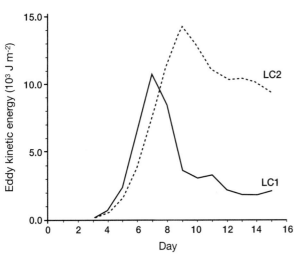

Figure 8.6 Eddy kinetic energy evolution for LC1 (solid) and LC2 (dashed). From Thorncroft et al. (1993). © Royal Meteorological Society.

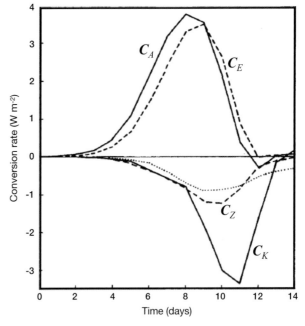

Figure 8.7 Time history of the conversions in the numerical integration of LC1 baroclinic waves. Symbols as in Fig. 8.1. The dotted line represents the dissipation of eddy kinetic energy. From Simmons and Hoskins (1978). © American Meteorological Society. Used with permission.

shown in Fig. 8.7 it is evident that the reduction of eddy kinetic energy in the LC1 waves from Day 7 to Day 9 is due to the barotropic conversion term C_K, in which the kinetic energy residing in the wave is transferred to the zonal flow. The process by which this collapse occurs is illustrated in Fig. 8.8, which shows SW–NE tilted ridges and troughs indicative of a northward transport of westerly momentum out of the subtropics and into the zonally averaged jet stream. As the waves collapse, the westerlies poleward of the storm

track accelerate. The LC2 wave does not experience a rapid collapse due to the barotropic energy conversion. It follows that most of the observed barotropic energy conversion

Figure 8.8 Streamfunctions of 300 hPa at 1 day intervals from Day 8 to Day 13 of a numerical simulation of LC1 baroclinic waves (zonal wavenumber 6). Latitude circles are drawn at 20° intervals extending out to the equator. From Simmons and Hoskins (1978). © American Meteorological Society. Used with permission.

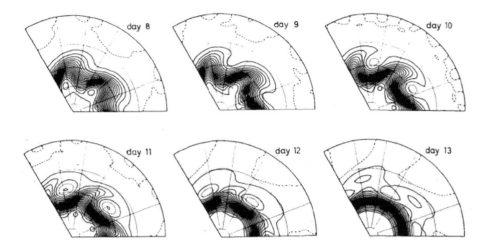

in Fig. 8.3 is associated with waves whose life cycles resemble LC1.

Apart from the greater strength of the negative C_K term in LC1, the energy conversions in the two life cycles are similar. The conversions C_A and C_E grow exponentially at first, with C_A peaking about a day earlier than C_E. Throughout the life cycle, zonal kinetic energy is being converted to zonal available potential energy ($C_Z < 0$) and eddy kinetic energy is being dissipated ($D_E > 0$). Some key points that bear on the distinctions between LC1 and LC2 are illustrated by Fig. 8.9, which shows distributions of potential temperature and wind on Days 5 and 7 of the simulations on a potential vorticity surface (2 PVU) that closely follows the tropopause in the meridional plane. On Day 5, near the end of the linear growth phase, the patterns in LC1 and LC2 are still quite similar: both are wavelike at all latitudes. The wave axes exhibit a SW–NE tilt to the south of the storm track and a SE-NW tilt to the north of it, indicative of an eddy transport of westerly momentum into the latitude belt of the storm track from both sides. The SW–NE tilt to the south of the storm track is more prominent in LC1 and the SE–NW tilt to the north of it is more prominent in LC2. By Day 7, the waves in LC1 and LC2 are both beginning to break. In LC1, the breaking involves an anticyclonic circulation on the southern flank of the storm track, in which most of the mass flux is concentrated in elongated, SW–NE tilted troughs that parallel the cold frontal zones trailing behind the cyclones at the Earth's surface. As time goes on, the troughs sharpen, become more elongated, and penetrate deeper into the subtropics.

In the LC2 life cycle, it is mainly the features to the north of the storm track that become more accentuated. The wave morphs into a series of cyclonic gyres separated by ridges. Irreversible wave-breaking occurs here as well, but in this case the resulting eddy pattern remains focused on the cyclonic side of the jet stream and does not develop a pronounced SW–NE tilt. What little tilt exists is in the opposite sense, indicative of an equatorward transport of westerly momentum. It is evident from the bottom panels of Fig. 8.5 that by Day 7 of the numerical integrations, the zonal

mean flow has already changed significantly in response to the meridional transports of westerly momentum. In LC1, a westerly jet has developed poleward of the storm track, extending all the way down to the Earth's surface, rendering the zonal mean flow in the vicinity of the storm track (at ~45°) anticyclonic. In LC2, a weaker jet has developed equatorward of the storm track, rendering the flow more cyclonic over the storm track. The changes in the zonal mean in the LC1 case serve to amplify the ambient horizontal shear at the latitude of the storm track that was imposed in the initial conditions.

The contrasting shapes of the waves in LC1 and LC2 can be interpreted in terms of the schematic of the flow shown in Fig. 8.10, which is applicable to all finite amplitude baroclinic waves irrespective of their LC1/LC2 categorization. The arrows A, B, C, and D are clearly discernible in Day 7 in Fig. 8.9. A and D are emphasized in LC1, whereas B and C are emphasized in LC2. But the distinction between LC1 and LC2 involves more than a difference in emphasis. Numerical experiments show they are discrete species of baroclinic waves, and not merely contrasting samples drawn from a continuum of life cycles. For each zonal wavenumber, there exists a critical value of cyclonic meridional wind shear of the zonal mean flow, below which the waves evolve in accordance with LC1 and above which they evolve in accordance with LC2.[7]

Whether a wave evolves in accordance with the LC1 or LC2 life cycle depends on the environmental meridional shear as well as on the wavelength. The longer waves with zonal wavenumber $k = 6$ are strongly inclined to evolve toward LC1 or (with sufficiently strong cyclonic environmental shear) LC2 structures. In contrast, shorter waves with zonal wavenumber $k = 9$ tilt in the opposite sense and transport westerly momentum equatorward, even in the absence of a barotropic component of cyclonic shear of the zonal mean flow.[8] In general, the axes of waves with

---

[7] Hartmann and Zuercher (1998).
[8] See Fig. 4 of Simmons and Hoskins (1978).

Figure 8.9 Fields of potential temperature and wind on the potential vorticity 2 PV surface on (top panels) Day 5 and (bottom panels) Day 7 in numerical simulations of LC1 (left panels) and LC2 (right panels) life cycles. Contour interval 5 K. Adapted from Figs. 7 and 10 of Thorncroft et al. (1993). © Royal Meteorological Society..

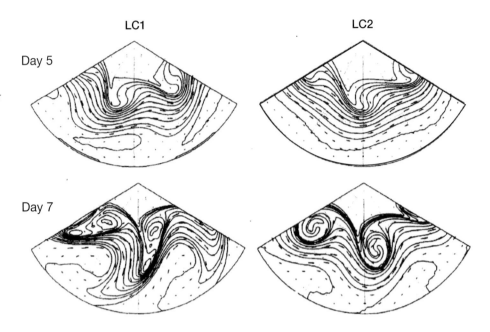

LC1   LC2

Day 5

Day 7

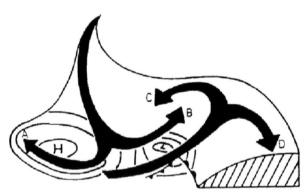

Figure 8.10 Schematic depicting the three-dimensional trajectories on sloping isentropic surfaces in a developing baroclinic wave in coordinates moving with the wave. Heavy solid arrows represent the flow along sloping isentropic surfaces. The sea level pressure pattern and fronts are also indicated. From Thorncroft et al. (1993). © Royal Meteorological Society..

the conversion C_A and giving it back through the conversion C_K. In simulations of the stratospheric planetary waves with general circulation models, these two conversions proceed at almost exactly the same rate, such that the energy cycle is a "merry go round" consisting of equal (but nonzero) conversions and flux divergences that nearly cancel, such that the net tendencies of A_Z, A_E, K_E, and K_Z are all nearly equal to zero.[9] By going through the laborious process of calculating all the conversions, one learns nothing about the sources and sinks of energy for the waves. The transformed Eulerian mean (TEM) formalism that will be introduced in the next section circumvents this problem.

higher zonal wavenumbers tilt in the same direction as the meridional wind shear and, hence, they tend to feed kinetic energy into the barotropic background flow at the expense of the eddy kinetic energy. These shallower waves never achieve sufficient amplitude to "break" in the sense that the LC1 and LC2 waves in Fig. 8.9 do.

8.1.3 The Mechanical Energy Cycle in the Winter Stratosphere

From the data shown in Figs. 1.21 and 5.14, it is evident that throughout the stratosphere in winter, the poleward heat transport is directed down the zonally averaged temperature gradient while the momentum transport (Figs. 1.19 and 3.11) is, for the most part, directed up the gradient of zonal momentum. Hence, like baroclinic waves, the stratospheric planetary waves are extracting energy from the mean flow by

---

Exercises

8.1   Based on data for the annual mean climatology shown in Chapter 1, compare the rates of generation of kinetic energy C_Z by the Hadley and Ferrel cells and show that they nearly cancel one another.

8.2   (a) By comparing meridional profiles of $[u]$ with $[v]$, and $[u^*v^*]$ with $[v^{*2}]$ at the tropospheric jet stream level, show that the first term on the top line of Eq. (8.2) is much larger than the first term on the second line. (b) Give a physical interpretation of these two terms.

8.3   Starting with the First Law of Thermodynamics in the approximate form $\frac{\partial \widehat{\alpha}}{\partial t} = -u\frac{\partial \widehat{\alpha}}{\partial x} - v\frac{\partial \widehat{\alpha}}{\partial y} - \sigma\omega + Q$ and decomposing $\widehat{\alpha}$ into zonally symmetric and eddy components, $[\widehat{\alpha}]$ and $\widehat{\alpha}^*$, respectively, and assuming that $d\langle\widehat{\alpha}\rangle/dt = 0$, derive Eqs. (8.4) and (8.5).

8.4   Show that for perturbations that vary sinusoidally with longitude, the ratio K_E/A_E is directly pro-

---

[9] Plumb (1983).

portional to the square of the two-dimensional wavenumber $K^2 = k^2 + l^2$, where k is the zonal wavenumber and l is the meridional wavenumber.

8.5   Figure 2.10 shows the structure of an amplifying baroclinic wave before it achieves finite amplitude – the so-called *Charney mode*. A more complete selection of fields for the more idealized, but mathematically much simpler *Eady mode* is shown in Fig. 9.12 of Vallis (2017) and in Fig. S.8.1 of the *Solutions Manual*. Using these figures as a reference, describe how the structure of the observed waves documented in this section departs from Charney's and Eady's solutions for the fastest growing wave.

8.6   Contrast the time histories of the various energy conversion terms in LC1 and LC2.

8.7   Show how a wave that is originally sinusoidal is distorted to the breaking point by the horizontal shear of the zonal mean westerly flow.

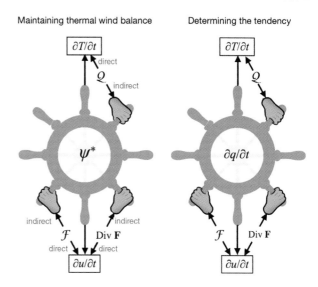

Figure 8.11 The schematic Fig. 7.5 revisited from a TEM perspective. The response to diabatic heating and friction is unchanged, but the eddy forcing is expressed as a single term involving the divergence of the Eliassen–Palm flux, which appears only in the zonal momentum equation (8.7). The Eulerian mean meridional circulation is replaced by the TEM residual circulation.

8.2 THE TRANSFORMED EULERIAN MEAN FORMALISM

The development of the transformed Eulerian mean (TEM) formalism was arguably one of the most important advances in dynamical meteorology in the late twentieth century. It provides

- an estimate of the Lagrangian mean meridional circulation based on grid point data that closely resembles the circulation computed in isentropic coordinates and interpolated onto a pressure grid,
- a representation of the eddy forcing of the $[u]$ field that takes into account the direct forcing by the momentum transports $[u^*v^*]$ and the indirect forcing by the eddy heat transports $[v^*T^*]$ by way of the mean meridional circulations, and
- a metric that can be used to trace the dispersion of waves in the meridional plane from regions of generation to regions of dissipation.

The equations that govern the evolution of a zonally symmetric circulation can be transformed by decomposing the vertical velocity $[\omega]$ into a component that induces an adiabatic temperature tendency that just balances the adiabatic temperature tendency by the eddies, plus a residual vertical velocity $[\omega]^*$ that is driven by diabatic heating. In Cartesian geometry, we can write Eqs. (7.6) and (7.7) as

$$\frac{\partial [u]}{\partial t} = f\,[v]^* \; + \; \nabla \cdot \mathbf{F} + [\mathcal{F}_x] \tag{8.7}$$

$$\frac{\partial [\alpha]}{\partial t} = \sigma\,[\omega]^* + [\mathcal{Q}], \tag{8.8}$$

where

$$[v]^* \equiv [v] - [v]_B + \frac{\partial}{\partial p}\frac{[v^*\alpha^*]}{\sigma} \quad \text{and}$$

$$[\omega]^* \equiv [\omega] - [\omega]_B - \frac{\partial}{\partial y}\frac{[v^*\alpha^*]}{\sigma} \tag{8.9}$$

are referred to as the *transformed Eulerian mean* (TEM) or *residual meridional circulation*, $[v]_B$ and $[\omega]_B$ represent the boundary-forced component, and

$$\mathbf{F} = -[u^*v^*]\,\mathbf{j} - f\frac{[v^*\alpha^*]}{\sigma}\,\mathbf{k} \tag{8.10}$$

is the Eliassen–Palm flux.[10] For the sake of clarity, we have reintroduced the explicit notation for zonal averaging and we continue to use thickness (specific volume) in place of temperature. Recalling the continuity equation, the TEM residual circulation in Eq. (8.9), exclusive of the boundary-forced term, can be rewritten in terms of TEM and Eulerian streamfunctions ψ^* and ψ,

$$\psi^* = \psi + [v^*\alpha^*]/\sigma. \tag{8.11}$$

The TEM circulation defined by Eq. (8.9) takes into account (i) the Eulerian mean meridional circulation, (ii) the eddy-induced Stokes drift, and (iii) a component that cancels the Stokes drift through the bottom boundary, as required to satisfy the conservation of mass. It is an estimate of the Lagrangian meridional motions of tagged air parcels, derived from Eulerian measurements: hence the label "TEM."

Away from the bottom boundary, the evolution of the zonally symmetric flow, symbolized by the steering of the ship in the schematic Fig. 7.5, looks different when viewed from a TEM perspective. The eddy forcing terms $\mathcal{G}$ and $\mathcal{H}$ are combined in a single eddy forcing term $\nabla \cdot \mathbf{F}$, which appears only in the zonal momentum equation (8.7), as depicted in Fig. 8.11. The "dynamical co-captain" is using both hands, while the "thermodynamic co-captain" is using only one hand.

---

[10] Note that the unit vector $\mathbf{k}$ is positive downward.

The Eliassen–Palm flux has another, equally fundamental interpretation in terms of wave activity and group velocity concepts. The "generalized Eliassen–Palm" relation can be expressed in the form[11]

$$\frac{\partial A}{\partial t} + \nabla \cdot \mathbf{F} = \mathcal{D}, \qquad (8.12)$$

where A is the *wave activity* and $\mathcal{D}$ represents the sources and sinks of wave activity. The sources include the mechanical mixing associated with flow instabilities as well as orography and zonally varying thermal forcing; the sinks include both frictional drag and thermal damping. If the flow is quasi-geostrophic and not too wavy,

$$A = \frac{1}{2} \frac{\partial [q_g]}{\partial y} [\eta^{*2}],$$

where η^* is the meridional displacement of air parcels in the eddies from their zonally averaged latitude and $\partial [q_g]/\partial y$ is the meridional gradient of the zonally averaged quasi-geostrophic potential vorticity. Hence, the distribution of A can be inferred from the distribution of potential vorticity.

The Eliassen–Palm flux is related to wave activity by way of the group velocity:

$$\mathbf{F} = \mathbf{c_g} A, \qquad (8.13)$$

where $\mathbf{c_g} = c_{gy} \, \mathbf{j} + c_{gp} \, \mathbf{k}$ is the group velocity in the meridional plane. Hence, $\mathbf{F}$ traces the flow of wave activity from the region(s) in the meridional plane in which the eddies are generated to regions where they are dissipated. If there is no generation or dissipation ($\mathcal{D} = 0$) and if waves are not changing in amplitude with time ($\frac{\partial A}{\partial t} = 0$), it follows from Eq. (8.12) that $\nabla \cdot \mathbf{F} = 0$ and the waves pass through the atmosphere without affecting the zonal averaged circulation. The dual identity of $\nabla \cdot \mathbf{F}$ as the divergence of the wave activity flux and the eddy-forced westerly acceleration implies that the eddies transport zonal momentum and wave activity in opposing directions. Combining Eqs. (8.12) and (8.13) we obtain the conservation law, applicable in the absence of sources and sinks of wave activity

$$\frac{\partial A}{\partial t} + \nabla \cdot \mathbf{c_g} A = 0, \qquad (8.14)$$

which is much simpler than the corresponding expression involving mechanical energy.[12, 13]

The applicability of the TEM formulation is contingent upon the validity of neglecting the terms involving the vertical eddy heat flux in Eq. (5.16). At stratospheric levels the zonally averaged static stability is sufficiently strong that this approximation is well justified, but in applying the TEM formulation to baroclinic waves in the next section, it should be borne in mind that it is a crude approximation: the neglected terms involving $[\omega^* T^*]$ are up to one-third as large as the term involving $[v^* T^*]$.

The EP flux formalism is applicable to gravity waves induced by flow over rough terrain and deep convection. In fact, the EP flux formulation was first developed for studying stationary mountain waves.[14] It can be used to diagnose waves on scales large enough to be resolved by the analysis and it has guided the development of parametrization schemes to represent the forcing of the large-scale flow by waves on scales too small to be resolved.

8.2.1 Further Interpretation

The Eulerian mean meridional circulation features thermally direct tropical Hadley cells and weaker thermally indirect midlatitude Ferrel cells, as discussed in previous chapters. By contrast, the TEM (mean meridional) circulation is thermally direct, not only in the tropics but all the way from equator to pole, like the isentropic mass circulation shown in Fig. 5.12. It is thermally direct because the Stokes drift is stronger than the Ferrel cell and thus determines the sense of the overturning in the meridional plane.

To understand why the Stokes drift is overwhelmingly poleward at the jet stream level requires consideration of the Eliassen–Palm (EP) flux $\mathbf{F}$. It can be inferred from Eq. (8.7) that above the boundary layer $\nabla \cdot \mathbf{F}$ is, in effect, a westerly momentum source. It follows that the $\mathbf{F}$ vectors, with sign reversed, trace the flux of westerly momentum in the meridional plane. As a consequence of the strong poleward heat transport at the Earth's surface associated with baroclinic waves, there exists a net convergence of $\mathbf{F}$ within the atmosphere as a whole, which tends to be concentrated in the vicinity of the jet stream level, where the waves break and the $\mathbf{F}$ vectors converge. In accordance with Eq. (8.7), in the climatological mean ($\partial / \partial t = 0$), the westward acceleration due to the excess convergence of $\mathbf{F}$ must be balanced by an eastward Coriolis force. It follows that the TEM circulation must be poleward and strongest in the vicinity of the jet stream level.

An additional, boundary-related component of the TEM circulation is needed in order to ensure that the TEM circulation satisfies the boundary condition for the continuity of mass, which requires that the vertical velocity vanishes at the Earth's surface. Poleward of the storm track where $[v^* \alpha^*]$ is largest, $\mathcal{H} > 0$. Hence, it follows from Eq. (8.9) that the Stokes drift is downward at the bottom boundary, so there must exist an upward boundary layer-related component that balances it. A similar cancellation must exist equatorward of the storm track. The boundary-related contribution to the TEM circulation is rarely mentioned in the literature on stratospheric and middle atmosphere dynamics. Because of the stable stratification, the associated equatorward boundary-related mass flux is concentrated in a shallow layer just above the Earth's surface. In the vertical integral it cancels the poleward Stokes drift, thus satisfying

---

[11] Andrews and McIntyre (1976, 1978). See also Andrews et al. (1987) and Edmon et al. (1980).

[12] Extended to finite amplitude in Nakamura and Zhu (2010).

[13] A conservation law for wave activity combined with mechanical energy is derived in Takaya and Nakamura (2001).

[14] Eliassen and Palm (1961).

the requirement there be no net mass flux across latitude circles.

For the atmosphere as a whole, the westerly momentum imparted by the eddy-induced poleward Lagrangian mass flux across midlatitudes is lost by frictional drag at the Earth's surface. The upward EP fluxes carry the westerly momentum downward from the jet stream level, where the poleward mass flux is strongest, into the boundary layer where it is subject to strong frictional drag.

In the troposphere, the Stokes drift is upward equatorward of the storm track, poleward across the storm track, and downward poleward of the storm track; that is, in opposition to and stronger than the upward, poleward, and downward branches of the eddy-forced Eulerian Ferrel cell.

For nondivergent eddies in thermal wind balance, the net eddy forcing of the zonal mean flow is

$$\nabla \cdot \mathbf{F} = -\frac{\partial}{\partial y}[u^* v^*] - f_o \frac{\partial}{\partial p}\frac{[v^* \alpha^*]}{\sigma} = \left[q_g^* v^* \right], \qquad (8.15)$$

where q_g^* is the quasi-geostrophic potential vorticity

$$q_g^* = \zeta^* - f_o \frac{\partial}{\partial p}\frac{\alpha^*}{\sigma}. \qquad (8.16)$$

The identity between the divergence of the EP flux and the net eddy (or wave) forcing of $\partial[u]/\partial t$ (Eq. (8.7)) applies to gravity and inertio-gravity waves as well as Rossby waves.

8.2.2 The TEM Version of the Baroclinic Wave Life Cycle

Figure 8.12 shows the distribution of $\mathbf{F}$ in the simulation of LC1 baroclinic waves. The packet of arrows proceeds

upward toward the tropopause level and then bends equatorward (panels a, b, and c). Throughout the life cycle, $\mathbf{F}$ vectors stream upward, diverging out of the boundary layer in the vicinity of the "storm track." The region of convergence disperses upward and equatorward at the leading edge of the lengthening packet of arrows. Averaged over the complete life cycle (panel d), the convergence is strongest just equatorward of the jet stream. Hence the waves are extracting zonal momentum from the mean flow where the westerlies are relatively strong and imparting it to the weaker westerlies in the midlatitude boundary layer.

By summing the conversions C_K in Eq. (8.3) and C_A in Eq. (8.5) and making use of the thermal wind equation $\partial[\alpha]/\partial y = f\partial[u]/\partial p$, it is readily verified that the net energy conversion from the mean flow to the eddies is $\mathbf{F} \cdot \nabla[u]$. Except for a small region just equatorward of the jet stream near the tropopause level, $\mathbf{F}$ is directed up the gradient of $[u]$ in the meridional plane – eddies are extracting energy from the mean flow. The eddy energy in this numerical simulation is lost to frictional dissipation in the boundary layer and to numerical damping of the mesoscale features that develop in association with fronts.

As noted above, in the absence of transience and the generation or dissipation of the eddies by diabatic heating or friction, Eq. (8.12) reduces to the statement that $\nabla \cdot \mathbf{F}$, the net forcing of the mean flow by the eddies, is equal to zero, which is referred to as the *non-acceleration theorem*.[15] It follows that the important wave–mean flow interactions in the meridional plane can only occur in those places where

---

[15] Charney and Drazin (1961). See also section 10.4 of Vallis (2017).

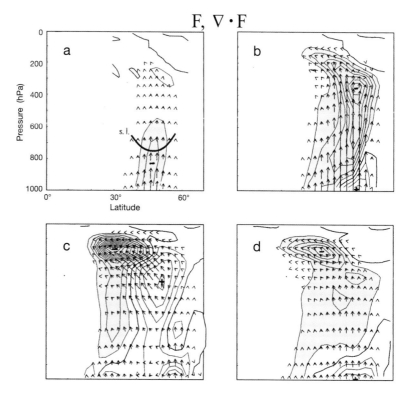

Figure 8.12 Meridional cross sections of Eliassen–Palm fluxes $\mathbf{F}$ (arrows) in pressure coordinates as obtained from a simulation of the life cycle of baroclinic waves shown in Figs. 8.7 and 8.8. (a) Linear calculation for the initial perturbation representative of the early stages of the simulation; (b), (c) successive instantaneous cross sections for later (nonlinear) stages of the simulation; (d) time average over the entire life cycle. Contours indicate the divergence of the flux, yellow positive, blue negative. The heavy contour in (a) represents the "steering level," where the phase velocity of the waves is equal to the background zonal flow. Near this level, the EP flux is not a measure of the group velocity. From Edmon et al. (1980). © American Meteorological Society. Used with permission.

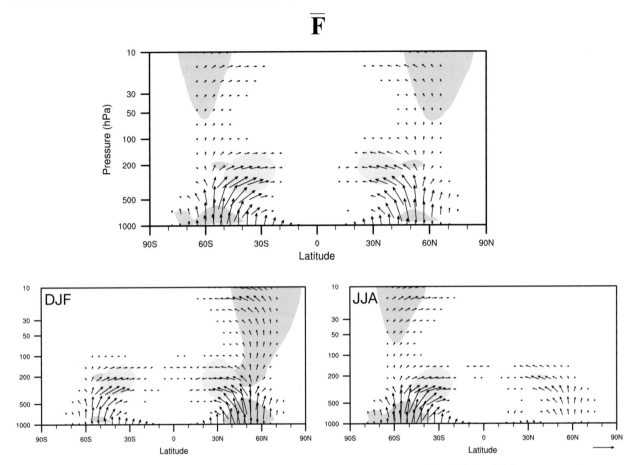

Figure 8.13 Climatological mean poleward eddy heat transports (gold shading) and eddy momentum transports (blue shading) and EP flux vectors: annual mean at top and DJF and JJA as indicated. Poleward heat transports greater than 15 K m s^{-1} are indicated by gold shading; equatorward momentum fluxes greater than 35 m^2 s^{-2} by blue shading. The vectors have been scaled by the inverse square root of the density in order to make the EP fluxes visible in the stratosphere. Courtesy of Ying Li.

one or more of these conditions does not apply. For example, baroclinic waves are generated near their "steering level" near 700 hPa (3 km) and they are dissipated as they approach 20°N/S at the jet stream level. Both regions are "critical zones" in which the zonal wind speed is comparable to the phase speed of the waves – about 10 m s^{-1}. As waves propagate into such regions in the meridional plane, their Doppler-shifted frequency approaches zero and frictional dissipation and thermal damping have more and more time to act. It is in these same regions that the Lagrangian particle displacements in the waves become irreversible, as indicated by the wrapping up of the cyclonic and anticyclonic gyres in the bottom panels of Fig. 8.9.

8.2.3 TEM Circulation and the Climatological Mean Eddy Forcing

Figure 8.13 shows annually and seasonally averaged climatological mean EP fluxes.[16] To remind the reader of

the definition of the EP flux, the patches of blue and gold shading highlight regions of strong poleward momentum and heat transports, respectively. As in the baroclinic wave simulations, the arrows stream upward out of the boundary layer ~50°N/S. They are stronger in the winter hemisphere. In the upper troposphere, the arrows split into two streams. Most of the lower stream bends equatorward, passes through the tropospheric jet stream, and converges in the subtropics. The observed equatorward branch of the EP flux at the jet stream level is consistent with that seen in the life cycle of the LC1 baroclinic wave in previous figures. There is just the barest hint of a poleward turning of the upward pointing arrows poleward of the storm track. The upper stream of arrows in the stratosphere is present only in the winter hemisphere. It is directed upward toward the polar night jet and some of it eventually bends equatorward.

The amount of bending that the EP fluxes and the associated group velocity vectors experience is determined by the potential vorticity gradient of the mean flow and by the spectrum of zonal wavenumbers and Doppler-shifted phase speeds of the waves. A distribution of refractive index can be computed for each zonal wavenumber and ray tracing can be

---

[16] The EP flux vectors shown in Fig. 8.13 were calculated using full spherical coordinates and a log p vertical coordinate, rendering the units of **F** to be kg s^{-2} (see Eq. 9.5).

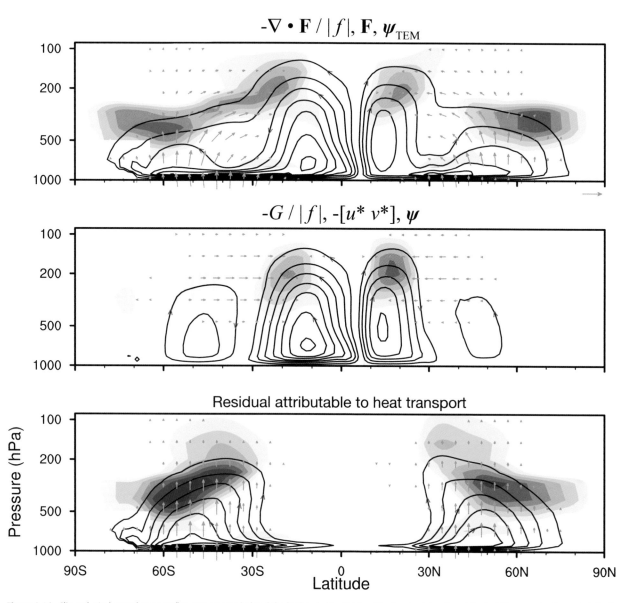

Figure 8.14 Climatological annual mean EP flux convergence (colored shading) scaled by dividing it by the absolute value of the Coriolis parameter f at each latitude. It can be interpreted as the poleward eddy-induced TEM circulation $[\bar{v}]^*$. Values range as high as 1 m s^{-1} poleward (orange shading) and as low as -0.4 m s^{-1} equatorward (blue shading). The EP flux itself is indicated by the gray vectors and the contours indicate the TEM circulation (contour interval 16×10^9 kg s^{-1} (16 Sv)). (Top) total TEM streamfunction, EP flux, and EP flux convergemce; (middle) the contribution attributable to the transports of zonal momentum, and (bottom) the contribution attributable to the heat transports. Courtesy of Ying Li.

performed in the meridional plane.[17] Under some conditions the EP flux vectors are bent into westerly jet streams, as is evidently happening around the tropopause level in the LC1 simulation. The EP flux vectors in the upper stream are usually refracted equatorward as they approach the stratospheric polar night jet, but even rather subtle changes in the

shape or latitudinal position of the jet can enable the waves to propagate upward into the jet, in which case they can trigger a stratospheric sudden warming event as discussed in Section 9.4.[18]

Figure 8.14 shows the annual mean climatology of the EP flux divergence and the TEM meridional circulation. The flux divergence is scaled by dividing it by the Coriolis parameter so that it represents the TEM meridional circulation $[v]^*$ required to balance it under steady state

[17] Fundamental papers on this topic include Charney and Drazin (1961) which deals with the effects of vertical wind shear, and Dickinson (1968). For a comprehensive discussion, see Holton (1987).

[18] See also Chapter 6 of Andrews et al. (1987).

conditions (see Eq. 8.7). (The tropical belt equatorward of 5°N/S, where the pattern is ill defined and noisy, is masked out.) The TEM meridional circulation shown in the top panel consists of a pair of tropical cells centered ~15°N/S and a pair of shallower extratropical cells centered ~50°N/S, both thermally direct, with little overlap between them. That the poleward flow in the upper branches of both pairs of cells coincides with regions of convergent EP flux illustrates the critical role of wave-driving in accommodating the poleward flow in the mean TEM meridional circulation.

Figure 8.14 also shows the components of the EP flux divergence and the TEM circulation partitioned into components attributable to the eddy transports of westerly momentum (middle panel) and heat (bottom panel). By construction, the component forced directly by the momentum transports corresponds to the Eulerian mean meridional circulation (e.g., as shown in Fig. 1.27), and the remainder, forced indirectly by the eddy heat transports corresponds to the Stokes drift. The tropical cells in the top panel are seen to be forced (or accommodated) mainly by the momentum transports and the extratropical cells by the heat transports. The forcing by the momentum transports in the poleward branch of the tropical cell coincides with the region of anticyclonic wave-breaking in LC1 (Fig. 8.12d). Heat transports in both LC1 and LC2 waves drive the extratropical cell, with transports in the poleward branch of the cell associated with cyclonic wave-breaking in LC2 and the shorter wavelength ($k \sim 9$) baroclinic waves. This interpretation is confirmed by a hemispheric analysis of the locations and orientations of potential vorticity streamers on isentropic surfaces.[19] In the cross section of EP fluxes in LC2 (not shown) the arrows do not curve equatorward in the upper troposphere as they do in LC1: they point slightly poleward and get shorter with increasing height above ~400 hPa, consistent with the tilt of the wave axes in Fig. 8.9. The longer waves with $k \sim 6$ are mainly responsible for the forcing of the Hadley cells, whereas the smaller scale, shallower waves with $k \sim 9$ are mainly responsible for forcing the poleward circulation across the extratropical storm tracks.

Exercises

8.8 In the stratospheric literature Eq. (8.11) is often referred to as the TEM meridional circulation or the *residual* mean meridional circulation, even though it is the sum of, rather than the difference between the conventional Eulerian mean circulation and the Stokes drift. Explain.

8.9 Prove that the eddy-driven components of $\boldsymbol{\Gamma}$ in Eq. (7.21) can be expressed as $\frac{\partial (\mathbf{F}/f)}{\partial y}$.

8.10 Prove that the eddy geopotential flux vector $\mathbf{W}$ and the Eliassen–Palm flux vector $\mathbf{F}$ based on the zonal

---

[19] Martius et al. (2007).

---

momentum and heat transports by the geostrophic wind are related by

$$\mathbf{W} \equiv [v^*\Phi^*]\,\mathbf{j} + [\omega^*\Phi^*]\,\mathbf{k} = ([u] - c)\,\mathbf{F}, \quad (8.17)$$

where c is the wave phase speed. Give a dynamical interpretation of this result.

8.11 Starting with the linearized, quasi-geostrophic potential vorticity equation with an exponential growth or damping term

$$\frac{\partial q_g^*}{\partial t} = -[u]\frac{\partial q_g^*}{\partial x} - v^*\frac{\partial [q_g]}{\partial y} + \gamma q_g^*,$$

derive Eq. (8.12)

$$\frac{\partial A}{\partial t} + \nabla \cdot \mathbf{F} = \mathcal{D},$$

where

$$A = \frac{1}{2}\frac{[q_g^{*2}]}{\partial [q_g]/\partial y} = \frac{1}{2}[\eta^{*2}]\frac{\partial [q_g]}{\partial y} \quad \text{and} \quad \mathcal{D} = \frac{\gamma [q^{*2}]}{\partial [q_g]/\partial y}. \quad (8.18)$$

8.12 Use Eqs. (8.7), (8.8), the continuity equation, geostrophy, and hydrostatic balance to derive the TEM form of the tendency equation for the quasi-geostrophic potential vorticity (c.f., (7.26)):

$$\frac{\partial [q_g]}{\partial t} = \frac{\partial}{\partial t}\left(\frac{1}{f_o}\frac{\partial^2 [\Phi]}{\partial y^2} + f_o\frac{\partial}{\partial p}\frac{1}{\sigma}\frac{\partial [\Phi]}{\partial p}\right)$$
$$= -\frac{\partial}{\partial y}\nabla \cdot \mathbf{F} - \frac{\partial [\mathcal{F}_x]}{\partial y} - f_o\frac{\partial}{\partial p}\left(\frac{[Q]}{\sigma}\right). \quad (8.19)$$

8.13 The accompanying figure shows the annual mean Eulerian vertical velocity $[w]$ superimposed upon the northward eddy heat transport in the equatorial belt in the ERA5 Interim Reanalyses. (Vertical velocity is expressed here in (x, y, z) coordinates, but that has no bearing on the question that is posed here.) Note that the contour interval for $[v^*T^*]$ is much smaller than in previous figures in the book showing the heat transports. The maximum at the 100 hPa level is due to the equatorial stationary waves that will be discussed in Section 14.5. The Eulerian upwelling directly over the equator exhibits a local minimum at all levels. The upwelling in the TEM cross section shown in Figs. 1.29 and 9.7 also exhibits an equatorial minimum, but it is evident only at the 50 and 70 hPa levels and it is not as pronounced as at 100 hPa. Explain the difference, making use of the TEM formulation.

8.14 (a) By adding Eqs. (8.7) and (8.12), derive an expression for the time rate of change of $[u] + A$ and derive an analogous expression for the rate of change of $[u] - A$. (b) Consider the time variations in these two quantities at the tropospheric jet stream level on a latitude circle near the center of the storm track. Bearing in mind the non-interaction theorem, which relates wave–mean flow interaction to transience and dissipation, how would you expect the two spectra to be different?

$$\overline{w},\ \overline{[v*T*]}$$

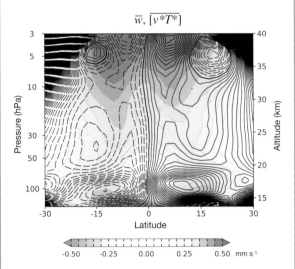

Figure for Exercise 8.13. Annual mean, zonally averaged Eulerian vertical velocity (colored shading) and northward heat transport, contour interval 0.25 K m s^{-1}, solid contours denote northward, the zero line is bolded. Courtesy of Hamid A. Pahlavan.

8.3 THE BALANCE REQUIREMENTS REVISITED

Let us reconsider the balance requirements for heat, westerly momentum, and pseudo-potential vorticity in light of the TEM formalism, with emphasis on the role of baroclinic waves in mediating the eddy transports and the mean meridional circulations.

8.3.1 The Eddy Heat Transport

From the viewpoint of the balance requirement for a poleward energy flux, the poleward eddy heat transports in baroclinic waves are most important in the lower troposphere, where they are not strongly compensated by an equatorward transport of geopotential. At these levels, the heat transport is down-gradient and resembles turbulent diffusion. The trajectory slopes in the waves in the meridional plane are about half as steep as the slopes of the isentropes. Hence, the mixing of potential temperature by baroclinic waves acts to rapidly flatten the isentropes, opposing the influence of the diabatic heating gradient in steepening them. In effect, the existence of baroclinic instability imposes a limit on the steepness of the isentropes. Whenever it becomes supercritical with respect to this limit, the waves quickly amplify and cause it to relax back to the limiting value, much as dry convection prevents the vertical temper-ature gradient from appreciably exceeding the dry adiabatic lapse rate.[20]

8.3.2 The Momentum Balance

The poleward transport of westerly momentum by the eddies is seen to be a consequence of the equatorward flux of wave activity from the source region for baroclinic wave activity in the midlatitude lower troposphere to the regions of wave-breaking equatorward of the tropospheric jet stream. Because the wave-breaking region for LC1 baroclinic waves is in the subtropics, there is a systematic equatorward flux of wave activity and hence a poleward transport of westerly momentum across $30°$N/S, which maintains the trade winds and the midlatitude westerly wind belts against frictional drag. [Recall that the westerly momentum transport by eddies follows the path $-\mathbf{F}$.]

That the waves must transport westerly momentum out of latitude belts in which waves are dissipated, and into the latitude belts in which they are generated can also be deduced from the local balance requirement for the r.m.s. amplitude of the eddy component of the relative vorticity field ζ^* in a two-dimensional flow, which follows from the relative vorticity tendency equation

$$\frac{\partial \zeta^*}{\partial t} \approx -[u]\frac{\partial \zeta^*}{\partial x} - v^*\frac{\partial (f + [\zeta])}{\partial y} + \gamma\zeta^*, \qquad (8.20)$$

where the last term represents wave generation and dissipation. The coefficient γ is positive in the storm tracks, where baroclinic waves are amplifying, and negative in the subtropical upper troposphere, where irreversible mixing is taking place in the waves as they approach their critical latitude. Multiplying Eq. (8.20) by ζ^* and zonally averaging, we obtain

$$\frac{1}{2}\frac{\partial [\zeta^{*2}]}{\partial t} = -[v^*\zeta^*]\frac{\partial (f + [\zeta])}{\partial y} + \gamma[\zeta^{*2}]. \qquad (8.21)$$

Since we are interested in climatological mean conditions, we can ignore the time derivative term. We recognize $[v^*\zeta^*]$ as $\mathcal{G}$, the zonally averaged transport of relative vorticity by the waves. Hence, Eq. (8.21) reduces to

$$\mathcal{G}\frac{\partial (f + [\zeta])}{\partial y} = \gamma[\zeta^{*2}]. \qquad (8.22)$$

The absolute vorticity gradient and enstrophy ζ^{*2} are positive definite. Hence, algebraic signs of $\mathcal{G}$ and γ must be the same, from which it follows that the waves induce an eastward acceleration in generation regions and a westward acceleration in dissipation regions.[21]

It is instructive to revisit the celebrated exchange of letters published in the *Journal of Meteorology* in 1949 concerning the relative importance of the eddies and the

---

[20] Stone (1978).
[21] This explanation is adapted from Held (1975), which was published just a year before the TEM formulation first appeared in Andrews and McIntyre (1976). Held refers back to similar arguments in earlier papers of Kuo (1951) and Green (1970).

mean meridional circulation in the poleward transport of angular momentum across midlatitudes in light of the TEM formulation. In his "Comment on Rossby and Starr" (1949), Palmén argued that a cell with poleward flow at the jet stream level and an equatorward return flow in the lower troposphere could account for the required transport of angular momentum across midlatitudes.[22] Starr countered by pointing out that the observed eddy transports were sufficient to satisfy the balance requirement, so it was not necessary to attribute the transport to a hypothetical circulation cell, whose existence could not be verified on the basis of observations. With the benefit of hindsight, it can be said that Palmén's hypothesized mean meridional circulation cell was indeed in the same sense as the observed TEM circulation cell. However, it is clear from Fig. 8.14 that the poleward flow across 30°N/S cell is associated with the shallow response to the poleward heat transport at the Earth's surface. More generally, the positive $f[v*]$ term is indeed of first order importance in the zonal momentum balance, but not for the reasons articulated by Palmen in his letter.

8.3.3 The Potential Vorticity Transport

The tropospheric pseudo-potential vorticity balance is dominated by the thermal component related to the static stability. Above the boundary layer the poleward heat transport across the storm track decreases with height, and therefore acts to decrease the static stability on its poleward flank and increase it on its equatorward flank. It follows that the transport of the thermal component of the pseudo-potential vorticity within the atmosphere is equatorward and downgradient. The much smaller dynamical (vorticity-related) transport, which is strongest at the tropospheric jet stream level, is poleward at the latitude of the storm tracks (the genesis region for baroclinic waves) and equatorward at subtropical latitudes (where baroclinic waves encounter their critical latitude). The poleward eddy heat transports across the storm tracks at the Earth's surface result in an additional equatorward transport of potential vorticity within the reservoir at the bottom boundary, as explained in Section 7.4.3. The poleward gradient of potential vorticity is maintained by the diabatic heating gradient at the Earth's surface, which is acting to increase the static stability poleward of the storm track and to decrease it equatorward of the storm track.

8.3.4 The Climatological Mean Meridional Circulation

Figure 8.15 shows a schematic of the climatological-mean TEM meridional circulation, also referred to as the *diabatic circulation*, whose adiabatic temperature tendencies $\sigma[\omega]^*$ balance the observed distribution of diabatic heating rate Q. The meridional flow is entirely poleward, and that is a valid description of the observed TEM meridional circulation,

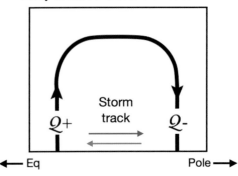

Figure 8.15 Schematic showing the interpretation of the extratropical cells in the climatological mean TEM circulation, which straddle the storm tracks. The poleward eddy heat transports in baroclinic waves (colored arrows: red the poleward flow of warmer air and blue the equatorward flow of colder air) drive the temperature field out of balance, inducing diabatic cooling on the equatorward flank and warming on the poleward flank of the storm track, which is balanced by the diabatic circulation.

apart from the weak equatorward TEM mass flux across the storm track in the boundary layer that exists by virtue of the upward rate of increase of the poleward heat transport near the surface (1000–800 hPa).

The diabatic circulation, as represented in pressure coordinates, is not a closed cell because it does not include the equatorward mass flux associated with the diabatically-induced poleward transport of specific volume $[v^*\alpha^*]$, i.e, the component induced by the heating (expansion) and cooling (contraction) of air parcels during their equatorward and poleward excursions in the eddies. When represented in isentropic coordinates, as in Fig. 5.12, the diabatic circulation is a closed cell.[23]

The observed diabatic circulation shown in the top panel of Fig. 8.14 involves the superposition of a tropical cell, which closely resembles the Eulerian Hadley cell and a weaker, shallower thermally direct cell centered at the latitude of the storm tracks (~50°N/S). The extratropical cell is associated with a belt of lower and mid-tropospheric diabatic heating centered ~40°N/S in combination with a band of cooling at subpolar latitudes, as shown in Fig. 5.18. This heating/cooling couplet is induced by the poleward eddy heat transports in baroclinic waves. At latitudes ~40°N/S, the predominance of cold advection (i.e., $\mathcal{H} < 0$ in Fig. 5.17) maintains temperature below radiative equilibrium, inducing radiative heating and (more importantly) enhancing the sensible and latent heat fluxes from the underlying oceans.

8.3.5 Toward a More Holistic Treatment

Thus far, we have treated the forcings $\mathcal{F}_x$, $\mathcal{G}$, $\mathcal{H}$, and $\mathcal{Q}$ as if they were free agents, driving the zonally averaged zonal flow $[u]$, but largely independent of it. In Chapter 7, we showed that under steady state conditions, the combined forcing field must be nondivergent and that the eddy-induced

---

[22] His justification is more clearly articulated in Section 1.5 of Palmén (1956). See also Rossby and Starr (1949), Palmén (1949) and Starr (1949).

[23] For a more detailed discussion of the interpretation of the lower branch of the adiabatic circulation, see Held and Schneider (1999).

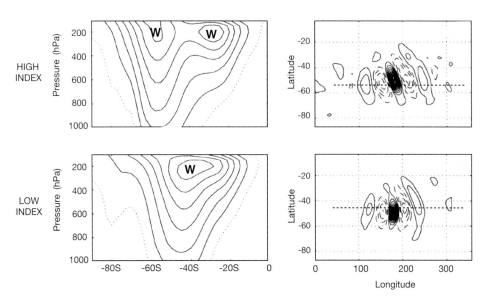

Figure 8.16 The Southern Hemisphere annular mode (SAM). (Left panels) "High index" and "low index" composite meridional cross sections of zonally averaged zonal wind. The zero contour is dotted: contour interval 5 m s^{-1}. (Right panels) The zonal average of the corresponding one-point correlation maps for 300 hPa 2.5–10d bandpass filtered relative vorticity anomalies for reference grid points along 50° S. The dotted lines indicate the latitudes of the zonally averaged jet stream. "High index" and "low index" nomenclature can be interpreted as relating to the latitude of the westerly jet in the lower to mid-troposphere. Contour interval 0.1; negative contours are dashed. Data for all seasons are included in the composites. Adapted from Figs. 8 and 10 of Hartmann and Lo (1998). © American Meteorological Society. Used with permission.

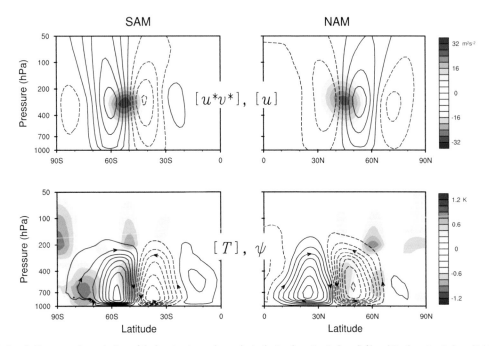

Figure 8.17 Various fields regressed upon indices of the barotropic annular modes in the Southern Hemisphere (left) and Northern Hemisphere (right). The indices are the standardized leading principal components (PCs) of the daily, vertically averaged [u] field in the domains 1000–200 hPa, 20–70° latitude in the respective hemispheres. (Top panels) [u] (contour interval 0.5 m s^{-1} : ..., -0.25, $+0.25$, $+0.75$, ...) and anomalous northward eddy transports of westerly momentum (colored shading) observed in association with an anomaly of 1 standard deviation in the standardized index. (Bottom panels) corresponding temperature anomalies (colored shading) and the Eulerian mean meridional circulation anomalies. Contour interval 0.5 × 10^9 kg s^{-1}. Courtesy of Ying Li.

frictional drag $\mathcal{F}_x$ is instrumental in enforcing this constraint in the boundary layer. In this chapter we have shown that not only does $\mathcal{F}_x$ limit the strength of the surface westerlies: it also determines the strength of the vertically integrated poleward mass flux in the diabatic circulation. Were it not for the frictional drag at the Earth's surface, the meridional gradient

$$[u^*v^*], \; [u]$$

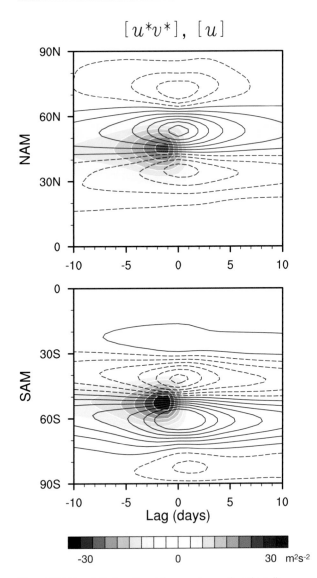

Figure 8.18 Time–latitude section of the anomalous northward zonally averaged eddy transport of westerly momentum (colored shading) and zonally averaged zonal wind anomalies (contours) in the barotropic annular modes, derived by performing regression analysis at various time lags relative to the (top) SAM and (bottom) NAM indices. Contour interval 0.7 m s^{-1}: contours at ..., $-0.35 + 0.35, + 1.05$.... Courtesy of Ying Li.

of diabatic heating could not be balanced by the adiabatic temperature tendencies associated with low latitude ascent and high latitude descent, which in turn, requires a poleward mass flux. In a frictionless atmosphere, the equator-to-pole temperature gradient and the atmospheric angular momentum would increase without limit. We have also shown how the poleward eddy heat transports in baroclinic waves reshape the diabatic circulation. The interactions between the wave forcings, $\mathcal{G}$ and $\mathcal{H}$ and the mean flow are more clearly apparent in the transient variability considered in the remainder of this chapter and in the next two chapters.

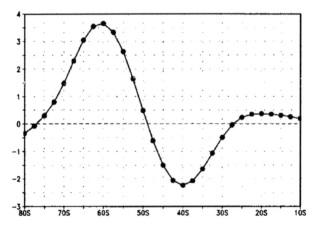

Figure 8.19 The leading EOF of vertically averaged [u] in the Southern Hemisphere (10-80°S) based on daily data. From Lorenz and Hartmann (2001). © American Meteorological Society. Used with permission.

8.4 THE BAROTROPIC ANNULAR MODES

In the distribution of annual mean of zonally averaged zonal wind [u] in the Southern Hemisphere (Fig. 1.19) the strongest westerlies in the tropospheric jet stream are observed ∼30°S, on the poleward flank of the Hadley cell in the so-called *subtropical* jet. The strongest westerlies at the Earth's surface are around 50°S, the latitude at which the poleward eddy heat transport is a maximum. At times, this higher latitude feature, referred to as the *eddy driven jet*, is discernible all the way up to the jet stream level. Subtropical and eddy-driven jets co-exist in aquaplanet models with rotation rates similar to that of Earth's (Fig. 2.13). The *barotropic annular mode* can be described as an alternating merging and separating of the subtropical and eddy-driven jets through the agency of the eddy transports of zonal momentum $[u^*v^*]$. It is the leading mode of variability of the [u] field in the Southern Hemisphere.[24] Although an eddy-driven jet is not discernible in the Northern Hemisphere zonal mean zonal wind climatology (Fig. 1.19), these modes of variability exist in both hemispheres and are referred to respectively as the *Northern Hemisphere annular mode* (NAM) and the *Southern Hemisphere annular mode* (SAM). The NAM was discovered over a century ago,[25] while the SAM was discovered much more recently.[26] Both the NAM and the SAM are recoverable using EOF analysis, in which they emerge as the leading modes of variability of the geopotential height field in their respective hemispheres. The prefix "barotropic" is used to distinguish the phenomenon discussed in this section from a more recently discovered pair of *baroclinic annular modes* discussed in the next section.

Figure 8.16 summarizes the structure and dynamics of the barotropic annular modes, using the SAM as an example. The left panels show zonally averaged zonal wind [u] composites for contrasting polarities of an index of the SAM

---

[24] Lorenz and Hartmann (2001).
[25] See Ivanov and Evtimov (2014) and Visbeck et al. (2003).
[26] Kidson (1988a,b).

Figure 8.20 (Top four panels) power spectra and autocorrelation functions for PC1 of vertically averaged [u] in the Southern Hemisphere (here labeled z) and the associated pattern of eddy forcing $\mathcal{G}$ (here labeled m). (Bottom) cross-correlation functions for eddy forcing (m) and the first PC of zonally averaged [u] (z). From Lorenz and Hartmann (2001). © American Meteorological Society. Used with permission.

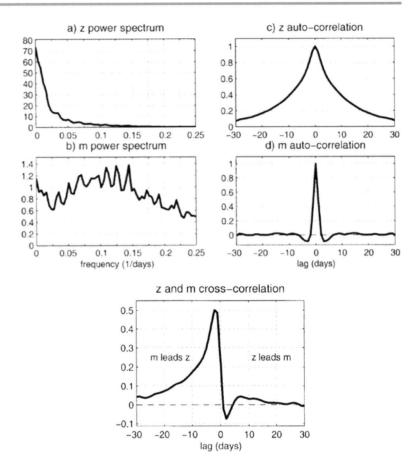

defined on the basis of the leading principal component (PC) of the Southern Hemisphere [u] field. In the "high index" composite, which corresponds to positive values of the PC, the subtropical and higher latitude eddy-driven jets are clearly separated, whereas in the "low index" composite the jets are merged. The right panels provide an indication of the contrasting horizontal structures of the leading modes of day-to-day variability under these contrasting zonal mean conditions: they are zonal averages of one-point correlation maps of the 300 hPa vorticity field for all grid points on the 50°S latitude circle. The waves in the high index composite are mainly equatorward of (i.e., on the anticyclonic side of) the jet, and accordingly, exhibit a pronounced SE–NW tilt indicative of a strong poleward transport of westerly momentum, reminiscent of the LC1 baroclinic wave life cycle described in Section 8.1.2. In contrast, those in the low index composite are located close to or somewhat poleward of the jet and accordingly exhibit a weaker tilt and are more bowed eastward in the middle so as to produce a convergence of westerly momentum transport into the storm track. The eddy structure in the low index phase is more like that of LC2 and the shorter wavelength (k ~9) baroclinic waves that don't develop sufficient amplitude in the upper troposphere to be refracted equatorward. In both cases, the transports of westerly momentum are in the sense as to maintain the jet in its present configuration: displaced poleward during

high index conditions and equatorward during low index conditions.

The vertical structures of the SAM and its Northern Hemisphere counterpart, the NAM, are shown in Fig. 8.17. They are derived by regressing the respective fields upon the leading principal component (PC1) of vertically averaged [u] in the respective hemispheric domains. We will refer to the *high index* polarity of the annular mode as matching the sign convention in this figure. The high index polarity of the SAM is characterized by an enhanced eddy-driven jet at ~60°S, a weakening of the westerlies at ~40°S, and an enhancement of the poleward eddy transport of westerly momentum at ~50°S (top left panel). The NAM (top right panel) exhibits similar features, but shifted equatorward by ~5°. The strongest anomalies in $[u^*v^*]$ precede those in $[u]$ by about a day, but in lowpass filtered data such as weekly or monthly means, the perturbations in $[u^*v^*]$ and $[u]$ appear to be virtually simultaneous.

The zonal wind $[u]$ perturbations in these modes are driven by the perturbations in the poleward transport of zonal momentum $[u^*v^*]$. The $[u^*v^*]$ anomalies are concentrated at the jet stream level, whereas the $[u]$ anomalies that develop in response to them exhibit a deep, barotropic structure. The mean meridional circulations spread the zonal wind response to the eddy forcing in the vertical, as illustrated

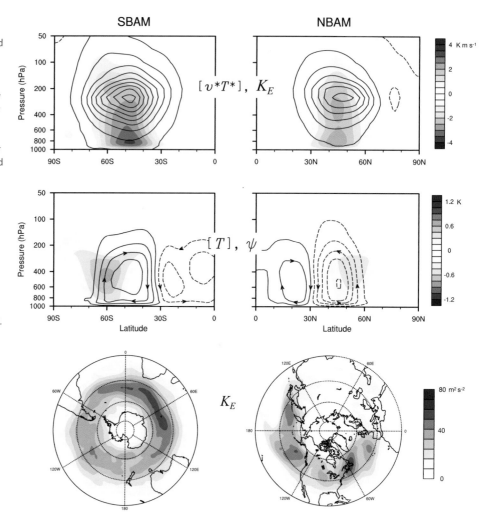

Figure 8.21 Various fields regressed upon indices of the (left) Southern and (right) Northern Hemisphere baroclinic annular modes: the standardized leading PCs of the daily zonally averaged eddy kinetic energy field in the domain 1000–200 hPa, 20–70° latitude. (Top panels) Eddy kinetic energy per unit mass (contour interval 6 m^2 s^{-2} : +3, +9 ···) and the northward eddy heat transport (colored shading) observed in association with an anomaly of 1 standard deviation in the index. (Middle panels) The corresponding temperature anomalies (colored shading) and Eulerian mean meridional circulation anomalies, contour interval 0.5×10^9 kg s^{-1}. (Bottom panels) The corresponding anomalies in eddy kinetic energy at the 300 hPa level. Courtesy of Ying Li.

schematically in Fig. 7.2, and in the process they induce $[T]$ anomalies as required for thermal wind balance.

The evolution of the SAM and NAM-related $[u^*v^*]$ and $[u]$ perturbations is shown in Fig. 8.18 based on lag regressions with their respective time-varying indices. As noted above, the anomalous eddy transports are strongest about a day before the extrema in the $[u]$ anomalies. The $[u^*v^*]$ anomalies build up to that level over an interval of about a week and collapse within a day or so after their peak. The corresponding $[u]$ anomalies amplify and decay more gradually. The long memory of the barotropic annular mode derives from the existence of a positive feedback between the forcing of the vertically averaged $[u]$ profile by the eddies and the influence of the same $[u]$ profile upon the meridional structure of the eddies, as would result from a prevalence of LC1 in high index and LC2 in low index states.

Let us consider the nature of this feedback in further detail. Following the protocol in Lorenz and Hartmann (2001), we define a time-varying index z representing the amplitude and polarity of the annular mode in the zonal wind field and a second index m representing the forcing by the eddy transport of westerly momentum $[u^*v^*]$. The index z is

the leading principal component (PC1) of vertically averaged $[u]$ and m is obtained by projecting the spatial pattern of the eddy forcing $\mathcal{G}$ upon the leading EOF of the vertically averaged $[u]$ shown in Fig. 8.19. By means of cross spectrum analysis (not shown) it can be shown that to a very close approximation,

$$\frac{dz}{dt} = m - \frac{z}{\tau},\qquad(8.23)$$

where τ is the decay timescale ($\sim$9 days) ; that is, variations in z are driven by variations in m and linearly damped.

Figure 8.20 shows power spectra and autocorrelation functions for z and m based on observations. Consistent with Fig. 8.18, the characteristic timescale of the variations in m is much shorter than those of z. In the cross-correlation function between z and m, shown in the bottom panel of Fig. 8.20, the strongest correlations are observed with m leading z by a day or two, consistent with Eq. (8.23). The negative lag correlations with z leading m by a day or two are a consequence of the negative side-lobes in the autocorrelation function of m. Of greater importance are the positive lag correlations for z leading m extending from

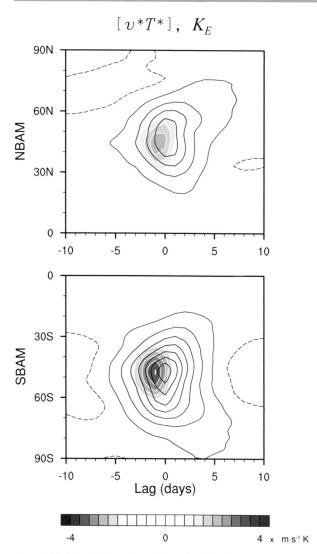

$[v^*T^*]$, K_E

Figure 8.22 Time–latitude section of the northward eddy heat transport at the 850 hPa level (colored shading) and eddy kinetic energy at the 300 hPa level (contours) in the Northern and Southern Hemisphere baroclinic annular modes, derived by regressing the fields upon the indices defined in the text. Contour interval 6 m^2 s^{-2} : +3, + 9, . . . Courtesy of Ying Li.

stronger Northern Hemisphere climatological mean stationary waves. However, the structure is analogous to that of the SAM and the dynamical interpretation is the same. Longitudinally dependent features of the NAM, such as those seen in the bottom right panel in Fig. 8.17, will be discussed in Section 13.1.2.

Exercises

8.15* Describe the impacts of the SAM upon the climate of Southern Australia.
8.16* Describe how the NAM (a.k.a. Arctic Oscillation) modulates energy usage by the nations of the Northern Hemisphere on a week-to-week basis.

8.5 THE BAROCLINIC ANNULAR MODES

The *baroclinic annular modes* involve temporal variations in the zonally averaged poleward heat transport across the climatological storm tracks and zonally averaged eddy kinetic energy.[28] Here we will focus mainly on the Southern Hemisphere baroclinic annular mode, which is more prominent and more straightforward to diagnose than its Northern Hemisphere counterpart. As an index, we will use the time series of the leading PC of the zonally averaged of eddy kinetic energy K_E in the domain 1000–200 hPa, 20–70°S.

The spatial structure shown in the bottom panels of Fig. 8.21. is, to first order, zonally symmetric and consequently well represented by the meridional cross sections of zonally averaged fields shown in the top and middle panels. Positive anomalies of the index are marked by enhanced K_E throughout the hemisphere: the latitude–height structure exhibits a broad peak at the jet stream level at 50°S. Enhanced hemispheric K_E is accompanied by enhanced poleward eddy heat transports with a meridional and vertical structure very similar to the transient eddy heat transport climatology, which is dominated by baroclinic waves. Enhanced hemispheric K_E is accompanied by anomalous tropospheric warmth at subpolar latitudes, indicative of a weakened meridional temperature gradient across midlatitudes, and by an enhancement in the strength of the Ferrel cell.

The relationships in Fig. 8.21 can be interpreted in terms of the mechanical energy cycle shown in Fig. 8.1. Enhanced energy conversion C_A is associated with enhanced baroclinic waves, and this interpretation is borne out by Fig. 8.22, which shows positive anomalies in poleward eddy heat transport occurring in the same latitude belt as the positive anomalies in K_E and preceding them by about a day.

The weakening of the meridional temperature gradients across midlatitudes by the burst of $[v^*T^*]$ renders the westerlies at the jet stream level around 50°S supergeostrophic, inducing an anomalous thermally indirect circulation, in

4 days out to about 20 days. If the variations in z were simply a passive response to variations in m, there is no way that z could anticipate an incipient burst in m so far in advance. It follows that positive z must be conducive to positive m and vice versa; that interactions with the eddies exert a positive feedback upon variations in z, enhancing the low frequency variability. Experiments with this highly simplified model of the SAM confirm that suppressing the feedback eliminates the positive cross correlations with m lagging z and reduces the variability of both z and m at periods longer than 20 days.[27]

The corresponding patterns for the NAM are not quite as annular because of the complicating effects of the much

[27] Lorenz and Hartmann (2003).

[28] Thompson and Barnes (2014).

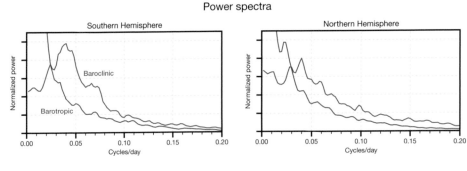

Figure 8.23 Power spectra for the indices of the barotropic and baroclinic annular modes, as defined in the text, based on the period of record 1979–2015. (Bottom) 10-day lowpass filtered time series for the Southern Hemisphere modes starting on 1 November 1982. Courtesy of Simchan Yook.

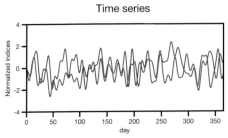

effect strengthening the climatological mean Ferrel circulation, which acts to restore the temperature gradient while weakening the westerlies. To fully restore the temperature gradient to the level that triggered the burst of anomalously strong baroclinic wave activity requires an extended interval of differential heating.

The distinctions between the time-dependence of the barotropic and baroclinic annular modes are clearly evident in their power spectra shown in the top panels of Fig. 8.23 and it is pronounced enough to be discernible in the raw time series of the indices in the bottom panel. Just as the positive feedback of the eddy forcing upon the zonally symmetric flow is responsible for the redness of the power spectra of the barotropic modes, a negative feedback appears to be responsible for the enhanced variability of the baroclinic modes at periods of about 3 weeks.

Exercises

8.17 The mean meridional circulations shown in the middle panel of Fig. 8.21 are Eulerian means. Sketch what the corresponding TEM (or isentropic) circulations should look like.

8.18 Contrast the barotropic and baroclinic annular modes (a) in terms of the mechanical energy cycle and (b) in a representation in terms of the EP fluxes.

Part IV The Stratospheric General Circulation

The prominence of the stratosphere in general circulation research derives from

- its role as a reservoir for dangerous radioactive debris from nuclear explosions and reflective sulfate aerosols from volcanic eruptions,
- its role in mediating ozone photochemistry, which shields life on Earth from harmful effects of solar ultraviolet radiation,
- its distinctive low frequency variability, which is seen as a possible vehicle for extending the range of useful weather prediction,
- its relative simplicity as a dynamical system in which the interactions between waves and the zonally symmetric flow play a central role.

Part IV on the stratospheric circulation is made up of two chapters, numbered 9 and 10. Chapter 9 documents and interprets the stratospheric jet streams, the planetary waves in the winter hemisphere, and the pole-to-pole TEM circulation, referred to as the *Brewer–Dobson circulation*, which ventilates the stratosphere with tropospheric air that enters by ascending through the tropical tropopause and exits at higher latitudes. We will consider the seasonally varying climatology and the non-seasonal variability on the year-to-year timescale, as well as so-called *stratospheric sudden warmings* induced by episodic planetary Rossby wave-breaking events in the winter hemisphere. Chapter 10 is mainly concerned with phenomena in the tropical stratosphere: the quasi-biennial oscillation (QBO) in zonally averaged zonal wind $[u]$ and the family of equatorially trapped planetary waves and smaller scale gravity and inertio-gravity waves that drive it. The main motivation for separating the topics in this manner is that the phenomena in Chapter 9 can be understood in terms of the simplified TEM formulation presented in Chapter 8 which captures the dynamics of Rossby waves, whereas the topics considered in Chapter 10 require a formulation generalized to include the momentum transports by vertically propagating gravity and inertio-gravity waves.

9 The Global Stratospheric Circulation

The middle atmosphere encompasses the stratosphere and the mesosphere. Its geometric midpoint at ~ 50 km corresponds roughly to the stratopause, the top of the stratosphere and the level of strongest heating (per unit mass) due to the absorption of solar ultraviolet radiation by ozone molecules. The volumes of the stratosphere and mesosphere are comparable, but more than 99% of the mass of the middle atmosphere resides in the stratosphere. The climatology and most of the non-seasonal variability of the zonally averaged, middle atmospheric circulation is wave-driven and can be understood in terms of the TEM formalism introduced in Chapter 8. In the stratosphere it is dominated by Rossby waves – baroclinic waves at lower levels and orographically and thermally forced planetary waves at higher levels – whose upward dispersion is proportional to the poleward eddy heat transport, in accordance with Eq. (8.10). In the mesosphere, the driving is dominated by gravity waves, whose amplitudes increase in inverse proportion to the square root of density until they break. Above the 30 km (10 hPa) level, the meridional gradient of the ozone heating also plays an important role in the dynamics. The zonally symmetric circulation in the meridional plane ($[v], [w]$), referred to as the Brewer–Dobson *circulation*, ventilates the middle atmosphere with air that formerly resided in the troposphere, carrying water vapor, carbon monoxide and other tropospheric tracers upward and carrying ozone that originated in the stratosphere downward.

The chapter begins with a section documenting the distributions of ozone and other tracers that provide insight into stratosphere–troposphere exchange. This is followed by a section focusing on the exchange processes themselves. It introduces the Brewer–Dobson circulation (BDC), in which air enters the stratosphere by way of the *tropical tropopause layer*. A tracer transport model is used to illuminate the dispersion and transport of a hypothetical tracer due to the poleward mass transport in the BDC and meridional mixing by eddies. It is shown how stratospheric air bearing high concentrations of ozone and potential vorticity can reenter the troposphere and descend to the Earth's surface within a day in episodic *tropopause folding events*.

The third section documents the seasonally varying climatology of the stratosphere in greater depth than in Section 1.4 and it includes a brief discussion of the mesospheric

circulation as well. It discusses the BDC in greater detail, drawing a distinction between its shallow branch, which is driven by baroclinic waves dispersing upward in the vicinity of the tropospheric jets, and its deep branch, which is driven by planetary-scale Rossby waves dispersing upward at higher latitudes of the winter hemisphere beneath the polar night jet. It offers explanations of why:

- the BDC is stronger in the Northern Hemisphere than in the Southern Hemisphere,
- the Antarctic stratospheric wintertime polar cap region is colder than its Arctic counterpart, rendering the air more susceptible to the catalytic destruction of ozone,
- the strongest upward mass fluxes in the annual mean BDC are in the outer tropics and subtropics, rather than directly over the equator,
- the tropical cold point tropopause is colder during DJF than during JJA, and
- on all timescales, a strong compensation is observed between high and low latitude temperature perturbations.

The final section describes and interprets the *stratospheric sudden warming* phenomenon, in which temperatures over the winter polar cap region may rise by 50 K or more, accompanied by a weakening, or even the disappearance of, the polar night jet over the course of a week or two.

Exercise

9.1 Why is the horizontal gradient of diabatic heating in the troposphere dominated by the equator-to-pole contrasts, while the gradient in the ozone-related heating at the stratopause level is dominated by the contrast between winter and summer hemispheres?

9.1 OZONE AND OTHER STRATOSPHERIC TRACERS

Stratospheric ozone (O_3) is formed by the so-called *Chapman reactions* involving the photodissociation of O_2 into atomic oxygen, followed by collisions involving O, O_2, and a third molecule. Ozone is photochemically active in the upper stratosphere, where it is photolyzed by ultraviolet radiation and the liberated O atoms quickly recombine with O_2 in an exothermic three-body reaction. The heat imparted to the air in these reactions is responsible for the temperature maximum that defines the stratopause. Lower stratospheric ozone is shielded from the ultraviolet radiation by the absorption that takes place higher in the air column; i.e., it resides below the level of unit optical depth for ultraviolet radiation. Hence, rather than being in photochemical equilibrium, its concentration is determined by the recent history of where it resided. Total column (i.e., vertically integrated) ozone is dominated by this lower stratospheric "reservoir" in which ozone behaves as a passive tracer. It is this vertical dependence of the photochemical lifetime of ozone, which ranges from less than a day in the upper stratosphere, to a month or longer in the lower stratosphere, that makes it such a valuable tracer.

Total column ozone has been monitored at a network of ground stations since the 1920s and from satellite-borne, downward-viewing sensors since 1979. The three-dimensional structure and variability of ozone has been revealed by limb-scanning sensors (i.e., radiometers looking toward the horizon and detecting vertical gradients in ozone concentration) on satellites since 1991. Figure 9.1 shows the zonally averaged ozone distribution. In the left column ozone abundance is expressed in terms of mixing ratio and in the

right column in terms of partial pressure, which is a direct measure of mass concentration per unit volume. The highest mixing ratios are in the middle and upper stratosphere in the low latitude photochemical source region. The level of peak mixing ratio slopes upward slightly with latitude, mirroring the rise in the level of unit optical depth with increasing solar zenith angle. The distribution of ozone partial pressure is quite different. The lower stratospheric reservoirs over the polar cap regions show up clearly, and there is a sharp discontinuity at the tropopause level. Concentrations tend to be higher in winter than in summer, especially in the Northern Hemisphere.

The spallation of cosmic rays in the stratosphere gives rise to radioisotopes ^{14}C, ^{3}H, and [7]Be, with half-lives of 5700 years, 12 years, and 53 days, respectively. In situ measurements of these species have been used in paleo-climate and other geophysical studies. Trace species originating in the troposphere, most notably water vapor, have played an important role in elucidating the ventilation of the stratosphere with air entering from below (i.e., the pathways by which it enters the stratosphere) and how it mixes with the drier stratospheric air. CO and CH_4, which are also monitored by limb-scanning instruments, are also used as tracers of tropospheric air that enters the stratosphere.

Figure 9.1 Climatological annual and zonal-mean ozone concentrations based on limb-scanning measurements from polar orbiting satellites expressed as (left) volume mixing ratio and (right) partial pressure: annual mean, DJF and JJA as indicated. Courtesy of Ying Li.

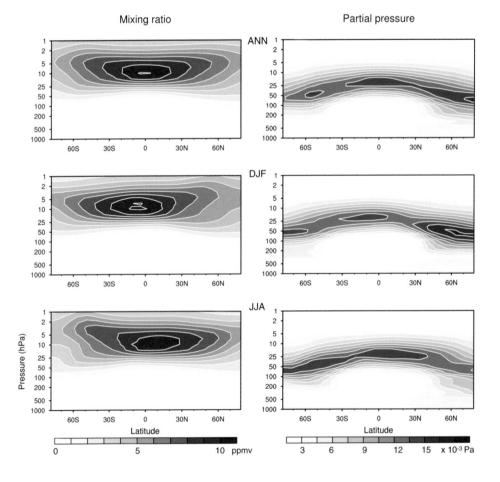

9.2 STRATOSPHERE–TROPOSPHERE EXCHANGE PROCESSES

Tropospheric air enters the stratosphere in the tropical tropopause layer (TTL) and much of the return flow occurs in so-called tropopause folding events in higher latitudes.

9.2.1 Entry of Tropospheric Air into the Stratosphere

As an air parcel enters the stratosphere by ascending through the TTL, nearly all of its water vapor is condensed and falls out as ice crystals. Below the 200 hPa level most of the condensation occurs in the anvils of cumulonimbus clouds, but above that level most of it takes place in the thin, cirriform cloud layers. As air parcels ascend through the 200–100 hPa layer layer, their mixing ratio decreases by about an order of magnitude, as shown in the left panel of Fig. 9.2. The high relative humidity and extensive cloud coverage of the layer of air at and just below the cold

point and the dryness of the air immediately above it, as shown in the right panel, supports the notion that air parcels ascend through the cold point. Proof that the stratosphere as a whole is ventilated by air that ascended in the TTL rests on that fact that the dewpoint of stratospheric air is substantially lower than the temperature of the extratropical tropopause.

The TTL is defined as extending upward from the level of zero radiative heating ~150 hPa, through the cold point tropopause, and up to the level of the highest overshooting convective cloud tops ~70 hPa. The chemical makeup of air at the base of the TTL is typical of tropical tropospheric air, while that at the top of the TTL is typical of stratospheric air.

The vertical profile of annual mean upward mass flux $-\omega$ averaged over the tropics (20°N–20°S) is shown in Fig. 9.3, together with the corresponding geometric vertical velocity w profile. The vertical velocity is upward through the entire depth of the TTL, but it decreases by more than an order of magnitude (in terms of mass transport ω) from the base to the top. The mass flux through the cold point in the ERA-I Reanalysis is ~6 kg m^{-2} per day, which is equivalent to a geometric rate of ascent, w, of ~30 m d^{-1}. The corresponding values for 70 hPa, at the top of the TTL and the minimum in the w profile, are ~2 kg m^{-2} d^{-1} and 20 m d^{-1}, respectively. (The corresponding ascent rates in ERA5 are about 40% smaller.) Above the 70 hPa level, geometric vertical velocity increases slowly with height while vertical mass flux continues to drop off, but more slowly than in the layer below.

It can be inferred from Fig. 1.24 that the potential temperature at the top of the TTL (70 hPa) is ~70 K higher than at the bottom (200 hPa). It is readily verified (e.g., from the diagram accompanying Exercise 9.5) that if an air parcel were lifted dry adiabatically from the bottom of the TTL to the top, it would have had to experience diabatic heating sufficient to raise its temperature by ~45 K in order to emerge at the observed temperature. Different processes contribute to the warming at different levels.

Figure 9.2 Climatological annual-mean (left) temperature, water vapor mixing ratio, and (right) relative humidity and cloud fraction averaged from 20°N–20°S. The blue-white boundary of the shading denotes the level of the tropical cold point tropopause. Courtesy of Ying Li.

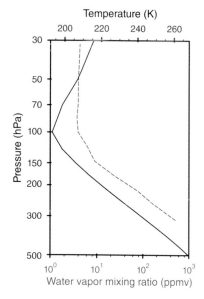

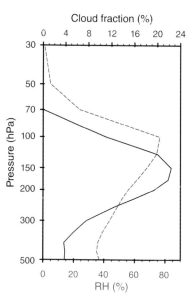

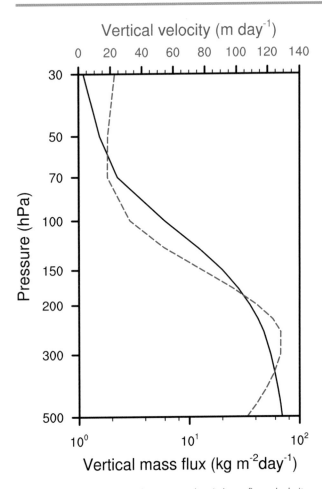

Figure 9.3 Climatological annual-mean upward vertical mass flux and velocity averaged 20°N–20°S. Courtesy of Ying Li.

In the vicinity of the cold point (100 hPa, 17 km), rising air parcels are warmed mainly by radiative transfer, as evidenced by the fairly close match between the adiabatic cooling rate $s\omega$ and the diabatic heating rate inferred from a radiative transfer model.[1] Given the slow ascent rates in the upper part of the TTL, even quite modest heating rates are sufficient to account for the observed warming of ascending air parcels.

9.2.2 Transport of Tracers within the Stratosphere

The first quantitative estimate of the mass transport in the Brewer–Dobson circulation (BDC) was by Murgatroyd and Singleton, who inferred the vertical velocities from heating rates and derived the meridional velocities from the zonally averaged continuity equation as in Eqs. 5.22 and 5.23.[2] Subsequent analysis of the tracer transports using the conventional Eulerian formalism yielded a more complicated picture, with eddies and the zonally averaged overturning circulations contributing to the mass transport in

equal measure, and often working at cross purposes to one another.[3]

The true Lagrangian mean mass transports in the meridional plane in a flow with eddies are difficult to visualize. Figure 9.4 shows results from recent numerical simulation that are helpful in distinguishing between the roles of the eddies and the mean meridional circulations. It is in the spirit of Kida's experiment described at the end of Chapter 5, but based on a tracer transport model in which a "cloud" of air parcels is advected by the three-dimensional velocity field in the ERA-I Reanalysis.[4] Air parcels released in the tropical boundary layer in a month-long pulse are tracked as they are advected by the time-varying three-dimensional wind field in ERA-I Reanalysis. After the pulse ends, any tagged parcel that is remaining within the tropical boundary layer or returns to it is immediately terminated. Note that the vertical coordinate in Fig. 9.4 is potential temperature θ, rather than $ln\ p$, so that vertical velocity is determined exclusively by the heating rate which is dominated by latent heat release in the lower and middle troposphere and radiative heating at higher levels. Owing to the flatness of the θ surfaces (e.g., see Fig. 1.24), showing the tracer concentrations in this format does not seriously distort the patterns. The spreading of the cloud of tagged air parcels and the shifting of its centroid during the experiment reveals the true Lagrangian mean meridional circulation.

At the end of the month-long release period, almost all of the tracer is still in the troposphere, where it has spread out until it extends from pole to pole due to meridional mixing by the eddies. In contrast, six months after the release, the mixing ratios of the tracer are highest in the lower stratosphere. Much of the remaining tracer has ascended in the shallow branch of the BDC in the tropics, and been mixed and transported laterally so that it extends from pole to pole, arching downward toward high latitudes like the distribution of ozone mixing ratio in Fig. 9.1. By this time, the sink in the tropical boundary layer has eroded the concentration of the tracer that remains in the troposphere, so that the lower stratosphere is serving as a reservoir, much as it did for the debris from the atomic weapons tests conducted circa 1960. By 12 months after the release, tropospheric air has become almost tracer free because of the sink in the tropical boundary layer, and concentrations in the lower stratosphere are beginning to drop as it becomes progressively more ventilated with "cleaner" tropospheric air. The highest tracer concentrations at this time are at the 600 K (~30 hPa) level, carried upward in the deep branch of the BDC. Three years after the release, the remnants of the tracer are confined to the high latitudes of the upper stratosphere, which has not yet been fully ventilated with cleaner, post-release air.

In this numerical simulation, the spread of the tracer from its source in the tropical boundary layer reveals both the lateral diffusion by the eddies on isentropic surfaces and the

---

[1] Lin et al. (2013).
[2] Murgatroyd and Singleton (1961).

[3] Newell (1963).
[4] Ploeger and Birner (2016). The "cloud" consists of a month-long pulse of air parcels released into the tropical boundary layer.

Figure 9.4 Dimensionless "mixing ratio" of tagged air parcels released in the tropical boundary layer (gray hatching) in a month-long pulse January 2000 and advected by the three-dimensional wind field in the ERA Interim Reanalyses. Panels show mixing ratio at the end of the month of the release (0) and at 6, 12, and 36 months after their release. Parcels that remain in or return to the tropical boundary layer after the end of the month of their release are terminated. The black solid line indicates the climatological location of the tropopause. The darker the shading, the higher the mixing ratio. Adapted from Ploeger and Birner (2016). Courtesy of Felix Ploeger.

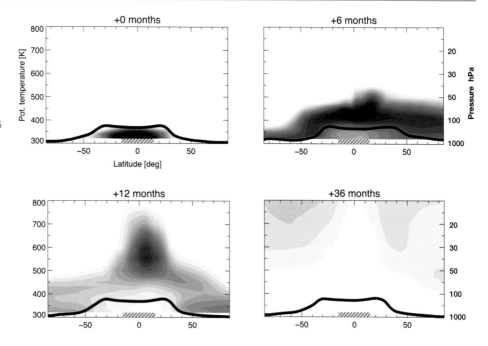

upward and subsequent poleward transport of the centroid of the "cloud" of tracked air parcels by the Lagrangian mean meridional circulation. The ascent rate of the centroid of the cloud is determined by the zonally averaged diabatic heating rate.

An estimate of the Lagrangian mean meridional circulation based on the TEM formalism in Eqs. (5.22) and (5.23) is shown in the bottom panel of Fig. 9.5. The vertical velocity component was inferred from the distribution of diabatic heating rate [Q] shown in the top panel, based on radiative transfer calculations. It is an update of the Murgatroyd and Singleton calculations, extended upward to the level of the mesopause.[5] Below 10 hPa, the heating is strongest in the tropics of the summer hemisphere, and the inferred circulation is consistent with the TEM estimate based on ERA-I, shown in Fig. 5.25. Above 10 hPa the strongest heating shifts from the tropics to the summer polar cap region and the mean meridional circulation accordingly exhibits a stronger summer pole to winter pole component.

The Lagrangian circulation cells that advect the tracers poleward in the stratosphere are analogous in some respects to the idealized steady state circulation cell in the right panel of Fig. 2.8, which is also driven by the horizontal gradient of diabatic heating. However, unlike that idealized cell, it is situated on a rotating, spherical planet, where the conservation of angular momentum imposes a strong restriction upon the strength of the poleward flow in its upper branch. In the words of Brewer, "there is considerable difficulty to account for the smallness of the westerly winds in the stratosphere, as the rotation of the Earth should convert the slow poleward movement into strong westerly winds."[6]

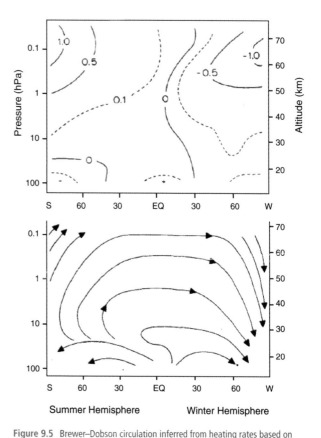

Figure 9.5 Brewer–Dobson circulation inferred from heating rates based on radiative transfer calculations. (Top) the vertical velocities whose adiabatic temperature tendencies balance the radiative heating rates and (bottom) the corresponding streamlines, whose meridional component is inferred from the continuity equation. W refers to the winter pole and S to the summer pole. From Dunkerton (1978). © American Meteorological Society. Used with permission.

[5] Murgatroyd and Singleton (1961).
[6] Brewer (1949).

By acting as an angular momentum sink for poleward-moving rings of air, the eddies mediate the poleward mass flux and thereby regulate the strength of the BDC.

The seasonality of the TEM circulation in Fig. 5.25 is reflected in the distribution of ozone mixing ratio shown in Fig. 9.1. The storage of ozone in the polar reservoirs increases during winter in response to the poleward and downward transport by the BDC. The reservoir over the Arctic contains more ozone than the one over the Antarctic, reflecting the greater strength of the Northern Hemisphere branch of the BDC, augmented by the impact of CFCs.

9.2.3 Intrusions of Stratospheric Air into the Troposphere

Fallout of radioactive debris from the numerous US and USSR nuclear weapons tests that were conducted from the late 1950s through the early 1960s turned out to be much more serious than anticipated. Most of the surface contamination occurred in association with discrete, regional events that occurred up to a year or so after the tests. Atmospheric dynamicists discovered that these radioactive aerosols could be traced back to recent intrusions of stratospheric air into the troposphere. The stratosphere was serving as a long-term reservoir for the radioactive atmospheric aerosols by sequestering them from scavenging by precipitation processes that limit the lifetimes of tropospheric aerosols to a week or two. Widespread public concern about these findings led to the 1963 nuclear test ban treaty.

Intrusions of stratospheric air into the troposphere involve a process referred to as *tropopause folding* on the cyclonic flank of the jet stream during the development of an intense upper air low. The ageostrophic circulation in the plane perpendicular to the section deforms the isentropes, as shown in the left panel of Fig. 9.6. Stratospheric air on the cyclonic (cold) side of the jet is subducted and stretched along a tilting axis, forming a fold, with tropospheric air both above and below it. The air within the fold is characterized by strong cyclonic vorticity and high static stability and hence, high PV, the distinctive marker of air that has recently resided in the stratosphere. It also exhibits concentrations of ozone characteristic of stratospheric air, sometimes reaching levels considered by regulatory agencies to be hazardous to human health. In some cases the ageostrophic deformation field is strong enough to advect the fold, in effect an upper level frontal zone, all the way down to the Earth's surface, or at least close enough to it for the air within it to be irreversibly mixed with ambient tropospheric air in convection or shear-induced turbulence. In the schematic shown in the right panel of Fig. 9.6, air parcels that passed through the fold descend all the way down into the boundary layer behind the cold front. It was a synoptic situation like this one in which radioactive debris from the nuclear tests, which had been sequestered in the stratosphere for many months, were brought back down to the Earth's surface in a matter of less than a day.

Tropopause folding occurs most frequently during winter and early spring, the season of strongest baroclinic wave activity. These events account for on the order of half the mass flux from the stratosphere back into the troposphere in the extratropics.[7] The remainder is presumably due to a slow, downward diabatic circulation.

Exercises

9.4 Show that the time required for the stratosphere to be ventilated by tropospheric air ranges from months in the lower stratosphere up to years in the upper stratosphere.

9.5 Show that a hypothetical air parcel entering the TTL at the 200 hPa level and cooling by adiabatic expansion as it ascends would need to be warmed by about 4K, 17 K and 45 K to be consistent with the observed temperatures at the 150, 100, and 70 hPa levels, respectively. (b) Would the required warming be appreciably less if the reference parcel ascended moist adiabatically rather than dry adiabatically?

9.6 The ascent rate of air at the equatorial cold point at ~100 hPa averaged over the tropics is about 30 m per day. (a) Estimate the heating rate in K d^{-1}, assuming that at the cold point the lapse rate is isothermal. (b) If the temperature were determined by Newtonian cooling with a radiative relaxation time of 30 d, estimate the departure from the radiative equilibrium temperature at the level of the cold point.

9.7 In the Eulerian equation for the time rate of change of ozone concentration averaged over the tropical lower stratosphere at, say, the 70 hPa level, what is the term that balances the negative tendency associated with the vertical advection by the mean meridional circulations?

9.8 What would the distribution of potential vorticity as defined in Eq. (4.15) look like for the wind and temperature fields in the vertical cross section shown in the left panel of Fig. 9.6?

9.3 THE STRATOSPHERIC CIRCULATION

The last section was mainly concerned with mass transport into, within, and out of the stratosphere. This section provides further documentation and dynamical interpretation of the seasonally varying climatological mean TEM circulation that is responsible for that transport.

9.3.1 The Seasonally Varying TEM Circulation

Meridional cross sections extending up to the 10 hPa level for annual mean, DJF and JJA zonally averaged zonal wind [u] and temperature [T] are shown separately in Figs. 1.19

[7] Jaeglé et al. (2017).

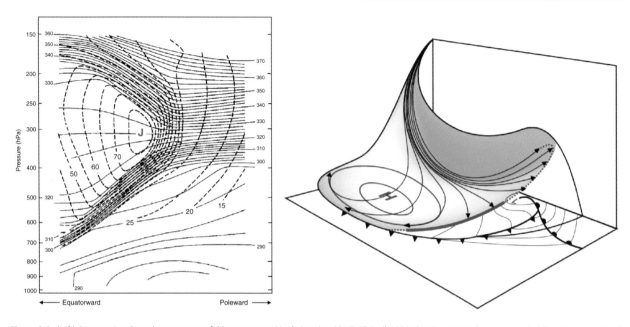

Figure 9.6 (Left) Cross section through a tropopause fold over western North America, 00 UT, 17 April, 1976, showing potential temperature (solid contours) in units of K and wind speed, out of the section (dashed contours) in units of m s^{-1}. Adapted from Fig. 1(a) of Shapiro (1981). © American Meteorological Society. Used with permission. (Right) Isometric showing idealized three-dimensional air trajectories in a vigorous extratropical cyclone. The stream of descending stratospheric air enters the troposphere in a tropopause fold below the jet stream and becomes entrained into the boundary layer behind the cold front. Some of this dry, ozone-rich air becomes entrained into the circulation around the cyclone, where it is interleaved with and eventually mixes with streamers of moist tropospheric air. The red band marks the end points of the trajectories of the air parcels that passed through the frontal zone. The isometric embodies some of the features of the mature phase of baroclinic waves represented schematically in Fig. 8.10. From a diagram created by E. F. Danielsen that appears in the Project Springfield Report, U.S. Defense Atomic Support Agency, NTIS 607980 (1964).

Figure 9.7 Meridional cross section extending from the summer pole (left) to the winter pole (right) constructed by averaging the DJF fields and the JJA fields, after transposing the latter in latitude. The climatological fields shown are the departure from the global mean temperature at each level (colored shading), the TEM circulation (arrows) and zonal wind (contour interval 5 m s^{-1}). The longest horizontal arrows correspond to 0.85 and 3 m s^{-1} above and below 100 hPa respectively, while vertical arrows correspond to 2 and 8.8 mm s^{-1}. The circulation at 100 hPa and above is referred to as the Brewer–Dobson circulation. With this scaling, only the Hadley cell is clearly discernible in the troposphere. Courtesy of Hamid A. Pahlavan.

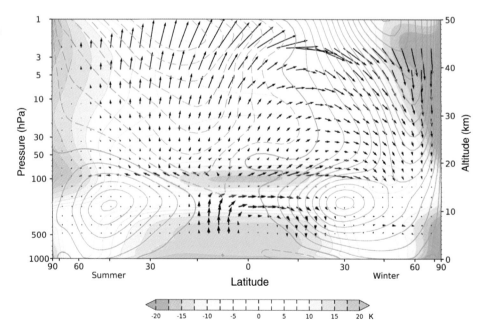

and 1.21 and, in combination, superimposed upon the TEM circulation in Fig. 1.29. Another version of the superimposed fields is shown Fig. 9.7, in which the JJA and DJF fields are combined to create a single winter hemisphere vs. summer hemisphere section extending all the way up to the stratopause level ∼1 hPa. The zonal flow is seen to be dominated by two matched pairs of jets: the tropospheric jet streams at the base of the stratosphere, and the westerly polar night jet of the winter hemisphere in combination with the subtropical easterly jet in the summer hemisphere, both of which achieve their peak amplitudes near the stratopause level. The temperature field can be inferred from the [u] field,

with which it is in thermal wind balance. The layer immediately above the tropospheric jet streams and extending up to ∼50 hPa is marked by easterly vertical wind shear and hence by temperature increasing with latitude from the tropics out to ∼60°N/S. Temperature increases by ∼25K from the tropics to subpolar latitudes. The inverse spatial correlation between zonally averaged temperature and radiative heating in Fig. 5.25 is indicative of radiative damping, from which it follows that the poleward temperature gradient in this layer must be maintained by thermally indirect circulations. Above the 50 hPa level, the temperature gradient is increasingly dominated by the contrast between the summer and winter polar cap regions, which mirrors the diabatic heating contrasts shown in Fig. 9.5. Around the 0.1 hPa (65 km) level in the middle mesosphere, the temperature gradient reverses and at the mesopause level the summer pole is more than 60 K colder than the winter pole.

Upon close inspection, the TEM circulation shown in Fig. 9.7 is seen to consist of a *shallow branch* extending from the tropopause up to about 50 hPa and a *deep branch* extending upward into the mesosphere. The ascent in the shallow branch is almost symmetric about the equator and the flow is poleward year-round in both Northern and Southern Hemispheres. The ascent in the deep branch is biased toward the summer hemisphere and virtually all of the descent occurs over the winter polar cap. More than two-thirds of the mass of air that enters the stratosphere through the tropical cold point flows poleward and returns to the troposphere in the shallow branch without ever reaching the 50 hPa level.

The flow in the shallow branch of the BDC roughly parallels the isotachs of climatological mean zonally averaged zonal wind $[u]$ above the cores of the tropospheric jet streams, with ascent on the cold, equatorial flanks and descent on the warm, poleward flanks. It is maintained by the strong meridional gradient of diabatic heating (Fig. 5.25), which is continually relaxing the temperature profile toward the radiative equilibrium profile, in which temperature decreases with latitude. The radiative heating gradient drives the zonally averaged zonal wind and temperature fields $[u]$ and $[T]$ out of thermal wind balance, inducing a thermally indirect mean meridional circulation that maintains the thermal contrast. Its strength is thus inversely proportional to the radiative relaxation time, which is determined by the concentrations of radiatively active trace gases H_2O, O_3, and CO_2 and thin cirrus clouds. In the poleward flow above the jet streams, the eastward Coriolis force is balanced by the westward force due to the EP flux convergence. This circulation cell could not exist in the absence of an angular momentum sink (i.e., a convergence of the EP flux) in its poleward branch, which derives mainly from the breaking of planetary-scale Rossby waves but also includes the breaking of gravity and inertio-gravity waves.

From the middle stratosphere upward into the lower mesosphere, the ascent in the BDC occurs mainly in the tropics and lower middle latitudes of the summer hemisphere, while the descent occurs in the high latitudes of the winter hemisphere. Hence, the cell is thermally direct, with rising

of warmer air and sinking of colder air, but it could not exist in the absence of wave-breaking, which enables zonally symmetric rings of air to move meridionally in the presence of strong inertial stability. In the upper stratosphere, most of the wave-breaking is associated with the upward dispersion of planetary waves in the winter hemisphere, whereas at mesospheric levels it is mainly associated with gravity and inertio-gravity waves, which produce a drag force on the zonally averaged zonal wind as explained in Section 7.1.

9.3.2 The Breaking of Planetary-Scale Rossby Waves

Rossby waves exhibit appreciable amplitudes only at latitudes and levels where the background flow $[u]$ is from the west. Westerly jet streams serve as *waveguides*; regions in which wave activity tends to be focused or "ducted", as evidenced by the coincidence between the regions of maximum eddy kinetic energy in Figs. 1.30 and 1.32 and the westerly jet streams in Fig. 1.19. The tropospheric jet streams serve as waveguides for Rossby waves with zonal wavenumbers ranging from $k = 5$ to about 10, which are generated mainly by baroclinic instability, while the stratospheric polar night jet serves as a waveguide for *planetary-scale Rossby waves*, with zonal wavenumbers $k = 1$ and 2.[8] The latter include stationary waves forced by flow over mountain ranges and continent–ocean thermal contrasts, as will be explained Chapter 11. The Rossby waves generated by baroclinic instability are thus largely confined to the lower stratosphere and to the shallow branch of the BDC, while planetary-scale Rossby waves dispersing upward from the troposphere are dominant in the deep branch.

Figure 9.8 shows the EP flux $\mathbf{F}$ and the EP flux divergence $\nabla \cdot \mathbf{F}$ in the same format as the previous figure. To apply the TEM formulation to a layer extending over a depth of multiple scale heights, it is necessary to make some adaptations. Pressure is replaced by a height-like vertical coordinate derived from the pressure field

$$z \equiv -H \ln \left(\frac{p}{p_s} \right), \tag{9.1}$$

where H is the prescribed scale height, RT/g, typically set to ∼7 km for stratospheric applications, and p_s is the surface pressure, typically set to 1000 hPa. Vertical velocity $w = dz/dt$ in this coordinate system resembles geometric vertical velocity. It follows from Eq. (9.1) that $\omega/w \approx e^{-z/H}$, where ω is vertical velocity in pressure coordinates. Hence, w is not attenuated the way ω is as one proceeds upward through a depth of the multiple scale heights. We have already made use of this in plotting Figs. 1.29 and 5.25.

The counterparts to Eqs. (8.7)–(8.9) in spherical coordinates are

---

[8] Charney and Drazin (1961) explain why the higher wavenumbers associated with baroclinic waves are unable to disperse upward into the polar night jet.

$$\frac{\partial [u]}{\partial t} = \left(f - \frac{1}{R_E \cos\phi} \frac{\partial}{\partial \phi} [u] \cos\phi \right) [v]^* - [w]^* \frac{\partial [u]}{\partial z}$$
$$+ \frac{\nabla \cdot \mathbf{F}}{\rho_o R_E \cos\phi} + [\mathcal{F}_x], \tag{9.2}$$

$$\frac{\partial [T]}{\partial t} = [w] \, (\Gamma_d - \Gamma), \tag{9.3}$$

$$[v]^* = [v] - \frac{1}{\rho_o} \frac{\partial}{\partial z} \left(\rho_o \frac{[v^* T^*]}{\Gamma_d - \Gamma} \right),$$
$$[w]^* = [w] + \frac{1}{R_E \cos\phi} \frac{\partial}{\partial \phi} \left(\frac{[v^* T^*]}{\Gamma_d - \Gamma} \cos\phi \right), \tag{9.4}$$

where $\Gamma_d - \Gamma = g/c_p - d\overline{T}/dz$ is the height-dependent static stability and $\mathbf{F}$ is given by (c.f., Eq. (8.10))

$$\mathbf{F} = \rho_o R_E \cos\phi \left(-[u^* v^*] \mathbf{j} + \left(f - \frac{1}{R_E \cos\phi} \frac{\partial}{\partial \phi} [u] \cos\phi \right) \frac{[v^* T^*]}{\Gamma_d - \Gamma} \mathbf{k} \right), \tag{9.5}$$

where ρ_o is a reference density, which is prescribed to decrease exponentially with height and $\mathbf{j}$ and $\mathbf{k}$ are unit vectors northward and upward, respectively. We have already made use of this scaling in plotting Fig. 8.13. Note that with the inclusion of the reference density, the EP flux vectors $\mathbf{F}$ have units of kg s^{-2}.

Figure 9.8 extends the fields shown in Fig. 8.13 upward into the stratosphere, making use of this height-like vertical coordinate. However, instead of dividing $\nabla \cdot \mathbf{F}$ by f, as in Fig. 8.14, a nonlinear color scale is used to bring out the weaker, but dynamically important low latitude features. As in Fig. 8.14, the lower branches of arrows in both hemispheres, which represent the flux of baroclinic wave activity, curve equatorward into the tropospheric jet streams and converge

on their equatorward flanks. In this section, a deep branch representing the flux of planetary-scale Rossby wave activity is also clearly discernible, ascending into the polar night jet, curving equatorward, and converging on its equatorward flank. The breaking of these waves, indicated by the broad, blue-shaded band of EP flux convergence, is the principal momentum sink that enables the poleward flow in the winter hemisphere in the deep branch of the BDC. The EP flux convergence is strongest in midlatitudes, but convergence strong enough to enable substantial poleward flow extends to within 15° of the equator.

9.3.3 The Cross-equatorial Mass Flux in the BDC

It remains to be explained why the ascent in the deep branch of the BDC is centered in the subtropics of the summer hemisphere rather than in the winter hemisphere, where the EP flux convergence associated with the breaking of planetary-scale Rossby waves is occurring. This bias is in part a reflection of the thermally direct cell driven by the cross-equatorial radiative heating gradient between summer and winter hemispheres, but that is not the whole story. Numerical simulations indicate that wave-breaking in the winter hemisphere can, in fact, induce substantial upwelling in the summer hemisphere tropics and subtropics, provided that it extends to within 15° of the equator. The inertial stability, which determines the meridional extent of the response, is directly proportional to the meridional gradient of zonally averaged angular momentum m, which vanishes on or very close to the equator where it changes sign. If the breaking of planetary-scale Rossby waves in the winter hemisphere extends deep enough into the tropics, there is nothing to prevent the cell from extending across the equator into the

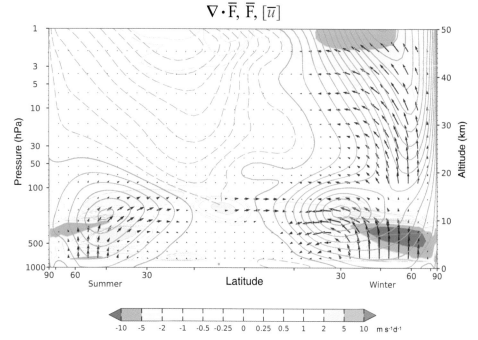

Figure 9.8 Meridional cross section extending Fig. 9.7, showing the summer pole to the winter pole, constructed from DJF and JJA fields, as described in the caption of Fig. 9.7, showing the divergence of the EP flux estimated using Eq. (9.5) (colored shading), the EP flux (arrows) and zonal wind (contour interval 5 m s^{-1}). The vector field at the 100 hPa level and above is stretched by a factor of 4.4 relative to the troposphere. Fields below 800 hPa are not shown. Courtesy of Hamid A. Pahlavan.

$$\nabla \cdot \overline{\mathbf{F}}, \; \overline{\mathbf{F}}, \; [\overline{u}]$$

summer hemisphere. Cross-equatorial flow, in turn, advects air with lower angular momentum onto the equator, inducing a westward acceleration. A weakening of the westerlies in the vicinity of the equator reduces the inertial stability, and thus serves as a positive feedback.[9]

Because wave-breaking in the winter hemisphere induces upwelling in the outer tropics and subtropics of both hemispheres, the annual mean upwelling exhibits peaks in the subtropics of both hemispheres, with a relative minimum over the equator (not shown). In the observations, this double-peaked structure is most clearly discernible at the 70 and 50 hPa levels. A related feature is the subtropical easterly jet in the summer hemisphere (Fig. 9.8), which coincides with the equatorward flow in the deep branch of the BDC (Fig. 9.7).

9.3.4 Northern versus Southern Hemisphere Asymmetries

The wintertime stationary waves are much stronger in the Northern Hemisphere than in the Southern Hemisphere: their r.m.s. amplitudes are stronger (Fig. 1.30) and they produce much stronger poleward heat transports (Fig. 5.16), indicative of a stronger upward EP flux into the middle and upper stratosphere. The poleward TEM circulation is accordingly stronger during DJF than in the Southern Hemisphere during JJA. The stronger upward EP flux (Fig. 8.13) results in a stronger angular momentum sink for the poleward-flowing air in the BDC, accommodating a stronger TEM circulation. A stronger TEM circulation, in turn, favors a weaker equator-to-pole temperature difference and hence a weaker polar night jet.

Another factor comes into play in accounting for the greater strength of the Northern Hemisphere wintertime circulation. Theory and numerical experiments indicate that excessively strong westerlies inhibit the dispersion of Rossby waves.[10] Hence, while the interhemispheric differences must ultimately be attributable to the stronger Northern Hemisphere boundary forcing, they are enhanced by a positive feedback. The stronger Southern Hemisphere polar night jet refracts the upward dispering Rossby waves equatorward, so that they serve as less of a brake on the polar night jet.

9.3.5 The Annual Cycle in the Strength of the Shallow Branch of the BDC

It is evident from Figs. 1.22 and 1.23 that the temperature of the tropical lower stratosphere exhibits an annual cycle over the equator, strongest just above the cold point: the lower stratosphere is colder in the Northern Hemisphere winter.[11] Just below the tropical cold point tropopause, where relative humidities are generally high and water vapor is often condensing in thin cloud layers, the saturation vapor

pressure varies in phase with the annual cycle in cold point temperature. As alternating dry and moist layers ascend into the lower stratosphere, they retain their identities, because above the cold point water vapor behaves as a passive tracer, whose mixing ratio tends to be conserved. The ascending moist and dry layers have been likened to rhythmic sounds on a virtual *tropical tape recorder*,[12] which can be "heard" for a year or longer as they ascend in the BDC, as shown in Fig. 9.9. This figure also serves to confirm the existence of the ascending branch of the BDC and that the ascent rate weakens with height in its shallow branch.

It was shown earlier in this section that the remarkable coldness of the tropical lower stratosphere is maintained by adiabatic cooling in the shallow branch of the BDC. It follows that the annual cycle in lower stratospheric temperature must be the reflection of an annual cycle in the strength of the shallow branch of the BDC. Why should the shallow branch be stronger during the boreal winter (DJF) than during the boreal summer (JJA)? There are two ways of looking at it, both of which are substantiated by the climatological mean statistics shown in Chapter 1 and on the companion web page.

The first explanation is from the perspective of the balance requirement for the conservation of angular momentum. The stronger tropical upwelling during DJF is consistent with and in some sense can be viewed as a consequence of the stronger poleward Lagrangian mass flux across subtropical and midlatitudes in response to the stronger wave driving. This line of argumentation, in which the vertical velocity at some point in the atmosphere – the tropical cold point in this instance – is viewed as being forced by the wave-induced meridional circulation in the layer above is referred to as the *downward control principle*.[13] Given that zonally averaged baroclinic wave activity is strong year-round in the Southern Hemisphere, but substantially stronger in DJF than in JJA in the Northern Hemisphere, it follows that the globally averaged wave-breaking at, say, 30°N/S must be stronger in DJF, when both hemispheres are making strong contributions.

In the second explanation, the role of thermodynamics is more clearly apparent. From the perspective of thermal wind balance, it can be argued that the coldness of the tropical lower stratosphere is proportional to the strength of the globally averaged tropospheric jet streams. Given that the Southern Hemisphere jet stream is strong year-round, whereas the Northern Hemisphere jet stream is much stronger during DJF than during JJA, it follows that the global average of the tropospheric jet stream strength must be higher in DJF. The Northern Hemisphere, with its much larger continents, warms up more during JJA, resulting in reduced globally averaged baroclinity, which is reflected in the weaker tropospheric jet stream, the reduced planetary wave driving, and the weaker shallow branch of the BDC

[9] Zhou et al. (2006).
[10] Charney and Drazin (1961); see also Section 11.2.1.
[11] Reed and Vicek (1969).
[12] Mote et al. (1996).
[13] Haynes et al. (1991).

Water vapor 12°N–12°S

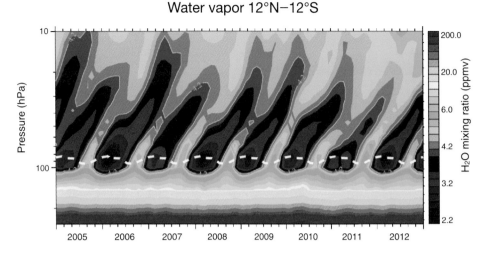

Figure 9.9 Time–height section of zonally averaged water vapor mixing ratio in the equatorial belt (12°N–12°S) as sensed by the microwave limb sounder (MLS) instrument. The dashed line marks the cold point tropopause. Courtesy of William Randel.

during boreal summer.[14] The two explanations offered here are equally valid and both derive from the seasonal cycle in the bottom boundary conditions.

9.3.6 Compensating Tropical and Extratropical Temperature Perturbations

The BDC and the related eddy heat transports redistribute energy meridionally, but they do not directly affect the globally averaged energy budget at any given level. Hence if the tropical lower stratosphere is colder in DJF than in JJA, as discussed in the previous subsection, then the extratropical lower stratosphere must be warmer, and indeed it is warmer, as shown in Fig. 9.10. That the extratropical lower stratosphere is warmer during DJF than during JJA is seen in Fig. 1.29 to be attributable to the coldness of the Antarctic (relative to the Arctic) winter stratosphere.

The compensation between tropical and extratropical temperature perturbations in the lower stratosphere is not limited to the climatology; the year-to-year temperature anomalies shown in Fig. 9.11 exhibit a similar compensation. Notable exceptions are positive temperature anomalies in both tropics and extratropics following the eruptions of El Chichon (1982) and Mt. Pinatubo (1991), both of which injected sulfate gases into the stratosphere that condensed to form long-lasting layers of sulfate aerosols that induced global warming at these levels by absorbing solar radiation.

Exercises

9.9 The shallow branch of the Brewer–Dobson circulation is diabatically forced, and yet it is characterized by the rising of colder air and the sinking of warmer air, as in a thermally indirect circulation. Explain.

9.10 Speculate on how the TEM circulation and associated jet streams in the stratosphere and above might

[14] Jucker and Gerber (2017).

Lower stratosphere temperature

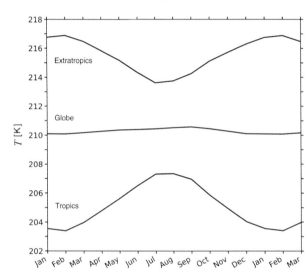

Figure 9.10 Climatological mean lower stratospheric temperature, as inferred from satellite-borne microwave sounders averaged over specified latitude belts as indicated. *Tropics* denotes 30°N–30°S and *extratropics* the remainder of the globe. Courtesy of Hamid A. Pahlavan.

be different from the observed in the absence of an ozone layer.

9.11* (a) Are tropical tape recorder signals observed for phenomena other than the annual cycle? (b) Are tropical tape recorder signals observed for trace species other than water vapor?

9.4 SUDDEN WARMINGS

In the winter hemisphere, there are short intervals during which the BDC is much more intense than usual, causing temperature to rise and column ozone to increase rapidly in its descending branch over the polar cap region and opposing tendencies in its ascending branch in the tropics. To define

Lower stratospheric temperature

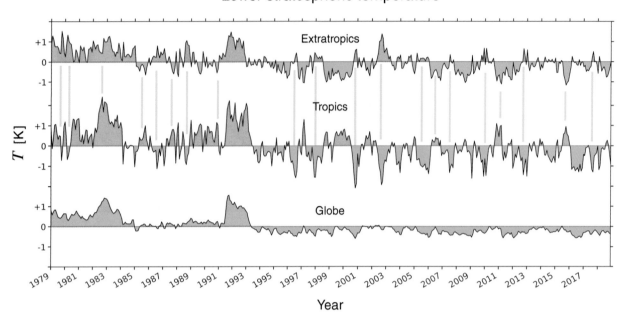

Figure 9.11 Time series of lower stratospheric temperature anomalies, as inferred from satellite-borne microwave sounders averaged over the specified latitude belts as indicated. *Tropics* denotes 30°N–30°S and *extratropics* the remainder of the globe. Courtesy of Hamid A. Pahlavan.

these events, we use the time series of daily 10 hPa geopotential height anomalies (i.e., departures from the seasonally varying climatology) averaged over the Northern Hemisphere polar cap region 60°–90°N. When a stratospheric sudden warming (SSW) event occurs, this index rises rapidly in response to the warming in the underlying layer. Table 9.1 lists the onset dates of the 14 SSW events included in the composite sections and maps shown in this section. All SSW events are in the winter months DJF and early March, even though no seasonal restriction was imposed on the selection of events.

Figure 9.12 shows time–height sections of selected zonally averaged variables for the SSW composite, lagged relative to the onset date on which the index rises above two standard deviations (Day 0). Temperature over the polar cap region begins rising around Day −20, coincident with a strengthening of the poleward eddy heat transport and it rises abruptly by as much as 20 K at the uppermost levels on or around Day −2, coincident with a sharp spike in the transport. At the peak of the warming ∼Day +1, the polar cap region is as warm as the midlatitudes. Radiative cooling acts to damp the positive temperature anomalies. Hence, it is clear that the temperature perturbations are dynamically induced and radiatively damped. The damping is more rapid at the higher levels, where the radiative relaxation times are shorter. The anomalously high ozone concentrations over the polar cap that develop during the warming persist even longer than the temperature anomalies. The meridional extent of the SSW signature is revealed by Fig. 9.13, which shows that the onset of the warming in high latitudes is accompanied by low latitude ascent and cooling that extends deep into the tropics.

The amplified eddies observed during sudden warmings assume a variety of shapes. Some warmings are dominated

Table 9.1 *Onset dates (Day 0) of the events included in the sudden stratospheric warming (SSW) composite shown in Figs. 9.12, 9.13, and 9.15. The onset dates are defined as the dates on which the daily Z_{10} anomaly averaged over 60–90°N rises two standard deviations above its climatological mean value.*

1	2 Mar 1984
2	31 Dec 1984
3	24 Jan 1987
4	6 Dec 1987
5	21 Feb 1989
6	7 Jan 1998
7	15 Dec 1998
8	17 Feb 2001
9	25 Dec 2001
10	18 Jan 2003
11	3 Jan 2004
12	21 Jan 2006
13	23 Jan 2009
14	6 Jan 2013

by zonal wavenumber $k = 1$, some by $k = 2$, and some by the mixture of the two. Figure 9.14 shows a warming dominated by $k = 2$: the top panels show the evolution of the geopotential height field at the 10 hPa level. As the warming proceeds, the polar vortex shrinks as the (bolded) zero-wind line intrudes into the polar cap region. Close to the zero-wind line the air parcel trajectories are non-sinusoidal, and strong

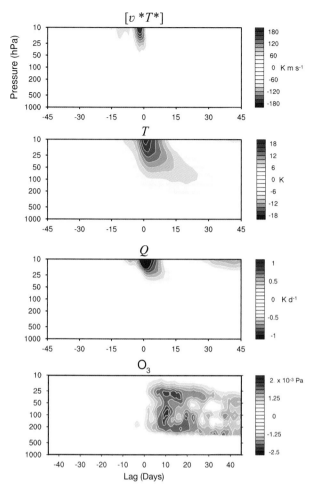

Figure 9.12 Time–height sections of selected zonally averaged variables averaged over the polar cap region poleward of 60°N during stratospheric sudden warmings (SSWs) based on a composite of the events listed in Table 9.1. Day 0 is the onset date at the 10 hPa level. (Top) Eddy heat transport anomalies; (second) temperature anomalies; (third) diabatic heating anomalies; (bottom) ozone partial pressure anomalies. Courtesy of Ying Li.

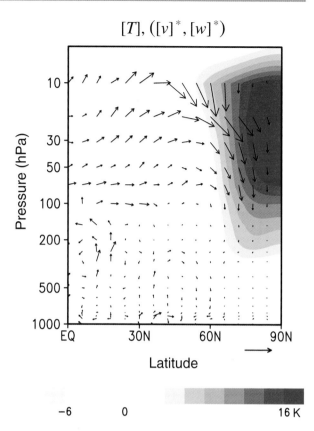

Figure 9.13 The transformed Eulerian mean (TEM) meridional circulation superimposed on temperature based on composites for the difference between SSW events and quiescent periods. Results are based on anomalies about the seasonal cycle. The reference arrow denotes 0.001 m s^{-1} (vertical) and 0.4 m s^{-1} (meridional) at and above the 100 hPa level and 0.005 m s^{-1} (vertical) and 2 m s^{-1} (meridional) below that level. The SSW event composites are based on the 10-day intervals before each of the onset dates listed in Table 9.1 and the quiescent periods correspond to intervals when geopotential height anomalies over the polar cap are two standard deviations lower than normal. Courtesy of Ying Li.

deformations develop, as indicated by the wind and potential vorticity fields on the nearby 850 K isentropic surface shown in the bottom panels of Fig. 9.14. Even in this planetary-scale flow, clusters of (formerly) neighboring air parcels are drawn out into thin filaments that lose their identities as the air within them mixes irreversibly with the surrounding air. Clusters of air parcels that recently resided at low latitudes, as evidenced by their low values of potential vorticity, are entrained into the polar vortex, and air parcels that formerly resided in the vortex are injected into the flow at much lower latitudes. These exchanges of air parcels weaken, and, in the most extreme cases, they completely reverse the strong meridional gradients in potential vorticity and other conservative tracers that existed prior to the warming.

9.4.1 The Role of Wave Driving

The dynamics of sudden warmings can be understood in terms of the governing equations for a geostrophically

balanced, zonally symmetric flow, as discussed in Chapter 7, prescribing the time-varying eddy forcing in accordance with the observed EP fluxes.

The layer below 200 hPa experiences only a slight increase in the upward flux of wave activity during sudden warmings. The big increases relative to quiescent periods are in the layer above, where the upward EP fluxes increase dramatically while they are in progress, as shown in Fig. 9.15. Much of the wave activity is directed almost straight upwards, where wave-breaking occurs, decelerating the polar night jet. The breaking occurs first up around the 10 hPa level, but as the jet progressively weakens the planetary waves encounter their critical levels sooner and sooner, and the wave-breaking occurs at successively lower levels until it reaches the polar tropopause, as shown in Fig. 9.12. The waves disperse equatorward and break at low latitudes, enhancing the upwelling in the Brewer–Dobson circulation and cooling the tropical stratosphere, as shown in Fig. 9.13.

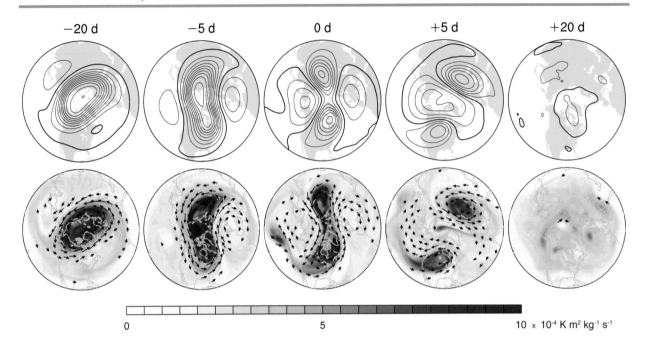

Figure 9.14 (Top) 10 hPa geopotential height and (bottom) potential vorticity and horizontal winds on the 850K surface for the January 23, 2009 event (event #13 in Table 9.1) for lags relative to the onset date as indicated in the titles. In the top panel the contour interval is 250 m and the 30750 m contour is bolded. Courtesy of Ying Li.

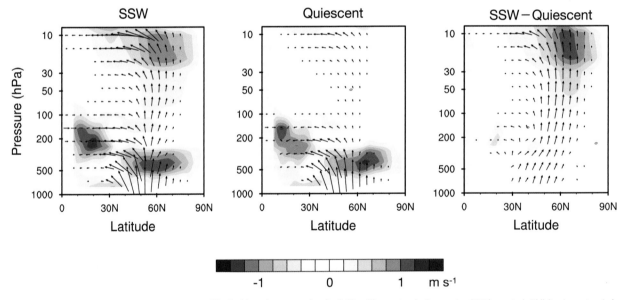

Figure 9.15 The EP flux (vectors) and its convergence (shading) based on composites for (left) sudden stratospheric warming (SSW) events, (middle) quiescent periods, and (right) the difference. The EP flux convergence is divided by the Coriolis parameter f at each latitude and expressed in units m s^{-1}; positive values denote a westward torque and vice versa. The SSW event composites are based on the 10-day intervals before each of the onset dates listed in Table 9.1 and the quiescent periods correspond to intervals when geopotential height anomalies over the polar cap are two standard deviations lower than normal. Courtesy of Ying Li.

During quiescent periods, energy circulates clockwise around the mechanical energy cycle (Fig. 8.1). Despite the strong poleward heat transport, the polar vortex doesn't weaken because K_Z is being replenished by the counter-gradient, poleward transport of westerly momentum. During sudden warmings the momentum fluxes weaken or reverse and energy is rapidly converted from the zonally symmetric flow into the eddies.

9.4.2 Climatology and Timing

Sudden stratospheric warmings can occur at any time during the winter season. By commonly used definitions, events strong enough to qualify as stratospheric sudden warmings occur, on average, about six times per decade in the Northern Hemisphere. Midwinter warming events occur much less frequently in the Southern Hemisphere,

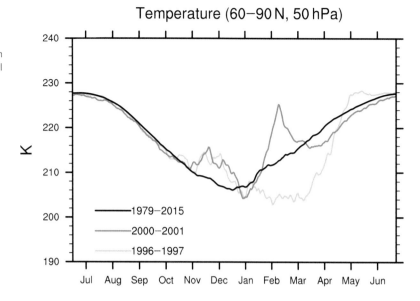

Figure 9.16 The annual march of 50 hPa temperature averaged over the Northern Hemisphere polar cap region 60–90° N: the climatology and two contrasting individual winter seasons, one with and one without a major midwinter warming event. Courtesy of Ying Li.

presumably because of the weaker orographic wave forcing. When a sudden warming event occurs in midwinter, the polar vortex recovers, at least partially, whereas if it occurs close to the end of the winter season, it is referred to as a "final warming." Northern Hemisphere final warmings typically occur within 6 weeks after the spring equinox, whereas Southern Hemisphere final warmings typically occur later, 6–12 weeks after the equinox, by which time the polar vortex has weakened in response to the heating of the ozone layer, rendering it less resistant to the breaking of planetary waves dispersing upward from the troposphere. The austral winter season often ends precipitously with a major warming in which polar temperatures rise from near their lowest values of the year to their highest values of the year in a matter of a few weeks.

Much of what appears to be interannual variability in stratospheric temperature in Fig. 9.11 is related to the occurrence or non-occurrence of stratospheric sudden warmings. Winters without strong midwinter warmings tend to be colder than normal over the polar cap region, as shown in Fig. 9.16.

What are the necessary and sufficient conditions for sudden warming events to occur? The strength of the planetary wave forcing appears to be an important factor. Numerical experiments with a simplified dynamical model of the stratospheric circulation with an idealized steady state forcing suggest that if the forcing is weak, the time-dependent solution is steady. When the forcing is increased, the model yields a quasi-periodic, vacillating solution suggestive of a sequence of minor (i.e., reversible) SSW events interspersed with quiescent periods.[15] That midwinter SSW events occur much more frequently in the Northern Hemisphere, with its stronger orography, than in the Southern Hemisphere attests to the importance of the wave forcing. It has recently been hypothesized that it is the ratio of the finite amplitude wave activity to the (zonal wind) speed of the jet that determines

the threshold beyond which a major (irreversible) warming becomes inevitable.[16]

The arguments about necessary and sufficient conditions do not address the question why SSW events occur when they do. In the SSW composites shown in Fig. 9.12, enhanced heat transports and warming set in about two weeks in advance of the peak of the warming. It is suggestive of a preconditioning of the flow that sets the stage for the major event that is to follow, and indeed, this has been one of the recurrent themes in the SSW literature. Anticipating SSWs and identifying their precursors has proven difficult because their development is influenced by the amplitude and configuration of both the wave forcing and the zonal mean flow, and because positive feedbacks render their development and evolution highly sensitive to the initial conditions.

Exercises

9.12 Describe the Eulerian mean meridional circulations during a sudden stratospheric warming event.

9.13 Interpret the dynamics of stratospheric sudden warming events from a potential vorticity perspective.

9.14 Describe how the mechanical energy cycle is perturbed by a sudden warming. Take into account the weakening of the polar night jet, the short-lived burst in eddy kinetic energy, and the impact of the dynamically induced temperature changes upon the radiative heating. Assume that the warming is just strong enough to obliterate the polar night jet.

9.15 Compare winters with and without sudden warmings with respect to time mean (a) diabatic heating rates over the polar cap region and (b) the strength of the Brewer–Dobson circulation.

---

[15] Holton and Mass (1976).

[16] Nakamura et al. (2020).

10 Wave–Mean Flow Interaction in the Tropical Stratosphere

with Hamid A. Pahlavan

From a global perspective, the dynamics of wave–mean flow interaction in the stratosphere is dominated by Rossby waves, as described in the previous chapter. However, the tropical stratosphere is a notable exception, which merits a chapter of its own. Almost all the wave activity converging into the tropics from higher latitudes is absorbed by the time it reaches 15°N/S. In the absence of Rossby waves, the main source of wave activity in the tropical stratosphere is planetary waves and gravity and inertio-gravity waves dispersing upward from below. The upward EP fluxes are smaller than the lateral fluxes associated with Rossby waves at higher latitudes, but they are nonetheless locally important because of the smallness of the Coriolis parameter. Interactions between these upward dispersing waves and the mean flow give rise to the remarkable *quasi-biennial oscillation* (QBO), which dominates the zonal wind climatology of the tropical stratosphere.

The first section of this chapter documents the structure and time dependence of the QBO, as represented in a state-of-the-art reanalysis dataset, ERA5. The second section presents observational evidence of the existence of planetary waves in the equatorial belt. They can be seen propagating zonally – some eastward and others westward – in time–longitude sections and systematically downward in time–height sections. The waves are continually dispersing upward from the troposphere. The prevalent wave species is shown to be dependent on whether the zonal mean flow is westerly or easterly. The reader is then introduced to the linear theory of *equatorially trapped planetary waves* on an equatorial beta-plane, as represented by a simplified set of governing equations referred to as the *shallow water wave equations*. Dispersion curves and structures for the observed waves are compared with those in the solutions.

The third section extends the TEM formulation in Sections 8.2 and 9.3.2 to include the eddy transports by equatorially trapped planetary waves – Rossby, Kelvin, mixed Rossby–gravity waves, inertio-gravity waves – and the smaller scale, higher frequency "pure" gravity waves, all of which play important roles in wave–mean flow interaction in the tropical stratosphere.

In the fourth section it is shown how wave–mean flow interaction in the tropical stratosphere gives rise to the QBO. The basic theory for how it works, which dates back 50 years,[1]

sounded fanciful to some at the time it was proposed.[2] It was not until recently that the spatial resolution of the observations has become sufficient to verify that the theory really works. The final section describes the QBO's imprint upon the general circulation beyond the confines of the tropical stratosphere by way of its influence on the wintertime stratospheric circulation at higher latitudes.

10.1 THE ZONAL WIND CLIMATOLOGY

The annual mean zonal wind in the tropical stratosphere (Fig. 1.19) is from east to west, increasing with height, reflecting the easterly jets centered at ~15° latitude in the summer hemisphere. It exhibits maxima at ~15°N/S with speeds on the order of 10 m s^{-1} at the 30 hPa level. In the equatorial belt, seasonal variations in zonal wind are small; rather than conforming to the rhythm of the seasons, the equatorial, zonally averaged zonal wind marches to its own, less regular drum beat, the QBO.

Figure 10.1 shows a time–height section of lightly low-pass filtered zonal wind data for a grid point over the equator. This particular point is on the Date Line, but sections for other equatorial grid points are virtually identical, apart from the timing of the weak perturbations on the day-to-day timescale. The feature of interest is the downward propagating westerly shear zone, which is a reflection of low-frequency variations in the zonally averaged zonal wind component [u]. The higher frequency oscillations mark the passage of equatorial waves discussed in the next section.

To document the slow evolution of the equatorial [u] field, Fig. 10.2 shows an equatorial time–height section extending over a 20-year interval. Fluctuations with zero-to-peak amplitude range up to 28 m s^{-1} at the 25 km (25 hPa) level. Successive easterly and westerly wind regimes, separated by strong shear zones, propagate downward at an average rate of ~1 km per month. The westerly shear zones are sharper and propagate downward somewhat faster than the easterly shear zones. Apart from a short lapse in 2016–2017, the variations have been quasi-periodic with similar vertical structures recurring at intervals of ~27 months: hence

---

[1] Lindzen and Holton (1968); Holton and Lindzen (1972).

[2] A biographical memoir of James R. Holton. National Academy of Sciences, 2014. www.nasonline.org/member-directory/deceased-members/54439.html

$$u$$

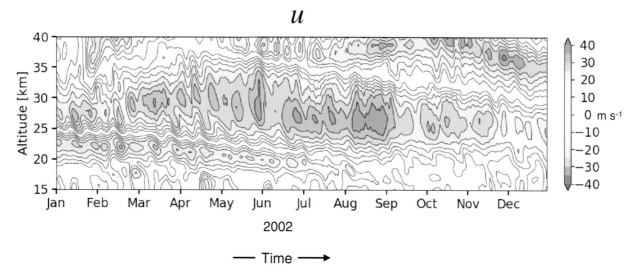

Figure 10.1 Time–height section of lightly lowpass filtered zonal wind for a grid point over the equator. This reference grid point is on the Date Line but sections for other equatorial grid points are virtually identical, apart from the timing of weak perturbations on the day-to-day time-scale. The feature of interest is the downward propagating westerly wind regime, preceded by a westerly shear zone with easterlies below and westerlies above, and followed by an easterly shear zone. These features are associated with low-frequency variations in the zonally averaged zonal wind component $[u]$. The higher frequency features are a reflection of the passage of equatorial waves.

$$[u]$$

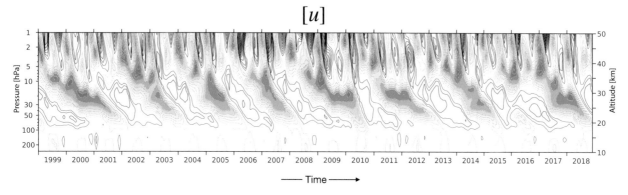

Figure 10.2 Time–height section of monthly mean zonally averaged zonal wind $[u]$ showing the quasi-biennial oscillation (QBO). From Pahlavan et al. (2021a). © American Meteorological Society. Used with permission.

the name quasi-biennial oscillation.[3] This phenomenon is responsible for the large r.m.s. amplitude of the $[u]$ perturbations in the equatorial belt in Fig. 1.31. It is the dominant mode of $[u]$ variability throughout the tropical lower stratosphere. The downward propagating easterly and westerly wind regimes are rapidly attenuated as they approach the cold point tropopause $\sim$17 km (100 hPa). Above 35 km, the QBO coexists with a semiannual oscillation (SAO). Most westerly shear zones of the QBO originate as leading edges of westerly wind regimes of the SAO. This relationship is reflected in the histogram of the periods of individual QBO cycles, which exhibits peaks centered on 24 and 30 months, with a pronounced dip in between.[4]

Not long after the discovery of the QBO, it was hypothesized that the observed downward propagation of easterly

and westerly shear zones over the equator is due to vertical advection, but it is now well established that the mean vertical velocity in the tropical lower troposphere is upward, not downward, in association with the rising branch of the Brewer–Dobson circulation. The descending wind regimes in the QBO coexist with the ascending moist and dry layers of the "tropical tape recorder" (Fig. 9.9).[5] Since the downward propagation is not associated with downward mass transport, it must be in response to downward propagating sources and sinks of westerly momentum.

Meridional cross-sections of zonally averaged zonal wind and temperature are shown in Fig. 10.3 for descending westerly and easterly shear zones of the QBO. On the timescale of the QBO, the temperature tendency term $d[T]/dt$ in the thermodynamic energy equation is negligibly small, so the adiabatically and radiatively induced temperature tendencies must balance,

---

[3] The QBO was discovered simultaneously by Veryard and Ebdon (1961) and Reed et al. (1961).
[4] See Fig. 2a of Kuai et al. (2009).

[5] See Fig. 2 of Pahlavan et al. (2021a).

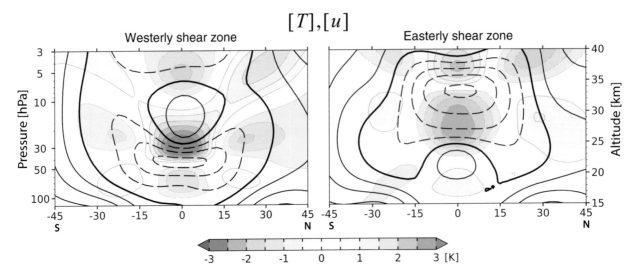

Figure 10.3 Composite meridional cross-sections showing the structure of the QBO at the times when westerly and easterly shear zones are descending through the 25 hPa level. Colored shading indicates temperature anomalies. Contours indicate the zonal wind (the total field, not the anomalies), contour interval 7.5 m s^{-1}, westerlies solid, easterlies dashed, zero contour bolded. From Pahlavan et al. (2021a). © American Meteorological Society. Used with permission.

$$[Q_R] = [w]\,(\Gamma_d - \Gamma)\,. \qquad (10.1)$$

That the QBO-related $[T]$ and geometric vertical velocity $[w]$ perturbations in Fig. 10.4 are linearly congruent and out of phase with one another indicates that the temperature perturbations are adiabatically induced and radiatively damped.

The QBO-related temperature anomalies are concentrated in the shear zones between the easterly and westerly wind regimes. The easterly regimes extend all the way to $\sim28°$N/S, whereas the westerly regimes are confined to within $\sim15°$ of the equator. The temperature $[T]$ and (not shown) geopotential height $[Z]$ anomalies are strongest on the equator and exhibit weaker centers of opposing sign at $\sim20°$N/S. The $[u]$ and $[T]$ fields are in quadrature with one another (i.e., a quarter of a wavelength out of phase), with the warmest air below the westerly maxima, as required for thermal wind balance, which prevails to within a few kilometers of the equator.[6]

On the basis of the continuity equation, it can be inferred that the QBO-related mean meridional circulations consist of downward propagating cells with equatorward flow in the westerly wind regimes and poleward flow in the easterly regimes, and that is, indeed, the case. The vertical velocity perturbations observed in association with the QBO, which range up to a few tenths of a millimeter per second (or tens of meters per day), are as large as the mean rate of ascent in the BDC at these levels (e.g., see Fig. 9.3). Hence, at the times when westerly shear zones are descending through the lower stratosphere, the shallow and deep branches of the BDC are separated by a layer in which the air is stationary or even slowly subsiding, as shown in the bottom panels of Fig. 10.4. Under these circumstances, tropospheric tracers entering the lower stratosphere in the shallow branch of the BDC are

deflected around the descending westerly shear zone. The QBO-related shear zones weaken as they approach the cold point ~17 km, allowing air in the shallow branch of the BDC to break through and ventilate the middle and upper stratosphere. Hence, the strength and structure of the BDC varies not only seasonally, but also from one year to the next, and so does the chemical makeup of the air in the tropical stratosphere.

Exercises

10.1 How could it be that the QBO was discovered on the basis of data from just one or two equatorial stations?

10.2 (a) Show that the thermal wind equation on the equator is of the form
$$\frac{\partial[u]}{\partial z} = -\frac{R}{H\beta}\frac{\partial^2[T]}{\partial y^2},$$
where R is the gas constant, H is the scale height and $\beta = df/dy$. (b) The QBO-related zonal wind perturbations are $\sim15°$ of latitude in half-width, the shear zones are ~5 km deep and the difference in zonal wind speed between easterly and westerly wind regimes is 35 m s^{-1}. Estimate the amplitude of the QBO-related temperature perturbations and compare the result with Fig. 10.3.

10.3 Prove that the zonal wind and temperature fields in the QBO are in thermal wind balance in the free atmosphere to well within a degree of the equator.

10.4 Show that the greater meridional width of the $[u]$ anomalies in the easterly wind regimes relative to westerly regimes is consistent with the annual mean zonal wind climatology shown in Fig. 1.19.

[6] Reed et al. (1961).

Figure 10.4 As in Fig. 10.3 but including additional fields. (Top panels) QBO-related [T] and TEM mean meridional circulation anomalies and (bottom panels) the total TEM circulation, including the climatological mean. The longest horizontal components of the arrows correspond to 0.3 m s^{-1}, while the longest vertical components correspond to 1 mm s^{-1}. Total zonal mean zonal wind [u] is contoured in the bottom panels; contour interval 7.5 m s^{-1}. Negative contours are dashed and the zero contour is bolded. From Pahlavan et al. (2021a). © American Meteorological Society. Used with permission.

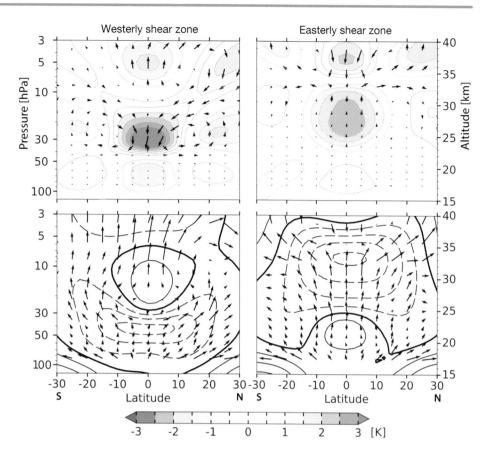

10.5 The meridional cross sections in Fig. 10.3 show the total fields. Describe how sections showing anomaly fields; that is, departures of [u] from the seasonally varying climatology, are different in appearance.

10.6 Estimate the ratio of the zonal kinetic energy K_Z to the zonal available potential energy A_Z in the QBO.

10.7 Show that the QBO must be wave driven through the energy conversion term C_K in the mechanical energy cycle.

10.8 Explain why the QBO does not show up prominently in time series of atmospheric angular momentum or in time series of globally averaged temperature at stratospheric levels.

10.2 EQUATORIALLY TRAPPED PLANETARY WAVES

Throughout the early history of large-scale atmospheric dynamics, the term *wave* was used almost exclusively in reference to *Rossby wave*. That changed toward the end of the 1960s with the discovery of a family of equatorially trapped planetary waves in the lower stratosphere.[7] It was recognized almost immediately that these waves are most

[7] Yanai and Maruyama (1966); Wallace and Kousky (1968); Takayabu (1994).

prominent within layers of strong vertical shear of the zonally symmetric background flow and that they play an important role in wave–mean flow interaction in connection with the QBO.

Figure 10.5 shows time–height sections of the eddy or wave (*) component of the zonal wind u and temperature T fields over the equator during a 3-month interval that corresponds to a time when a westerly shear zone with strong, positive $\partial[u]/\partial z$ was descending through the lower stratosphere. The wave fields are lightly smoothed using a 5-day running mean filter. It is evident that downward propagating waves with periods on the order of 10–15 days are present in both fields, and that they are particularly well organized within the westerly shear zone.

Figure 10.6 shows time–longitude sections of the eddy component of T over the equator at the 50 hPa (20 km) level for selected seasons in successive calendar years, one in which the 50 hPa level is in a westerly shear zone (left) and the other when it is an easterly shear zone (right). Both sections are dominated by bands of color that slope upward toward the right, indicative of eastward propagating waves with periods on the order of 15 d, zonal wavelength $k = 1, 2$, and eastward phase speeds on the order of +30 m s^{-1}. That the phase speed is independent of frequency indicates that the waves are nondispersive. Corresponding sections for the meridional wind component v_{50} are shown in Fig. 10.7. These waves have a zonal wavenumber $k \sim 4$ and a period of 4–5 d. The waves in the westerly shear zone

u^*, $[u]$

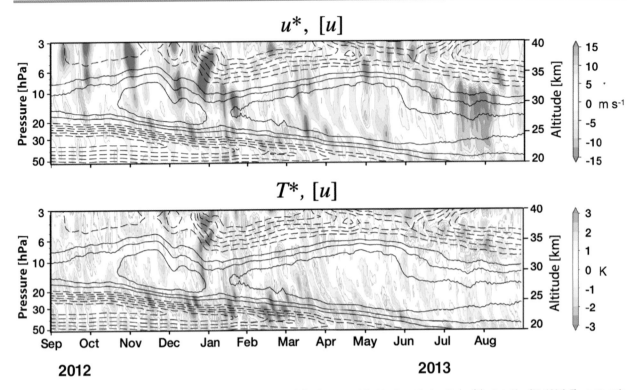

T^*, $[u]$

Figure 10.5 Time–height sections of eddy zonal wind u^* and temperature T^* for the equatorial grid point at the longitude of the Date Line (0°, 180°). The contours in both panels indicate zonally averaged zonal wind $[u]$, contour interval 5 m s^{-1}; positive values (i.e., westerlies) solid, negative values (easterlies) dashed, zero contour not shown. The colored shading indicates the eddy (or wave) component of (top) zonal wind u^*; (bottom) temperature T^*. From Pahlavan et al. (2021a). © American Meteorological Society. Used with permission.

Figure 10.6 Equatorial time–longitude sections of departures of temperature from the zonal average at 50 hPa (T_{50}) in (left) westerly and (right) easterly shear zones of the QBO at the 50 hPa level. These sections are dominated by eastward propagating Kelvin waves with phase speeds ~30–35 m s^{-1}. From Pahlavan et al. (2021b). © American Meteorological Society. Used with permission.

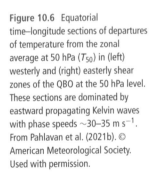
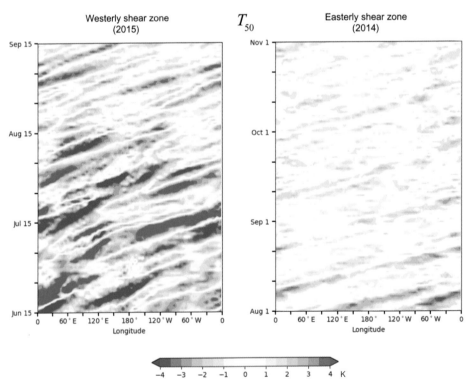

(left panel) do not have a preferred direction of propagation, whereas those in the easterly shear zone (right panel) are westward propagating with a phase speed of about 25 m s^{-1}.

Understanding the features in these sections will require consideration of the linear theory of the equatorial wave modes introduced in the next section.

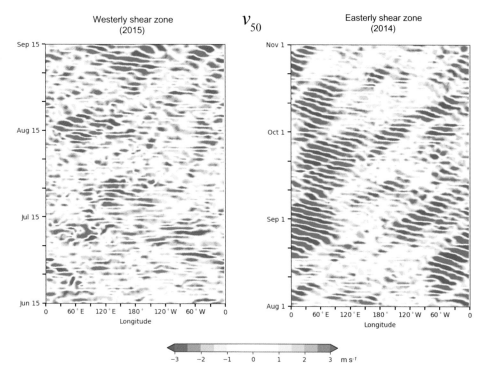

Figure 10.7 Equatorial time–longitude sections of departures of v_{50} from the zonal average in (left) westerly and (right) easterly shear zones of the QBO. The section for the easterly shear zone of the QBO is dominated by westward propagating mixed Rossby–gravity waves with a westward phase speed of $\sim$27 m s^{-1}. From Pahlavan et al. (2021b). © American Meteorological Society. Used with permission.

10.2.1 The Shallow Water Wave Equations

Equatorially trapped planetary waves are solutions of Laplace's linearized, shallow water wave equations on a rotating sphere, which may be written in Cartesian form as

$$\frac{du}{dt} = -g\frac{\partial h'}{\partial x} + fv \tag{10.2}$$

$$\frac{dv}{dt} = -g\frac{\partial h'}{\partial y} - fu \tag{10.3}$$

$$\frac{\partial h'}{\partial t} = -h_o \nabla \cdot \mathbf{V}, \tag{10.4}$$

where h_o is the resting depth of the fluid and h' refers to the perturbations about it.

For application to equatorial waves, the Eqs. (10.2)–(10.4) can be simplified by writing $f = \beta y$; that is, by representing the spherical geometry in terms of an *equatorial beta-plane*. The solutions, referred to as *equatorially trapped wave modes*, are sinusoidal in longitude with zonal wavenumber k, the number of zero crossings on a latitude circle, and in latitude by Hermite polynomials, bearing the index n, the number of zero crossings in the pole-to-pole meridional profile of the v perturbations. Odd-numbered modes, in terms of n, are equatorially symmetric in the variables u^*, T^*, and Z^*, and even-numbered modes are equatorially antisymmetric.[8]

---

[8] The most influential of the theoretical papers on equatorially trapped waves is Matsuno (1966). Other deviations published around the same time are those of Rosenthal (1965) and Rattray and Charnell (1966).

10.2.2 The Shallow Water Wave Solutions

The family of waves supported by Eqs. (10.2)–(10.4) includes equatorially trapped synoptic and planetary-scale *Rossby waves*, higher frequency *inertio-gravity waves*, as well as a *Kelvin wave* and a *mixed Rossby–gravity wave*. For waves with frequencies much higher than f, the solutions are indistinguishable from pure *gravity waves* with horizontal wavelengths ranging over several orders of magnitudes, propagating horizontally with phase speed

$$c = \sqrt{gh_o}. \tag{10.5}$$

Figure 10.8 shows the dispersion diagram (i.e., a plot of frequency ω versus zonal wavenumber k) for the wave solutions. The resting depth plays a critical role in the scaling of the ω and k axes. In this plot the axes are nondimensional, scaled as described in the figure caption. Positive values of k correspond to eastward propagating waves and negative values to westward propagating waves. Straight lines radiating out from the origin (i.e., lines of constant ω/k) represent specific phase speeds: the steeper the slope the higher the phase speed. The gravest mode, which corresponds to the $n = -1$ solution, is the *Kelvin wave*, which behaves as an eastward propagating gravity wave with phase speed $\sqrt{gh_o}$. The next gravest $n = 0$ solution is referred to as the *mixed Rossby–gravity* (MRG) wave. It has a dual identity, behaving like a westward propagating Rossby wave at the lower frequencies and like an eastward propagating gravity wave at the higher frequencies. At the higher frequencies ($\omega \gg f$), the eastward propagating segment of the $n = 0$ mode behaves as an eastward propagating gravity wave and the Kelvin wave is a member of that family too. IG

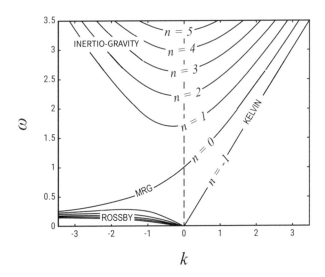

Figure 10.8 Dispersion diagram for equatorially trapped wave modes. Zonal wavenumber on the *x* axis is nondimensionalized by dividing it by $(c/\beta)^{1/2}$ and frequency on the *y* axis is nondimensionalized by dividing it by $(c\beta)^{1/2}$. For $c = 50$ m s^{-1} and $\beta = df/dy$ on the equator equal to 2.29×10^{-11} m^{-1} s^{-1}, $k = 1$ corresponds roughly to 9300 km or zonal wavenumber 4.3 at the equator, and $\omega = 1$ to a period of ∼3 days. Adapted from Matsuno (1966) and Cane and Sarachik (1976).

waves may propagate zonally in either direction, whereas Rossby waves only propagate westward. Kelvin waves are equatorially symmetric with *v* identically equal to zero, and MRG waves are antisymmetric in the u, T, and Φ fields and symmetric in the *v* field.

10.2.3 Vertical Structure

To properly interpret the vertical structure of these waves, it is necessary to understand the distinctions between what are referred to as *external* and *internal wave modes*. In an incompressible fluid, the horizontal pressure gradients in external waves are exclusively due to gradients in the height of the free surface, whereas those in internal waves like the ones considered in this section are due to horizontal density gradients in a continuously stratified fluid. Imposing a rigid lid suppresses the external modes but not the internal modes, while specifying that density be constant has the opposite effect.

External modes feel the full effect of the Earth's gravity *g*, whereas the internal modes feel a "reduced gravity" $g\rho'/\rho_o$, where ρ' refers to the density perturbations in the waves and ρ_o is the mean density at that level. Because of the much weaker restoring force, the internal wave modes exhibit much slower horizontal phase speeds than external wave modes. Unlike external modes, which are inherently barotropic and propagate only horizontally, internal modes are baroclinic and propagate both horizontally and vertically. Their horizontal structures and horizontal dispersion characteristics are well described by the same shallow water

equations, but with fluid depth h_o replaced by an equivalent depth h_e, which is smaller than h_o.

The shallow water wave equations are also applicable to waves in a compressible fluid, but some modifications are required. There exists an analogous distinction between *external* and *internal* modes, but for the external mode of a compressible, non-rotating atmosphere, the mean depth h_o in Eqs. (10.2)–(10.4) is replaced by γH, where $\gamma = c_p/c_v = 1.4$ and $H = RT/g$ is the scale height (∼8 km). It follows that the external gravity wave in a compressible fluid travels at the speed of sound in an isothermal atmosphere.[9] The perturbation density field in the external mode is three dimensional, but it is constrained to be equivalent barotropic; that is, $\rho'(x, y)$ is proportional to $p(x, y)/RT$. To match the 30–35 m s^{-1} phase speed of the Kelvin waves in Fig. 10.6, which is about one-tenth of the speed of sound, h_e would have to be on the order of 100 m, ∼1% of that of the external mode.

Whereas for an incompressible fluid there exists an infinite set of discrete vertical modes with well-defined equivalent depths, for an atmosphere – a compressible fluid in a semi-infinite domain – there exists a continuous spectrum of internal modes, whose distribution of equivalent depths is determined by the properties of the medium through which they propagate and by the depth scale of the heating that forces them.[10] For a given pressure perturbation, the vertical velocity perturbations are much larger in an internal wave than those in an external wave and the vertical transports of westerly momentum are accordingly much larger.

10.2.4 Wavenumber–Frequency Spectra

To relate the observations to the dispersion diagram, it is informative to perform spectrum (Fourier) analysis of scalar fields such as *u*, *v*, or *T* defined on a latitude circle. The following protocol yields a two-sided wavenumber-frequency (k, ω) spectrum in the same format as Fig. 10.8.[11] At each time step, grid-point values of a variable are first decomposed into symmetric (like-signed) and antisymmetric (opposite signed) components about the equator by summing or differencing the values at corresponding pairs of latitudes and dividing by two. The time series of these values along a given latitude band are then converted into eastward and westward zonal wavenumbers (*k*) by means of a complex Fourier transform, and a further complex Fourier transform of these transformed coefficients in the time domain is then used to create a two-sided wavenumber spectrum for each frequency, whose right side (*k*+) represents power of eastward propagating waves and whose left side (*k*−) represents power of westward propagating waves.

The preferred modes of variability in the two-sided spectrum can be accentuated by dividing the two-sided spectrum

[9] Vallis (2017).
[10] Salby and Garcia (1987); Wu et al. (2000).
[11] Hayashi (1974) refined by Takayabu (1994) and Wheeler and Kiladis (1999).

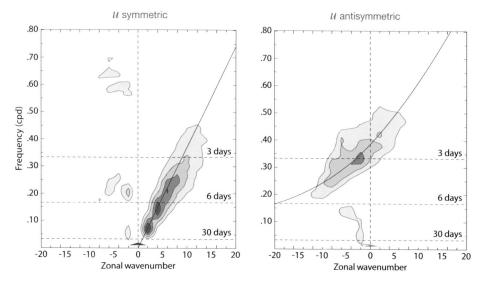

Figure 10.9 Two-sided wavenumber–frequency spectra for the 50 hPa zonal wind component: (left) symmetric and (right) antisymmetric about the equator, computed on each latitude circle and averaged over the domain 15°N to 15°S. The values plotted are dimensionless ratios of the raw spectrum to the smoothed spectrum. Positive zonal wavenumbers correspond to eastward propagating waves and negative ones to westward propagating waves. The black line in the left panel is the dispersion curve for Kelvin waves and the one in the right panel is for MRG waves, both in a shallow water wave equation model with a 120 m equivalent depth ($c = 34$ m s^{-1}). Contour interval 0.1 starting at 1.2.

at each wavenumber and frequency by a highly smoothed "background spectrum" constructed from the same data.[12] These spectra are often averaged over a latitude band. For example, Fig. 10.9 shows two-sided spectra for symmetric and antisymmetric equatorial zonal wind at the 50 hPa level u_{50} in the band 15°N–15°S. The equivalent depth for the dispersion curves shown in the diagram is chosen empirically, to align them as well as possible with the observed variance pattern in the (ω, k) domain. The theoretical curves in Fig. 10.8 fit the lower stratospheric data best for $h_e = 120$ m. This value is used in scaling the theoretical dispersion curves in Fig. 10.9.

10.2.5 Kelvin and Mixed Rossby–Gravity Waves

The two gravest modes, the Kelvin and MRG waves are set apart from the higher modes of IG waves in the sense that the zonal wind component is in geostrophic balance with the pressure field. In the Kelvin wave there is no meridional wind component and in the MRG wave, the relationship between meridional wind component and the pressure field is analogous to that in IG waves. For the planetary waves (i.e., the waves with low zonal wavenumbers), the frequency of the $n = -1$ Kelvin wave, whose zonal wind component is in geostrophic balance, is substantially lower than that of the corresponding unbalanced $n \geq 1$ IG wave in which neither wind component is in geostrophic balance. The same is true of the MRG wave, whose phase speed is intermediate between those of Rossby and IG waves. These two modes are clearly discernible in the pressure, wind, and temperature fields by virtue of their relatively low frequencies and their distinctive structure on the equator.

The spectrum of the equatorially symmetric component of zonal wind at the 50 hPa level, shown in the left panel of Fig. 10.9, exhibits enhanced variance relative to the background spectrum along a line of constant phase speed $c = 34$ m s^{-1}, in line with the dominant eastward phase speed of the Kelvin disturbances shown in Fig. 10.6. The spectrum of the corresponding antisymmetric component (right panel) exhibits enhanced variance along the dispersion curve for the MRG wave. The band of enhanced variance follows the dispersion curve for $n = 0$ toward higher frequencies, well into the domain in which they behave as eastward propagating IG waves. However, it is evident from Figs. 10.7 and 10.9 that the lower frequency, westward propagating MRG waves are dominant. Unlike Kelvin waves, the $n = 0$ waves are dispersive with a positive group velocity $\partial \omega / \partial k$, indicative of eastward energy dispersion, as evidenced by the fact that their dispersion curve cuts across lines of constant phase speed. Idealized depictions of equatorially trapped Rossby, Kelvin, MRG, and IG waves as they appear on pressure surfaces and in vertical cross sections can be found in papers, textbooks, and lecture notes dating back to Matsuno.[13]

Figure 10.10 shows the structure of the Kelvin wave: in the top and bottom left panels based on linear theory and in the middle and bottom right panels constructed empirically by regressing various fields upon a designated reference time series (see Appendix F for details). The Kelvin wave is the most energetic of the equatorially trapped modes. Its horizontal velocity field involves only the zonal wind component: v^* is everywhere equal to zero. Geopotential height, zonal wind, and (upward) vertical velocity vary in-phase with one another, indicative of an upward flux of westerly

[12] Wheeler and Kiladis (1999).

[13] Matsuno (1966).

Figure 10.10 Horizontal structure of the equatorially trapped Kelvin wave: (top) based on theory, where contours represent geopotential and arrows represent horizontal velocity, and (middle) as reconstructed empirically at 50 hPa. (Bottom) Vertical structure in the equatorial plane as viewed looking toward the north: (left) based on theory – HIGH and LOW refer to Z^* extrema, black arrows to extrema in u^* and w^*, and colored shading to temperature; wide gray arrows indicate the direction of the phase propagation (downward and eastward); (right) as reconstructed empirically based on regression analysis, showing temperature only. The bottom left panel is also applicable to pure gravity waves.

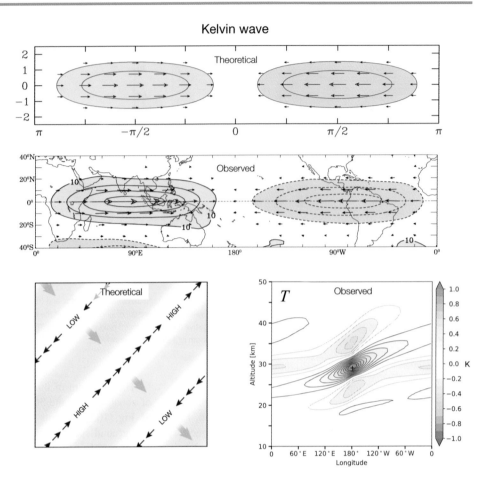

momentum, geopotential and wave activity. The zonal wind field is in geostrophic balance with the height field. The temperature and height perturbations are in quadrature, with highest T^* quarter wavelength to the east of and below the highest Z^*, u^*, and vertical velocity w^*. It is evident from Fig. 10.6, that Kelvin waves are nondispersive and propagate eastward with a phase speed $\sim 30–35$ m s^{-1}, which corresponds to a period of ~ 15 d for zonal wavenumber $k = 1$ and correspondingly shorter periods for the higher zonal wavenumbers. Like all gravity and inertio-gravity waves that disperse upward in a stratified fluid, their wave fronts propagate downward, as shown in the bottom left panel of Fig. 10.10.

Figure 10.11 shows the structure of the MRG wave, the next most energetic of the equatorially trapped modes. In the horizontal plane it consists of a series of alternating westward propagating clockwise and counterclockwise gyres, centered on the equator. The v^* perturbations are equatorially symmetric, while the u^*, w^*, Z^*, and T^* perturbations are equatorially antisymmetric, with u^*, w^*, and Z^* in phase with one another and in quadrature with T^*. Hence, as in Kelvin waves, the flux of westerly momentum and geopotential in the MRG wave are upward. Along their outer flanks, the flow is nearly geostrophic, whereas equatorward of the strongest Z^* perturbations, the flow is dominated by the v^* component, which varies in quadrature with Z^*. The T^* and

v^* perturbations vary in-phase, with highest temperatures in the poleward flow. Hence, the MRG wave produces a poleward eddy heat transport. Since the flow in the waves is nearly adiabatic, the temperature tendencies induced by the meridional convergence of the eddy heat transport must be cancelled by the tendencies induced by the Stokes drift. By imagining the trajectory of an air parcel in the meridional plane as a wave passes by it, moving from east to west, it is readily verified that the mean meridional circulation is in the same sense as in the Ferrel cell, with air parcels ascending at a higher latitude in the highs and descending at a lower latitude in the lows in the geopotential field. As in the Ferrel cell, this is an effective mechanism for the downward transport of westerly momentum: in fact, it is stronger than the upward eddy flux $[u^*w^*]$ and thus determines the direction of the net vertical transport by the MRG wave. The wave fronts propagate downward, in agreement with theory, as shown in the bottom left panel of Fig. 10.11.

10.2.6 Rossby Waves

The family of equatorially trapped, westward propagating Rossby waves comprises meridional modes with $n = 1$ and higher. In all Rossby waves, the wind and geopotential height fields are in geostrophic balance: v^*, T^* and Z^* vanish on the equator, while u^* is large in the equatorial belt and out

Figure 10.11 Horizontal structure of the westward propagating mixed Rossby–gravity wave. (Top left) Based on theory, where contours represent Z, arrows represent horizontal flow, and the patches of color represent temperature, gold warm and blue cool. (Top right) Reconstructed empirically, by regressing Z^*_{50}, u^*_{50}, v^*_{50}, and T^*_{50} fields upon the time series of v^*_{50} over the equator at $60°$W. The bottom panels show the vertical structure north of the equator (as viewed looking toward the north). (Left) Based on theory – HIGH and LOW refer to Z^* extrema, arrows to extrema in u^* and w^*, $\otimes$ and $\odot$ to the strongest poleward and equatorward winds, and colored shading to temperature. (Right) As reconstructed empirically based on regression analysis, showing temperature only. The thick gray arrows indicate the direction of phase propagation (downward and westward).

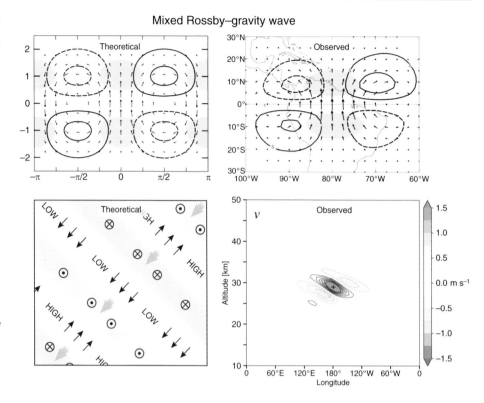

Mixed Rossby–gravity wave

of phase with Z^*. The odd-numbered modes $n = 1, 3, \ldots$ are equatorially symmetric, while the even-numbered modes are antisymmetric. When forced from below, their axes tilt westward with height such that they produce a poleward heat transport analogous to their extratropical counterparts. Because their zonal and meridional wind components are both in geostrophic balance, they exhibit lower frequencies and slower phase speeds and group velocities than the other members of the family of equatorially trapped waves. They are not clearly discernible in the lower stratosphere, likely because convective heating is less efficient at producing them and because their slower upward dispersion renders them more susceptible to damping.[14]

10.2.7 Gravity Waves and Inertio-Gravity Waves

State-of-the art reanalyses with hourly time steps are capable of resolving inertio-gravity waves and some of the smaller scale gravity waves in the topmost segment of the dispersion diagram (Fig. 10.8). The composite fields shown in Figs. 10.12 and 10.13 provide an indication of the three-dimensional structure of these waves. They were constructed by compositing the fields in the ERA5 Reanalysis anomalies following "downwelling events" in time series of tropospheric vertical velocity (the sum of standardized w_{300} and w_{700}) at reference grid points along the equator at $1°$ longitude increments. (A justification for basing the composite on downwelling events in the tropospheric vertical velocity field is provided in the extended caption of Fig. 10.12 in Appendix

F.). Figure 10.12 shows the geopotential, wind, and vertical velocity fields at the 50 hPa level 8 hours after the events. The vertical structure is identical to that of the Kelvin wave shown in the bottom left panel of Fig. 10.10, with in-phase perturbations in vertical and radial motion and geopotential. The patterns in this composite exhibit almost perfect axial symmetry, but there are places and times when the outward phase propagation is biased toward one direction. Figure 10.13, based on the temperature field, provides further information on how the waves develop and propagate. The left panels show lagged-composite maps for T at the 50 hPa level and the right panels show longitude–height sections. The waves can be seen to radiate outward from the respective reference grid points, initially as disks and later as rings. The wave axes slope outward with increasing height, so that as the waves propagate outward, they also propagate downward. The group velocity is upward and outward, causing the sloping bands of positive and negative anomalies in the section to lengthen with time. Amplitude decreases rapidly with lag time as the waves disperse outward and upward from their source. The radially outward phase speed of the waves is roughly 1.6 degree of longitude per hour, which is equivalent to ~ 50 m s^{-1}.

Inertio-gravity (IG) waves on a rotating planet also exhibit an azimuthal velocity component, with anticyclonic flow following the divergent, radially outward flow by a quarter cycle such that the wave-related horizontal wind vector rotates anticyclonically with time. In contrast to the flow in Rossby waves, in which the nondivergent component dominates the horizontal wind field, in IG waves it is the irrotational component that dominates.

[14] Holton (1972); Salby and Garcia (1987).

Figure 10.12 Composites showing gravity waves at the 50 hPa level radiating out from downwelling events in the tropospheric vertical velocity field. The reference time series used in creating these composites is $(\omega_{700} + \omega_{300})$, where the time series are standardized before they are averaged so they have equal weight. "Downwelling events" are defined as the hours that fall within the bottom 10% of the frequency distribution of (upward), vertical velocity. The analysis is performed independently at 360 equally spaced equatorial grid points spaced one degree of longitude apart, and the patterns are shifted in longitude to bring the reference grid points into alignment, thus creating a single composite with $(360 \times N)/10$ members, where N is the number of hourly samples in the record that is analyzed, in this case 10 years long (2010–2019). The fields displayed are geopotential (colored shading), horizontal wind (vectors, the longest correspond to 0.1 m s^{-1}), and vertical velocity (contour interval 0.4 mm s^{-1}). For justification and further specifics concerning the data and the data processing, see he extended caption in Appendix F. From Pahlavan et al. (2022). © American Meteorological Society. Used with permission. This preliminary version has been accepted for publication in the *Journal of Atmospheric Sciences* and may be fully cited. The final typeset copyedited article will replace the EOR when it is published.

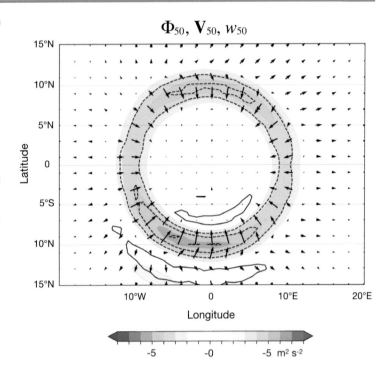

$\Phi_{50}, \mathbf{V}_{50}, w_{50}$

Figure 10.13 As in Fig. 10.12 but for temperature. The left panels show the composite 50 hPa temperature fields 2, 4, and 8 hours after the downwelling events. The right panels show the corresponding equatorial longitude–height cross sections. From Pahlavan et al. (2022). © American Meteorological Society. Used with permission. This preliminary version has been accepted for publication in the *Journal of Atmospheric Sciences* and may be fully cited. The final typeset copyedited article will replace the EOR when it is published.

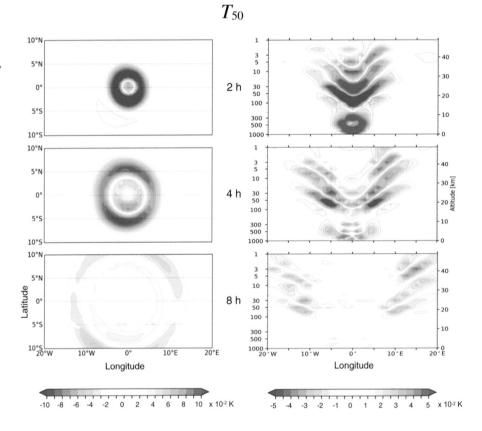

T_{50}

To see IG waves more clearly in the two-sided spectra, it is convenient to use (log k, log ω), rather than (k, ω) as horizontal and vertical coordinates, taking care to multiply power or spectral density by $k \times \omega$ to preserve the property that (power $\times$ area) be proportional to variance anywhere on the plot. When k and ω are plotted on logarithmic scales, nondispersive waves such as Kelvin waves and gravity waves, for which ω/k is a constant, all appear as straight

Figure 10.14 Dispersion diagram as in Fig. 10.8 scaled to an equivalent depth of 54 m to match a phase speed of 23 m s^{-1}, plotted on logarithmic ω and k scales but linear in the range $k = -2$ to $+2$. Rossby waves do not appear in this plot because their frequencies are < 0.1 cycles per day. From Pahlavan et al. (2022). © American Meteorological Society. Used with permission. This preliminary version has been accepted for publication in the *Journal of Atmospheric Sciences* and may be fully cited. The final typeset copyedited article will replace the EOR when it is published.

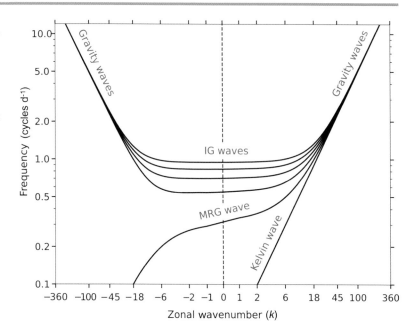

lines with a slope of one power of 10 increase of ω per power of 10 increase of k. Differences in phase speed are reflected, not as differences in slope, but as vertical displacements of the phase speed line; the higher the line, the faster the phase speed. For reference, Fig. 10.14 shows dispersion curves plotted on log k–log ω axes for gravity waves and equatorially trapped planetary waves. The equivalent depth that yields a gravity wave phase speed $c = \sqrt{gh_e} = 23$ m s^{-1} has been used in order to match the observed spectra shown in the next figure as closely as possible. In this representation, the curves for the IG waves quickly asymptote to the respective gravity wave phase speed at frequencies higher than ∼2 cycles per day. Rossby wave curves are not visible in this plot because their highest frequencies are less than 0.1 cycles per day. The Kelvin wave and gravity wave phase speeds are the same, but the two waves correspond to different ranges of frequencies. The spectral density (or power) scale in the observed spectra also spans several orders of magnitude. If a linear scale were used to represent it, most of the plot would be empty. Hence, it is convenient to use a logarithmic scale for that as well. One can think of these two-dimensional spectra as an assemblage of one-dimensional frequency spectra, one for each zonal wavenumber k, plotted on a log–log scale.

Figure 10.15 shows two-sided spectra for u, v, w, and T at the 50 hPa level averaged from $10°$N to $10°$S. All of them exhibit a clearly distinguishable gravity and inertio-gravity wave signature extending along the broad swaths with phase speeds on the order of 30 m s^{-1}, consistent with Fig. 10.13. In this chapter, we focus on the broad swaths of enhanced variance, and defer discussion of the more subtle features embedded within them until Section 15.2.

The next figure is derived from spectrum and cross spectrum analysis involving the three wind components, temperature, and geopotential at the 50 hPa level. The vertical flux

of geopotential $[w^*\Phi^*]$ shown in the top left panel of Fig. 10.16 is upward for both eastward and westward propagating waves, consistent with the structure of gravity and inertio-gravity waves. The in-phase relationship between w^* and Φ^* is indicative of an upward flux of mechanical (kinetic plus potential) energy from below. The vertical flux of eastward momentum $[u^*w^*]$ (top right) is characterized by an upward flux of eastward momentum for eastward propagating waves and an upward flux of westward momentum for westward propagating waves, consistent with the structure of vertically propagating gravity waves in which air parcels ascend and descend along slanted trajectories parallel to the advancing wave fronts. Evidence that these are internal, rather than external, waves is afforded by the fact that Φ and T are in quadrature with T leading Φ, as indicated in the bottom left panel of Fig. 10.16, rather than in-phase. In the quadrature spectrum shown in the bottom right panel of Fig. 10.16, the poleward meridional wind component v_p leads the zonal component for both eastward and westward propagating waves, indicative of clockwise rotating perturbation wind vectors in the Northern Hemisphere and counterclockwise rotating vectors in the Southern Hemisphere, consistent with the structure of IG waves.

Exercises

10.9 (a) Using the bottom left panel of Fig. 10.11, infer the temperature perturbations that a hypothetical air parcel undergoes during the passage of a westward propagating MRG wave and interpret them in terms of the thermodynamic energy equation. Assume that the wave motions are adiabatic and the displacements that the air parcels undergo in association with the passage of the wave are

Figure 10.15 Two-sided wavenumber–frequency spectra for variables, as indicated, at the 50 hPa level, extending out to frequencies of 12 cycles per day on a logarithmic scale for wavenumber and frequency, and for spectral density in units of variance per unit area on the plot. This plot represents an average of 81 spectra computed at $0.25°$ latitude intervals for latitudes ranging from $10°$N to $10°$S. The contour interval corresponds roughly to a factor of 3. The higher pair of sloping reference lines correspond to phase speeds of $\pm 49\,\mathrm{m\,s^{-1}}$ and the lower pair to $\pm 23\,\mathrm{m\,s^{-1}}$. Between $k = -2$ and $+2$ the wavenumber scale is linear. From Pahlavan et al. (2022). © American Meteorological Society. Used with permission. This preliminary version has been accepted for publication in the *Journal of Atmospheric Sciences* and may be fully cited. The final typeset copyedited article will replace the EOR when it is published.

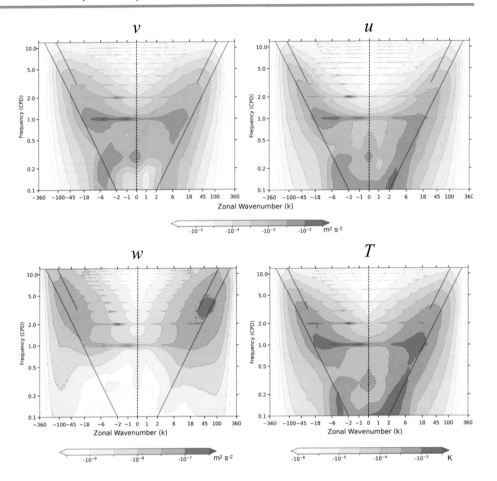

vanishingly small. (b) Perform the same analysis for the zonal and meridional wind components.

10.10 Based on their idealized structures in the longitude height plate, prove that Kelvin and MRG waves must be downward propagating.

10.11 Enumerate the similarities and differences between equatorially trapped Kelvin waves and gravity waves.

10.12 Why is the quadrature spectrum shown in the bottom right panel of Fig. 10.16 calculated on the basis of the poleward meridional wind component v_p rather than on the northward component v?

10.3 THE GENERALIZED ELIASSEN–PALM FLUX

The TEM formalism developed in Sections 8.2 and 9.3.2 includes only the leading terms important for extratropical wave–mean flow interactions. To diagnose the wave–mean flow interaction in the QBO it will be necessary to use the generalized TEM formulation, which takes into account the contributions of vertically propagating Kelvin waves, MRG waves, IG waves, and smaller scale gravity waves, as well as equatorially trapped Rossby waves. The general expression for the EP flux $\mathbf{F}$ in $z = -H\,ln(p/p_s)$ coordinates is (c.f., Eq. (9.5))

$$\mathbf{F} = F^{(y)}\mathbf{j} + F^{(z)}\mathbf{k} \tag{10.6}$$

where

$$F^{(y)} = \rho_o R_E\,cos\phi\left(\frac{\partial[u]}{\partial z}\frac{[v^*T^*]}{\Gamma_d - \Gamma} - [u^*v^*]\right),$$

$$F^{(z)} = \rho_o R_E\,cos\phi\left(\left(f - \frac{1}{R_E\,cos\phi}\frac{\partial}{\partial\phi}[u]cos\phi\right)\frac{[v^*T^*]}{\Gamma_d - \Gamma} - [u^*w^*]\right), \tag{10.7}$$

where ρ_o is a reference density, which decreases exponentially with height, and $\mathbf{k}$ is the vertical (upward) unit vector.

The wave forcing term $\nabla \cdot \mathbf{F}$ in Eq. (9.2) can be written as the sum of the direct forcing by the eddy transports of zonal momentum

$$-\nabla \cdot \mathbf{F}_M = \frac{\rho_o}{cos\phi}\frac{\partial}{\partial\phi}\left([u^*v^*]cos^2\phi\right) + R_E\,cos\phi\frac{\partial}{\partial z}\left(\rho_o\,[u^*w^*]\right) \tag{10.8}$$

and the indirect forcing associated with the mean meridional circulations induced by the poleward eddy heat transports,

$$-\nabla \cdot \mathbf{F}_H = -\frac{\rho_o}{cos\phi}\frac{\partial}{\partial\phi}\left(cos^2\phi\frac{\partial[u]}{\partial z}\frac{[v^*T^*]}{\Gamma_d - \Gamma}\right)$$
$$- R_E\,cos\phi\frac{\partial}{\partial z}\left(\rho_o\left(f - \frac{1}{R_E\,cos\phi}\frac{\partial}{\partial\phi}[u]cos\phi\right)\frac{[v^*T^*]}{\Gamma_d - \Gamma}\right). \tag{10.9}$$

When written in this generalized form, the EP flux is still in the direction opposite to the flux of westerly momentum,

Figure 10.16 Products derived from two-sided cross spectrum analysis of 50 hPa fields. (Top left) the upward flux of geopotential $[w^*\Phi^*]$, given by the cospectrum (Co) between w^* and Φ^*; (top right) the upward flux of westerly momentum $[u^*w^*]$ given by the cospectrum between u^* and w^*; (bottom left) the quadrature spectrum (Q) between geopotential height Φ^* and temperature T^*, and (bottom right) the quadrature spectrum between the zonal wind component u^* and the poleward meridional wind component v_p^*, which is indicative of the turning of the wind with the passage of inertio-gravity waves. Sloping reference lines as in Fig. 10.15. From Pahlavan et al. (2022). © American Meteorological Society. Used with permission. This preliminary version has been accepted for publication in the *Journal of Atmospheric Sciences* and may be fully cited. The final typeset copyedited article will replace the EOR when it is published.

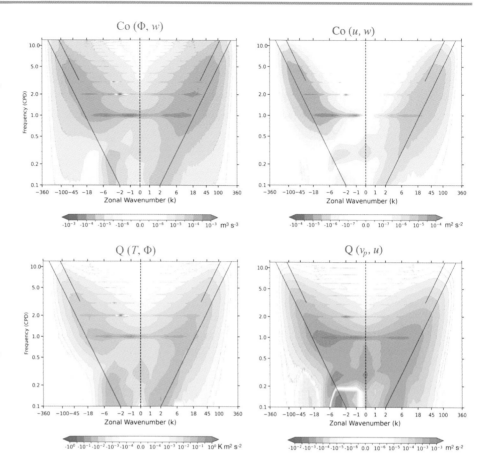

but it is not quite as simply related to the flux of wave activity as it is in the diagnosis of Rossby waves. For Kelvin waves and eastward propagating IG and gravity waves, the transport of both westerly momentum and wave activity is upward, as evidenced by the observed in-phase relationships between u, w and Z, yet it is evident from Eq. (10.7) that they produce a downward EP flux despite their upward group velocity. Hence, in the case of eastward propagating Kelvin and IG waves the vertical fluxes of wave activity and westerly momentum are in the same direction; i.e., they are both upward.

> **Exercise**
>
> **10.13** Offer a dynamical interpretation of the terms that appear in Eq. (10.7) that do not have counterparts in Eq. (9.5).

10.4 WAVE–MEAN FLOW INTERACTION IN THE QBO

Having documented the structure and evolution of the QBO in the $[u]$ field (Section 10.1), described the theory of small-scale gravity waves and equatorially trapped planetary waves (Section 10.2), and extended the expression for the wave-forcing of the zonal flow (Section 10.3), we now explain

how wave–mean flow interaction gives rise to the QBO. The top panel of Fig 10.17 shows an equatorial time–height section of the forcing required to account for the downward phase propagation of the QBO in the presence of the upwelling associated with the Brewer–Dobson circulation. It is qualitatively similar to the observed acceleration $\partial[u]/\partial t$ (not shown) but the accelerations are somewhat larger because they need to account for the downward phase propagation. The middle panel shows the forcing by the waves resolved by the reanalysis, as given by the divergence of the EP flux, estimated from Eqs. (10.8) and (10.9), and the bottom panel shows the contribution of waves with zonal wavenumbers k higher than 20. In all three panels, the acceleration and forcing fields are superimposed upon the $[u]$ field, the same as shown in Fig. 10.2.

The wave forcing resolved by the reanalyses accounts an appreciable fraction of the required forcing. The fit is particularly impressive for the westerly shear zones. It is notable that the wave forcing accounts for the composite QBO structure as well as the peculiarities of the individual QBO cycles such as the hiatuses in the descent of the shear zones and the brief suspension of cyclic behavior in 2016–2017. Waves with zonal wavenumbers $k > 20$ behave in a similar manner and their contribution to the observed accelerations is not inconsequential.

The concepts of wave activity and group velocity applied to extratropical waves in Section 8.2 also apply to equato-

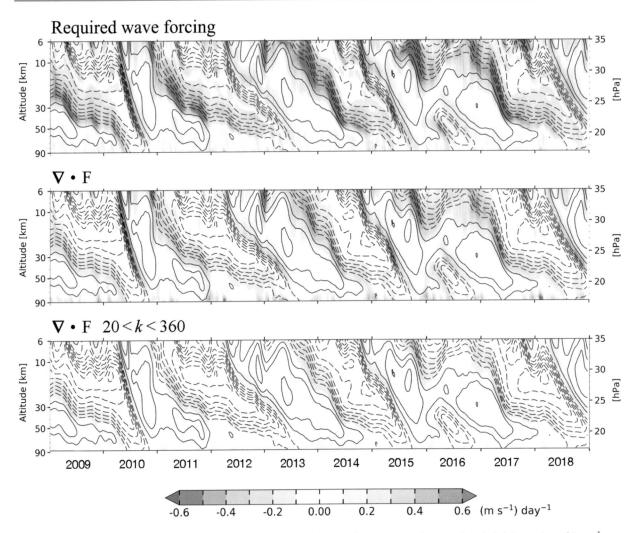

Figure 10.17 Time–height sections averaged from 5°N to 5°S. The contours in all three panels represent zonally averaged zonal wind. Contour interval 5 m s^{-1}, westerlies are solid, easterlies are dashed, and the zero contour is omitted. Colored shading indicates (top) the required forcing; that is, the observed acceleration estimated using a 2-month centered difference minus the vertical advection $-[w]\partial[u]/\partial z$; (middle) the resolved wave forcing $\nabla \cdot \mathbf{F}$; and (bottom) the contribution of waves with zonal wavenumbers $k > 20$ to the forcing. All fields are based on 6-hourly ERA5 data, smoothed with a 30-day running mean. From Pahlavan et al. (2021a). © American Meteorological Society. Used with permission.

rially trapped waves. The vertical component of the group velocity c_g, a measure of the rate at which wave activity disperses upward from below, is proportional to the Doppler-shifted phase speed $[u] - c$. It follows that eastward propagating Kelvin waves, eastward propagating IG and gravity waves disperse upward much more rapidly through easterly regimes than through westerly regimes and that their wave activity thus tends to accumulate in westerly shear zones, where their group velocity is decreasing rapidly with height, causing them to break and transfer their westerly momentum to the zonal flow. It is the eastward acceleration due to this wave forcing that causes the westerly shear zones to descend. In a similar manner, westward propagating IG and mixed Rossby–gravity (MRG) waves tend to accumulate and break in easterly shear zones, causing them to descend. The bottom panel of Fig. 10.17 shows that it is not only the planetary-scale waves that interact with the QBO-related $[u]$ field.

Theory predicts that eastward propagating waves of all species should disperse upward more readily during the easterly wind regimes of the QBO, when $[u]$ and c are of opposing sign so that the Doppler-shifted phase speed $[u] - c$ is large, and the reverse should be true of westward propagating waves. Two-sided wavenumber-frequency spectra of w^* during westerly and easterly phases of the QBO are shown in Fig. 10.18, together with the difference spectrum. It is evident that the eastward propagating waves in the vertical velocity field are indeed more vigorous during the easterly phase, in agreement with theory.

The contributions of the individual terms in Eqs. (10.8) and (10.9) have been computed and the forcing has also been decomposed into the contributions from the various wave modes by partitioning the variances and covariances in the u^*, v^*, w^*, and T^* time series into frequency bands that match those of the respective wave species and partitioning the perturbations in each band into contributions from east-

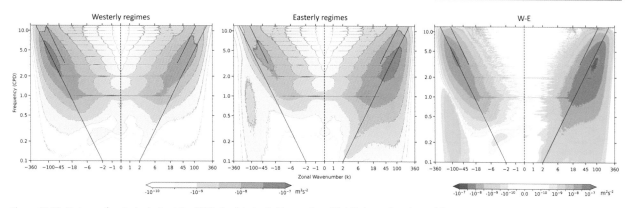

Figure 10.18 Variance of vertical velocity at the 50 hPa level during (left) westerly and (middle) easterly regimes of the QBO, as represented by two-sided spectra of w^*. (Right) the flux during westerly regimes minus the flux during easterly regimes. The sloping reference lines correspond to phase speeds of ± 49 m s^{-1} (higher pair) and ± 23 m s^{-1} (lower pair).

ward and westward propagating waves. The results based on ERA5 reanalyses can be summarized as follows.[15]

- Resolved waves with zonal wavenumbers k ranging up to 360 account for slightly more than half of the wave forcing required to account for the observed zonal wind accelerations in the QBO. The remainder is presumably due to IG and gravity waves with shorter wavelengths and higher frequencies than those resolved by the reanalysis.
- On the equator, the resolved forcing is dominated by the divergence of vertical transports of westerly momentum $[u^*w^*]$, whereas in the latitude bands from $5°$ to $15°$N/S, the mean meridional circulations induced by the poleward heat transports $[v^*T^*]$ in MRG waves and the momentum transports $[u^*v^*]$ in extratropical Rossby waves also make appreciable contributions.
- The forcing by the resolved waves is roughly evenly divided between waves with zonal wavenumber k greater or less than 20. The former are mainly inertio-gravity waves with periods on the order of a day or less, while the latter are lower frequency Rossby waves, Kelvin waves, and MRG waves.
- The planetary wave contribution to the forcing comprises Kelvin, MRG, and IG waves. Kelvin waves are dominant in the westerly shear zones, while westward propagating IG waves are the most important contributors in the easterly shear zones, with a secondary contribution from MRG waves.
- Eastward propagating IG waves contribute to the descent of westerly wind regimes and westward propagating IG waves contribute to the descent of easterly wind regimes, as is evident from the bottom panel of Fig. 10.17.

These recent observational results serve to validate a mechanism, proposed over 50 years ago, for explaining why the QBO cycles back and forth between easterly and westerly wind regimes.[16] The forcing in a simple, mechanistic model was prescribed as consisting of upward-

[15] Pahlavan et al. (2021a,b).
[16] Lindzen and Holton (1968); Holton and Lindzen (1972).

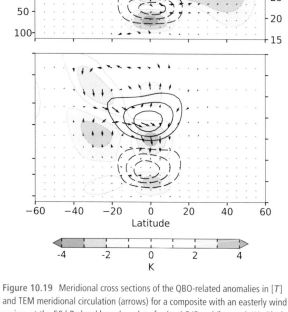

Figure 10.19 Meridional cross sections of the QBO-related anomalies in $[T]$ and TEM meridional circulation (arrows) for a composite with an easterly wind regime at the 50 hPa level based on data for (top) DJF and (bottom) JJA. Black contours denote the $[u]$ anomalies, contour interval 7.5 m s^{-1}; zero contour omitted. The longest arrows correspond to 0.4 m s^{-1} for $[v]$ and 0.6 mm s^{-1} for $[w]$. From Pahlavan et al. (2021a). © American Meteorological Society. Used with permission.

dispersing gravity waves with a continuum of phase speeds, both eastward and westward, ranging up to a maximum of $\pm c_{max}$. The waves give up their zonal momentum to the zonal flow when they reach a level in the atmosphere at which their phase speed matches the zonal wind speed, at which point their group velocity goes to zero and the momentum that they carry with them is imparted to the

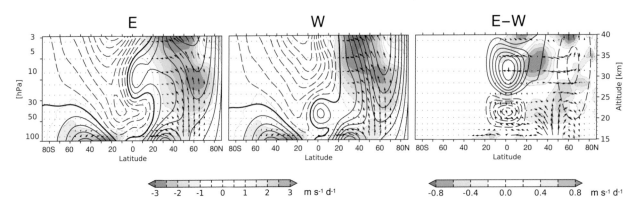

Figure 10.20 Composite fields of EP flux divergence (colored shading), EP flux (vectors), and zonally averaged zonal wind [u] (contours) for DJF. (Left and middle panels) Easterly (E) and westerly (W) phases of the QBO at the 50 hPa, and (right) differences between the left and middle panels. The longest vectors in the left and middle panels correspond to 10^{11} kg m^{-2} for the meridional component and 1.5×10^{10} kg m^{-2} for the vertical component and the vectors in the right panel scaled up by a factor of 15 to make them clearly visible. Contour interval 5 m s^{-1}; negative values are dashed and the zero contour is bolded.

zonal flow, causing the shear zone to propagate downward. Waves propagating in the opposite direction are unaffected by this shear zone and are able to disperse upward without absorption until they encounter a mean flow that matches their phase speed. Successive easterly and westerly shear zones continue to propagate downward until they reach the wave source, which is presumed to lie at 19 km, just above the equatorial cold point. Integrations of the model yielded alternating, downward propagating easterly and westerly wind regimes with wind speeds ranging up to $\pm c_{max}$.

Further evidence that the QBO could be generated by wave–mean flow interaction was provided by laboratory experiments with a non-rotating annulus filled with salt-stratified water. An oscillating bottom boundary with sinusoidal perturbations with a zonal wavenumber k and frequency ω was imposed, which induces a standing wave response with fixed nodes and antinodes. [A standing wave is equivalent to waves with phase speeds ω/k propagating in opposite directions.] The interactions between the waves and the mean flow give rise to robust, low frequency, quasi-periodic, downward propagating fluctuations in the azimuthal flow.[17] A phenomenon analogous to the QBO but with a quasi-quadrennial period has been identified in the equatorial stratosphere of Jupiter.[18]

To simulate the wave–mean flow interactions that give rise to the equatorial stratospheric QBO in a GCM, it is necessary to resolve the shear zones, which are only a few km deep. Some of today's state-of-the-art models are capable of producing realistic simulations.[19]

> **Exercise**
>
> 10.14 How is it that radially propagating gravity waves produce vertical fluxes of zonally averaged zonal momentum in association with the QBO?

10.5 INFLUENCE OF THE QBO ON THE WINTERTIME POLAR VORTEX

The patterns in the meridional cross sections based on year-round data shown in Fig. 10.4 are nearly equatorially symmetric. In contrast, when the data are separated by season, as shown in Fig. 10.19, the midlatitude centers of action in the QBO-related temperature field and the mean meridional circulation anomalies are largely confined to the winter hemisphere. A likely explanation for this pronounced seasonality is that poleward of 15°N/S, the existence of mean meridional circulations is contingent upon the presence of "wave driving" (i.e., the divergence of the EP flux), which is appreciable only during the winter season when the background flow at these levels is westerly, as documented in Fig. 8.13. This constraint applies to the climatological mean as well as to the QBO-related variations about the mean.

The QBO modulates the strength of the Northern Hemisphere wintertime stratospheric polar vortex. Figure 10.20 shows composite fields of EP flux divergence, EP flux vectors and zonal wind during easterly and westerly phases of the QBO at the 50 hPa level in DJF. The fields are similar, but the difference field shown in the right panel exhibits an interesting structure. In the easterly phase, the polar vortex is weaker and the flux of wave activity into the stratosphere from below is stronger. Owing to the stronger wave-driving, the Brewer–Dobson circulation is also enhanced (not shown). A secondary feature in this section is the redirection of the high latitude EP flux into the polar vortex in the easterly composite, reflecting a higher frequency of

[17] Plumb and McEwan (1978).
[18] Friedson (1999); Flasar et al. (2004).
[19] Richter et al. (2020).

incidence of stratospheric sudden warmings. This is a pathway by which the QBO exerts an (albeit subtle) influence upon extratropical tropospheric weather patterns, as will be documented in Fig. 13.40.[20] At the time of this writing, there is considerable interest in the question of whether another pathway might exist. Several recent observational studies have reported the QBO may influence tropical convection and/or convectively coupled tropical waves discussed in Part VI of this book.

[20] For further discussion, see, for example, Garfinkel et al. (2012).

Exercises

10.15 The Northern Hemisphere polar vortex is most sensitive to the phase of the QBO in midwinter, whereas in the Southern Hemisphere the sensitivity is greatest a few months later, around the time of the final warming. Explain.

10.16* Explain the observed semiannual oscillation (SAO) in zonal wind in the upper stratosphere.

10.17* Are equatorial jets observed in atmospheres other than the Earth's?

Part V The Zonally Varying Extratropical Tropospheric Circulation

Parts II, III, and IV are exclusively concerned with the zonally averaged circulation. All representations of the eddies and the transports that they produce are based on zonally averaged statistics. Interest in the zonally varying circulation dates back to theoretical studies of the extratropical stationary waves in the 1950s. However, it was not until the advent of archived, gridded, global datasets in the mid-1970s that the zonally varying circulation became amenable to quantitative analysis based on variance and covariance statistics.

Diagnosing the zonally varying circulation has required the development of new metrics for characterizing the structure and evolution of the eddies, such as one-point correlation (or covariance) maps, teleconnectivity maps, and empirical orthogonal functions (EOFs), all of which can be derived from the temporal covariance matrix. These metrics may be viewed as tools for describing the spatial distribution and characteristic shapes of the eddies. The transports of momentum or wave activity in the horizontal plane are intimately related to the anisotropy of the eddies, that is, the tendency for them to be elongated in a preferred direction in the horizontal plane. It follows from the dynamics of Rossby wave propagation and dispersion that the structure and evolution of the eddies exhibits a pronounced frequency dependence that can be exploited to separate baroclinic waves from other kinds of variations.

To understand the zonally varying general circulation it is necessary to transcend the "channel vision" that dominated the early dynamics literature, and conceive of wave propagation and dispersion as occurring along preferred waveguides shaped by the background flow. Even in the simplest case of a steady state forcing in the presence of a background flow consisting of pure superrotation, wave activity disperses not along latitude circles, but along great circles. In Part V, we will focus on the zonally varying extratropical circulation, mapping the observed fields on polar stereographic projections rather than global Hammer plots like the ones shown in Chapter 1.

In Parts II–IV the structure of the eddies was considered only to the extent that it is necessary to understand the zonally averaged eddy transports. In the remainder of this book, the longitudinal and frequency dependence of the eddy statistics is taken into consideration. Making use of an expanded array of three- (as opposed to two-) dimensional diagnostics, it is possible to document the structure and evolution of the eddies more clearly and to interpret them using terms and concepts derived from the synoptic and dynamical meteorology literature, such as the climatological mean stationary waves, extratropical cyclones with their attendant frontal features, zonally varying jet streams and storm tracks, blocking, and wave-breaking. The effects of the continent–ocean configuration and the major mountain ranges in shaping the extratropical general circulation become more clearly apparent.

As in previous chapters, we will be concerned with the relationships and interactions between the waves and the background flow, but rather than the various terms being partitioned into zonal mean versus eddy components, they are partitioned into time mean versus transients, as in the right panel of Fig. 6.6. Zonally varying extratropical and tropical phenomena will be considered separately in Parts V and VI, respectively.

Part V is divided into three chapters. The first (Chapter 11) is concerned with the time-mean, zonally varying circulation; that is, the *stationary waves*, various representations of which are shown in Figs. 1.4 and 1.30. Whereas in Parts II–IV the emphasis is on understanding how the waves affect the zonally averaged flow, the emphasis in Chapter 11 is on understanding how the zonally varying flow is forced by the land–sea geometry, mountains, and the transports of heat and momentum by the transients. The results of experiments with general circulation models are incorporated into the discussion in order to clarify the role of the various forcings. Chapters 12 and 13 are focused on the structure and evolution of the transients: the former is mainly concerned with baroclinic waves and the latter with a variety of lower frequency phenomena.

11 The Northern Hemisphere Winter Zonally Varying Climatology

with Rachel H. White

In Chapter 1, we presented a survey of the general circulation encompassing both Northern and Southern Hemispheres and winter and summer seasons. In this chapter, we focus on the Northern Hemisphere winter season DJF, which arguably exhibits the most distinctive patterns in terms of zonally varying jets, storm tracks, and climatological-mean stationary waves. Figure 11.1 shows a time–longitude section of 500 hPa height on the 50°N latitude circle extending through one arbitrarily chosen but representative winter. Such displays, often referred to as *Hovmöller diagrams*, were frequently used by mid-twentieth century longrange weather forecasters to project how the longwave pattern was likely to evolve over the next week. The preferred longitudes of warm and cool colors indicate the positions of the climatological mean *stationary wave* ridges and troughs. The stripes of stronger and weaker color, which slope upward toward the right with a phase speed on the order of 10 m s^{-1}, are the signature of baroclinic (Rossby) waves.

For the phase speed of a Rossby wave to be zero it is required that the westward propagation due to the advection of planetary vorticity $\beta_* \equiv \beta - \partial^2[u]/\partial y^2$ by the wave exactly cancels the eastward advection of the vorticity perturbations in the waves by the climatological mean zonally symmetric flow $[u]$, that is, that in the equation for the vorticity tendency, $-[u]\partial\zeta/\partial x = \beta_* v^*$. To derive the formula for the phase speed of a Rossby wave, the horizontal flow is represented in terms of a velocity streamfunction ψ whose Laplacian is the relative vorticity field. It follows that the zonal phase speed of a Rossby wave with zonal wave number k and meridional wavenumber l in a background zonal wind U is

$$c = \frac{\omega}{k} = U + c_o, \tag{11.1}$$

where

$$c_o = -\beta_*/K^2 \tag{11.2}$$

is the westward Doppler-shifted phase speed due to the advection of planetary vorticity by the meridional wind component v^* in the wave, and $K^2 = k^2 + l^2$.

Setting c equal to zero in Eq. (11.1), substituting c_o from Eq. (11.2), and solving for K yields

$$K_s \equiv \sqrt{\frac{\beta_*}{U}}, \tag{11.3}$$

where K_s is the so-called stationary wavenumber, the two-dimensional wavenumber whose phase speed is zero. Ignor-

ing the contribution of the zonal wind field to the meridional gradient of absolute vorticity and setting l equal to zero for a one-dimensional Rossby wave propagating along a latitude circle yields the expression $k_s = \sqrt{\beta/U}$ in the foundational paper of Rossby.[1]

Based on Fig. 1.19, the mean wintertime zonal wind speed in the mid-troposphere is on the order of 15 m s^{-1}, which corresponds to a stationary wavenumber K_s of around 4 in Table 11.1. Since the planetary-scale waves observed in the extratropics are not strongly anisotropic, $K_s \sim 4$ corresponds to a stationary zonal wavenumber $k_s \sim 4/\sqrt{2}$ or between 2 and 3.

This chapter documents the structure of the Northern Hemisphere wintertime circulation; that is, the combined zonally symmetric circulation and stationary waves. Reanalysis products will be used for documenting these features in the observations. The first section compares the observed DJF wintertime stationary waves with the waves simulated in a free running, state-of-the-art atmospheric general circulation model (AGCM), the Whole Atmosphere Community Climate Model (WACCM) version 6, run with SST and sea ice prescribed to match the DJF observations.

The second section explores the role of mountains, making use of an additional simulation with the WACCM. The model code is altered to eliminate the variations in the level of the bottom boundary as well as parameterizations related to orographaphically induced drag. The simulated planetary waves are thus indicative of those that would exist on an otherwise Earth-like planet with no mountains, and thus the difference between the simulations with and without mountains provide insight into the role of mountains. To provide an indication of the planetary wave response to orography with the zonally symmetric flow held fixed, experiments conducted with simple linearized models are also shown. The results of these experiments also serve as a demonstration of Rossby wave dispersion on a spherical planet in the presence of a zonally symmetric background flow. The third section shows the response to diabatic heating, also based on the results of experiments with linear models.

The fourth section, which is largely based on observations, illuminates the role of the transient eddies (strictly speaking, transients) in the maintenance of the stationary

---

[1] Rossby (1939).

Table 11.1 *Doppler-shifted phase speed as a function of zonal wave number k for a one-dimensional Rossby wave along 55° latitude.*

k	c_o (m s^{-1})
1	−250
2	−62
3	−27
4	−15
5	−10
6	−7
10	−3

waves. In the fifth section it is shown that while linear models with prescribed zonally symmetric flows have proven useful for diagnosing the shapes of the zonally varying flow patterns that develop in response to the various prescribed forcings, they give a false impression of the relative importance of the forcings because they do not take into account the nonlinear interactions, particularly those involving the zonally symmetric flow.

The final section, in way of an historical perspective, explains why similar numerical simulations conducted 40 years ago using the models available back at that time, yielded such a different impression of the relative importance of orography and diabatic heating in accounting for the observed stationary waves.

11.1 OBSERVED AND SIMULATED STRUCTURE IN DJF

The reader was introduced to the climatological mean stationary waves back in Chapter 1. The egg-shaped Hammer plots shown there and in subsequent chapters reveal various aspects of the zonally varying structure of the general circulation. In this section, we will view some of these same fields again, but in a polar stereographic projection, which emphasizes their zonal dependence. In these figures (11.2–11.7, 11.10, 11.13, and 11.19), observations, as represented in ERA-I, are shown in the left panels, while the center panels show results from the WACCM control (CTL) simulation, in which WACCM is driven by the observed seasonally varying climatological mean SSTs (1982–2001). The right panels show results from the WACCM No Mountain (NM) experiment, which will be discussed in Section 11.2. The fidelity of the control simulation is evidenced by the strong similarity between the paired ERA-I and WACCM CTL plots. The positions of the storm tracks are shown in Fig. 11.2, where they are represented by the standard deviation of the 700 hPa vertical velocity field based on unfiltered daily data. They are slightly stronger in ERA-I than in CTL and in CTL than in NM, but their positions in the three panels are similar.

Figure 11.3 shows the DJF climatological mean mid-tropospheric (500 hPa) vertical velocity field. The subsidence around the periphery of the map (20–30°N) is a reflection of the sinking branch of the Hadley cell. Prominent zonally varying features include patches of ascent upstream

and descent downstream of the mountain ranges of central and eastern Asia and North America. Embedded in the planetary-scale orographic signature are bands imprinted by the individual mountain ranges. Over the open oceans there are patches of ascent at the eastern end of the storm tracks.

Figure 11.4 shows the distributions of SLP and SST/skin temperature. At the higher latitudes, land, with its low heat capacity, tends to be much colder than the ocean at the same latitude. The subpolar Aleutian and Icelandic lows are prominent features of the SLP field: an analogous ring of low

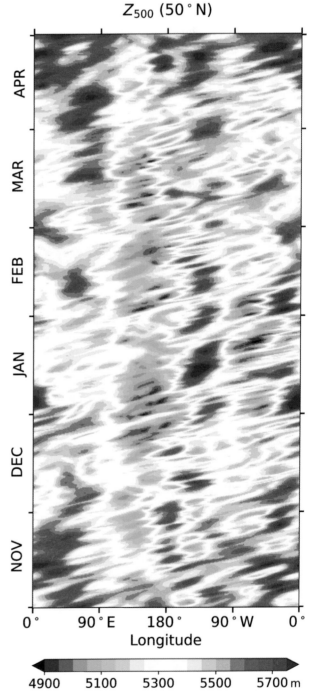

Figure 11.1 Time–longitude section of 500 hPa height on the 50°N latitude circle during the 2012–2013 winter.

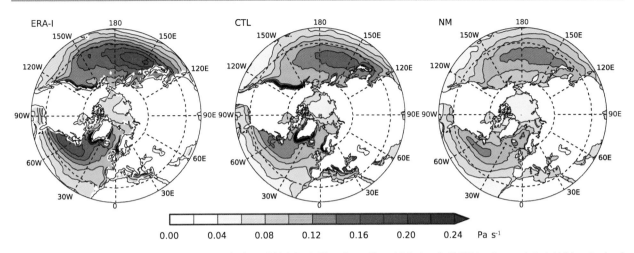

Figure 11.2 The standard deviation of the 700 hPa vertical velocity field during DJF based on unfiltered daily data. (Left) ERA-Interim reanalysis; (middle) as simulated in the control run of the Whole Atmosphere Community Climate Model (WACCM) with realistic orography (CTL) and (right) in a run of the same model in which the orography is entirely flattened and the parameterized frictional drag is the same as it would be on a planet with the same continent–ocean configuration, but with all land surfaces at sea level (NM). Unless otherwise noted, the domain in these polar stereographic projections is poleward of 20°N.

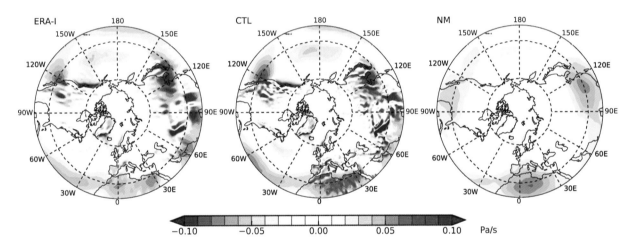

Figure 11.3 Vertical velocity at 500 hPa during DJF; three panel format as in Fig. 11.2.

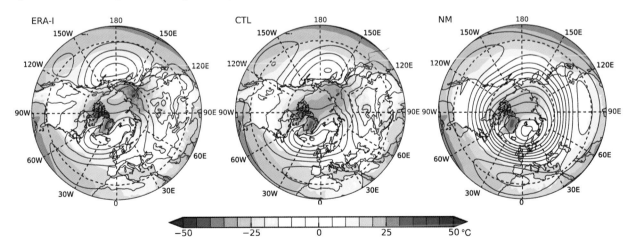

Figure 11.4 Skin temperature (colored shading) and SLP, contour interval 5 hPa during DJF in the same three panel format as in Fig. 11.2.

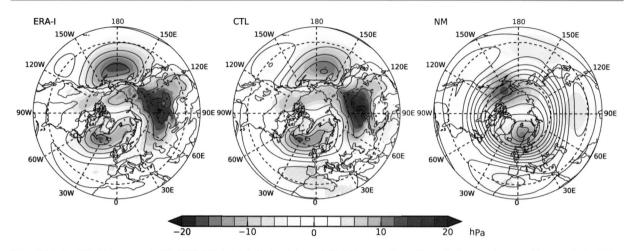

Figure 11.5 The DJF eddy component of the SLP field (colored shading) and the total SLP field, contour interval 5 hPa, in the same three panel format as in Fig. 11.2.

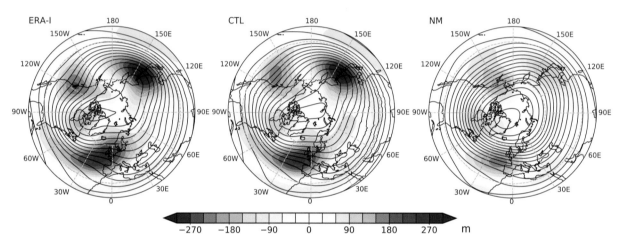

Figure 11.6 The DJF eddy component of the Z_{300} field (colored shading) and the total Z field, contour interval 100 m, in the same three panel format as in Fig. 11.2.

SLP is observed over the subpolar seas surrounding Antarctica (not shown). The Northern Hemisphere near-surface subpolar features assume the form of closed cyclones rather than a series of troughs in the westerlies because the axially symmetric flow at subpolar latitudes is almost zero, as evidenced by the strong correspondence between the SLP contours and the colored shading showing the eddy component of the SLP field (Fig. 11.5).

The corresponding Z_{300} patterns, shown in Fig. 11.6, exhibit pronounced ridges over the west coasts and troughs over the east coasts of the continents. The westerlies are particularly strong along the base of the troughs, where the height contours are pinched together. The stronger of these sectoral jet streams, shown in more detail in Figs. 1.4 and 1.5, extends across North Africa and South Asia, and into the western Pacific; the weaker one extends east-northeastward from central North America into the North Atlantic. The storm tracks shown in Fig. 11.2 are located along the poleward flanks of these jets and somewhat downstream of them.

The boreal wintertime planetary waves disperse upward into the stratosphere, but it is only the lowest zonal wavenumbers $(k = 1, 2)$ that are capable of remaining

stationary in the presence of the strong wintertime polar night jet (Fig. 1.19).[2] Figure 11.7 shows the pattern in the Z and Z^* fields at the 30 hPa level. The climatological mean ridge centered over Alaska is a persistent feature of the wintertime stratospheric polar vortex and it evidently plays a role in the midwinter warming documented in Fig. 9.14.

The patterns shown in Figs. 11.5, 11.6, and 11.7 are slices through a three-dimensional structure that tilts westward with height, as illustrated by the longitude–height cross sections at 60°N shown in Fig. 11.8. The amplification of the waves and the transition toward lower zonal wavenumbers with increasing height are also evident. The continuous westward tilt with height is consistent with the observed poleward heat flux by the stationary waves in DJF (Fig. 5.16) and the upward EP fluxes in the stationary waves shown in Fig. 11.9. The EP fluxes split into two branches – one dispersing equatorward into the subtropics at the tropospheric jet stream level (~200 hPa) and the other one dispersing straight upward and converging into the polar night jet. The regions of EP flux convergence, indicated by blue shading,

[2] Charney and Drazin (1961).

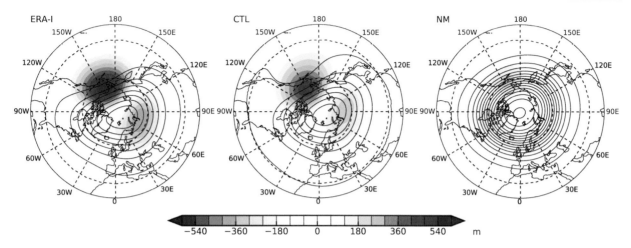

Figure 11.7 The DJF eddy component of the Z_{30} (colored shading) and the total Z_{30} field, contour interval 200 m, in the same three panel format as in Fig. 11.2.

Figure 11.8 Longitude–height sections along 60°N showing the departure of the geopotential height from the zonal mean during DJF. (Top) based on ERA-I, (middle) the WACCM control run with realistic orography (CTL), and (bottom) the WACCM "no mountains" (NM) run.

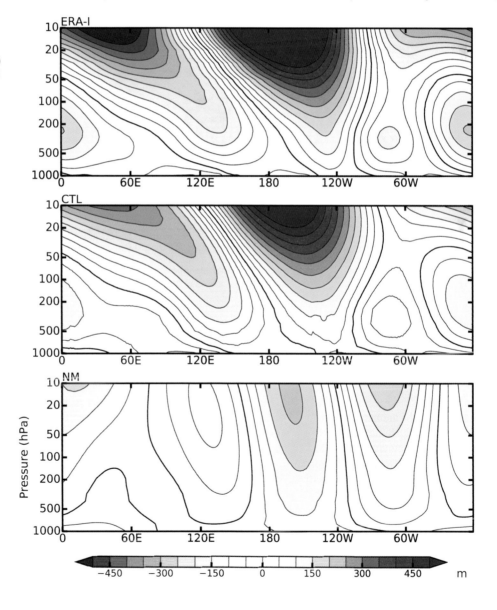

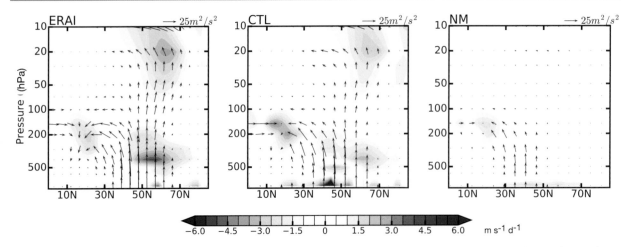

Figure 11.9 Meridional cross sections of DJF Eliassen–Palm fluxes by the stationary waves (arrows) and its divergence (colored shading) in the same three panel format as in Fig. 11.2. The vectors have been scaled by the inverse square root of density to make the EP fluxes in the stratosphere clearly visible.

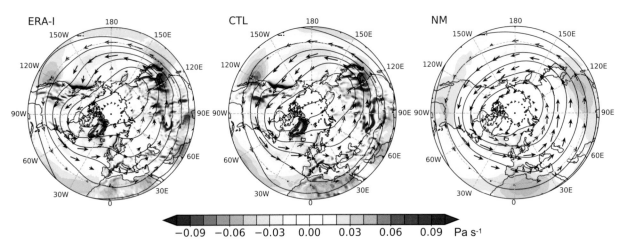

Figure 11.10 The DJF 500 hPa vertical velocity fields (colored shading) superimposed on 1000–300 hPa layer averaged wind vectors (the longest arrows correspond to 25 m s^{-1}) and 1000-300 hPa thickness (contour interval 200 m), an indicator of the mean temperature of the layer. The three panels are arranged as in Fig. 11.2.

are characterized by breaking planetary waves, which extract westerly momentum from the zonal mean flow in accordance with the properties of EP fluxes as discussed in Section 8.2.

11.1.1 Energetics

Making use of the fields shown in Fig. 11.10, it is possible to infer the algebraic signs of C_A and C_K in the stationary waves. The contoured field in Fig. 11.10 is the thickness of the 1000–300 hPa layer, a measure of tropospheric temperature and the vectors show the vertically averaged wind field in the same layer. Consistent with the prevalence of poleward heat transports, a strong positive correlation between the stationary wave fields $\bar{v}^*$ and $\bar{T}^*$ is clearly evident. Meridional temperature advection is acting to amplify the stationary waves while weakening the meridional gradient of zonally averaged temperature, thereby converting zonal to eddy available potential energy, just as it does in amplifying baroclinic waves. Comparing the thickness and mid-tropospheric vertical velocity fields in Fig. 11.10, it is evident that air is

rising in the ridges over the eastern oceans and sinking in the troughs over East Asia and (albeit less distinctly) over the eastern seaboard of North America. The correspondence is not perfect, but the marked tendency for the rising of warmer air and the sinking of colder air is indicative of a conversion from potential to kinetic energy, also analogous to that in baroclinic waves. The energy conversion in the winter stratospheric planetary waves (not shown) is in the same sense.

The sign of the barotropic energy conversion C_K in the stationary waves can be inferred from the EP fluxes in Fig. 11.9. The stream of equatorward arrows centered at the 200 hPa level is indicative of a poleward, mostly countergradient transport of westerly momentum analogous to the one observed in the later stages of the life cycle of baroclinic waves (Fig. 8.12). Hence, the barotropic energy conversion is from the stationary waves back to the zonally symmetric flow. The same is true of the higher stream of equatorward arrows at the level of the wintertime polar night jet, where the energy conversion is much larger owing to the stronger gradient of zonally averaged zonal wind.

Although the conversions in the mechanical energy cycle in the stationary waves are qualitatively similar to those in baroclinic waves, the dynamical interpretation is somewhat different. Rather than playing the central role in maintaining the waves and being responsible for their existence, the conversion from zonal to eddy available potential energy C_A in the stationary waves can be viewed as a positive feedback that serves to amplify the response to whatever wave forcing is present. To understand why the stationary waves exist and why they assume the form that they do, it is necessary to consider the responses to the various forcings: orography, diabatic heating, and the transports by the transients. In the next three sections, we will consider the responses individually and in the final section of this chapter we will attempt a synthesis of these results, taking nonlinearity into account.

Exercises

11.1 It is evident from Fig. 1.30 that at tropospheric levels, most of the kinetic energy in the climatological mean stationary waves resides in the zonal wind component. How is this anisotropy reflected in the structure of the stationary waves in the 300 hPa height field shown in the left panel of Fig. 11.6?

11.2 In Figure 1.30 the maxima in the r.m.s. amplitude of $\overline{v^{*2}}$ lie in latitude belts that correspond to minima in the r.m.s. amplitude of $\overline{u^{*2}}$ and vice versa. Why is this the case?

11.3 Making use of Fig. 1.30, compare the r.m.s. amplitudes of the stationary waves in the boreal winter with those in the boreal summer, austral winter, and austral summer.

11.4 Using the figures for this section in the mini-altas on the companion web page, document the structure of the stationary waves in both Northern and Southern Hemispheres in both DJF and JJA.

11.5 Based on the global fields shown in Figs. 1.4 and 1.5, relate the wintertime jet streams at the 300 hPa level to the Z_{500} and to the storm tracks, as inferred from the variance of v'_{700} (Fig. 11.2).

11.6 Relate the zonally varying 850 hPa temperature pattern shown in the bottom left panel of Fig. 1.10 to the boreal wintertime geopotential height field at the jet stream level, shown in Fig. 11.6. Show that the patterns are hydrostatically consistent with the westward tilt of the ridges and troughs with height.

11.7 Relate the storm tracks in Fig. 11.2 to the global distribution of precipitation shown in Figs. 1.14 and 1.15.

11.2 OROGRAPHIC FORCING

To investigate the role of mountains in forcing the stationary waves, we compare a control run (CTL) with realistic land–sea geometry and orography and a "no mountains (NM)" run with the same land–sea geometry and SST distribution but lacking orographic forcing.

11.2.1 Impact on the Zonally Symmetric Flow

Figure 11.11 shows a set of meridional cross sections of climatological mean zonally averaged zonal wind $[\overline{u}]$ during the boreal winter months DJF. The left panel is based on observations and the middle panel is from a control run of the WACCM that has both orographic and thermal forcing, and includes all parameterizations of subgrid-scale processes. Note the close agreement. The right "no mountain" (NM) panel is based on the same model, but the land surface is completely flattened and the orographic forcing is entirely eliminated in both the resolved dynamics and in the parameterizations that involve gravity wave drag. The westerlies are substantially stronger in the NM run than in the control run, especially in the Northern Hemisphere. The strengthening is most pronounced at the uppermost levels but it is discernible all the way down to the Earth's surface.

In a set of auxiliary experiments, the influence of the orography in reducing the amplitude of the stratospheric polar night jet was partitioned into the drag from the EP flux convergence by resolved waves, estimated on the basis of Eq. (10.6), and the drag associated with subgrid scale orographic features that are parameterized in the model, including the

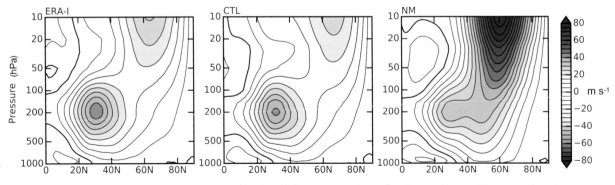

Figure 11.11 Meridional cross sections of DJF zonally averaged zonal wind $[u]$, contour interval 5 m s^{-1}, in the same three panel format as in Fig. 11.2.

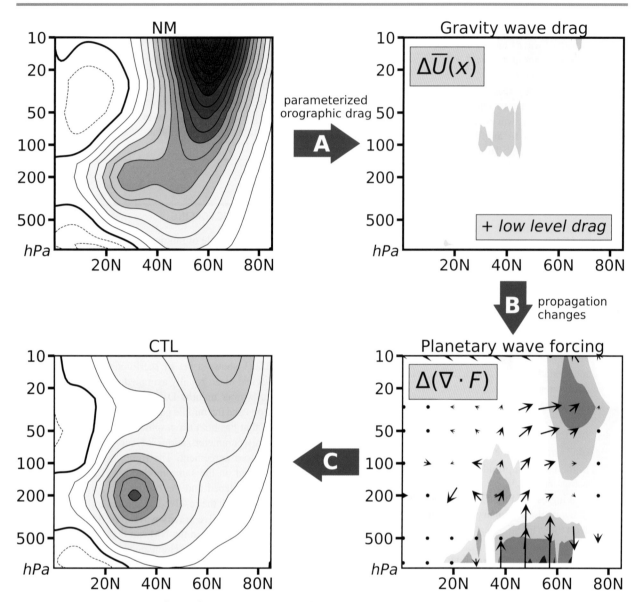

Figure 11.12 Schematic showing the influence of orography on DJF NH stratospheric zonally averaged zonal wind [$\overline{u}$]. (Top left) NM [$\overline{u}$]. (Top right and arrow "A") Incremental forcing of [$\overline{u}$] due to unresolved gravity waves alone. (Bottom right and arrow "B") Subsequent change in EP flux and EP flux divergence initiated by the weakening of the westerlies in "A." The enhanced wave forcing leads to a further weakening of the westerlies at stratospheric levels, as reflected in the climatology of the control experiment CTL (bottom left and arrow "C"). From White et al. (2021). © American Meteorological Society. Used with permission.

drag at stratospheric levels due to the breaking of small-scale (parameterized) gravity waves that are generated in response to the roughness of the surface. Neither the forcing by the resolved or the unresolved waves, acting alone, can account for more than a small fraction of the marked weakening of the jet in the control experiment CTL as compared with the No Mountain Experiment NM. The much larger impact of the combined forcings can be understood in terms of the schematic shown in Fig. 11.12. The direct effect of drag from unresolved gravity waves slightly reduces the strength of the westerlies just above the tropospheric jet stream (the blue-shaded region in the top right panel). The weakening of the westerlies at this level renders the stratospheric polar night jet more susceptible to the breaking of planetary

waves dispersing upward from the troposphere, resulting in enhanced EP flux and EP flux convergence (bottom right panel). It is this positive feedback between the forcing by resolved and unresolved orography that makes the whole so much greater than the sum of the parts.

Since the Southern Hemisphere's high mountain range in the latitude belt of the westerlies, the Andes, is much more limited in longitudinal extent than the ranges on the Northern Hemisphere continents, and the Andes achieve their highest altitudes equatorward of the westerly wind belt, comparing control (CTL) and No Mountain (NM) simulations in a climate model can be considered as analogous, in some respects, to comparing Northern and Southern Hemisphere wintertime climatologies, using the latter as an indicator

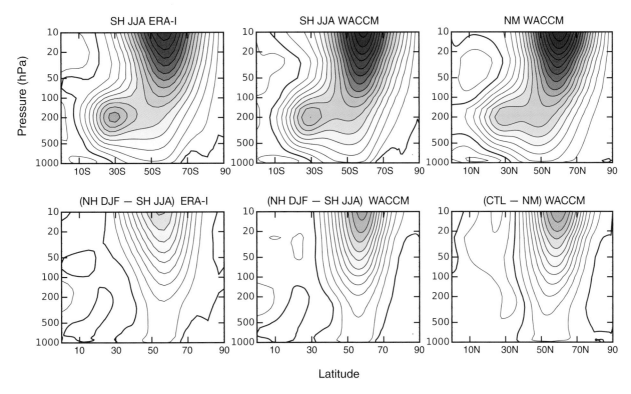

Figure 11.13 Meridional cross sections of zonally averaged zonal wind [*u*]. (Top left) Southern Hemisphere JJA, horizontally flipped so that the South Pole is on the right-hand side, based on ERA-I; (top middle) same but as simulated by the WACCM and; (top right) WACCM no mountain (NM) simulation for Northern Hemisphere DJF. (Bottom left) Southern Hemisphere winter (horizontally flipped) minus Northern Hemisphere winter based on ERA-I; (bottom middle) same but as simulated by WACCM; (bottom right) WACCM control run (CTL) minus no mountain (NM) simulation for Northern Hemisphere DJF. Contour interval 5 m s^{-1}, zero contour is bolded. In top panels gold highlights indicate westerlies and blue highlights indicate easterlies. In bottom panels, blue highlights indicate negative extrema where the westerlies are significantly weakened by the presence of orography.

of what the zonally symmetric circulation would be like on a planet without mountains; that is, as a surrogate for the NM simulation. It is evident from Fig. 1.30 that the Southern Hemisphere wintertime stationary waves are much weaker than their Northern Hemisphere counterparts, and from Fig. 5.16 that the poleward heat transport by the Southern Hemisphere stationary waves, an indicator of the upward flux of wave activity, is much weaker than that associated with the Northern Hemisphere stationary waves. Figure 11.13 contrasts the zonal mean zonal wind in the Northern and Southern winters in (left) observations and (middle) the WACCM CTL simulation; the right panel shows WACCM CTL minus NM for the Northern Hemisphere winter. The similarity of the patterns in the three panels confirms that the zonally symmetric circulation in the NM experiment mimics the Southern Hemisphere wintertime zonally symmetric circulation, and that the Andes range has little impact on the zonal mean zonal wind.

11.2.2 Impact on the Stationary Waves

Having dealt with the influence of the mountains on the zonally symmetric flow, let us now consider their impact on the stationary waves. It is evident from a comparison of the middle and right panels of Figs. 11.4–11.7 that

including mountains in the WACCM simulations substantially increases the prominence of the stationary waves. It does so for two reasons: (1) it increases the r.m.s. amplitude of the waves, as evidenced by the more robust color shading in the eddy fields in the CTL than in NM and (2) it weakens the zonally symmetric flow, thereby rendering the total (zonally symmetric plus eddy) flow more wavy. The greater waviness is most clearly evident at the Earth's surface (Fig. 11.4), where the Aleutian and Icelandic lows and the Siberian high are barely discernible in the total SLP field of the NM simulation.

That the upper level Z^* fields are stronger in the CTL than in the NM experiments (Fig. 11.7), yet similar in shape, indicates that the response to orographic forcing (CTL-NM) projects positively upon the NM pattern, from which it can be inferred that the weakening of the zonally symmetric flow due to the inclusion of mountains amplifies the stationary-wave response to the planetary wave forcing as a whole. The strong influence of the mountains on the boreal winter general circulation also derives from the symbiotic relationship between the mechanical and thermal influences that exists by virtue of the particular positioning of the mountain ranges on the Northern Hemisphere continents. For example, a numerical simulation of the general circulation of a planet with the same geomorphology as Earth, but rotating in the

Figure 11.14 Velocity streamfunction field in the eddies induced by flow over a localized circular mountain at 30°N, 90°E longitude as inferred from a nondivergent shallow water wave equation model (Eq. (11.4)) linearized about two different zonally symmetric background flows: (top) pure superrotation (constant angular velocity $[\bar{u}]\cos\phi$), with westerly flow everywhere and a value of 20 m s^{-1} on the equator, and (bottom) a more realistic zonal mean flow. The prescribed linear damping is sufficient to prevent the Rossby wave train from dispersing all the way back to the mountain and thereby causing the perturbations to amplify with time. From Held et al. (2002). Published (2002) by the American Meteorological Society.

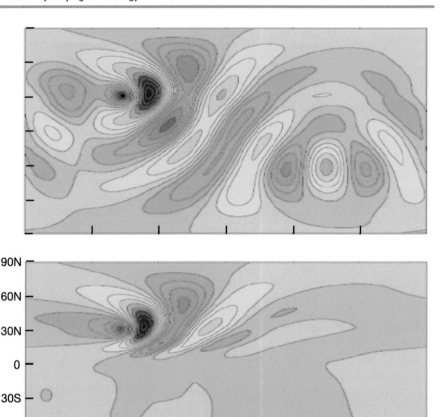

opposite direction yields much weaker Northern Hemisphere stationary waves, presumably because the relationship in this case is not as symbiotic.[3]

All things considered, the NM simulation shows that the climate of a hypothetical planet with Earth-like, but flat continents would be more like that of an aquaplanet model than like the observed climate. The presence of orography and, in particular, the extensive Northern Hemisphere mountain ranges greatly enhances the zonal asymmetries in the boreal wintertime climate at all levels, deepening the oceanic, subpolar lows at the Earth's surface, enhancing the east–west temperature contrasts along latitude circles, and weakening the polar night jet and thereby rendering it more susceptible to midwinter sudden warmings.

11.2.3 Insights Derived from Experiments with a 2D Model

Since orography influences the stationary waves indirectly, by slowing down the zonally symmetric flow, and thereby enabling the stationary waves to disperse upward more readily, the differences between the CTL and NM simulation can-

not be interpreted as a linear wave response to the orographic forcing. To show what a linear response looks like, we rely on results of numerical experiments conducted back in the 1980s to investigate how the major Northern Hemisphere orographic features (Tibet and the East Asian Mountains, the Rocky Mountains, and the Greenland ice sheet) perturb the climatological mean flow. These experiments closely followed the discovery that in an atmosphere in solid body rotation, the eastward dispersion of Rossby waves from a local source region is not along latitude circles, as assumed in earlier theoretical studies of the stationary waves, but along great circles.[4] A mountain range perturbs the vorticity field in a shallow water wave equation model by deflecting the flow around the mountain range or, equivalently, acting as a mass source/vorticity sink upstream of the mountain, and the opposite downstream of it.

Let us consider first the nondivergent, steady state Rossby wave response to an idealized circular mountain at 30°N in the Northern Hemisphere, using the shallow water wave equations described in Section 10.2.1. For nondivergent barotropic flow, $dh/dt = 0$ and thus height h in Eqs. (10.2)

---

[3] White et al. (2021).

[4] Hoskins et al. (1977).

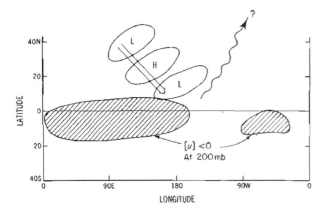

Figure 11.15 Schematic of the Rossby wave train generated by the East Asian orography dispersing into the tropics. The DJF mean zonal flow in the upper troposphere is easterly within the hatched areas. The boundary of these regions acts as a critical line for the stationary Rossby waves. From Held et al. (1983). © Academic Press, Inc. Published by Elsevier Ltd. All rights reserved.

and (10.3) is identical to the streamfunction for the horizontal wind field ($\mathbf{V} = \mathbf{k} \times \nabla h \equiv \mathbf{V}_h$) and the two momentum equations reduce to the barotropic vorticity equation:

$$\frac{\partial \zeta}{\partial t} = -\mathbf{V}_h \cdot \nabla (\zeta + f) - r\zeta, \tag{11.4}$$

where $\zeta = \nabla^2 h$ is the vertical component of relative vorticity and we have included a linear damping term $-r\zeta$ to prevent the wave amplitude from increasing indefinitely in response to the sustained forcing. This equation can be integrated forward in time until the flow reaches a steady state. When the basic state upper tropospheric flow is prescribed to be pure superrotation (i.e., constant angular velocity, with westerly flow), the streamfunction field for the Rossby wave response, shown in the top panel of Fig. 11.14, consists of a wave train that disperses eastward along a great circle: southeastward across the equator to the antipodal point in the Southern Hemisphere, and then back toward the latitude of its source. The wave train is attenuated along the way by the prescribed linear damping in the model. When the experiment is repeated using a more realistic zonal mean zonal flow, as shown in the bottom panel, the wave train is absorbed when it encounters the zero zonal wind (the critical line for stationary waves) in the zonal mean flow, at $\sim 10°N$. In the real world, the situation is a bit more complicated because the critical line is interrupted over the eastern Pacific, where westerlies extend across the equatorial belt in the DJF mean climatology at upper tropospheric levels (Figs. 11.15 and 16.4). In that limited sector, waves originating in the Northern Hemisphere are able to disperse across the equator.[5]

Now let us consider the response to more realistic orographic forcing, a highly smoothed version of the Northern Hemisphere orography as depicted in the left panel of Fig. 11.16. The meridional profile of climatological zonal mean zonal wind at the Earth's surface in DJF $[\bar{u}_s]$, shown in the

right panel, flows over the smoothed terrain. The mountain range perturbs the basic state vorticity field $f - \partial[\bar{u}]/\partial y$ (which is prescribed to match that observed at 300 hPa), either by producing horizontal divergence on the upslope side and convergence on the downslope side, or by deflecting the flow around the mountain range. This gives rise to Rossby wave trains to the east of each of the major mountain ranges, as shown in the middle panel of Fig. 11.17. This pattern resembles the observed 300 hPa geopotential height field shown in the left panel of Fig. 11.6. As in the idealized experiment, the amplitude of the waves and the rate at which they are attenuated as they disperse are controlled by the prescribed damping rate. Since the model is linear, the simulated response is the sum of the responses to the individual mountain ranges. The responses to the mountains in the eastern hemisphere (the east Asian orography) and the western hemisphere (mainly the Rocky Mountains), shown in the left and right panels of Fig. 11.17, resemble the idealized wave trains in Fig. 11.14. When a perfectly reflecting wall is placed at the equator, the tilt is almost entirely eliminated, as shown in the right panel of Fig. 11.18.

11.3 DIABATIC FORCING

In this section, we will briefly examine two lines of evidence concerning the influence of the distribution of diabatic heating upon the climatological mean extratropical general circulation, the first based on observations and the second based on experiments with a linearized primitive equation model.

11.3.1 The Observed Heating and its Relation to the Storm Tracks

The distributions of DJF diabatic heating based on ERA-I and as simulated by WACCM with and without mountains are shown in Fig. 11.19. The boundary layer heating shown in the bottom panels is strongest off the eastern seaboards of Asia and North America, where the ocean surface is much warmer than the air flowing off the continents during wintertime. They are enhanced by the existence of the warm, wind-driven western boundary currents in these regions, the Gulf Stream in the Atlantic and the Kuroshio–Oyashio current system in the Pacific.

The sharpness of these features is revealed in Fig. 11.20, which shows the speed of the surface current in the left panel and the Laplacians of the SST and SLP fields in the center and right panels, respectively. Viewing the Laplacians, rather than the fields themselves, reveals the disproportionally strong influence of subtle features in the SST field with scales of 100 km or less, oriented in the direction transverse to the Gulf Stream, upon the SLP field. The core of the Gulf Stream is warmer than its immediate surroundings and is thus marked by an elongated band of strong negative Laplacian of SST. Sensible heat fluxes at the air–sea interface convey this excess warmth to the atmospheric boundary

---

[5] Webster and Holton (1982). For further discussion, see Section 19.3.

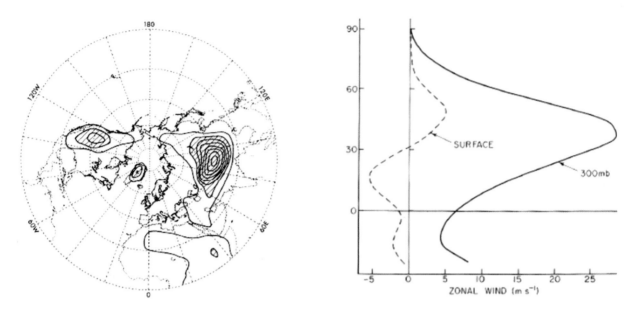

Figure 11.16 Experimental setup for a numerical experiment with a linearized barotropic shallow water wave equation model. The left panel shows the smoothed mountain ranges used to perturb the vorticity field. The right panel shows meridional profiles of the prescribed zonally averaged DJF zonal surface winds that blow over the mountain ranges (dashed line) and the 300 hPa background, zonally averaged zonal wind (solid line), both based on observations. The domain in this plot and in the next two plots is poleward of the equator, rather than 20°N. From Held et al. (1983). © Academic Press, Inc. Published by Elsevier Ltd. All rights reserved.

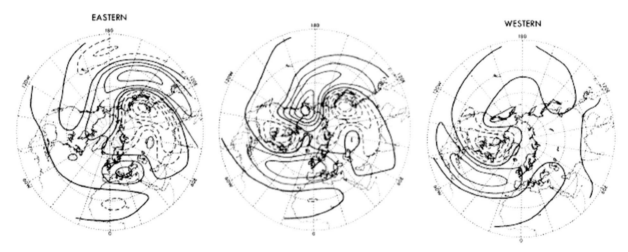

Figure 11.17 Simulations of the DJF orographically-forced geopotential height field at the 300 hPa level with a linearized barotropic shallow water wave equation model with a 20 day damping time: (center) all mountains; (left) mountains of the eastern hemisphere; (right) mountains of the western hemisphere. Contour interval 50 m. From Held et al. (1983). © Academic Press, Inc. Published by Elsevier Ltd. All rights reserved.

layer, inducing a hydrostatic response in the SLP field with a narrow band of low pressure directly over the Gulf Stream. Moisture fluxes are also enhanced over the Gulf Stream, increasing the convective available potential energy (CAPE) of the boundary layer air that moves across it. The fluxes are particularly strong on winter days with strong northwesterly flow from the cold continent flowing over the much warmer water, rendering the boundary layer highly unstable. Because this band of low pressure along the core of the Gulf Stream is thinner than the Rossby radius of deformation,[6] it induces

boundary layer convergence. The low-level convergence, in turn, induces a narrow "chimney" of ascent, which slopes eastward with height, as shown in the left-hand panel of Fig. 11.21. Upper tropospheric divergence (middle panel) and an enhanced frequency of pixels with low OLR (right panel), indicative of enhanced deep convection and rain rate, are clearly evident, shifted slightly to the east of the band of low-level convergence. The frequency of incidence of lightning is also enhanced in this band.[7]

---

[6] See section 3.9.2 of Vallis (2017).

[7] For example, see Fig. 21.2 and Figs. 3 and 4 of Virts et al. (2013).

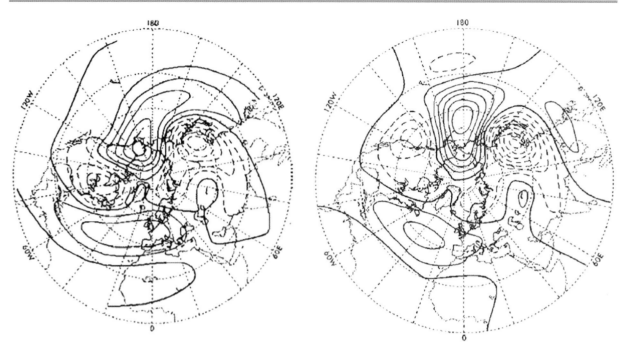

Figure 11.18 (Left) 300 hPa height response to orographic forcing derived from the linearized model, repeated from the middle panel of Fig. 11.17. (Right) The same experiment rerun with a perfectly reflecting wall at the equator. From Held et al. (1983). © Academic Press, Inc. Published by Elsevier Ltd. All rights reserved.

Numerical experiments with atmospheric models forced by prescribed SST distributions indicate that sharpness of the SST gradients along the western and poleward flanks of the western boundary currents contributes to the prominence of the oceanic storm tracks. Representing the sharpness of these features and resolving their week-to-week variability accurately has been shown to have a beneficial impact on the skill of long-range weather prediction,[8] though there remain questions about the dynamical mechanisms that are responsible for it.

11.3.2 Impact on the Stationary Waves

The stationary wave response to the diabatic forcing has been estimated using a linearized primitive equation model, prescribing the eddy component of the diabatic heating in accordance with the observations or with the climatology of an AGCM run with realistic boundary conditions. Figure 11.22 shows the response partitioned into tropical and extratropical contributions, where the boundary between tropics and extratropics is set at $25°$N.[9] The tropical contribution emanates mainly from the strong heat source over the Maritime Continent and disperses poleward and then eastward across North America. The extratropical heating yields a response with wave troughs upstream and ridges downstream of the storm tracks, consistent with linear theory. Both tropical and extratropical components of the response project positively on the observed stationary waves. We will

reserve discussion of the amplitude of these features until Section 11.5.

11.4 FORCING BY THE TRANSIENTS

In discussing the forcing by the transient eddies, it is convenient to treat the time-mean, zonally varying flow, including both the stationary waves and the climatological mean zonally symmetric flow as a single entity. The analysis is less cumbersome, but no less revealing in terms of the forcing of the stationary waves. The mechanical energy cycle is reformulated in terms of the time-mean circulation (M) and the transients (T), as shown schematically in the right panel of Fig. 6.6; there is no partitioning between zonal means and eddies in this framework. In the governing Eqs. (7.1)–(7.4), the continuity and thermal wind equations are generalized to three dimensions and, in place of the conservation of angular momentum, we make use of the conservation of vorticity about a vertical axis. In the three-dimensional time-mean formulation, the geostrophic flow is zonally varying and the ageostrophic flow that maintains it comprises zonal as well as meridional overturning circulations. With these substitutions and generalizations, the dynamics of the time-mean flow are well described by the familiar quasi-geostrophic tendency and omega equations, augmented with additional terms to take into account the forcing of the time-mean flow by the transients by way of their horizontal transports of heat and vorticity. For example, if, at a particular place in space, the transient fluctuations v' and T' are positively correlated in time, then there will be a northward heat transport by the transients ($\overline{v'T'} > 0$). Regions in which the two-dimensional heat transport vector ($\overline{u'T'}$, $\overline{v'T'}$) converges (i.e., where $\nabla\cdot$

---

[8] Roberts et al. (2020).
[9] Held et al. (2002). In this simulation, the forcing by the transient eddy heat transports is included as a part of the diabatic forcing.

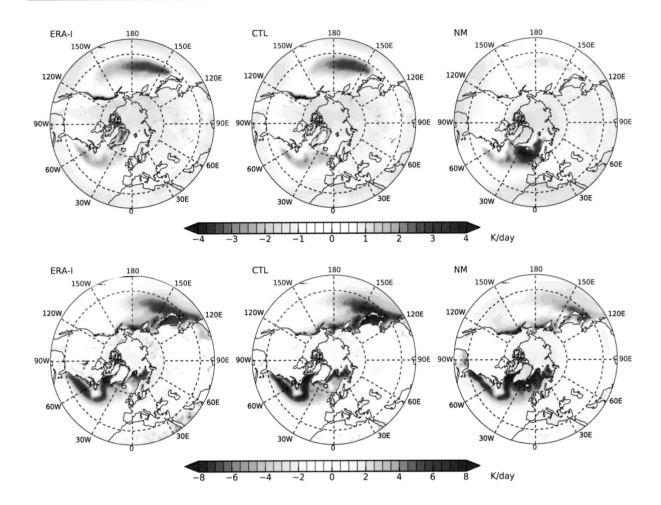

Figure 11.19 Vertically averaged DJF mass-weighted diabatic heating rate: (top panels) 850–500 hPa layer, (bottom panels) 1000–850 hPa layer. (Left panels) ERA-I, (middle panels) the WACCM control run (CTL), and (right panels) the WACCM "no mountains" (NM) simulation.

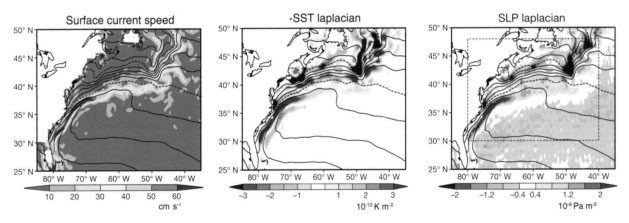

Figure 11.20 (Left panel) Surface geostrophic current speed showing the core of the Gulf Stream. (Center panel) Colored shading shows the Laplacian of the SST field with sign reversed; in effect, a spatially highpass filtered representation of it. (Right panel) Laplacian of the SLP field based on a ~4-year record of monthly mean, high resolution ECMWF operational analyses. The contours in all three panels are SST at 2°C intervals; the 10°C and 20°C contours are dashed. Reprinted by permission from Springer Nature Customer Service Centre GmbH: Springer Nature, Nature, Minobe et al. © 2008.

$\overline{V'T'} < 0$) experience heating. The equation for the local time rate of change of vorticity contains an analogous transient forcing term.

Evaluation of the transient terms is complicated by the fact that they contain strong nondivergent components, which do not contribute to the forcing by the transients: the

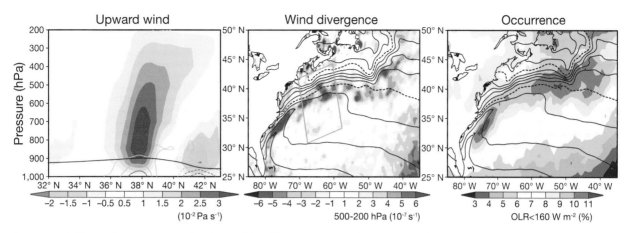

Figure 11.21 (Left) Vertical cross section, normal to the axis of the Gulf Stream and averaged along the length of the green box in the middle panel, showing vertical velocity in pressure coordinates based on the ERA operational analyses. (Middle) Horizontal divergence averaged over the 500–200 hPa layer from the same analyses. (Right) Fraction of the pixels with OLR < 160 W m^{-2}. The contours in the center and right panels are SST. Reprinted by permission from Springer Nature Customer Service Centre GmbH: Springer Nature, Nature, Minobe et al. © 2008.

Figure 11.22 Thermally forced stationary waves in the 300 hPa eddy streamfunction field as simulated by a linearized, steady state primitive equation model in which the zonal mean zonal wind is set to the observed climatological January flow, and the diabatic heating is prescribed on the basis of an AGCM run. "Tropical" denotes the contribution from heating equatorward of 25° latitude and "Extratropical" denotes heating poleward of that latitude. Contour interval 3 ×10^6 m^2 s^{-1}. From Held et al. (2002). Published (2002) by the American Meteorological Society.

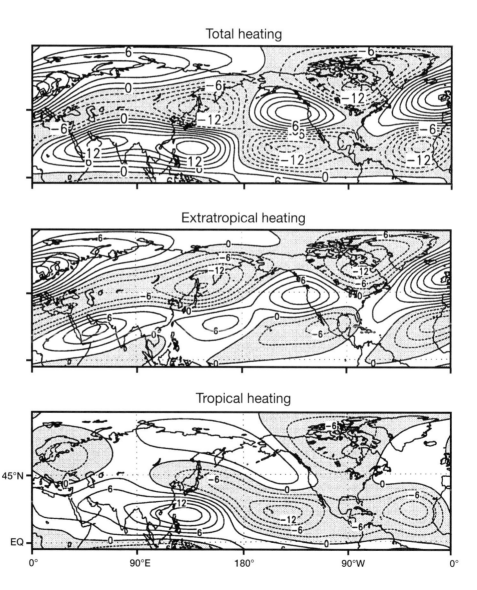

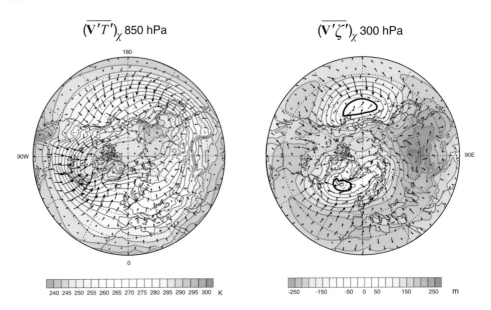

$$(\overline{\mathbf{V}'T'})_\chi \ 850 \ \text{hPa} \qquad (\overline{\mathbf{V}'\zeta'})_\chi \ 300 \ \text{hPa}$$

Figure 11.23 (Left) The DJF irrotational component of the eddy heat transport by the transients at 850 hPa superimposed upon the 850 hPa temperature field. The longest vector is on the order of 30 °C m s⁻¹ and the contour interval is 2.5 K. (Right panel) The DJF irrotational component of the 300 hPa vorticity transport by the transients superimposed upon the 1000 hPa height field. The 273 K (0°C) contour is bolded. The longest vector is on the order of 5×10^{-5} m s⁻² and the contour interval is 25 m.

heat transport vectors tend to parallel the contours of time-mean temperature and the vorticity transport vectors tend to parallel the contours of time-mean vorticity.[10] To make the forcing of the time-mean flow more explicit it is helpful to decompose the vectorial transport fields into irrotational and nondivergent components.

The left panel of Fig. 11.23 shows the irrotational transport of heat at the 850 hPa level, superimposed upon the time-mean temperature field. The transport is mainly poleward, but it exhibits a zonal component directed down the zonal temperature gradient. The poleward transport is larger over the oceanic storm tracks than over the continents, where the strongest meridional temperature gradients are observed. Hence, the transport cannot be characterized as being linearly proportional to the local temperature gradient, as it is in isotropic turbulence. The transport of relative vorticity at the jet stream level tends to be directed down the Z_{1000} gradient (Fig. 11.23, right panel) into the Icelandic and Aleutian lows, balancing the frictional sink in the vertically integrated, zonally varying vorticity budget.

The poleward vorticity transports are strongest over the oceanic storm tracks. In Exercise (3.24) the reader is invited to prove that if the horizontal wind field is assumed to be nondivergent, $G \equiv -\partial[u^*v^*]/\partial y = [\zeta^*v^*]$. The same operations that are applied to the eddies in that exercise can be applied to the transients here. Hence, the convergence of the transport of westerly momentum by the transients tends to accelerate the westerly flow in the vicinity of the storm tracks, and it is strongest at the jet stream level. It follows that in the storm tracks, the transients act to drive the time-mean flow out of thermal wind balance: the poleward heat transports weaken the meridional temperature gradient while the vorticity transports strengthen the vertical

wind shear. In analogy with the dynamics of the zonally symmetric circulation discussed in Chapter 7, this should tend to drive meridional circulation cells (in the longitudinal sectors of the storm tracks) that transport westerly momentum downward from the jet stream level into the boundary layer. That the poleward eddy transports of momentum and vorticity tend to be focused in the sectors of the storm tracks suggests that they are mainly attributable to baroclinic waves. We will confirm that this is, in fact, the case in Chapter 12.

The forcing of the time-mean flow by the transients (mainly the transient eddies) in the extratropics can be diagnosed quantitatively using the quasi-geostrophic formulation.[11] The geopotential tendency equation is written as

$$A\frac{\partial \Phi}{\partial t} = D + R_1, \tag{11.5}$$

subject to the bottom boundary condition that

$$C(p)\frac{\partial}{\partial p}\frac{\partial \Phi}{\partial t} = -\nabla \cdot \overline{\mathbf{V}'\theta'} + R_2, \tag{11.6}$$

where A is an elliptic operator that appears in the tendency equation, C is a function of pressure, D is the forcing due to the transients as detailed below, and R_1 and R_2 are collections of terms involving advection of vorticity and potential temperature by the time-mean flow. Since the tendency of the time mean flow is zero, it follows that the terms of the right-hand sides of Eqs. (11.5) and (11.6) must be equal and opposite. To get an indication of how the transients and their associated mean meridional circulations are forcing the time-mean flow, one can compute the geopotential tendency that they would produce if they were acting in isolation. The forcing by the transients is given by

$$D = D^{VORT} + D^{HEAT}, \tag{11.7}$$

---

[10] Lau and Wallace (1979).

[11] As adapted in Lau and Holopainen (1984).

where

$$D^{VORT} = -\nabla \cdot \overline{\mathbf{V}'\zeta'} \qquad (11.8)$$

and

$$D^{HEAT} = f\frac{\partial}{\partial p}\frac{\nabla \cdot \overline{\mathbf{V}'T'}}{\overline{S}}, \qquad (11.9)$$

where $\overline{S}$ is the hemispheric mean of the static stability, here defined as $-\partial\overline{\theta}/\partial p$. The transport terms in these expressions are analogous to the advection terms in the conventional quasi-geostrophic tendency equation, as discussed in Section 7.5. The geostrophic wind and temperature tendency fields induced by the transients can be deduced by differentiating the geopotential tendency in accordance with the geostrophic and hydrostatic balance equations.

The geopotential tendencies induced by the transients at the 300 and 1000 hPa levels are shown in Fig. 11.24. The transients induce negative geopotential tendencies over the Aleutian and Icelandic lows at the 300 hPa level. The stronger pattern at the 1000 hPa level consists of a series of dipoles with nodes centered along ~45°N, with negative tendencies to the north and positive tendencies to the south. The forced eastward wind tendency is particularly strong at the longitudes of the oceanic storm tracks, which are subject to boundary forcing by the strong poleward eddy heat transport at the Earth's surface, as described in Section 7.4.3. These tendencies are in the sense as to strengthen the geostrophic winds at the Earth's surface along ~45°N. When the forcing by the high and low frequency transients is examined separately, it is found that the robust pattern at the 1000 level is almost exclusively due to the forcing by the high frequency transients with periods shorter than a week. We will show in the next chapter that the high frequency transients are dominated by baroclinic waves.

In Fig. 11.25, the eddy-induced tendencies at the 300 and 1000 hPa levels are partitioned into the contributions from the heat transport and the vorticity transport. Consistent with the notion that the heat transport should act to reduce the baroclinicity of the time-mean flow, the geopotential tendencies induced by the heat transports are in the opposing directions at the two levels, so that to the north of ~45°N the 1000–300 hPa thickness is increasing (i.e., the transients are inducing warming) and to the south of that latitude it is decreasing. In contrast, the vorticity transports induce a barotropic response, which is of the same polarity at the two levels. At the upper level the tendencies induced by the heat and vorticity transport tend to cancel, whereas at the lower level they tend to reenforce one another. This explains why the 1000 hPa pattern in Fig. 11.24 is stronger and more robust looking than the 300 hPa pattern, particularly in the storm track entrance regions along the eastern seaboards of Asia and North America.

Fig. 11.26 offers a further dynamical interpretation of the geopotential tendencies induced by the transients. The heat transports induce a thermally indirect meridional circulation cell in the same sense as the zonally averaged Ferrel cell,

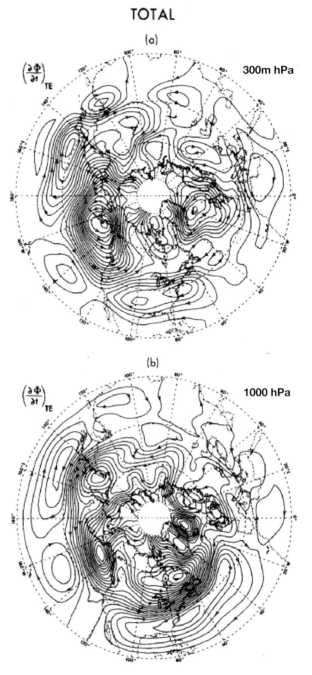

TOTAL

Figure 11.24 Geopotential tendencies induced by the transients in DJF at (top) 300 hPa and (bottom) 1000 hPa. Contour interval 5×10^{-4} m^2 s^{-3}. The arrowheads indicate the sense of the time-mean circulation forced by the transients. From Lau and Holopainen (1984). © American Meteorological Society. Used with permission.

but locally much stronger. The poleward flow in the lower branch of this cell induces a strong westerly acceleration at the Earth's surface at latitudes ~45°N. The vorticity transports are concentrated at the jet stream level, but by virtue of the elliptical operator in the geopotential tendency equation, they induce negative height tendencies poleward of the storm tracks and positive tendencies equatorward of

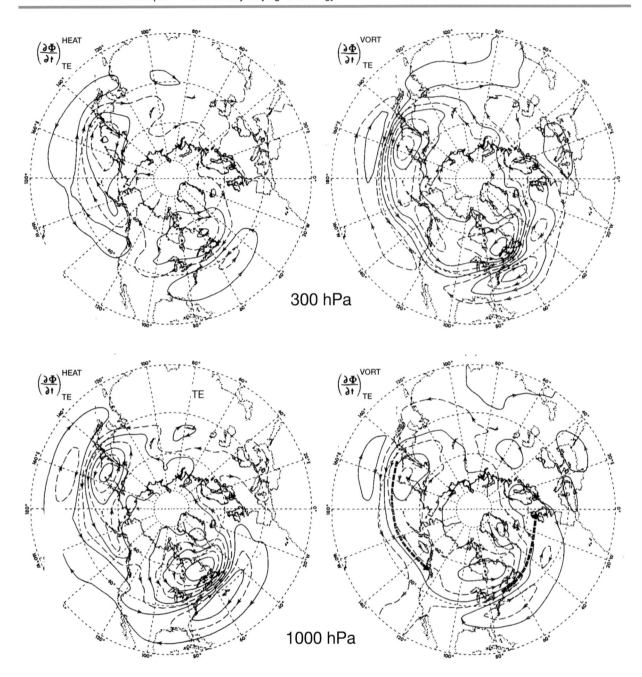

Figure 11.25 Geopotential tendencies induced by the high frequency transients with periods shorter than about a week partitioned into the contribution from (left) the heat transports and (right) the vorticity transports. Contour interval 5×10^{-4} m^2 s^{-3}. The arrowheads indicate the sense of the time-mean circulation forced by the transients. From Lau and Holopainen (1984). © American Meteorological Society. Used with permission.

them at both levels. The eddy-induced mean meridional circulation at the jet stream level is instrumental in spreading the tendencies through the depth of the troposphere, resulting in a more barotropic response to the poleward vorticity transports.

We will confirm in the next chapter that the strong dipole patterns in geopotential tendency, one centered over the western Pacific and the other centered over the western Atlantic, do, in fact, correspond to the climatological mean storm tracks and that the heat and vorticity transports by

the transients in these regions are, in fact, associated with baroclinic waves.

Exercises

11.8   Given the form of the barotropic vorticity equation (11.4), with absolute vorticity behaving as a passive tracer, one might ask whether the eddy transport can be parameterized as down-gradient

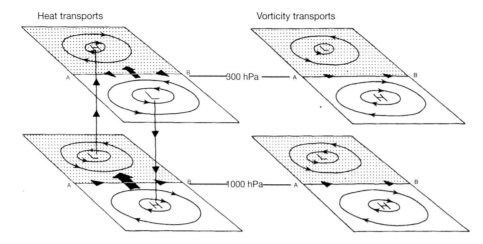

Figure 11.26 Schematic representation of the geopotential tendencies (contours) induced by the (left) heat transports and (right) vorticity transports at (bottom) 1000 hPa and (top) 300 hPa. The bold arrows in the left panel indicate the direction and relative strengths of the heat transports by the transients at the 1000 and 300 hPa levels. The dotted and undotted halves of each plane indicate convergence and divergence of the transports, respectively. Thin arrowheads in the horizontal planes indicate the directions of the transient-induced geostrophic wind tendencies. Arrowheads pointing in the vertical direction indicate the transient-induced vertical motion. From Lau and Holopainen (1984). © American Meteorological Society. Used with permission.

diffusion in zonally averaged general circulation models. Test this hypothesis on the basis of observational evidence.

11.9 In Cartesian plane geometry, u and v are the eastward and northward wind components of a purely horizontal flow and

$$\frac{\partial \overline{u}}{\partial t} = -\frac{\partial}{\partial x}\overline{u'u'} - \frac{\partial}{\partial y}\overline{u'v'}$$

and

$$\frac{\partial \overline{v}}{\partial t} = -\frac{\partial}{\partial x}\overline{u'v'} - \frac{\partial}{\partial y}\overline{v'v'}.$$

Prove that

$$\frac{\partial}{\partial t}\left(\frac{\partial \overline{v}}{\partial x} - \frac{\partial \overline{u}}{\partial y}\right) = \left(\frac{\partial^2}{\partial y^2} - \frac{\partial^2}{\partial x^2}\right)\overline{u'v'}$$
$$+ \frac{\partial^2}{\partial x \partial y}\left(\overline{u'^2} - \overline{v'^2}\right).$$

11.10 In the right panel of Fig. 11.26, the weak mean meridional circulation induced by the vorticity transports by the transients at the jet stream level is not shown. Indicate the sense of this circulation.

11.5 RESPONSE TO THE COMBINED FORCING

Orographic forcing, zonal variations in diabatic heating, and the heat and vorticity transports by transient eddies all contribute to maintaining the observed stationary waves, as evidenced by the fact that the responses to all three forcings project positively upon them. But the results presented in the previous sections do not provide a clear indication of which of the forcings is most important. One analysis protocol that has been used for this purpose is to compare the responses to

the three forcings in the same linearized primitive equation model. Figure 11.27 shows such a comparison based on the 300 hPa streamfunction.[12] The linear solution matches the observations quite well and the patterns for the responses to orographic forcing and the response to the vorticity transports by the transient eddies are generally consistent with the results presented in previous sections. Of the three forcings, diabatic heating induces by far the strongest linear response.

As it turns out, this comparison is misleading because the zonally symmetric background flow in the linear models is prescribed to match the observations, whereas in reality it is profoundly influenced by the presence of the wave forcings, and, in particular, by the orography. As described in Section 11.2, the breaking of resolved and unresolved orographically induced waves substantially reduces the strength of the westerlies poleward of ~40°N, thereby reducing the Doppler-shifted phase speed $c - [u] = -[u]$ of the stationary waves. Air parcels are accordingly exposed to the zonally varying diabatic heating and cooling for a longer time as they pass from west to east through the waves, inducing a stronger response to the heating than would be observed on a flat but otherwise Earth-like planet, even if the pattern of zonally varying diabatic heating were unaffected by the presence of mountains.[13]

In the presence of such strong nonlinearity, it really isn't meaningful to decompose the stationary waves into components attributable to the different forcings. But numerical experiments performed using today's state-of-the-art models that are capable of realistically simulating the observed stationary waves can nonetheless provide valuable dynamical insights. The "Control" versus "No Mountain" runs described in this chapter are but one

[12] Adapted from Figs. 1 and 2 of Held et al. (2002), p. 2125; see also Chang (2009).
[13] Held et al. (2002); Garfinkel et al. (2020).

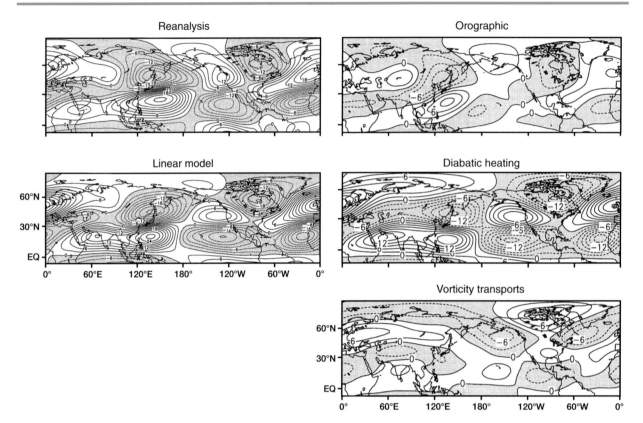

Figure 11.27 Stationary waves in the 300 hPa streamfunction field. (Top left) As represented in the NCEP/NCAR Reanalyses. (Bottom left) As simulated in a steady state primitive equation model, linearized about the climatological mean flow, in which the heating and the momentum and heat transports by the transients are prescribed on the basis of a GCM run. In the right panels the forcing is partitioned into the contributions from orography, diabatic heating including the convergence of sensible heat transports, and from vorticity transports. Contour interval 3×10^6 m^2 s^{-1}. Adapted from Held et al. (2002). Published (2002) by the American Meteorological Society.

example. Another approach is to prescribe various idealized continent and ocean distributions, for example, sectoral continents of prescribed widths with mountain ranges at various longitudes relative to the coastlines, or triangular- or trapezoidal-shaped continents whose shapes mimic the observed continents,[14] that serve as "building blocks" for identifying the various processes that gives rise to the zonally asymmetric circulation.

A deeper level of understanding of the Earth system can be achieved by modeling the distribution of SST rather than prescribing it. Atmospheric circulation models coupled to ocean models, in which the atmosphere exchanges energy and momentum with the ocean, are capable of illuminating the role of atmosphere–ocean interactions on the atmospheric general circulation. To cite two specific examples:

- In contrast to the results obtained using linearized models, experiments with a wide variety of general circulation models – some atmosphere only, some coupled to slab oceans, and some fully coupled – indicate that tropical SST gradients are far more influential in shaping and perturbing the extratropical stationary waves in the winter hemisphere than extratropical SST gradients. The teleconnections between the tropical oceans and the global atmosphere are in response to the upward mass fluxes in tropical convection, which force planetary-scale Rossby waves that disperse poleward along great circles, preferentially into the winter hemisphere, as will be discussed in Chapter 17.

- Results of numerical experiments in which the atmosphere is coupled to a slab ocean indicate that the amplitude of the stationary waves is dependent upon positioning of the mountain ranges relative to the land–sea geometry. For example, if the direction of the Earth's rotation is reversed so that the orography of East Asia and the Rockies are, in effect, on the opposite sides of their respective continents, the Northern Hemisphere wintertime stationary waves are not as strong as the observed. Evidently, in the Earth's atmosphere there exists something of a symbiotic relationship between the thermal and orographic forcing that exists by virtue of the geomorphology.[15]

Exercise

11.11* Discuss how general circulation models can be used to explore what the circulation might look like on a backward-rotating Earth.

---

[14] Brayshaw et al. (2009).

[15] White et al. (2021). See also Brayshaw et al. (2009).

11.6 A HISTORICAL NOTE

The general circulation models in use in the late 1970s with horizontal resolutions on the order of 5° latitude by 5° longitude, were surprisingly successful in capturing the structure of the zonally symmetric flow and the climatological-mean stationary waves. Not until much later was it recognized that the agreement was fortuitous. The higher resolution models that came into use during the early 1980s, which better resolved the baroclinic waves in the storm tracks, exhibited unrealistically strong surface westerlies in midlatitudes. It soon became apparent that this systematic bias was due to the neglect of the drag exerted on the atmosphere by the breaking of orographically induced gravity and inertio-gravity waves on horizontal scales far too small to be resolved by the models. In order to obtain a realistic simulation, it was found to be necessary to resort to a gravity wave drag parameterization scheme.[16] When this refinement was introduced, the fidelity of the models increased quite dramatically and continued to improve with further increases in resolution. In Section 11.2, the important role of orographically induced gravity wave drag was demonstrated by comparing a pair of simulations, one (CTL) with mountains and the other (NM) without them. Aquaplanet simulations (not shown) exhibit unrealistically strong westerlies much like those in the NM experiment, as does the observed Southern Hemisphere general circulation (Fig. 11.13).

Many of the pioneering studies in which CTL and NM simulations were used to diagnose the relative importance of diabatic heating and orography in forcing the climatological mean stationary waves were carried out around 40 years ago, during the period when the models were getting the right answer for the wrong reason – the eddy forcing of the westerlies and the gravity wave drag were both being underestimated, resulting in compensating errors.[17] The mountains in these models induced a stationary wave response, but in the absence of a gravity wave drag parameterization scheme, their presence did not have much of a braking effect on the westerlies. Hence, the "thermal" and "orographic" forcing inferred from the NM fields and the CTL – NM difference fields resembled the responses in linearized models with the prescribed zonal mean flow.[18] When the same experiments are rerun with a current state-of-the art general circulation model (Section 11.2), the orography assumes a more prominent role. Not only is the zonally symmetric flow weaker than in the NM experiments, the stationary waves and the sectoral jet streams and the oceanic storm tracks are more distinct. The zonally averaged poleward heat transport is stronger at all levels and especially in the stratosphere, where it is linked to a higher frequency of occurrence of stratospheric warmings.

In today's higher resolution simulations, orographically induced features in the rain rate climatology come into sharper focus (Fig. 11.3). Some of these features impact the climatology by way of their influence on the diabatic heating field. For example, the presence of coastal mountain ranges like those along the south coast of Alaska, which block the flow of cold, continental air that would otherwise flow over warmer coastal waters, locally reduces the heating of the boundary layer relative to what it would be on a flat planet with the same coastal geometry. The higher-resolution simulations also reveal features in the vertical velocity field that were not previously recognized, such as the bands of ascent immediately downstream of the sharp SST gradients along the landward edge of the western boundary currents (Fig. 11.20). The more vigorous transients in these simulations are also seen to play a more prominent role in forcing both the zonally symmetric flow and the stationary waves.

[16] Palmer et al. (1986).

[17] Manabe and Terpstra (1974); Held (1983).
[18] Held (1983).

12 The High Frequency Extratropical Transients

The realization that correlation statistics could provide useful information on the three-dimensional structure and evolution of the transients (i.e., variations about the seasonally varying climatological mean state) dates back al least 100 years, but at that time studies based on this methodology were largely restricted to the analysis of seasonal or annual mean time series at individual stations. Notable examples include studies of Exner, and Walker and Bliss.[1] From the late 1940s into the 1960s, V. P. Starr and his colleagues were making extensive use of variance and covariance statistics based on daily data from a sparse and irregularly spaced global network of rawinsonde stations in their research (see e.g., Fig. 1.1). It was only by resorting to zonal averaging that it was possible to extract useful information from these statistics. During the early 1970s, it became feasible to calculate three-dimensional fields of covariance statistics making use of newly available, objectively analyzed, gridded data-sets archived by several of the world's operational centers for numerical weather prediction. Despite their limited accuracy by today's standards, even the early operational analyses that were available prior to 1980 were capable of revealing the structure and evolution of the geopotential height and wind fields and qualitatively representing the gross features of the ageostrophic wind field. Coincidentally, it was at about the same time that the computational resources became available to process extended time sequences of large arrays of gridded data, including the memory required to efficiently perform the required calculations.

The reader has already been introduced to maps of the temporal variance and covariance field of atmospheric variables in Figs. 1.33 and 1.34. Here we will make extensive use of them to document the structure of the transients and their interactions with the climatological-mean basic state flow. The oceanic storm tracks and other features of the climatological mean general circulation show up more clearly in the wind, temperature and geopotential height fields when the data are pre-filtered before computing the variance and covariance statistics in order to separate the variability into day-to-day "highpass filtered" and week-to-week or longer "lowpass filtered" components.[2]

This chapter is divided into six sections. The first shows observational evidence of the frequency dependence and anisotropy of the transients. Perturbations with periods less than about a week tend to be organized in zonally oriented wave trains in which the patches of positive geopotential height or streamfunction anomalies are meridionally elongated, whereas those with periods longer than a week are anisotropic in the opposite sense. The second section offers a dynamical interpretation of these results based on considerations relating to phase speed of Rossby waves, taking into account their meridional structure as well as their zonal wavelength. The third section introduces the (temporal) velocity covariance tensor, which is helpful for quantifying the anisotropy of the transients. It is their anisotropy that causes them to behave as waves, rather than turbulent eddies. The fourth section introduces the extended EP flux vector, which is helpful for diagnosing the interactions between the transients and the zonally varying background flow. It is analogous to zonally averaged EP flux defined in Chapter 8, but it includes a zonal component that is related to the anisotropy of the transients.

The first four sections are focused mainly on the structure of the transients. In the fifth section, it is shown that the alternative interpretation of the EP flux, which relates it to the group velocity of Rossby waves, is applicable to the extended EP flux as well and it accounts for the observed "downstream development" of Rossby wave packets, as documented in the sixth section. The final section provides a very brief survey of the voluminous general circulation literature relating to baroclinic waves, which arguably dominate the high frequency transients in the extratropical troposphere.

12.1 FREQUENCY DEPENDENCE AND ANISOTROPY: OBSERVATIONAL EVIDENCE

In this section, we describe, diagnose, and interpret the anisotropy of the horizontal wind field, distinguishing the high frequency transients with periods shorter than about a week from the low frequency transients with longer periods. The partitioning of the transient variability in the frequency domain is accomplished through the use of temporal filters that are applied to the time series of each variable at each grid point. The results presented and discussed in this section are relatively insensitive to the design of the filter that is used in the partitioning; the cutoff can be anywhere between

---

[1] Exner (1913) and Walker and Bliss (1932).
[2] Klein (1951).

214

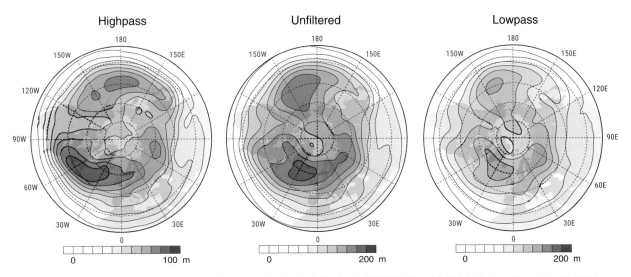

Figure 12.1 Standard deviation of the Northern Hemisphere DJF 500 hPa height field, $\sigma(Z_{500})$. (Middle) Unfiltered; (left) 7 day highpass filtered data; (right) 7 day lowpass filtered data. Contour interval 10 m for highpass and 20 m for lowpass filtered and unfiltered.

5 and 10 days. In the early literature on the transients, what we refer to in this chapter as the "high frequency transients" is subdivided into "bandpass" with periods from 2.5 to 6 or 7 days and "highpass" with periods shorter than 2.5 days.[3] The high frequency end of the spectrum is so red that the bandpass window dominates. Hence, we use the term "high frequency transients" and make no distinction between statistics based on "2.5–7 day bandpass" and statistics based on all periods less than 7 days. We focus on the Northern Hemisphere winter, but the results presented here are applicable to all seasons and to both hemispheres.

The frequency dependence of the structure of the transients is clearly evident in the variance maps for the 500 hPa geopotential height field, expressed as standard deviation "$\sigma(Z_{500})$" in Fig. 12.1. The variance map for the high frequency perturbations exhibits zonally elongated maxima suggestive of "storm tracks" located slightly poleward and downstream of the climatological-mean jet streams. These oceanic storm tracks are also clearly discernible in maps of the temporal standard deviation of unfiltered ω at the 700 hPa level shown in Figs. 1.34 and 11.2. Compared to the strongest high frequency variability, the maxima in variance of low frequency transients are farther downstream (to the east), extend farther poleward, and not as zonally elongated (cf. the left and right panels of Fig. 12.1). To explain what makes the variance patterns for the high and low frequency transients so different, we will need to make use of analysis techniques that reveal the structural distinctions between the low and high frequency fluctuations. The most basic of these tools is the *one-point correlation* (or regression) *map*.

One-point correlation maps are created by correlating the time series of a prescribed variable at a prescribed "reference grid point" with time series of the same (or another) variable at grid points extending over the entire hemisphere

or globe and mapping the resulting field. By construction, the correlation coefficient at the reference grid point is 1.0 and the correlations drop off with increasing distance from the reference grid point. In the case of isotropic turbulence, which has no preferred orientation, the correlations drop off at the same rate in all directions from the reference grid point, so the correlation contours are circular, like the rings surrounding the bullseye on a dart board. Beyond the outermost contour, the correlations are not statistically significant. If the correlations drop off more rapidly in one direction than along an axis perpendicular to it, that constitutes evidence of *anisotropy*, a systematic departure from isotropic behavior. A composite one-point correlation map (referred to in the objective analysis literature as a "horizontal structure function") may be generated by "registering" a number of individual one-point correlation maps (i.e., shifting them so that their reference grid points coincide) and averaging them. By displaying a series of one-point correlation maps for an orderly sequence of reference grid points, (e.g., points along a latitude circle at intervals of 1° of longitude) as an animation, it is possible to explore the geographical dependence of the anisotropy.

Rather than examining one-point correlation maps constructed separately for the zonal and meridional wind components, it is simpler and more instructive to infer the structure of the wind field from one-point correlation maps of a scalar representation of it. Given the dynamical constraints on the wind field, there are two possibilities: its nondivergent component $\mathbf{V}_{nd}$ can be represented by the horizontal streamfunction ψ field,

$$u_{nd} = -\frac{\partial \psi}{\partial y} \qquad v_{nd} = \frac{\partial \psi}{\partial x} \qquad (12.1)$$

or its geostrophic component $(\ )_g$ can be represented by the geopotential field,

$$u_g = -\frac{1}{f}\frac{\partial \Phi}{\partial y} \qquad v_g = \frac{1}{f}\frac{\partial \Phi}{\partial x}. \qquad (12.2)$$

[3] Blackmon (1976).

Highpass   Unfiltered   Lowpass

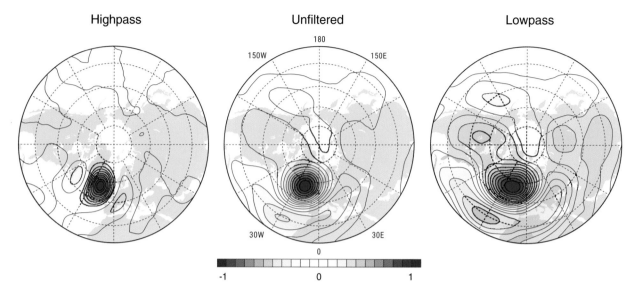

Figure 12.2 One-point correlation maps Z_{500} for the reference grid point (55°N, 20°W) based on DJF data. (Middle) unfiltered daily data; (left) 7 day highpass filtered data; (right) 7 day lowpass filtered data. Contour interval 0.1.

Since the wind field is both quasi-nondivergent and quasi-geostrophic, one-point correlation maps based on ψ and Φ are similar and either one gives a satisfactory representation of the total wind field for purposes of this discussion.

Figure 12.2 shows one-point correlation maps for the 500 hPa height field at a reference grid point to the west of the British Isles, at 55°N, 20°W. The high and low frequency transients are both quite anisotropic: the patch of high values surrounding the reference grid point in the former is meridionally elongated with strong zonal gradients, indicative of $\sigma(v') > \sigma(u')$ and its corresponding feature in the latter is zonally elongated with strong meridional gradients, indicative of $\sigma(u') > \sigma(v')$. The structure of the high frequency transients is suggestive of a train of waves with a wavelength (i.e., distance between centers of like sign) of about 4000 km. Based on inspection of daily sequences of synoptic maps or lag correlation maps constructed by regressing the geopotential height (or other) field upon the time series of the same variable at prescribed reference grid points a day earlier or a day later (not shown), it is evident that the high frequency transients are dominated by eastward propagating waves. In contrast, the low frequency pattern is larger in horizontal scale and its shape is suggestive of a north–south dipole configuration. The one-point correlation map based on unfiltered data, with its nearly circular contours (middle panel), is indicative of a more isotropic structure. That the high and low frequency transients exhibit contrasting forms of anisotropy is a reflection of fundamental properties of Rossby wave propagation, as discussed next.

12.2 PHASE VELOCITY VERSUS GROUP VELOCITY: THEORY

As explained in the preamble of Chapter 11, in the absence of a zonal flow, Rossby waves would propagate westward with phase speed

$$c = -\frac{\beta^*}{k^2 + l^2},\qquad(12.3)$$

where $\beta^* = df/dy - \partial^2[u]/\partial y^2$ is the meridional gradient of the absolute vorticity, and k and l are the zonal and meridional wavenumbers, respectively, expressed in reciprocal meters. The frequency of the zonally propagating waves is thus

$$\omega = -\frac{k\beta^*}{k^2 + l^2}.\qquad(12.4)$$

For waves of a given two-dimensional scale $\sqrt{k^2 + l^2}$, the phase speed is independent of the ratio of k to l, but the frequency depends on it because it takes waves of a given phase speed longer to propagate past a geographically fixed reference point if they are zonally elongated than if they are meridionally elongated. Even if the perturbations consisted of isotropic "red noise," the high frequency transients would contain a disproportionate share of meridionally elongated waves and vice versa.

This dynamical sorting mechanism might explain the very strong anisotropy to ocean currents. The top panel of Fig. 12.3 shows the climatology of the zonal component of the near surface currents. Some features, like the North Equatorial Countercurrent (NECC) extend all the way across their respective oceans while others are more sectoral, but all are remarkably narrow in meridional extent. The zonal striations are accentuated in the climatology of sea level, shown in the bottom panel, which has been subjected to an isotropic filter in the space domain that emphasizes the smallest resolved scales. The NECC is forced by the ITCZ-related pattern of wind stress, which is also strongly zonally elongated. The more subtle, yet statistically robust higher latitude striations in the bottom panel are likely due to dynamical sorting. That the climatology of sea level is much more anisotropic than that of SLP reflects the much smaller oceanic Rossby radius of deformation.

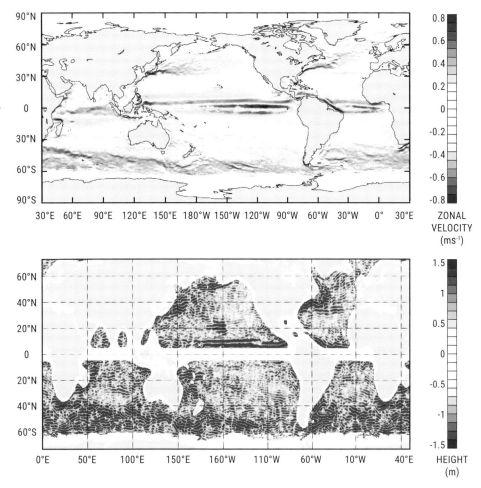

Figure 12.3 (Top) Climatological mean zonal surface velocity based on an analysis of surface drifter data (see Lumpkin and Johnson, 2013); (bottom) climatological mean surface height based on satellite altimetry (1993 - 2002). An isotropic filter in the space domain has been applied to emphasize the smallest resolved scales. (Top) courtesy of Gregory C. Johnson (NOAA/PMEL) and (bottom) adapted from Maximenko et al. (2008) © American Geophysical Union.

For Rossby wave propagation in the atmosphere, the background zonal wind field U needs to be taken into account, in which case Eq. (12.3) becomes

$$c = U - \frac{\beta_*}{k^2 + l^2}.$$ (12.5)

The waves are stationary when the two-dimensional wavenumber $K = \sqrt{k^2 + l^2} = \sqrt{\beta_*/U} = K_s$, and waves with scales close to that value propagate quite slowly and therefore exhibit low frequencies. For waves of a given two-dimensional scale, those that are meridionally elongated exhibit higher frequencies than those that are zonally elongated: hence the contrasting anisotropy of the observed high and low frequency transients. The dynamical sorting is not as dramatic as it is in the ocean, but it is sufficient to account for the frequency dependence in eddy anisotropy in Figs. 12.1 and 12.2.

Figure 12.4 shows one-point correlation maps of Z, u, and v at 500 hPa at the same reference grid point (55°N, 20°W) based on unfiltered DJF data. The pattern for Z, repeated from Fig. 12.2, is quite isotropic (i.e., the contours tend to be circular). In contrast, the centers of action in the pattern for v are meridionally elongated, as in the pattern

for the high frequency transients in Z, shown in the left panel of Fig. 12.2. The succession of positive and negative centers is suggestive of a zonally oriented wave train. In contrast, the centers of action in the pattern for u are zonally elongated like those in the pattern for the low frequency transients in Z.

The distribution of the variances $\overline{u'^2}$ and $\overline{v'^2}$ in the high frequency transients based on 500 hPa DJF data is shown in Fig. 12.5. In interpreting these fields, it should be kept in mind that u' and v' are nearly geostrophic and nondivergent. Both fields exhibit distinctive features associated with the storm tracks over the western Pacific and western Atlantic sectors. Because of the marked anisotropy of the high frequency transients, with meridionally elongated features in the streamfunction and geopotential fields, the variance of the meridional wind component is about three times as strong as that of the zonal wind component. The shapes of the u and v patterns are also quite different: patches of high variance of v are concentrated along the axes of the storm tracks, whereas those of u are less so. These distinctions can be understood in terms of the schematic in Fig. 12.6, in which the zonal wind perturbations derive from the meridional gradient of

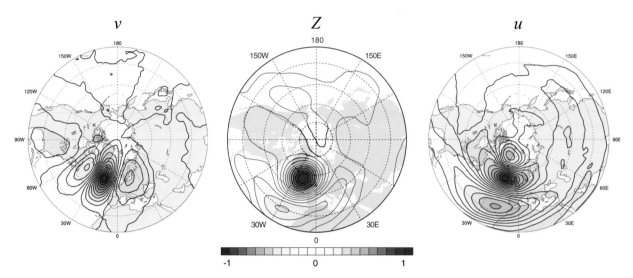

v Z u

Figure 12.4 One-point correlation maps of unfiltered (left to right) v, Z, and u at the 500 hPa level at the reference grid point (55°N, 20°W) based DJF data.

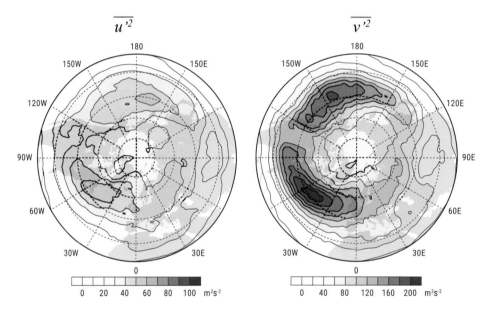

$\overline{u'^2}$ $\overline{v'^2}$

Figure 12.5 Temporal variances of 7-day highpass filtered fluctuations in the zonal and meridional wind components u and v at the 500 hPa level in DJF.

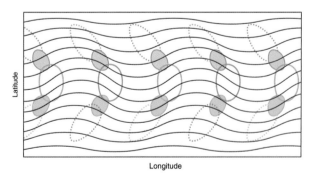

Figure 12.6 Idealized sinusoidal waves in westerlies confined within a channel, as represented by the geopotential height or streamfunction field. The waves exhibit a meridional tilt similar to those in the schematic in Fig. 3.10. Locations of extrema in the u and v perturbations are denoted by dashed and solid contours, respectively. Extrema in the zonal momentum transport $\overline{u'v'}$ are denoted by shading. Gold denotes positive and blue denotes negative values.

wave amplitude along the flanks of the idealized storm track.

The Doppler-shifted zonal group velocity of a Rossby wave is $\partial\omega/\partial k - U$. Using Eqs. (11.1) and (11.2), it can be written as

$$c_{gx} - U = \frac{\beta(k^2 - l^2)}{(k^2 + l^2)^2}. \tag{12.6}$$

It is evident that waves with $k > l$ (i.e., with centers of action elongated in the meridional direction, e.g., as in baroclinic waves) have an eastward Doppler-shifted group velocity and a westward Doppler-shifted phase velocity (Eq. (11.2)). So a wave packet of finite (zonal) extent that is embedded in an eastward zonal background or "steering flow" will disperse eastward more rapidly than the waves that comprise it; that is, new ridges and troughs will develop to the east of the existing ones. Conversely, zonally elongated features with $k < l$ have westward group velocity relative to the mean wind, but it is not as large as their phase velocity. Hence, for example, the group velocity of a stationary Rossby wave embedded in a westerly background flow is eastward, but smaller than U. Irrespective of the direction in which they are propagating, Rossby waves tend to disperse in the direction parallel to their minor axis: hence the tendency for them to be oriented along great circles (e.g., Fig. 11.14).

Exercises

12.1 Describe how one-point correlation and cross-correlation maps were generated from sparse station data only, before the advent of gridded analyses.

12.2 The accompanying figure is a snapshot of the streamfunction for sinusoidal waves propagating eastward in a channel, passing fixed grid points A and B. Sketch the one-point autocorrelation functions for the meridional wind component v' for a reference station at A and for the zonal wind component u' for a reference station at B. Assume that the wave is superimposed upon background "noise", which causes the autocorrelation to decline with distance. Why isn't it possible to construct a one-point autocorrelation map for u' at a reference station at A?

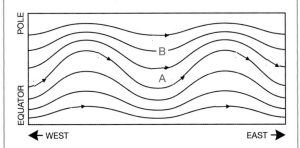

Figure for Exercise 12.2.

12.3 In the schematic Fig. 12.6, the amplitude of the zonal wind perturbations is zero along the axis of the storm track located along the central latitude in the figure and largest along the flanks, but in the observations (Fig. 12.5) there is no double maximum in the meridional profile of $\overline{u'^2}$ at the longitudes of the storm tracks. Explain this apparent discrepancy. Suggest an analysis protocol that might be capable of resolving a double maximum in observational data.

12.4 Show that, relative to the background zonal flow, the phase velocity of meridionally elongated Rossby waves is westward, while the group velocity is eastward.

12.3 THE HORIZONTAL VELOCITY COVARIANCE TENSOR

The horizontal velocity covariance tensor **C** can be divided into isotropic and anisotropic (trace = 0) components[4]

$$\mathbf{C} \equiv \begin{vmatrix} \overline{u'^2} & \overline{u'v'} \\ \overline{u'v'} & \overline{v'^2} \end{vmatrix} = \begin{vmatrix} K & 0 \\ 0 & K \end{vmatrix} + \begin{vmatrix} M & N \\ N & -M \end{vmatrix}, \quad (12.7)$$

where

$$K \equiv (\overline{u'^2} + \overline{v'^2})/2 \quad M \equiv (\overline{u'^2} - \overline{v'^2})/2 \quad \text{and} \quad N \equiv \overline{u'v'}. \quad (12.8)$$

The overbars represent climatological (time) means and the primes represent deviations from them.

The velocity covariance tensor conveys information about the statistically averaged structure of the transient variability. The kinetic energy is indicated by K, the relative prominence of the u and v components by M, and the degree to which the transient wind perturbations tend to be enhanced along a set of axes oriented at 45°/135° relative to the (x, y) axes by N. We already made implicit use of M in the previous subsection for diagnosing whether the transients in geopotential are elongated in the zonal ($M > 0$) or meridional ($M < 0$) directions. If the transients are elongated along an axis oriented at 45° relative to the (x, y) axes, then $M = 0$, and the anisotropy is expressed in a nonzero value of N, positive if the transient wind perturbations are elongated along a 45° axis.

Some simplification can be achieved by adopting a coordinate system oriented along the principal axes of the velocity covariance tensor, which lies at angle

$$\Psi \equiv (1/2) \tan^{-1}(N/M) \quad (12.9)$$

relative to the x axis, where $0 < \Psi < \pi/2$. Let $\widehat{u}$ and $\widehat{v}$ denote the transient velocity components in these rotated coordinates and $\widehat{M}$ and $\widehat{N}$ define the anisotropy. It is readily verified that K is not affected by this coordinate transformation and that

$$\widehat{M} \equiv \left(\overline{\widehat{u}^2} - \overline{\widehat{v}^2} \right)/2 = \sqrt{M^2 + N^2} \quad (12.10)$$

and

$$\widehat{N} = 0. \quad (12.11)$$

Hence, in these rotated coordinates, $\widehat{M}$ and Ψ convey all the essential information about the anisotropy of the horizontal wind field. In addition, it is useful to define the *coefficient of anisotropy*

$$\alpha \equiv \widehat{M}/K, \quad (12.12)$$

which may take on values from 0 to 1. For an isotropic horizontal wind field, it is zero; for motions constrained to one direction, as in Kelvin waves, $\alpha = 1$.

An ellipse provides a useful visual analog of the velocity covariance tensor. Ψ defines the angle of its major axis

[4] Hoskins et al. (1983).

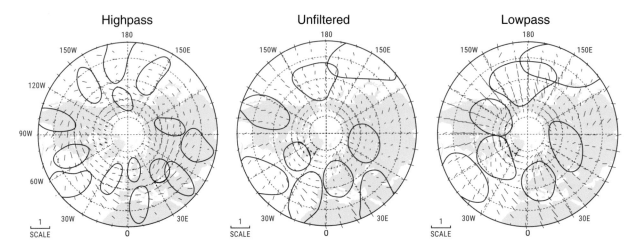

Figure 12.7 Statistics relating to the velocity covariance tensor. Line segments are linearly proportional in length to the coefficient of anisotropy α and they are oriented perpendicular to the angle Ψ, and parallel to the minor axis. The closed loops correspond to the 0.3 contours on one-point correlation maps of geopotential height for selected reference grid points. Based on DJF 300 hPa wind statistics. (Middle) All transients; (left) 6-day highpass filtered transients; (right) 6-day lowpass filtered transients. From Wallace and Lau (1985). © Academic Press, Inc. Published by Elsevier Ltd. All rights reserved.

and α is a measure of its ellipticity; for $\alpha = 0$ the ellipse degenerates into a circle and for $\alpha = 1$ it degenerates into a slit. The velocity covariance tensor can be calculated for the transient variability as a whole or for the transient variability in any particular frequency band that one can isolate through the use of temporal filters. Fig. 12.7 shows the distribution of α and Ψ over the Northern Hemisphere during wintertime for the transient variability as a whole (middle), for the high frequency transients (left) and for low frequency transients (right). The line segments are oriented perpendicular to the principal axis Ψ of the velocity covariance tensor and their length scaled to be proportional to α. The sharp contrast between the structure of the high and low frequency transients, which was pointed out in the previous section, is evident in Fig. 12.7. The east–west oriented line segments in the left-hand panel are indicative of a north–south elongation of the perturbations in the streamfunction field in the high frequency fluctuations, while the north–south orientation of many of the line segments in the right panel is indicative of an east–west elongation of the perturbations in the low frequency transients. The partial cancellation between the contributions from the high and low frequencies results in an overall lower level of anisotropy for the transient variability as a whole (middle panel).

The anisotropy of the flow has implications for the energetics. In Cartesian coordinates the barotropic energy conversion from the time-mean flow into the transients (C_K) is given by

$$C_K = -\underbrace{\overline{u'^2}\frac{\partial \overline{u}}{\partial x}}_{(1)} - \underbrace{\overline{v'^2}\frac{\partial \overline{v}}{\partial y}}_{(2)} - \overline{u'v'}\left(\underbrace{\frac{\partial \overline{v}}{\partial x}}_{(3)} + \underbrace{\frac{\partial \overline{u}}{\partial y}}_{(4)} \right), \quad (12.13)$$

where the terms on the right-hand side represent the conversion of KE from the mean flow to the transients by (1) the zonal transport of eastward momentum down the zonal gradient of eastward momentum, (2) by the meridional transport of northward momentum down the meridional gradient of northward momentum, (3) by the zonal transport of northward momentum down the zonal gradient of northward momentum and (4) by the meridional transport of eastward momentum down the meridional gradient of eastward momentum. If the mean flow is nondivergent, this expression can be rewritten as

$$C_K = -2M\frac{\partial \overline{u}}{\partial x} - N\left(\frac{\partial \overline{v}}{\partial x} + \frac{\partial \overline{u}}{\partial y}\right). \quad (12.14)$$

Now for realistic climatological mean atmospheric flows, which are characterized by narrow, zonally oriented jet streams, $\partial \overline{u}/\partial y \gg \partial \overline{v}/\partial x$ and therefore, to within about 10%, the conversion reduces to

$$C_K = \mathbf{E} \cdot \nabla \overline{u}, \quad (12.15)$$

where

$$\mathbf{E} = -2M\,\mathbf{i} - N\mathbf{j} = (\overline{v'^2} - \overline{u'^2})\,\mathbf{i} - \overline{u'v'}\,\mathbf{j} \quad (12.16)$$

and $\mathbf{i}$ and $\mathbf{j}$ are unit vectors in the zonal and meridional directions.[5] The "E vector" is referred to as the extended Eliassen–Palm flux.

12.4 THE EXTENDED ELIASSEN–PALM FLUX

Equations (12.15) and (12.16) serve to introduce what is referred to as the *extended Eliassen–Palm flux vector*, or

---

[5] A more complete and rigorous derivation, taking into account the terms associated with the spherical geometry, is given in Simmons et al. (1983).

Figure 12.8 Relationship between the velocity covariance tensor **C**, whose principal axis is aligned with the major axes of the ellipses, and the extended Eliassen–Palm flux vector **E**, indicated by the arrows. (a) $M < 0$, $N = 0$; (b) $M < 0$, $N = -M/2$; (c) $M < 0$; $N = -M$; (d) $M = 0$; $N > 0$; (e) $M > 0$, $N = 0$; (f) $M = 0$. $N < 0$; and (g) $M = N = 0$.

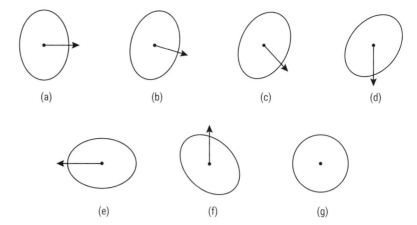

(a)   (b)   (c)   (d)

(e)   (f)   (g)

"E vector" for short, where "extended" connotes taking into account the zonally varying structure of the transients. Like their two-dimensional (zonally averaged) counterparts v (denoted **F**; see Eq. (8.10)), the E vectors trace the flow of zonal momentum and (with some restrictions) wave activity.

12.4.1 The Barotropic Component

We will consider first the horizontal flow at one level, for which the **E** can be regarded as horizontal, a function of (x, y). It can be shown[6] that if the second spatial derivatives $N_{xx} \ll N_{xy}$, as in the vicinity of the zonally elongated storm tracks, the acceleration of the mean flow induced by the transients through the momentum (or vorticity) fluxes can be approximated by the divergence of **E**,

$$\frac{\partial \overline{u}}{\partial t} \simeq \overline{v'\zeta'} \simeq \nabla \cdot \mathbf{E}. \qquad (12.17)$$

Like variances and covariances, **E** may be partitioned into contributions from various frequency bands; its orientation may be frequency dependent.[7]

E is related to the velocity covariance tensor **C** in the manner described in Fig. 12.8, in which the velocity covariance tensor is represented as an ellipse, elongated along its principal axis, as described in section 12.3. For $M = N = 0$, the transients are isotropic and the ellipse assumes the form of a circle, as in panel (g). For $N = 0$, the E vectors point eastward if $M < 0$ $(\overline{v'^2} > \overline{u'^2})$ as in panel (a), or westward if $M > 0$ $(\overline{u'^2} > \overline{v'^2})$ as in panel (e). Under either of these conditions and more generally, provided that $M \gg N$ in absolute value, **E** is an indicator of the group velocity relative to the mean flow, tracing the flux of wave activity like the zonally averaged Eliassen–Palm flux considered in Section 8.2. It is evident from Fig. 12.7 that $M \gg N$ over most of the hemisphere. In the low frequency transients, $M > 0$ and **E**

is westward, whereas in the high frequency transients $M < 0$ and **E** is eastward. Wherever N (i.e., the $\overline{u'v'}$ term) makes an appreciable contribution to the anisotropy of the transients, the orientations of **E** and the minor axis are different, as illustrated in panels (c), (d), and (f).

Figure 12.9 shows the climatological mean, DJF, 300 hPa E vectors for highpass filtered, unfiltered, and lowpass filtered fields superimposed on the distribution of $\overline{u}$. The three patterns are quite different so we will need to consider each one individually, but before doing so it is instructive to note the consistency of Figs. 12.7 and 12.9 in light of the relationships between the covariance tensor and E vectors illustrated in Fig. 12.8.

The high frequency transients in the left panel of Fig. 12.9 are dominated by eastward-pointing E vectors, strongest within and just downstream of the storm tracks, a reflection of the pronounced anisotropy $M < 0$ $(\overline{v'^2} \gg \overline{u'^2})$. The transport of westerly momentum in these regions by the high frequency transients is westward and it is evident from Fig. 12.9 that it is mainly countergradient. That the E vectors are strongly divergent in the storm tracks indicates that the eddy transports are imparting a strong westerly acceleration to the basic state flow. In terms of group velocity, the eastward-pointing E vectors in the storm tracks are indicative of eastward dispersion (i.e., group velocity) of the high frequency transients.

The low frequency transients in the right panel of Fig. 12.9 are dominated by the westward-pointing E vectors, strongest in the jet exit regions, directed up the gradient of $\overline{u}$. Bearing in mind that the E vectors point in the direction opposite to the flux of westerly momentum, this configuration is indicative of a down-gradient momentum flux and a barotropic conversion from the climatological mean flow into the low frequency transients $K_M \rightarrow K_T$ ($C_K > 0$, see Eq. (12.15)) in the energy cycle depicted in Figs. 6.6 and 8.1. Interpreted in terms of group velocity, the westward-pointing E vectors in the jet exit regions are indicative of westward dispersion of the low frequency transients.

The pattern of E vectors based on unfiltered $\overline{u'^2}$, $\overline{v'^2}$, and $\overline{u'v'}$ shown in the middle panel of Fig. 12.9, is a hybrid

---

[6] Hoskins et al. (1983).
[7] There are several different versions of the extended Eliassen–Palm flux. The one developed in this section is due to Hoskins et al. (1983). For alternative definitions, see Andrews (1983), Plumb (1985, 1986), and Trenberth (1986). Trenberth's paper contains a comparison of these various formulations.

Highpass

Unfiltered

Lowpass

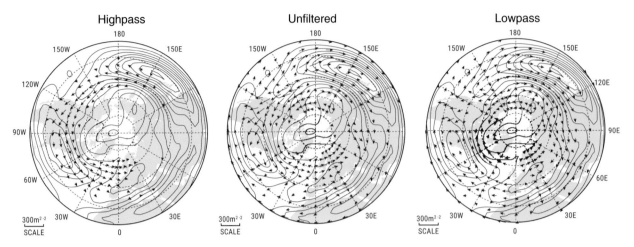

Figure 12.9 *E* vectors (arrows) based on observed 300 hPa winds, superposed upon the DJF climatological mean zonal wind field, indicated by the contours. (Middle) All transients; (left) high frequency (2.5 to 6 day period) transients; and (right) low frequency (>10 day period) transients. Contour interval for zonal wind 5 m s^{-1}. From Wallace and Lau (1985). © Academic Press, Inc. Published by Elsevier Ltd. All rights reserved.

of the contrasting patterns for the low and high frequency transients and is thus more difficult to interpret. There is some cancellation between the predominately westward *E* vectors in the low frequency component and the eastward *E* vectors in the high frequency component such that the meridional fluxes, in particular, the southward vectors at ~35°N over the United States and the Mediterranean are quite prominent, accounting for the pronounced maxima in $[u^*v^*]$ in Fig. 3.11.

It remains to be explained why a positive value of $M = (\overline{u'^2} - \overline{v'^2})/2$ should induce a zonal transport of westerly momentum. It is clear that $\overline{u'^2}$ must be producing an eastward transport of westerly momentum analogous to the way in which $\overline{u'v'}$ produces a northward transport of it. The westward transport of westerly momentum in the storm tracks is thus attributable to the meridional flux of meridional momentum $\overline{v'^2}$. Figure 12.10 provides a heuristic explanation of how the maximum in $\overline{v'^2}$ along the axis of the storm track induces an ageostrophic time-mean circulation that, in turn, induces a westerly acceleration at the upstream end of the storm track and an easterly acceleration at the downstream end, which amounts to a *de facto* westward transport of westerly momentum. The direct effect of the transport of meridional momentum is to produce a poleward acceleration of the air on the poleward flank of a storm track and an equatorward acceleration of the air on the equatorward flank, as shown in the top panel of Fig. 12.10. The Coriolis force induced by this diffluent meridional flow, in turn, induces an eastward ageostrophic circulation on the poleward flank of the storm track and a westward ageostrophic circulation on its equatorward flank. In order to conserve mass, the ageostrophic flow must assume the form of an anticyclonic gyre, with poleward flow at the upstream end of the storm track and equatorward flow at the downstream end, as shown in the bottom panel. These transverse ageostrophic circulations at the upstream and downstream ends of the storm tracks and the associated vertical velocities

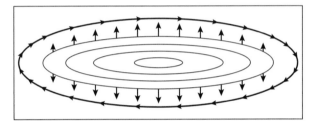

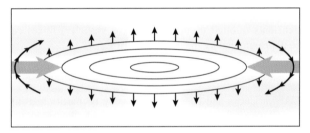

Figure 12.10 Response to the excess meridional transports of meridional momentum $\overline{v'^2}$ in the vicinity of an isolated storm track, denoted by contours of $\overline{v'^2} - \overline{u'^2}$. (Top) the direct response, a diffluent meridional circulation (vertically oriented vectors pointing away from the storm track) which is not, in and of itself, geostrophically balanced. The zonal component of the vectors circulating clockwise around the outermost, bolded contour is the ageostrophic circulation driven by this forcing. (Bottom) The zonal wind accelerations (thick gray arrows) and temperature tendencies (colored shading, blue cold and yellow warm) that develop in response to the ageostrophic circulation. The meridional component of the vectors circulating clockwise around the outermost, bolded contour is the ageostrophic circulation required to keep the zonal component of the geostrophic wind in thermal wind balance.

in the layer below them conspire to produce the pattern of geostrophically balanced zonal accelerations and temperature tendencies shown in the bottom panel of Fig. 12.10. In this manner, the dominance of $\overline{v'^2}$ in the high frequency transients in the storm track results in a diffluent meridional acceleration together with a confluent zonal acceleration, inducing a circulation that is quasi-nondivergent.

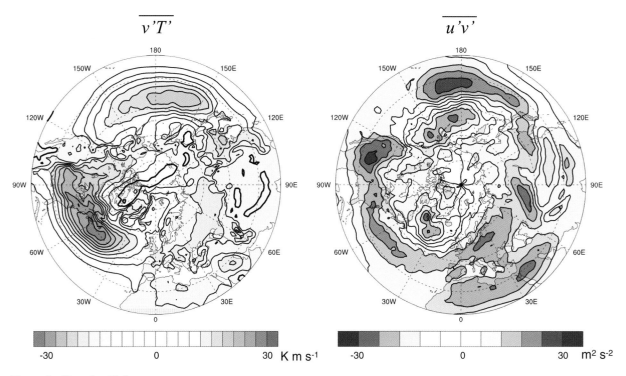

Figure for Exercise 12.5.

Consistent with the kidney bean-shape of the correlation contours in the left panel of Fig. 12.2, the E vectors streaming out of the storm tracks in the high frequency transients in Fig. 12.9 are diffluent, tracing the meridional dispersion of wave activity poleward and (mainly) equatorward out of the storm tracks, particularly toward their downstream ends. The associated meridional confluence of westerly momentum into the storm tracks is mainly countergradient with respect to $\partial \bar{u}/\partial y$. Hence, considering both the zonal and meridional transports of westerly momentum by the high frequency transients, the energy conversion is $K_T \to K_M$. It was shown in Section 8.1 that a conversion in this sense is consistent with the behavior of baroclinic waves in the later stages of their life cycle.

12.4.2 The Baroclinic Component

It was shown in Section 8.2 that a zonally averaged poleward heat transport is indicative of an upward flux of wave activity. In the three-dimensional formulation of the EP flux, the local vertical component of the E vector is directly proportional to $\overline{v'T'}$, so that E for the horizontal flow in Eq. (12.16) can be extended to three dimensions

$$E = \left(\overline{v'^2} - \overline{u'^2}\right) \mathbf{i} - \overline{u'v'} \, \mathbf{j} + \overline{v'\alpha'}/\sigma \, \mathbf{k}. \qquad (12.18)$$

As in Eq. (8.10), upward-directed E vectors are indicative of an upward flux of wave activity and a downward flux of westerly momentum. E vectors directed up the vertical gradient of $\bar{u}$ are indicative of energy conversions $A_M \to A_T$ in the energy cycle and vice versa. The upward-directed E

vectors in the storm tracks are the signature of amplifying baroclinic waves. They are consistent with the reenforcement of the surface westerlies in the storm tracks by the high frequency transients, as deduced from the geopotential tendency equation and summarized in Fig. 11.26.

Exercise

12.5 The accompanying figure shows the distributions of the poleward transports of heat and westerly momentum by the high frequency transients, the former at the 850 hPa level and the latter at the 500 hPa level. (a) Show that the distribution of the momentum fluxes is consistent with the pattern of extended EP fluxes for the high frequency transients shown in Fig. 12.9. (b) How does the distribution of heat transports relate to the EP fluxes?

12.5 PHASE VELOCITY VERSUS GROUP VELOCITY: OBSERVATIONS

In this section, we contrast the observed phase speeds of extratropical Rossby waves with the group velocity of wave packets and we demonstrate the influence of orography upon the phase propagation.

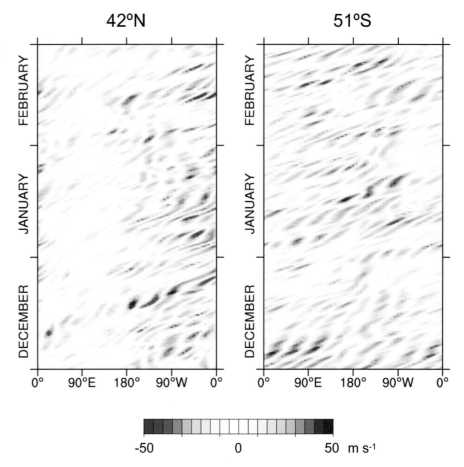

Figure 12.11 Time–longitude sections of v_{200} along (left) 42°N and (right) 51°S for the 2013/2014 winter.

12.5.1 Phase Velocity Versus Group Velocity

Figure 12.11 shows time–longitude sections of v_{200} on the 42°N and 51°S latitude circles for DJF 2013/14. Individual waves with zonal wavelengths on the order of 30-60 degrees of longitude and eastward phase speeds c on the order of 10 degrees of longitude per day are clearly evident in both sections. The waves are embedded in packets that disperse eastward with group velocities c_g on the order of 30 degrees of longitude per day. Individual waves evolve through their life cycles in a matter of a few days, but some of the wave packets can be traced for 10 days or longer. These properties are concisely summarized in the lag correlation plots shown in Fig. 12.12, which were constructed based on reference grid points near the Date Line. Similar patterns are obtained for reference grid points at other longitudes and for other latitude circles in the 30–60°N/S range. It is a fundamental property of Rossby wave dispersion that $c_g > c$. Hence, wave packets overtake the individual waves: as a packet moves eastward at a rate c_g, new waves are amplifying at the leading edge of the packet and mature waves are decaying as the packet leaves them behind. Synopticians refer to this phenomenon as "downstream development."[8]

Figure 12.13 provides a synoptic perspective on the individual waves in one specific wave packet. Letters designate specific trough lines that can be tracked from one panel to the next as they propagate eastward. Note how new trough lines (C and D), originally on the downstream end of the packet, become increasingly prominent, while A and B fall behind the advancing wave packet and dissipate. Figure 12.14 offers another perspective on the downstream development phenomenon – a series of one-point lag-correlation maps based on a reference time series for v_{300} on the Date Line. During this 2-day sequence, the center of the wave packet that is centered on the reference grid point on DAY 0 disperses eastward from ~150°E to 160°W, 50 degrees of longitude, while its primary positive center propagates eastward only ~20 degrees of longitude. Lagged-regression maps based on highpass filtered reference time series (not shown) exhibit a similar rate of phase propagation but the downstream dispersion of the wave packet is not as well-defined because the filtering artificially lengthens its longitudinal span. The downstream dispersion mechanism involves the flow in the upper troposphere, where $u > c$. It can be understood by considering the horizontal equations of the motion in a frame of reference moving with the phase velocity of the waves, bearing in mind that air parcels are moving through the waves from west to east, as in the schematic Fig. 5.22. The Coriolis force acting on the eastward ageostrophic wind

---

[8] Similar figures appear in Lee and Held (1993) where it is shown that this behavior can be simulated using a simple two-layer quasi-geostrophic model.

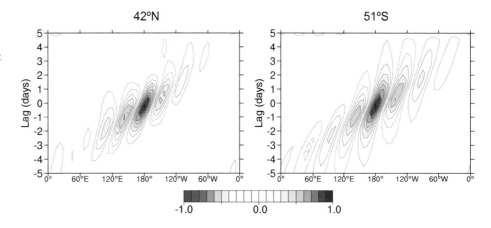

Figure 12.12 Lag-longitude sections of the correlation coefficient between v_{200} at the Date Line and v_{200} at other longitudes.

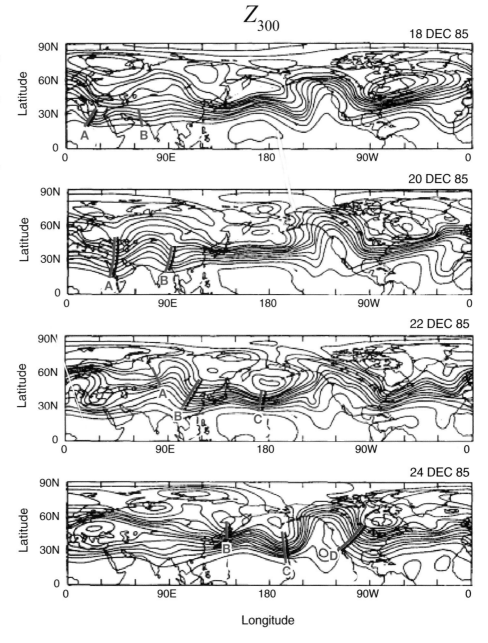

Figure 12.13 Geopotential height at the 300 hPa level at 2 day intervals as indicated. Letters designate specific trough lines that can be tracked from one panel to the next as they propagate eastward. New trough lines (C and D, in succession) develop to the east of the preexisting ones while A and B fall behind the advancing wave packet and dissipate. From Chang (1993). © American Meteorological Society. Used with permission.

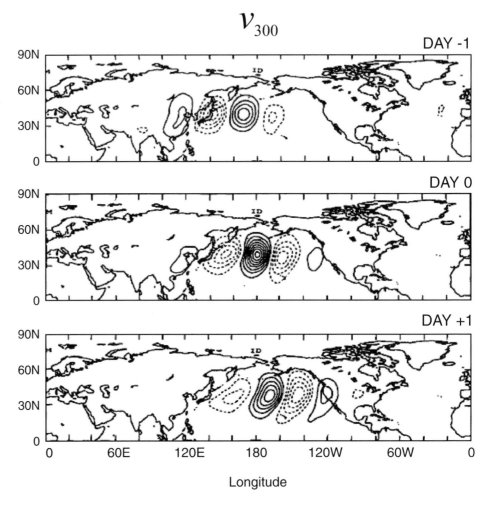

Figure 12.14 The unfiltered v field at the 300 hPa level regressed upon standardized v at 40°N,180°, lagged as indicated. Contour interval 2 m s^{-1}. Negative contours are dashed and the zero contour is omitted. From Chang (1993). © American Meteorological Society. Used with permission.

component in the ridges induces an equatorward acceleration $\partial v / \partial t = -f u_a$ and hence the development of a trough in the geostrophic wind field downstream of the ridge. In a similar manner, the subgeostrophic flow in the wave trough induces the development of a downstream ridge. That the eastward flow through the waves is supergeostrophic in the ridges and subgeostrophic in the troughs is indicative of an eastward flux of mechanical energy $\overline{u_a^*(\Phi^* + K)}$, in the same sense as the flux of wave activity.

12.5.2 Orographic Influences on Wave Propagation

At lower tropospheric levels the propagation of the transients is influenced by orography. Baroclinic waves and other wavelike features tend to propagate anticyclonically around regions of high terrain as indicated in Fig. 12.15. Two factors contribute to the tendency of low-level features to propagate anticyclonically around mountain ranges: one involves the geostrophic flow in Rossby waves, and the other involves the orographically induced ageostrophic flow. In Rossby wave dynamics the terrain slope acts as a waveguide

analogous to the β effect. To see how it works, consider a wave propagating along the eastern slope of a north–south oriented range like the Rockies and imagine that the flow is constrained by the presence of a rigid lid. Equatorward of the surface anticyclone, where the geostrophic flow is upslope, the vertical velocity is upward, decreasing with height, so the horizontal flow must be divergent. The flow in Rossby waves is governed by the barotropic vorticity equation, in which the divergence induces an anticyclonic vorticity tendency that causes the wave to propagate equatorward, parallel to the terrain contours. Now consider the ageostrophic flow. The upslope flow equatorward of the surface anticyclone tends to be blocked by the high terrain. The westward low into the barrier induces a back-pressure that slows it down, rendering it subgeostrophic and thereby forcing it to turn equatorward, creating what synoptic meteorologists refer to as a "barrier jet" that carries the wave along with it.[9]

---

[9] Colle and Mass (1995) argue that the equatorward ageostrophic flow in advance of the pressure surges over North America is the main reason they propagate equatorward.

Figure 12.15 Thick arrows indicate favored tracks of wintertime cold surges along the eastern slope of the major mountain ranges, which acts as a waveguide. Thin arrows indicate tracks of summertime cool surges that propagate poleward, trapped against the coastline, in which the coolness of the air derives from the onshore flow. From Garreaud (2001). © Royal Meteorological Society.

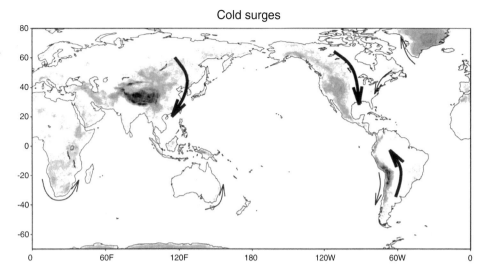

Cold surges

Exercises

12.6 Show that baroclinic waves are characterized by an eastward flux of geopotential at the jet stream level, consistent with their eastward group velocity.

12.7* Show examples of equatorward propagating cold surges to the east of mountain ranges and shallower, poleward propagating cold surges, referred to as "marine pushes" to the west of coastal mountain ranges.

12.6 CHARACTERISTICS OF BAROCLINIC WAVES

The high frequency transients are dominated by baroclinic waves, whose structure and evolution have been investigated in numerous observational studies based on several different approaches. In some studies, time series of wind, geopotential height, temperature, and other variables are linearly regressed upon a common reference time series. To make the waves clearly visible, the reference time series is either highpass filtered, or it is based on a variable that is known to exhibit a strong baroclinic wave signature, as in the last section. An alternative approach is to identify features on the basis of tracking algorithms applied to fields such as sea level pressure or potential vorticity.[10] The documentation of their structure and evolution presented in this chapter does not do justice to the vast body of literature relating to baroclinic waves based on these methodologies. Other published findings establish that baroclinic waves exhibit

• a westward tilt of the wave axes in the geopotential height field with height, strongest in the lower troposphere,

• an in-phase relationship between upward vertical velocity and temperature in the lower troposphere versus a quarter cycle phase difference between them in the upper troposphere,

• in most cases an upward development and a transition from a more baroclinic to a more barotropic vertical structure as the waves evolve through their life cycles while they grow and propagate eastward along the oceanic storm tracks,

• systematic departures from geostrophic balance, including supergeostrophic zonal flow in the ridges and subgeostrophic flow in the troughs, which account for the meridional accelerations experienced by air parcels passing through the waves,

• a tendency for cyclones at the Earth's surface to track poleward and anticyclones to track equatorward, a reflection of the eastward acceleration of the surface winds during the baroclinic wave life cycle,

• a positive correlation between zonal wind and geopotential height, indicative of an eastward transport of mechanical energy,

• the tendency for anticyclonic wave-breaking in regions where the climatological mean flow exhibits anticyclonic relative vorticity and vice versa,

• an enhancement of the generation of transient (or eddy) kinetic energy due to the release of latent heat, especially in the more intense systems,

• an upward displacement of the tropopause in the ridges of the waves relative to the troughs, as evidenced by the distribution of total ozone,

• well-defined patterns of cloudiness and cloud types in satellite imagery.[11]

---

[10] Hoskins and Hodges (2002).

[11] The literature pertaining to these bullet points is so voluminous that it is not practically feasible to include citations in the text.

The structure and evolution of baroclinic waves as inferred from these analyses is generally consistent with the picture that has emerged in the synoptic meteorology literature. It reflects the structure of the normal modes associated with baroclinic instability (Fig. 2.10), but it is also influenced by incidences of nonmodal behavior in the early stages of their development, and by nonlinearities that develop after they achieve finite amplitude.

Exercises

12.8　Compare and contrast the major storm tracks, as inferred from linear regression analysis, as in this chapter versus those inferred from cyclone tracking algorithms.

12.9*　Review the literature documenting the existence of a midwinter suppression of baroclinic wave activity in the North Pacific storm track and offering dynamical interpretations.

13 The Low Frequency Extratropical Transients

with Grant W. Branstator

The datasets and analysis tools for diagnosing the zonally varying general circulation that became available during the 1970s made it possible, for the first time, to clearly discern the signature of low frequency variations. This new capability sparked interest in phenomena that had been known to long-range weather forecasters dating back to the early twentieth century statistical studies of Exner and Walker, but had not hitherto been studied in the context of advancing our understanding of the general circulation. Research on what is now referred to as "climate variability" continues to expand as computational resources become available for performing extended numerical simulations.

The purpose of this chapter is to document the structure and evolution of the variability of the atmospheric circulation with periods ranging from about a week to a season or longer and, where possible, to offer a dynamical interpretation. Here the "high frequency cutoff" of about one cycle per week is intentionally chosen so as to exclude baroclinic waves, which were considered in the last chapter, and to focus on phenomena whose structure and evolution are not clearly revealed by instantaneous synoptic charts. As in the previous chapter, the emphasis is in the extratropical tropospheric circulation.

In the first section, it is shown that many of the phenomena of interest can be interpreted in terms of Rossby wave dispersion. Their structure and evolution can be described making use of linear regression methodology and simulated making use of linearized models. Some are quasi-one-dimensional, dispersing along westerly waveguides, while others are more two dimensional. The second section focuses on nonlinear phenomena, of which wave-breaking, blocking, and preferred states are examples. The final section is devoted to externally forced low frequency variability of the extratropical Northern Hemisphere winter circulation. The examples provided describe how stratospheric phenomena – sudden warmings, the quasi-biennial oscillation (QBO), and the ozone hole in the Antarctic – influence the low frequency variations about the mean circulation by virtue of their strong projections onto the barotropic annular modes. Discussion of external forcing from the tropics is reserved for Part V of this book.

13.1 ROSSBY WAVE DISPERSION

In this section, we will consider the structure and evolution of low frequency variability, as viewed from the perspective of Rossby wave dispersion, starting with one-dimensional wave trains trapped within westerly waveguides and then generalizing the treatment to include quasi-stationary, two-dimensional pulsing and shifting modes with geographically fixed nodes and antinodes that account for a disproportional share of the variance at these frequencies.

We will make extensive use of what we refer to as *teleconnectivity* maps for prescribed variables such as the 200 hPa meridional wind component v_{200} and the 500 hPa geopotential height Z_{500}. Teleconnectivity is a measure of the degree to which fluctuations of a variable at a given grid point, say x_i, are linearly related to fluctuations in x at distant grid points. Appendix C summarizes the various metrics used to quantify teleconnectivity and identify prominent teleconnection patterns: (a) the absolute value of the largest negative value on the one-point correlation map for x_i; (b) the area-weighted r.m.s. amplitude of the values on the one-point correlation map for that grid point; and (c) the area-weighted r.m.s. amplitude of the covariances between standardized x_i and non-standardized x at all grid points. For metrics b) and c) the r.m.s. amplitude is often evaluated based on the grid points that lie within a subdomain. We will refer to them as the *correlation teleconnectivity* and the *covariance teleconnectivity*, respectively.

In a field dominated by wavelike fluctuations like those in baroclinic waves, grid points derive their teleconnectivity in metric (a) from the grid points located one half wavelength upstream or downstream of them. Figure 13.1 shows covariance and correlation teleconnectivity based on the highpass filtered v_{200} field together with a map of the temporal standard deviation. Maps based on the two teleconnectivity metrics tend to be similar but the ones based on covariance place greater emphasis on the oceanic storm tracks by virtue of their high variance, whereas basing it on correlation places greater emphasis on the westerly waveguide over North Africa and Eurasia. In general, the more prominent the waves, the higher the teleconnectivity. In a field dominated by isotropic turbulence, the teleconnectivity is small everywhere.

In identifying and representing teleconnection patterns, we will also make use of empirical orthogonal function (EOF) analysis, a term coined by Lorenz (1956) to describe the application of a standard set of matrix operations to the data matrix comprising the time series of a prescribed variable for all the grid points within a prescribed spatial

229

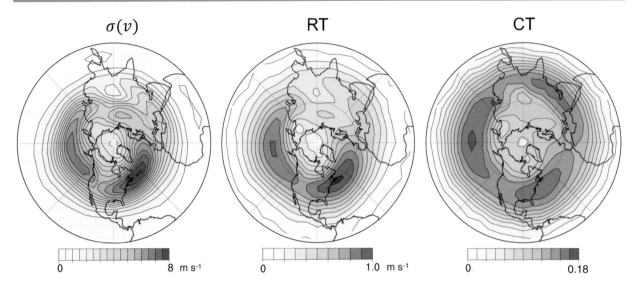

Figure 13.1 Statistics for the DJF high frequency transients in the v_{200} field as defined by 7-day highpass filtered data: (left) standard deviation; (middle) covariance teleconnectivity; and (right) correlation teleconnectivity.

domain. EOFs are a set of mutually orthogonal spatial patterns that account for maximal fractions of the temporal variance, summed over all the time series in the data matrix. Principal components (PCs) are a set of mutually orthogonal time series that define the polarity and amplitude of the respective EOFs at each of the times represented in the data matrix. For more specifics on EOF analysis, see Appendix C.

A related, but somewhat different method of identifying favored patterns of low frequency variability in observational data is to recover a set of patterns called *empirical orthogonal teleconnections* (EOTs), whose one-point correlation patterns exhibit the maximal covariance teleconnectivity, subject to the constraint that they be mutually orthogonal, as explained in Appendix C.[1] By construction, the leading EOT is identical to the one-point covariance pattern for the grid point with the strongest covariance teleconnectivity. The subsequent EOTs tend to be simpler, more robust, and easier to interpret than EOFs.

13.1.1 Dispersion Along Westerly Waveguides

Figure 13.2 shows one-point correlation maps for the meridional wind component v_{200} at a typical grid point along an extended westerly waveguide, which coincides with the axis of the wintertime jet stream that extends across North Africa, southern Asia and into the western Pacific. Patterns are shown for high, low, and very low pass filtered data, as defined in the caption. All are characterized by extended wave trains oriented along the waveguide. The wave trains for the low and very low frequency variability are almost circumglobal in extent. For all three frequency bands, when the reference grid point is shifted eastward or westward along the waveguide, the entire circumglobal wave train

[1] Van den Dool et al. (2000).

shifts with it (not shown). Zonal wavelengths range from zonal wavenumber $k \sim 8$ for the high frequency transients to $k \sim 5$ for the very low frequency transients.

The distribution of teleconnectivity for the low frequency transients in v_{200} exhibits a maximum along the axis of the climatological mean jet stream (Fig. 13.3, left panel). Teleconnectivity tends to be enhanced along the waveguides, as in the corresponding pattern for the high frequency transients (Fig. 13.1) but rather than being zonally uniform, it exhibits wavelike variations along the waveguide, with maxima spaced at intervals of $\sim 35°$ of longitude. These undulations in teleconnectivity are a weak reflection of the much stronger features in the teleconnectivity pattern for the very low frequency transients, shown in the right panel of Fig. 13.3. One-point correlation or regression patterns for grid points along the waveguide are all qualitatively similar and nearly circumglobal. The undulations in teleconnectivity are indicative of a preferred phase of the nearly circumglobal wave in the center and right panels of Fig. 13.2, which we will refer to as the *circumglobal teleconnection* (CGT) *pattern*. It is aligned with the zonally varying jet stream, and has a wavelength of zonal wavenumber $k \sim 5$.

The evolution of wave trains along the westerly waveguide is frequency dependent. Eastward phase propagation is dominant in the day-to-day variability, as shown in the previous chapter, but not in the low frequency variability. Eastward dispersion is clearly evident in the week-to-week variability, as illustrated in the lag-correlation plots shown in Fig. 13.4, in which features on the western end of the wave-train decay with time while features on the eastern end amplify. The corresponding lag-correlation patterns for the month-to-month variability (not shown) do not exhibit any notable phase propagation: as the lag is increased, they simply fade into obscurity without systematic changes in shape. The evolution on the month-to-month timescale

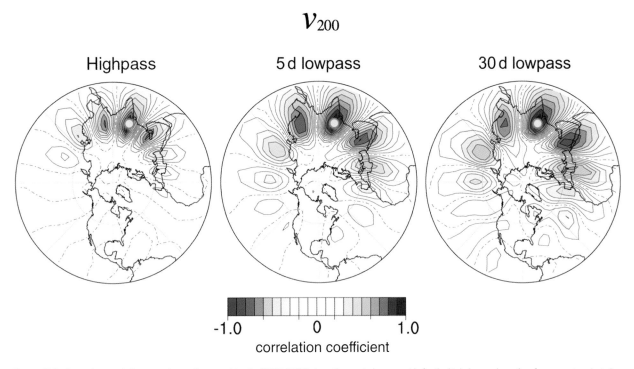

Figure 13.2 One-point correlation maps for a reference grid point (75°E, 30°N) along the westerly waveguide for the high, low, and very low frequency transients in DJF v_{200}. High, low, and very low frequency transients are defined here as daily data minus 5-day means, 5-day means minus 90-day means, and 30-day means minus 90-day means, respectively.

Figure 13.3 Correlation teleconnectivity maps for the low and very low frequency transients in DJF v_{200}, filtered as defined in the caption of Fig. 13.2. Adapted from Branstator and Teng (2017). © American Meteorological Society. Used with permission.

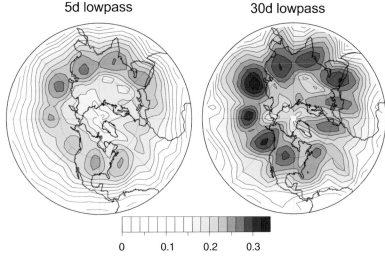

is uninteresting because the transient variability associated with Rossby wave propagation and dispersion does not extend to such low frequencies.

A comprehensive survey of the role of westerly wave-guides in the global general circulation is presented in Fig. 13.5, which shows the observed distributions of correlation teleconnectivity of v_{200} in both hemispheres during both DJF and JJA. A Southern Hemisphere counterpart of the CGT pattern, with centers of action anchored over the southern tip of South America and south of Africa and western Australia, is present year-round. Although CGT-like patterns

in the Northern Hemisphere summer tend to be weaker than during winter, there is evidence that they may contribute to the severity and persistence of simultaneous heat wave and flooding episodes over parts of Eurasia.[2]

Many aspects of the structure and evolution of low frequency atmospheric variability can be understood in terms of barotropic Rossby wave dispersion on a sphere in the presence of a basic state flow that includes strong westerly jets with well-defined exit regions. A dynamical quantity that

[2] Trenberth and Fasullo (2012).

$$v_{200}$$

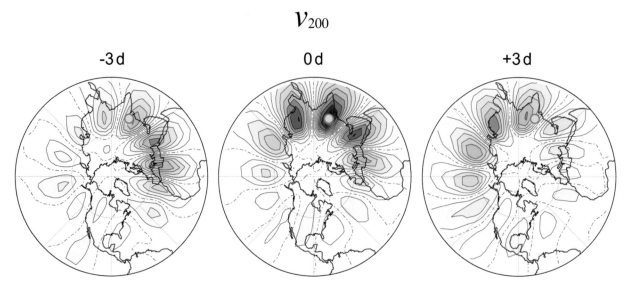

Figure 13.4 Simultaneous and lag correlation maps for the week-to-week variability in DJF v_{200} at a reference grid point (75°E, 30°N) along the westerly waveguide. Here *week-to-week variability* is defined as centered 5-day means minus centered 30-day means, the same as in the middle panel of Fig. 13.2. Warm colors denote positive and cool colors negative correlations. Contour interval 0.1.

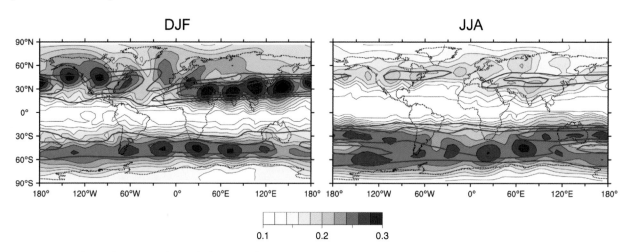

Figure 13.5 Correlation teleconnectivity for v_{200} for the very low frequency transients, as in the right panel of Fig. 13.3, but for both solstitial seasons and for a global domain. The climatological mean jet streams in u_{200} are highlighted with blue contours at wind speeds of 20, 30, and 40 m s^{-1}.

is instructive for understanding the locations of east–west waveguides is the stationary wave number K_s introduced in Eq. (11.3), which in simplified form is $K_s = (\beta_*/[u])^{1/2}$, where β_* is the meridional derivative of background absolute vorticity. K_s is the total wavenumber of stationary Rossby waves in the linear barotropic vorticity equation, where it acts like a refractive index for meridional propagation. As such, maxima in the meridional profiles of K_s identify regions in which low frequency Rossby waves are confined to propagate zonally. As depicted in Fig. 13.6, these maxima tend to occur in zonally oriented bands, often, but not always, aligned with the mean jets. The small red dots indicate the locations of prominent features in the distribution of teleconnectivity in Fig. 13.5. Indeed, most of them fall within the bands of high K_s. For a waveguide teleconnection pattern to exist, not only must its meridional propagation

be meridionally confined but it needs to be located in a region of high eastward group velocity so that its energy will disperse a long distance before it dissipates. For low frequency Rossby waves this can only happen in regions of strong mean westerlies, since the magnitude of the group velocity in the zonal direction of such waves is proportional to the local background zonal wind (see Exercise 13.2).

Simple solutions of linearized barotropic models provide examples of the waveguide behavior predicted by K_s. Figure 13.7 shows perturbations in the v field that develop in a linear barotropic model at times $t = 2$, 5, and 8 days after a small circular vorticity source is inserted into the flow in an atmosphere with zonally symmetric flow consisting of pure solid body superrotation ($[u] > 0$). Within a few days, wavelike circulation anomalies develop downstream of the source. In succeeding days additional centers appear along

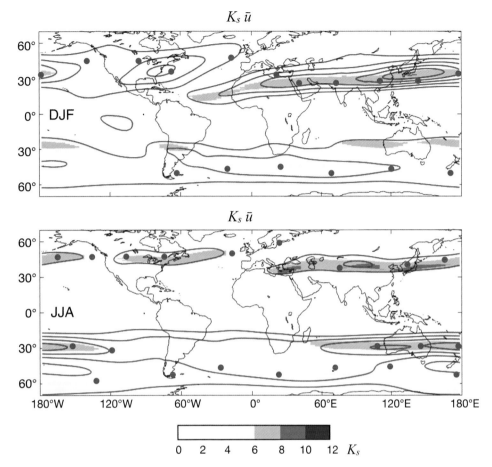

Figure 13.6 Two-dimensional stationary wavenumber for zonally propagating Rossby waves (colored shading) and zonal wind speed (contours at intervals of 10 m s^{-1} for $\bar{u} \geq 20$ m s^{-1}). Regions of high wavenumber and high wind speeds correspond to westerly waveguides. The red dots indicate centers of enhanced teleconnectivity transcribed from Fig. 13.5.

the leading edge of the flux of wave activity emanating from the source and remaining geographically fixed as the "signal" disperses downstream and crosses the equator into the Southern Hemisphere, following a great circle. Centers amplify until the convergence of wave activity is balanced by linear damping and a steady state is achieved. As time elapses the pattern becomes more and more analogous to the steady state pattern for the geopotential height field shown in the top panel of Fig. 11.14.

Figure 13.8 shows the corresponding steady state solution for a basic state flow in solid body rotation together with solutions for two flows with westerly jets that mimic the climatology. In the more realistic basic state flows in the two bottom panels, the strong poleward gradient of absolute vorticity along the axes of the climatological mean jets acts as a waveguide for low frequency Rossby waves of intermediate zonal scale ($k = 4 - 6$). Near these jets the low frequency disturbances are not able to propagate meridionally in arching patterns oriented along great circles as in the previous figure: their group velocity is constrained to be eastward.

From an inspection of the teleconnectivity patterns shown in Fig. 13.5, it is evident that all westerly jets serve as waveguides: the stronger the jet, the higher the teleconnectivity. Where the jets are weak or diffuse, as in the Southern Hemisphere during JJA near the Greenwich Meridian and in

the eastern North Pacific during DJF, the shape of the wave trains reverts to a great circle orientation. Where they are split, as in the Southern Hemisphere during JJA near the Date Line, the wave trains separate into northern and southern branches (Fig. 13.5 right panel).

The importance of the mean state is more quantitatively demonstrated by a set of experiments with a more realistic model in which the patterns develop in response to steady forcing that has no particular spatial organization. The three-dimensional primitive equations are linearized about the climatological mean basic state and perturbed with an array of thermal and vorticity forcings, which are prescribed to be random in three-dimensional space. A large ensemble of these experiments is performed and the family of steady state responses is analyzed just as if the individual responses were separate samples (in time) from the observational record. Figure 13.9 shows observed and simulated one-point correlation plots of v_{200} for two selected reference grid points. That the simulated patterns are remarkably realistic and the seasonality is well simulated confirms that it is not necessary to invoke any kind of special forcing to explain the existence of the CGT pattern: its structure derives from Rossby wave propagation and dispersion in the presence of the observed basic state flow. Further analysis shows that the zonal variations in the teleconnectivity of v_{200} along the

Transient solution. Solid body rotation

Steady state solutions

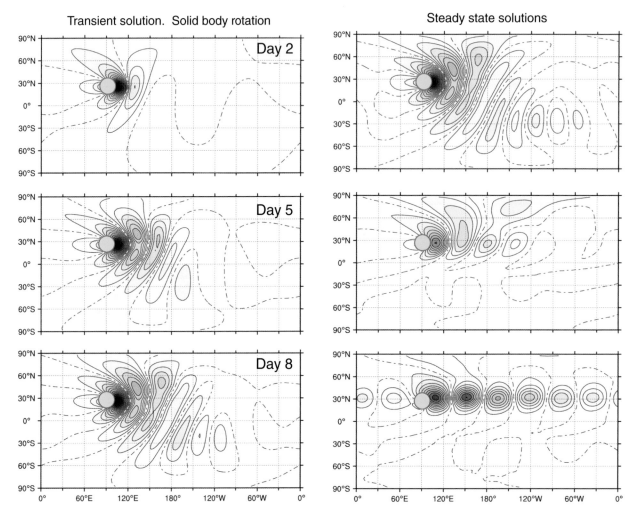

Figure 13.7 Solutions of the barotropic vorticity equation at Days 2, 5, and 8 as manifested in the meridional wind component v, when a positive, steady vorticity source is inserted into the flow at the position of the yellow circle on Day 0. The model is linearized about a background flow consisting of solid body rotation with $\overline{u} > 0$. Warm colors denote positive and cool colors negative correlations.

Figure 13.8 As in Fig. 13.7 but showing steady state solutions for three different prescribed basic state flows: (top) solid body rotation; (middle) with a basic state flow prescribed to match DJF $[u]_{300}$; and (bottom) $[u]_{300}$ set equal to the zonally averaged u_{300} over the sector 120°E to 150°W in DJF. Warm colors denote positive and cool colors negative correlations.

waveguide shown in Fig. 13.3 also derive from the zonal asymmetries in the structure of the waveguide itself.

13.1.2 Dispersion away from Westerly Waveguides

Away from the waveguides, the behavior of Rossby waves is quite different from that described in the previous section. Figure 13.10 shows lagged one-point correlation maps analogous to those in Fig. 13.4, but for three reference grid points outside the waveguides and for Z_{500}, in which the patterns tend to be more isotropic than in the v_{200} field. Away from westerly waveguides, Rossby waves are less constrained to propagate zonally and accordingly, they tend to propagate along great circles rather than along latitude circles.

The structure is frequency dependent, with teleconnection patterns most clearly evident at the very low frequencies.

Figure 13.11 shows a series of one-point correlation maps: the top row of panels for the very low frequency variability, represented by 30-day means and the bottom panels for the week-to-week variability, represented by 5-day means. The three columns are for reference grid points along the 30°N latitude circle at 85°W, 70°W, and 55°W, respectively. In the very low frequencies, a strong, arching pattern with centers over the Pacific and North America, is apparent in the map for 85°W (top left panel), but as the reference grid point is shifted eastward toward 70°W the pattern fades, and by the time it reaches 55°W, an Atlantic dipole pattern emerges. Shifting the reference grid point eastward yields a response like twisting the dial on an old-fashioned radio and hearing a strong station fade, followed by an interval without a strong signal, and then by the emergence of another strong station. In the corresponding sequence for the week-to-week variability, the pattern does not change

Figure 13.9 (Top) Observed and (bottom) simulated one-point correlation maps for DJF and JJA v_{200} for the reference grid points with the strongest correlation teleconnectivity, indicated by yellow dots. Observations are based on 30-day means minus 90-day means. The simulated teleconnectivity statistics are based on a large ensemble of steady state solutions in which a three-dimensional primitive equation model linearized about the climatological mean state is forced with randomly distributed vorticity and heat sources. Warm colors denote positive and cool colors negative correlations. Contour interval 0.1. Adapted from Branstator and Teng (2017). © American Meteorological Society. Used with permission.

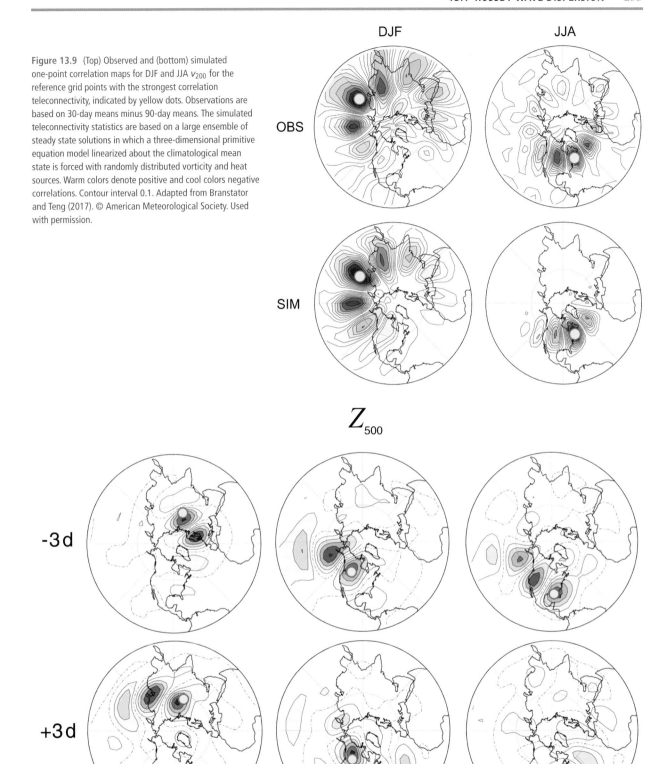

Figure 13.10 Lag correlation maps for the week-to-week variability in DJF Z_{500} at selected reference grid points. "Week-to-week variability" (5-day mean minus 90-day mean). Contour interval 0.1. Adapted from Figs. 3, 4 and 8 of Blackmon et al. (1984b). © American Meteorological Society. Used with permission.

85°W   70°W   55°W

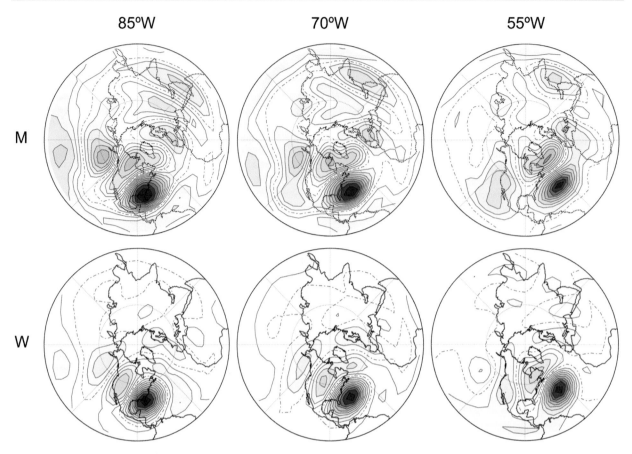

Figure 13.11 One-point correlation maps for the DJF low frequency transients in Z_{500} partitioned into week-to-week (W) variability, as represented by 5-day means minus 30-day means and month-to-month (M) variability, as represented by 30-day means minus 90d means for variations at the reference grid point along 30°N at (left) 85°W, (center) 70°W, and (right) 55°W. Warm colors denote positive and cool colors negative correlations. Contour interval 0.1. Adapted from Fig. 10 of Blackmon et al. (1984a). © American Meteorological Society. Used with permission.

appreciably: it simply shifts eastward with the reference grid point. The patterns for the week-to-week variability consist of zonally oriented wave trains not unlike those in the westerly waveguide, whereas the corresponding patterns in the very low frequency variability have more of a great circle orientation.

Figure 13.12 provides a hemispheric overview of the patterns of variability in the geopotential height field: a regression teleconnectivity map for the very low frequencies analogous to the one for v_{200} shown in the right panel of Fig. 13.3. The pattern is dominated by (1) an arching wave train extending from the subtropical Pacific along a great circle across the Gulf of Alaska, and the Yukon, across Florida, and into the tropical Atlantic, and (2) a dumbbell over the North Atlantic with a northern center over the southern tip of Greenland. The geographically fixed teleconnection patterns that are responsible for these features are revealed by the one-point correlation maps shown in Fig. 13.13: the arching wave train is the signature of the so-called *Pacific–North American* (PNA) pattern,[3] and the dipole in the center panel

corresponds to one of the preferred centers of action of the *North Atlantic Oscillation* (NAO).[4]

To demonstrate the dynamical plausibility of the existence of the PNA pattern, let us consider the results of three numerical experiments with a barotropic model

$$\frac{\partial \zeta}{\partial t} = -\mathbf{V} \cdot \nabla \left(\zeta + f \right) + S. \qquad (13.1)$$

The model is forced by a prescribed steady state vorticity source S and linearized about a zonally varying basic state flow – the observed January climatological mean streamfunction at the 300 hPa level. The basic state ψ_{300} field and the shape of the perturbation in the ψ_{300} field imposed on Day 0 are shown in Fig. 13.14 and the experimental results are shown in Fig. 13.15.

On Day 0 of the first experiment, the flow is perturbed locally at the upstream end of the Eurasian jet stream and the response can be followed as it disperses eastward and later emerges from the waveguide in the jet exit region. Were it not for the linear damping in the model, the disturbances

---

[3] Wallace and Gutzler (1981).

[4] Exner (1913); Walker and Bliss (1932); van Loon and Rogers (1978); Stephenson et al. (2003).

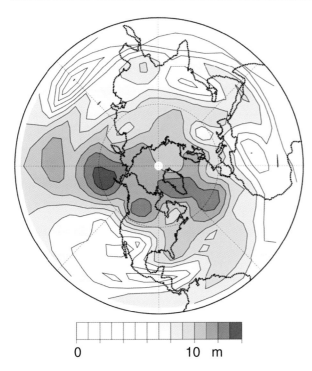

Figure 13.12 Covariance teleconnectivity maps in DJF Z_{500} for the very low frequency transients represented by 30-day means minus 90-day means.

would grow as they extract energy from the basic state flow, as explained below. By Day 2, the perturbations have dispersed into a wave train in the westerly waveguides, and by Day 6 the primary centers of action have emerged from the waveguide and become zonally elongated like those in the very low frequency transients. In the second experiment, the initial perturbation is placed over the Philippines and the wave train disperses along a great circle across the North Pacific. In both of these experiments (left and center rows), the dominant features in the patterns for Days 6 and 10 resemble the PNA pattern. In the third experiment, the initial perturbation is placed over Central America and the wave train that develops in response to it disperses northeastward across the North Atlantic and evolves into a north–south dipole over the eastern North Atlantic that resembles the observed Eastern Atlantic (EA) pattern in that sector. That patterns similar to these are obtained in many such experiments regardless of where the initial perturbation is placed is suggestive of a normal mode type behavior. Patterns that resemble PNA and EA patterns do, in fact, appear in different stages of the life cycle of the most unstable (or least strongly damped) normal mode of this model when it is linearized about this mean state.[5]

The PNA and EA patterns bear a unique relationship to the Northern Hemisphere winter climatological mean flow. Their upstream centers in the geopotential height field assume the form of dipoles straddling the two prominent jet exit regions. Perturbations in these patterns hence involve

extensions and retractions of the jet streams. The prominence of these patterns is largely responsible for the strong, westward pointing E vectors in the jet regions based on monthly mean wintertime 300 hPa wind statistics shown in Fig. 13.16. That the vectors point westward is indicative of an eastward, down-gradient flux of westerly momentum. Hence, it is evident that these modes extract kinetic energy from the climatological-mean flow through the barotropic conversion term $-u'u'\partial\overline{u}/\partial x > 0$. They also exhibit subtle phase shifts with height that are indicative of a baroclinic energy conversion.[6]

Patterns that resemble the NAO do not emerge either among the solutions of the linearized barotropic model or among the prominent one-point correlation patterns obtained when the more realistic multi-level linearized primitive equation model is subjected to stochastic forcing. The NAO also differs from the other patterns in that it exhibits a strong zonally symmetric component, with out-of-phase SLP fluctuations in polar and temperate latitudes. At very low frequencies, its time-varying index is so highly correlated with that of the Northern Hemisphere barotropic annular mode (NAM) described in Section 8.4 as to provoke a debate as to whether the NAM is the zonally averaged NAO or the NAO is the regional expression of the NAM.[7] Proponents of both interpretations agree that the positive feedback of the eddy fluxes in baroclinic waves upon the zonally averaged background flow, as described in Section 8.4, contributes to the prominence of this mode. Hence, whereas the PNA and EA patterns appear to be favored modes of variability because they are capable of extracting kinetic energy from the climatological mean stationary waves, and the waveguide patterns are prominent because of the trapping of energy by meridional gradients in the climatological mean jets, the NAO/NAM is favored by virtue of positive feedbacks involving interactions with baroclinic waves. The same positive feedbacks from the synoptic scale eddies that are responsible for the existence of the NAO/NAM also contribute to the maintenance of patterns like the PNA,[8] but these patterns would arguably exist even in the absence of feedbacks involving the high frequency transients.

The left panel of Fig. 13.17 shows composite 500 hPa charts for months in which the PNA is in its positive polarity and Fig. 13.18 shows regression patterns for temperature and precipitation. The climatological mean ridge over the Rocky Mountains is accentuated, along with the upstream and downstream troughs. Mild marine air masses prevail over southern Alaska, western Canada, and the Pacific Northwest of the contiguous United States, while coastal regions of southeast Alaska and British Columbia experience abnormally heavy precipitation. Temperatures in the downstream trough over the southeastern United States tend to be below normal. The positive polarity of the NAO (Fig. 13.19, left panel) is marked by a strengthening and northward shift

---

[5] Simmons et al. (1983).

[6] Kosaka and Nakamura (2006); Tanaka et al. (2016).
[7] See Exercise 13.6.
[8] Lau (1988); Cai and Mak (1990); Branstator (1992); Franzke et al. (2011).

$$Z_{500}$$

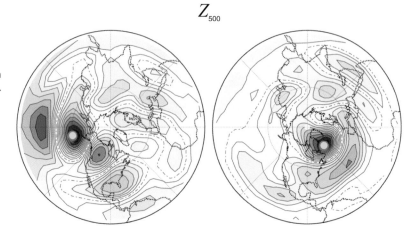

Figure 13.13 One-point correlation maps for the very low frequency transients in DJF Z_{500} at reference grid points near teleconnectivity maxima in Fig. 13.12. Warm colors denote positive and blue negative correlations with variations at the reference grid point. Contour interval 0.1.

Figure 13.14 Setup for a suite of initial value experiments in which the barotropic vorticity equation was linearized about the January Northern Hemisphere climatological mean state at 300 hPa and perturbed by adding localized vorticity perturbations on Day 0. The streamfunction for the mean state is shown in the left panel and the perturbation streamfunction in the right panel. The choice of reference grid point in the right panel is only for purposes of illustration. From Simmons et al. (1983). © American Meteorological Society. Used with permission.

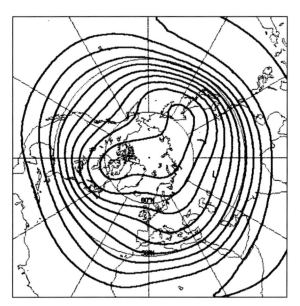

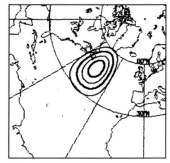

of westerlies across the North Atlantic, which favors mild, wet weather over northern Europe and drought over the Mediterranean (Fig. 13.20). The NAO's negative polarity is marked by a ridge over Greenland and a sharpening of the upstream trough and a southward shift of the downstream end of the Atlantic storm track: instead of being aimed at northern Europe, it extends toward the Mediterranean.

Now let us consider whether we find expressions of these same teleconnection patterns when we apply linear methods of analysis to other variables.

(a) *Zonal Wind at the Jet Stream Level*

Recalling the discussion of the anisotropy of the wind statistics in Section 12.2, we anticipate the leading patterns of unfiltered u_{200} will be dominated by the low frequency variability, just as the patterns of v_{200} are dominated by the high frequency variability. If this is indeed the case, then it is conceivable that the very low frequency patterns might be recovered without recourse to temporal filtering. The climatological-mean wintertime

(DJF) field of u_{250}, shown in the left panel of Fig. 13.21 is consistent with the 300 hPa height climatology shown in Fig. 11.6, with jet exit regions over the North Atlantic and North Pacific. The temporal variance (right panel) tends to be elevated over the jet exit regions. The top panels of Fig. 13.22 show the two leading EOFs of u_{250} in the Atlantic sector and the leading EOF in the Pacific sector. The leading Atlantic mode involves a meridional shifting of the jet exit region, reminiscent of a dog wagging its tail: the sign convention is chosen such that the positive polarity denotes a poleward shift and vice versa. In contrast, the other two modes both involve variations in wind speed along the axis of the jets in the jet exit regions; that is, extensions and retractions of the jets. The corresponding patterns in Z_{500}, shown in the bottom panels of Fig. 13.22, resemble the NAO, the EA pattern and the PNA patterns, respectively.

(b) *Storm Tracks*

Low frequency variations in the position and intensity of the storm tracks can be inferred from monthly maps

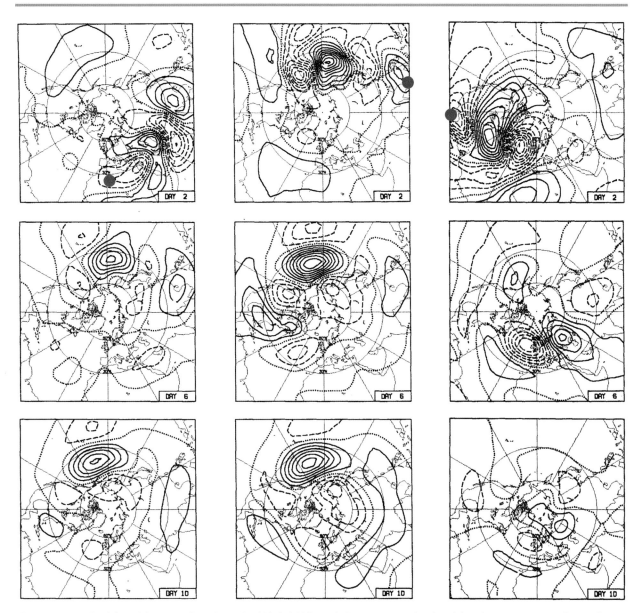

Figure 13.15 Results of three of the numerical experiments in which the initial perturbations were centered on the red dots. The contours represent the perturbation streamfunction field (top) 2, (middle) 6, and (bottom) 10 days into the integrations. From Simmons et al. (1983). © American Meteorological Society. Used with permission.

of the variance of highpass filtered fields such as the meridional wind component v, which exhibit distinctive baroclinic wave signatures. The top panels of Fig. 13.23 show the leading EOFs of the monthly averaged temporal variance of 6-day highpass filtered v_{300} based on year-round data for the Southern Hemisphere. The bottom panels show the variance regressed on indices of the baroclinic and barotropic annular modes, described in Sections 8.4 and 8.5. In all four panels, the climatological mean highpass filtered variance field, which represents the "storm track," the focus of baroclinic wave activity, is shown for reference. The leading EOF of the variance field (top left) involves a strengthen-

ing/weakening (i.e., a pulsing) of the storm track along its entire length and the second mode (top right) involves a poleward/equatorward shifting of the storm track. It is evident that the *pulsing mode* in the variance field resembles the baroclinic annular mode and the *shifting mode* resembles the barotropic annular mode (i.e., the SAM). During the high index polarity of the SAM both the jet (Fig. 8.16) and the storm track are shifted poleward of their climatological mean positions. Though it is not revealed by this figure, it is worth noting that the poleward shift in the storm track is not as great as the poleward shift of the jet. Hence, the basic state flow at the latitude of the storm track is more anticyclonic

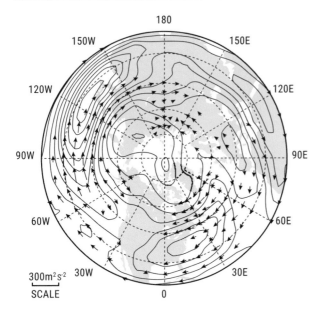

under high index of the SAM and more cyclonic under low index. Baroclinic waves whose life cycle resembles LC1 are consequently more prevalent in the high index state and those whose life cycle resembles LC2 are more prevalent in the low index state.

The close correspondence between the EOFs of the storm track variability and the Northern Hemisphere annular modes is more clearly revealed by separate analyses of the v'_{300} field for Pacific and Atlantic sectors shown in Fig. 13.24, which is also based on year-round data. Pulsing and shifting modes are clearly identifiable as leading EOFs in both domains and they collectively account for most of the variability of the Northern Hemisphere barotropic annular mode (i.e., the NAM). The patterns are more prominent in Fig. 13.25, which are based on DJFM data, where they are superimposed upon regression patterns for the 300 hPa geopotential height field Z_{300} based on monthly mean data. The patterns observed in association with the shifting modes coincide with the principal centers of action of the PNA and EA patterns and the pulsing of the Atlantic storm track lies along the node in the

Figure 13.17 Composite monthly mean 500 hPa charts for contrasting polarities of the Pacific–North American (PNA) pattern. Months included in the composites correspond to the top and bottom 10% of the months in the time series of EOT2 of the monthly mean DJF SLP field, whose reference grid point is indicated by the red dots.

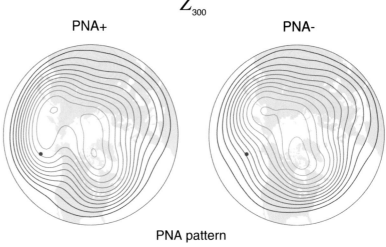

Figure 13.18 Surface air temperature and standardized precipitation regressed upon a standardized index of the PNA pattern. From Wallace and Hobbs (2006).

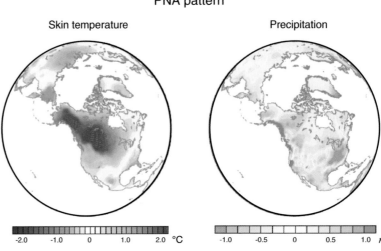

$$Z_{300}$$

NAO+ NAO−

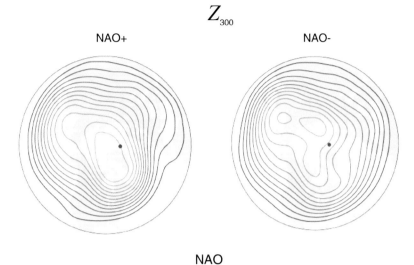

Figure 13.19 As in Fig. 13.17 but for the NAO, as defined by EOT1 of the DJF SLP field.

NAO

Skin temperature Precipitation

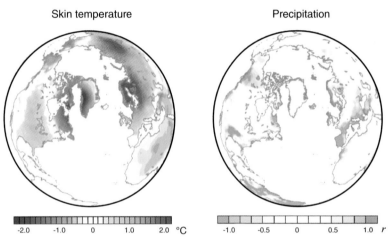

Figure 13.20 As in Fig. 13.18 but for the NAO. From Wallace and Hobbs (2006). © Elsevier Inc. All rights reserved.

-2.0 -1.0 0 1.0 2.0 °C   -1.0 -0.5 0 0.5 1.0 r

$$\bar{u}_{250} \qquad \sigma(u'_{250})$$

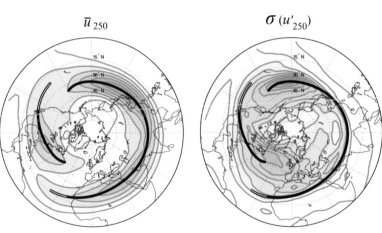

Figure 13.21 Wintertime 250 hPa zonal wind u_{250}: (left) the climatological mean, contour interval 10 m s^{-1} and (right) the day-to-day variance, contour interval 50 m^2 s^{-2}. The superimposed heavy lines indicate the positions of the climatological-mean jet streams. The black segments of these lines indicate wind speed along the jet axis exceeding 30 m s^{-1}, and the white segments in the jet entrance regions indicate wind speeds between 25 and 30 m s^{-1}. From Athanasiadis et al. (2010). © American Meteorological Society. Used with permission.

climatological NAO dipole in 300 hPa height. In the patterns for both pulsing modes, a strengthening of the storm tracks is accompanied by a strengthening of the westerly basic state flow and vice versa. In the patterns for both shifting modes, cyclonic anomalies in the basic state flow correspond to an equatorward shifting of the storm track and vice versa. Cyclonic anomalies in the basic state flow favor cyclonic wave-breaking and a southward shifted storm track. A southward shifted storm track, in turn, forces the background flow in the sense as to enhance the cyclonic shear. In essence, this is the same positive feedback mechanism that is responsible for the prominence of the barotropic annular modes.

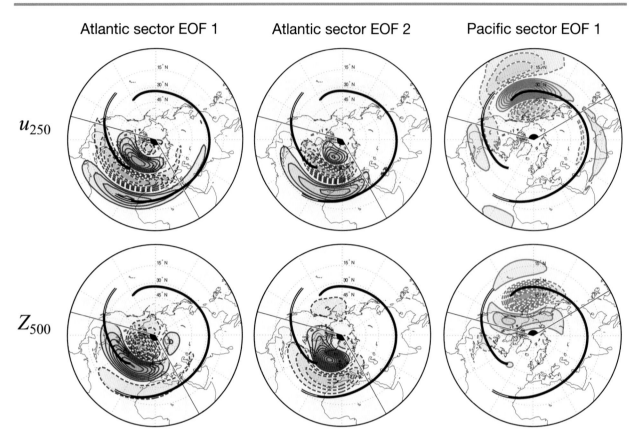

Atlantic sector EOF 1   Atlantic sector EOF 2   Pacific sector EOF 1

u_{250}

Z_{500}

Figure 13.22 (Top) u_{250} and (bottom) Z_{500} anomalies observed in association with variations in the amplitude and polarity of the EOFs of unfiltered daily wintertime u_{250} in the Atlantic and Pacific sectors, found by regressing these fields upon the standardized principal component (PC) time series of the first and second sectoral EOFs as indicated. Sector boundaries are indicated by the bolded meridians. The superimposed heavy lines indicate the positions of the climatological-mean jet streams. Contour interval (top) 2 m s^{-1} and (bottom) 12 m. From Athanasiadis et al. (2010). © American Meteorological Society. Used with permission.

Figure 13.23 Statistics based on averaged variance of the highpass filtered 300 hPa meridional wind component, v'^2_{300}. (Top) Its two leading EOFs (colored shading) superimposed upon its own annual mean (contours). (Bottom) Regressed on standardized indices of the Southern Hemisphere baroclinic and barotropic annular modes (colored shading) superimposed upon its own annual mean (contours). Contour interval 10 m^2 s^{-2} per standard deviation of the PC for the regression coefficients and 30 m^2 s^{-2} (30 m^2 s^{-2} bolded) for the mean fields. The annual mean variance field in the top and bottom panels are not quite the same because they are based on different datasets. Top panels adapted from Wettstein and Wallace (2010). © American Meteorological Society. Used with permission Bottom panels courtesy of Simchan Yook.

Southern Hemisphere v'^2_{300} field

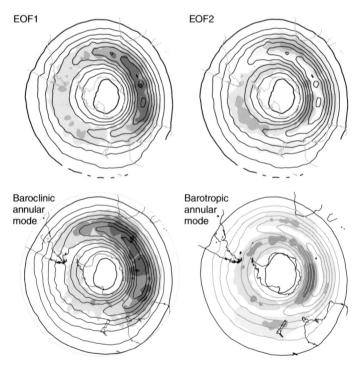

EOF1   EOF2

Baroclinic annular mode   Barotropic annular mode

Pacific sector

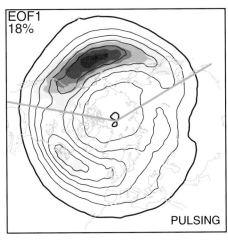
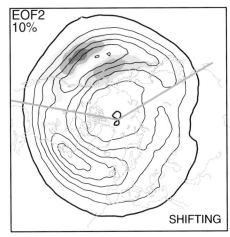

Figure 13.24 Leading EOFs of the monthly mean $v_{500}'^2$ field based on data for all calendar months in the Pacific (top) and Atlantic (bottom) sectors, as defined by the green outlines, warm colors positive, cool colors negative. Contours in all panels represent the variance of 6-day highpass filtered v_{300}'; contour interval 30 m² s⁻² and the 30 m² s⁻² contour is bolded. The printed numbers with percent symbols (%) indicate the fraction of explained variance. "Pulsing" and "shifting" pertain to the character of the modes of storm track variability. From Wettstein and Wallace (2010). © American Meteorological Society. Used with permission.

Atlantic sector

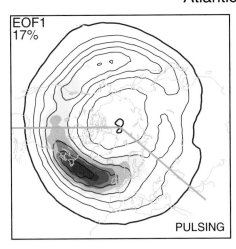
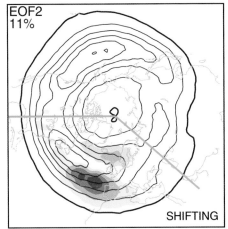

(c) *Sea Level Pressure*

Finally, let us consider the leading modes of variability of the boreal winter (DJFM) sea level pressure (SLP) field. In this case the analysis is hemispheric rather than sectoral, the daily data are strongly low-pass filtered by aggregating them into monthly means, and EOT analysis is used in lieu of EOF analysis. Patterns for the six leading EOTs of the SLP field are shown in Fig. 13.26 and the corresponding Z_{500} patterns are shown in Fig. 13.27. The NAO and the PNA pattern are clearly recognizable as EOT1 and 2, respectively. EOTs 2–6 are mainly monopolar in the SLP field, whereas the corresponding Z_{500} patterns are di- or multipolar.

EOT3 resembles the so-called Warm Arctic-Cold Eurasia pattern, which has been linked to Arctic sea ice variability. The Arctic temperature signature is barely discernible in this plot, which is based on monthly mean data, but it is more prominent in the variability on interannual and longer timescales.[9] EOT4 resembles the EA pattern and the pattern associated with the shifting mode of the Atlantic storm track in Fig. 13.24.

With the exception of EOT4, all the patterns exhibit strong temperature anomalies over land. The positive polarities of EOT's 3 and 6 are marked by adiabatically induced cold anomalies in upslope flow along the flanks of major mountain ranges. The damming of cold air by the major mountain ranges appears to be responsible for the peculiar three-dimensional structure of these patterns. They are obviously related to the high frequency perturbations propagating along orographic waveguides that give rise to cold surges (Fig. 12.15). What makes them evolve so slowly that they emerge as preferred patterns in monthly mean data is not clear. It may be that they are aliased signatures of high frequency

[9] Mori et al. (2014).

Pacific sector

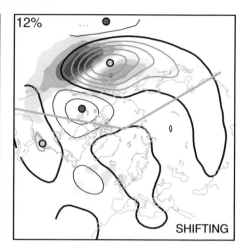

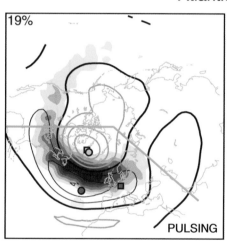

Figure 13.25 Monthly mean $\overline{Z_{300}}$ regressed on PCs 1 and 2 of $\overline{v'^2}_{300}$ based on data for DJFM only in the Pacific (top) and Atlantic (bottom) sectors: contour interval 20 m per standard deviation (black positive, gray negative, zero contour bold). Colored shading shows the EOFs of $\overline{v'^2}_{300}$. The percentage variance explained is noted in each panel. The small colored circles denote prominent centers of action of the leading EOFs of SLP in the same oceanic sectors. "Pulsing" and "shifting" pertain to the character of the modes of storm track variability. From Wettstein and Wallace (2010). © American Meteorological Society. Used with permission.

Atlantic sector

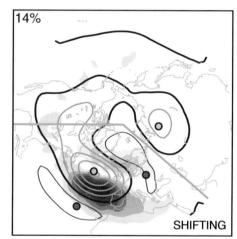

variability or that they are forced by slowly varying boundary conditions.

Figure 13.28 summarizes the locations and shapes of the primary centers of action of the nine leading EOTs in the Northern Hemisphere DJFM SLP field. EOT1, which occupies much of the polar cap region, is surrounded by a family of lesser "modes", arrayed in a horseshoe shape stretching from the North Atlantic eastward across northern Europe and Asia and across the Pacific, poleward of the jet stream. The mountain ranges of Asia form the southern boundaries of the centers of EOTs 3, 5, and 7, and the imprint of the Rockies is evident in EOTs 2, 6, and 9. The corresponding array of patterns for the Southern Hemisphere DJF (summer) SLP field is more symmetric about the pole, not unlike the patterns arising from sampling variability in numerical simulations with aquaplanet models. Hence, it its questionable whether they are representative of genuine teleconnection patterns.[10]

[10] Gerber and Vallis (2005); Gerber and Thompson (2017).

13.1.3 Retrograding Planetary-Scale Rossby Waves

Figure 13.29 contrasts time–longitude sections of Z_{500} on the 42°N and 60°N latitude circles. In the former, eastward propagating disturbances are prevalent at all frequencies, whereas in the latter, westward propagating planetary waves are relatively more prominent, especially at the lower frequencies.

The corresponding two-sided wavenumber–frequency spectra are shown in Fig. 13.30. The spectrum for 42°N exhibits a distinctive baroclinic wave signature. The spectrum for 60°N is different from the one for 42°N in three respects: (1) it peaks at zonal wavenumber $k = 2$ rather than at $k \geq 5$; (2) its frequency spectrum is much redder – waves with periods of 10 days or longer account for most of the variance of Z_{500}; and (3) the phase propagation is predominantly westward rather than eastward. The differences reflect the fact that 60°N is far poleward of the storm tracks with their high frequency, eastward propagating baroclinic waves and it is also far poleward of the jet stream and the associated westerly waveguide. The westerlies are

DJFM SAT, SLP

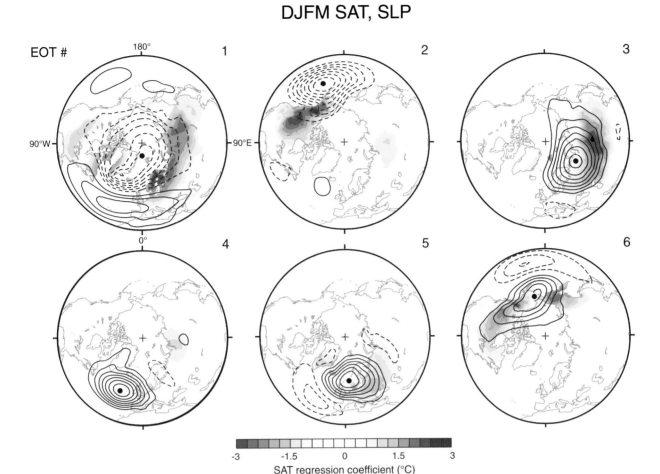

Figure 13.26 The leading EOTs of Northern Hemisphere wintertime monthly mean SLP, as represented by Z_{1000}. Contour interval 1 hPa per standard deviation of the corresponding PC time series, the zero contour is omitted, and negative values are dashed. Regressions of the surface air temperature field onto the respective standardized EOT expansion coefficient time series are indicated by the colored shading. From Smoliak and Wallace (2015). © American Meteorological Society. Used with permission.

so weak at these latitudes that westward propagation is prevalent (i.e., the waves are said to be retrograding) despite the weakness of the beta effect.

To verify the features in the spectrum for 60°N, we show in Fig. 13.31 one-point lag-correlation maps for the grid point at (64°N, 165°E). To emphasize the variability on the timescale of weeks, the analysis is performed upon 5-day means minus 30-day means. Consistent with the previous figure, the spatial patterns are dominated by zonal wavenumber $k = 2$ and the phase propagation is characterized by a prevalence of retrogression. That the spatial patterns for lags of +5 days and −5 days are nearly equal and opposite is indicative of a tendency for cyclic evolution with a period of about 20 days (i.e., four times the lag interval). In effect, two different spatial patterns, one resembling the simultaneous regression map and the other resembling the 5-day lag regression maps appear in quadrature with one another. Extrema of one pattern occur at the same time as extrema of the time derivative of the other pattern.

Exercises

13.1 In the vicinity of a zonal jet, the meridional gradient of absolute vorticity $\beta_* \approx -\partial^2[u]/\partial y^2$. For a Gaussian shaped jet with e-folding width of L, show that the stationary wavenumber K_s is maximum at the jet core with amplitude $K_s = \sqrt{2}/L$. Compare your results with the observed patterns shown in Fig. 13.6.

13.2 Derive the expression for the group velocity of a two-dimensional, stationary barotropic Rossby wave in a zonal mean flow $[u]$. When propagation is purely zonal, show that the group velocity is equal to $2[u]$.

13.3 One-point covariance maps, teleconnectivity maps, EOTs and EOFs can all be derived from the temporal covariance matrix $\overline{x_i x_j}$, where i and j refer to the ith and j^{th} grid points. (a) Describe the mathematical operations required to obtain each one of these fields or sets of fields. (b)

DJFM Z_{500}

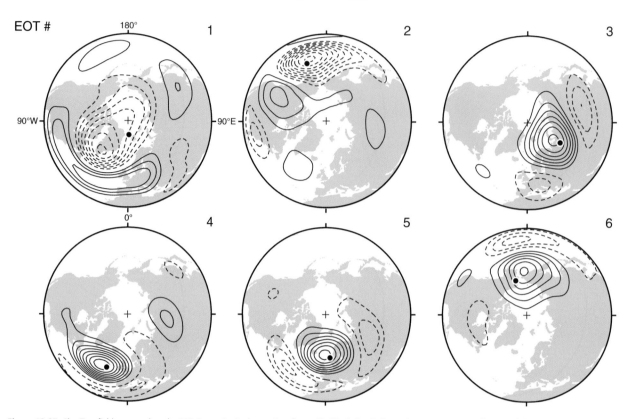

Figure 13.27 The Z_{500} field regressed on the EOT time series in the previous figure. The black dots indicate the primary centers of action of the SLP EOTs. From Smoliak and Wallace (2015). © American Meteorological Society. Used with permission.

DJFM SLP

Figure 13.28 Summary of the leading nine EOTs of monthly mean Northern Hemisphere wintertime (DJFM) SLP (left) and Southern Hemisphere summertime (DJFM) SLP (right) displayed as correlation coefficient contours (0.4, 0.6, and 0.8) between local SLP anomalies and the PC time series of the respective EOTs. From Smoliak and Wallace (2015). © American Meteorological Society. Used with permission.

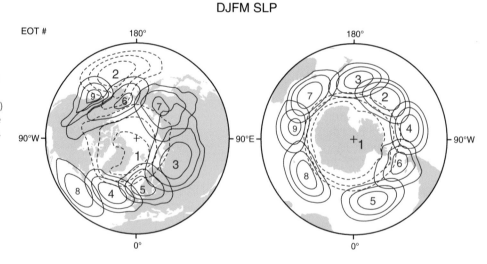

Relate the PCs derived from EOF analysis to the same temporal covariance matrix. (c) Describe two alternative operations that yield exactly the same EOFs.

13.4 Describe how the number of statistical degrees of freedom inherent in the spatial patterns derived from the covariance matrix relates to the number of grid points in the space domain and the number of (evenly spaced) samples in the time domain, taking into account the autocorrelation inherent in the field that is being analyzed.

13.5 The EOFs of random space/time variations of a scalar field in a spherical domain are spherical harmonics. How does the spatial autocorrelation

Figure 13.29 Time–longitude sections of Z_{500} (left) at 42°N and (right) at 60°N latitude circles during the 2012-2013 winter season.

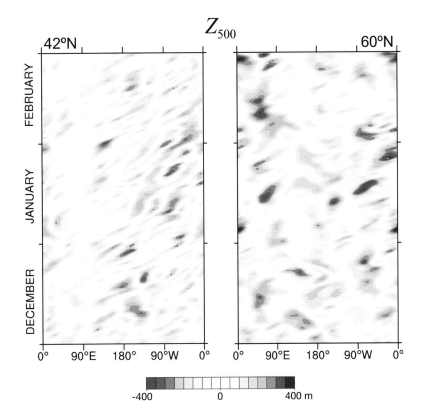

Z_{500}

42°N   60°N

inherent in the field that is being analyzed affect the associated eigenvalues?

13.6* (a) The Northern Hemisphere annular mode NAM discussed in Section 8.4 and the North Atlantic Oscillation NAO discussed in this section have so much in common that it could be argued that they are different ways of characterizing the same phenomenon. Do you have a distinct preference for one or the other, and if so, explain your reasoning. (b) Is the Southern Hemisphere annular mode SAM subject to an analogous ambiguity?

13.7 Is the strength (or prominence) of teleconnection patterns frequency dependent and, if so, why?

13.8 Relate the more prominent of the observed Northern Hemisphere wintertime teleconnection patterns mentioned in the literature, including the North Pacific Oscillation (NPO), to the more limited set of patterns discussed in the book.

13.9* Offer examples of the impacts of specific teleconnection patterns upon regional temperature and precipitation anomalies on timescales of a week or longer.

13.2 BLOCKING

In this section and the next, we will document and attempt to explain some aspects of the nonlinear behavior

of the hemispheric circulation. Figure 13.32 shows the climatological-mean distribution of skewness of the 6-day lowpass filtered Northern Hemisphere DJF Z_{500} field. Positive skewness (i.e., the existence of an elongated positive tail in the frequency distribution) is prevalent poleward of the jet stream and negative skewness is prevalent equatorward of it. Most of the values in the positive tail are associated with high amplitude ridges and so-called *blocking highs* (or *blocking anticyclones*) that exhibit one or more closed Z contours, with a patch of easterly flow on their equatorward flank that interrupts the westerly background flow in which they are embedded. In a similar manner, high amplitude negative values in the band of negative skewness to the south of the jet stream are associated with deep troughs or *cutoff lows* with a patch of easterlies on their poleward flank that "blocks" the westerlies along the axis of the jet stream, interrupting the eastward progression of baroclinic waves and their associated extratropical cyclones. When a blocking high and a cutoff low occur at the same longitude, the flow assumes a dipole configuration, and the blocking of the westerlies is particularly strong.[11] The water vapor imagery shown in Fig. 1.2 corresponds to one of these episodes. Such dipole blocking configurations resemble *modons* – steady state solutions of the barotropic vorticity equation.

Blocking episodes have been of interest to weather forecasters dating back to the 1940s because they account for a disproportional share of record-breaking warm and cold

[11] Berggren et al. (1949); Rex (1950a,b).

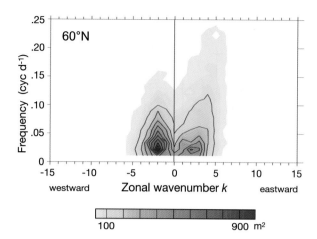

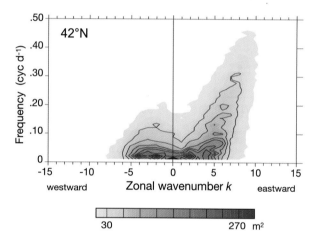

Figure 13.30 Two-sided wavenumber–frequency spectra for the DJF 500 hPa height field on the 42°N and 60°N latitude circles.

events and severe weather events. They tend to be long lasting for two reasons: (1) blocking flow configurations are marked by a near-balance between eastward advection of relative vorticity by the zonal flow and westward phase propagation due to the β effect, and (2) they are amplified and maintained by their interactions with the high frequency transients.[12] Blocks exhibit a strong barotropic component of the flow that extends from the Earth's surface upward to the lower stratosphere. They are characterized by a local reversal of the meridional gradient of potential vorticity on potential temperature surfaces. Figure 13.33 shows the distributions of Z_{250} and θ on the PV = 2 surface in a block characterized by anomalous anticyclonic flow over the North Sea and Scandinavia and cyclonic flow over much of Southern Europe. Note the close alignment of the θ contours with the geostrophic wind, which is indicative of a lack of horizontal advection, as required for a flow pattern to be quasi-stationary. An analogous dipolar configuration was situated over western Canada.

The processes that lead to the formation and maintenance of blocks are illustrated in Fig. 13.34, which is from a numerical simulation. In this experiment, a barotropic flow in a channel was initialized with a modon embedded in a uniform eastward flow and perturbed by a train of waves generated by an upstream wavemaker. The contoured field is quasi-geostrophic potential vorticity. As the waves approach the modon, they are meridionally stretched, zonally contracted, and twisted in such a way that anticyclonic blobs break off and circulate around the poleward flank of the anticyclonic gyre, re-enforcing it. In this manner, air parcels

---

[12] Mullen (1986); Nakamura and Wallace (1993); Nakamura et al. (1997); Berrisford et al. (2007).

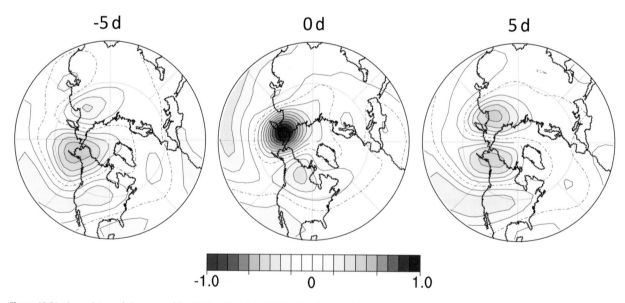

Figure 13.31 One-point correlation maps of the DJF Z_{500} hPa height field for the reference grid point (165°E, 64°N). (Middle) Simultaneous; (left) Z_{500} field leading the reference grid point by 5 days; and (right) lagging the reference grid point by 5 days. Contour interval 0.1.

that formerly resided at low latitudes are transported poleward and become sequestered within closed, anticyclonic eddies and vice versa. This interpretation has been verified in observational studies of blocking episodes.[13]

The initiation of blocking anticyclones is often associated with anticyclonic wave-breaking, as shown in the schematic Fig. 13.35. With a bit of twisting in either direction, the potential vorticity (PV) contours passing through a modon in a westerly flow assume a "yin-yang" shape.

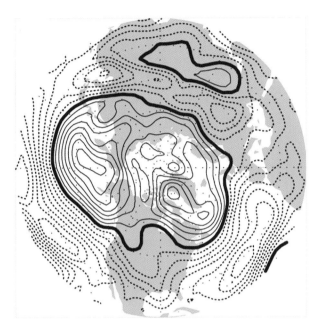

Figure 13.32 Moment coefficient of skewness of the 6-day low pass filtered Northern Hemisphere wintertime 500-hPa height field (contour interval 0.1; negative contours dashed, zero contour bold). From Rennert and Wallace (2009). © American Meteorological Society. Used with permission.

The direction of the twisting determines whether the contour loops first around the high latitude ridge and then around the low latitude trough or vice versa. In the former case, the wave-breaking is said to be anticyclonic and in the latter is said to be cyclonic. When a wave near the front of a Rossby wave packet breaks, it can interrupt the eastward dispersion of wave activity, leading to a long-lived block.[14] It is evident from the tilts of the wave in the PV contour in Fig. 13.35 that anticyclonic wave-breaking results in a poleward transport of westerly momentum while cyclonic wave-breaking results in an equatorward transport. Regardless of which way the waves break, they produce an equatorward transport of vorticity (and potential vorticity) along the axis of the developing block, weakening the zonally averaged zonal flow at that latitude. Anticyclonic wave breaking strengthens the anomalous anticyclonic flow poleward of the climatological mean jet and cyclonic wave breaking strengthens the anomalous cyclonic flow equatorward of it.

An alternative criterion for defining blocking that has been widely used in the literature is the occurrence of strong, long-lived geopotential height anomalies at fixed grid points. Figure 13.36 shows a blocking climatology based on the latter criterion. Blocking frequency tends to be highest during winter in the regions of high skewness over and near Greenland and the Gulf of Alaska, downstream of the climatological-mean upper tropospheric troughs, where the background flow is strongly diffluent. Another favored location is downstream of the oceanic storm tracks, which play a role analogous to the wave-maker in Fig. 13.34.

Figures 13.37 and 13.38 show blocking composites for grid points located in the regions of highest frequency over Greenland and the Gulf of Alaska, respectively. The amplitudes are even larger than those associated with the PNA- and NAO-related anomalies (Figs. 13.11 and 13.13). Over their roughly week-long lifetimes, blocking ridges in both regions exhibit a tendency for retrogression.

Figure 13.33 Fields associated with a prominent blocking event over northern Europe. (Left) Z_{250} (contour interval 1250 m) and (right) θ in units of K on the PV = 2 surface at 1200 UTC 21 Sep 1998. From Pelly and Hoskins (2003). © American Meteorological Society. Used with permission.

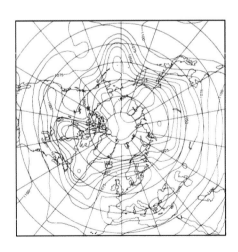

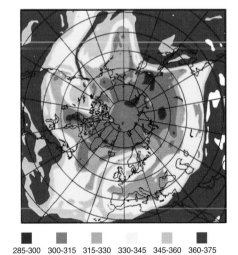

285-300 300-315 315-330 330-345 345-360 360-375

[13] Green (1977); Shutts (1983, 1986); Hoskins et al. (1985); Haines and Marshall (1987); Pelly and Hoskins (2003).

[14] Nakamura et al. (1997).

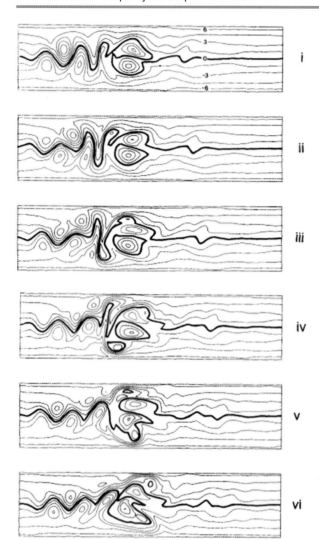

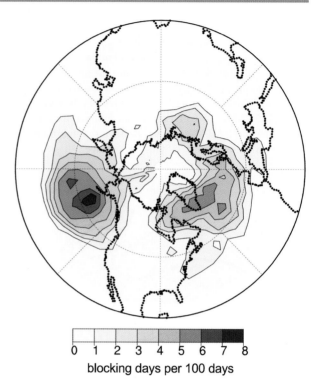

Figure 13.36 Climatology of blocking frequency. A day is considered to be a blocking day if it belongs to a sequence at least 7 days long during which $Z_{500} \times (\sin 45°/\sin \varphi) > 150$ m.

Exercise

13.10 (a) Are blocking and teleconnection patterns interrelated? (b*) Is the NAO a dynamical entity in its own right, or is it just the statistical signature of blocking an/or wave breaking events?

Figure 13.34 Contours of quasi-geostrophic potential vorticity at 2.5-day intervals from a numerical experiment with a shallow water wave equation model with large equivalent depth on a rotating planet. Initial conditions consist of a modon – a dipole – superimposed on a uniform zonal flow. The flow is perturbed by a train of waves generated by a wavemaker at the upstream left end of the channel. The modon induces a diffluent flow on its upstream side, which distorts the waves and causes them to break as they propagate downstream towards and around the modon. Blobs of low potential vorticity air from low latitudes are incorporated into the anticyclonic gyre to the north of the modon, while high potential vorticity air is incorporated into the cyclonic gyre to the south. Hence, rather than eroding the modon structure, the waves reenforce it. From Haines and Marshall (1987). © Royal Meteorological Society.

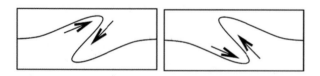

Figure 13.35 The morphology of wave-breaking in the presence of a westerly background flow. The contour is for a conservative tracer such as PV. The arrows indicate the sense in which the contour is being advected as the wave breaks. In the Northern Hemisphere the left panel would be described as anticyclonic wave-breaking and the right panel as cyclonic. Adapted from Berrisford et al. (2007). © American Meteorological Society. Used with permission.

13.3 FAVORED FLOW CONFIGURATIONS

In his classic nonlinear dynamics paper, Lorenz showed how a system consisting of three differential equations in three variables may exhibit two stable equilibrium states, around which it oscillates – sometimes around one and sometimes around the other, transitioning from one state to the other at irregular intervals.[15] The notion that the hemispheric circulation might exhibit analogous *multiple equilibria*, stimulated a flurry of observational studies aimed at detecting the presence of bi- or multimodality in general circulation statistics that would serve as proof of their existence.[16] The evidence thus far is inconclusive.[17] However, the notion that there exist favored circulation regimes persists

---

[15] Lorenz (1963).
[16] The first of these was the influential paper of Charney and DeVore (1979).
[17] See, for example, Hannachi and Iqbal (2019) and the references therein.

52W, 69N Blocking composite

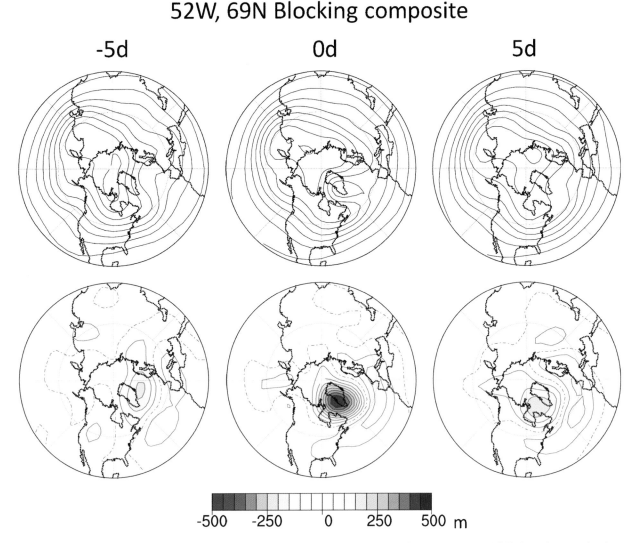

Figure 13.37 Composite 500 hPa height fields for 30 blocking events centered at 52°W, 69°N. (Top) total fields; (bottom) anomaly fields. (Center) At time of peak amplitude; (left and right) 5 d earlier and 5 d later. Contour interval 50 m.

and is well supported by observational evidence. The most prominent of these regimes are related to blocking or the absence of it.

The most widely used analysis technique for detecting favored circulation regimes is cluster analysis. If we define *distance* between two maps as the square root of the hemispherically integrated squared difference between corresponding grid point values, then a measure of goodness of fit is the ratio of the so-called "external variance" – the squared distance between the centroid (i.e., mean map) of a cluster and the centroid of the entire dataset – to the total variance (i.e., the mean squared distance between the individual maps in that cluster and the centroid of the entire dataset). Obviously, this variance ratio must increase as the maps in the dataset are partitioned among more and more clusters, attaining a value of unity for all clusters when the number of clusters equals the number of maps in the dataset. However, the robustness or reproducibility of the results, as

determined by performing the analysis on subsets of the data, declines rapidly with the number of clusters.

Figure 13.39 shows the centroids of clusters obtained when lowpass filtered DJF 500 hPa maps are partitioned among four clusters using the method of self organizing maps (SOM).[18] (The fourth cluster is not shown because it is in some sense just a residual.) All the clusters are highly reproducible in subsets of the data. The leading cluster (ranked in terms of the hemispherically integrated variance of its anomaly pattern) exhibits strong positive anomalies centered over the west coast of Greenland. Its anomaly pattern projects strongly upon the NAO, in its negative polarity. The corresponding total 500 hPa field shown in the upper left panel resembles a weak version of Greenland blocking. The second cluster exhibits a ridge along

---

[18] Kohonen (1990).

142W, 56N Blocking composite

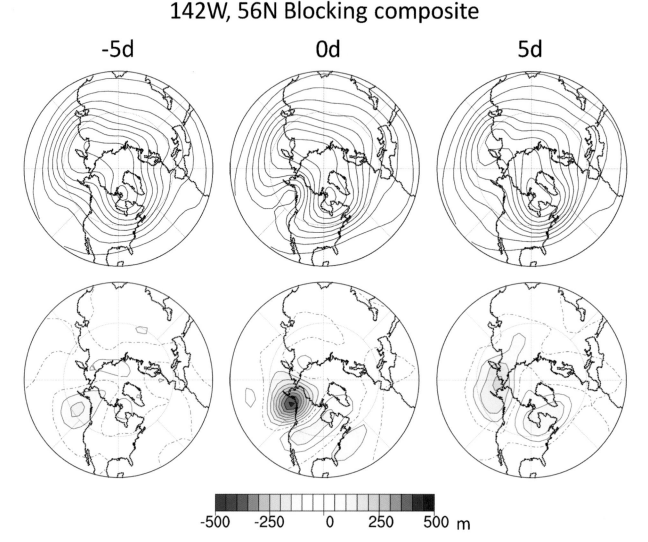

Figure 13.38 As in Fig. 13.37, but for blocking events centered at 56°N, 142°W.

~140°W, a sharpening of the climatological mean trough over Hudson Bay and a strengthening of the climatological mean westerlies across the North Atlantic into the British Isles. Its primary center of action, a high amplitude ridge, coincides with a region of frequent blocking over the south coast of Alaska. The third cluster exhibits a ridge just inland of the west coast of North America, with positive height anomalies over Canada and the northern United States, and the North Atlantic. Its two western centers of action project strongly upon the PNA pattern in its positive polarity and its eastern center upon the positive polarity of the EA pattern.

That the variance ratios for these patterns are modest, ranging from about 20% for the NAO-like cluster (left panel) to 12% for the PNA/EA-like cluster (right panel) testifies to the diversity of the hemispheric flow configurations assigned to a given cluster. The diversity can be reduced (or the homogeneity increased) by performing the analysis on the circulation in, say, the Atlantic sector, as opposed to the entire NH, but only at the expense of rendering the clusters and the dates assigned to them domain-dependent. Alternatively, the dates can be partitioned among a larger number of clusters, but only at the expense of robustness (i.e., stationarity): as the number is increased, the clusters and the dates assigned to them become more and more dependent upon the start and end dates of the record included in the analysis.

The frequency of occurrence of the three highest amplitude clusters shown in Fig. 13.39 has varied from zero in some winters to more than half the days in other winters. The range is particularly large for the leading cluster, which is observed in association with Greenland blocking (Fig. 13.37). It seems unlikely that the winter-to-winter variability would be as large if variations in blocking frequency were entirely stochastic. The other possibility is that the hemispheric circulation is, to some degree, being orchestrated by processes that are operative on a longer timescale such as externally forced variability.

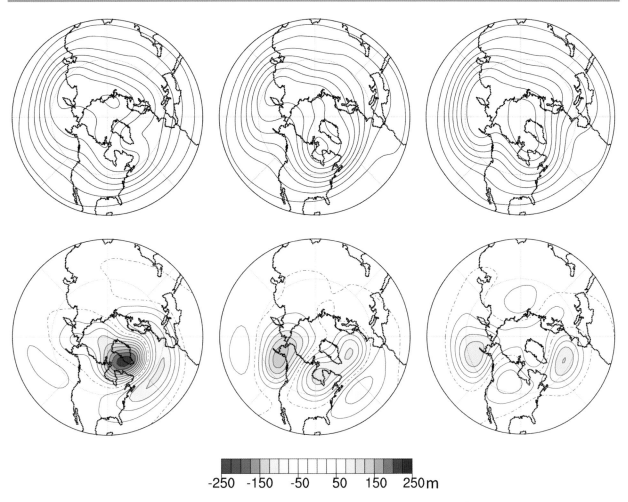

Figure 13.39 Composite maps corresponding to the three leading clusters of the Northern Hemisphere DJF 500 hPa height field poleward of 20°N based on 10-day lowpass filtered data. (Top panels) total fields, contour interval 100 m. (Bottom panels) anomaly fields, contour interval 25 m. The zero contour is dashed. Adapted from Bao and Wallace (2015). © American Meteorological Society. Used with permission. Courtesy of Ming Bao.

Exercise

13.11 How are favored flow configurations derived from cluster analysis different from teleconnection patterns?

13.4 EXTERNALLY FORCED LOW FREQUENCY VARIABILITY

Preferred patterns of variability play an important role in mediating externally forced climate variability and change. Just as a stretched violin string may emit a pure note – its normal mode of vibration – in response to white noise forcing, the extratropical circulation exhibits favored modes of variability, which assume the form of teleconnection patterns. The same positive feedbacks that maintain the patterns in the presence of frictional and thermal damping can serve to amplify the response to external forcings that project strongly onto them.

To illustrate the role of external forcing, we will demonstrate the influence of sudden stratospheric warmings (SSWs) and the quasi-biennial oscillation (QBO) in equatorial stratospheric zonal wind upon the Northern Hemisphere barotropic annular mode, the NAM. The forcing in this case is not external to the atmosphere, but it may be regarded as external as far as the extratropical tropospheric circulation is concerned. Figure 13.40 (left) shows the pattern of anomalies in surface air temperature averaged over the 60 days following stratospheric sudden warming (SSW) events, during which time the polar vortex tends to be anomalously weak, not only at stratospheric levels, but all the way down to the Earth's surface. The similarity between the SSW-related surface air temperature pattern in the left panel of Fig. 13.40 and its counterpart for EOT1 in Fig. 13.26 (with reversed sign) attests to the strong projection of the SSW signature upon the NAO/NAM.

The frequency of occurrence of SSWs, in turn, varies from one year to the next in accordance with the phase of the QBO, as documented in Fig. 10.20. During the

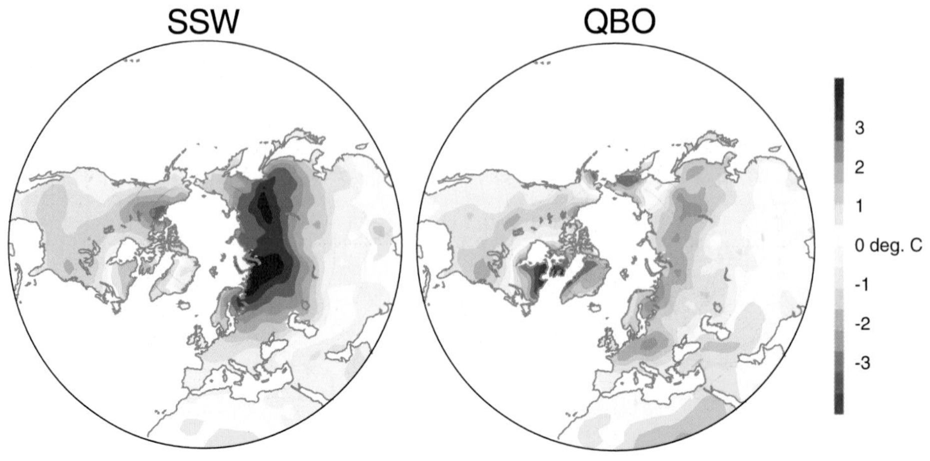

SSW   QBO

Figure 13.40 Northern Hemisphere surface air temperature anomalies. (Left) Averaged over the 60 JFM days following sudden stratospheric warming (SSW) events; (right) seasonally averaged difference between winters in the westerly and easterly phases of the equatorial stratospheric QBO at the 50 hPa level. Adapted from Thompson et al. (2002). © American Meteorological Society. Used with permission.

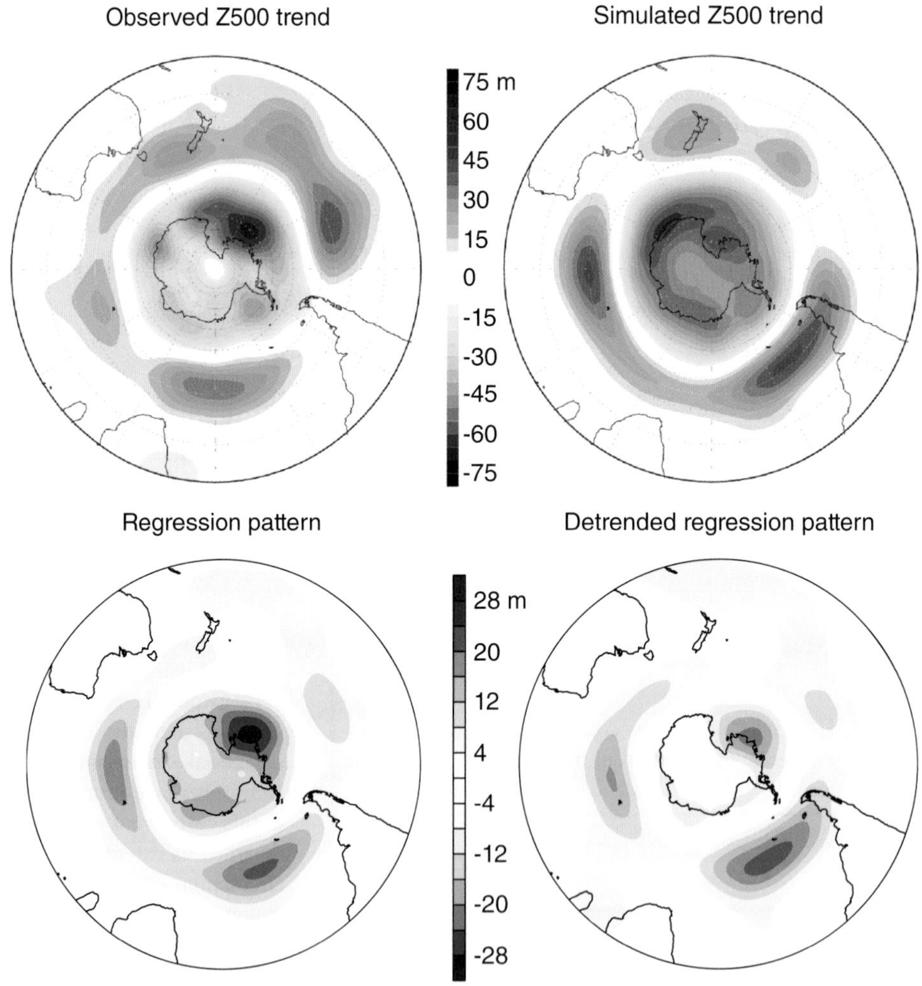

Observed Z500 trend   Simulated Z500 trend

Regression pattern   Detrended regression pattern

Figure 13.41 Observed and simulated ozone-related 500 hPa height (Z_{500}) perturbations during December-February. (Top left panel) Observed differences from the period 1995–2009 (excluding the 12-month period immediately following the September 2002 sudden warming) to 1979–1985. The results indicate the strengthening of the polar vortex during the development of the Antarctic ozone hole. Reprinted by permission from Springer Nature Customer Service Centre GmbH: Springer Nature, *Nature Geoscience*, Thompson et al. © 2011. (Top right) The simulated response of Z_{500} to prescribed ozone losses. From Gillett and Thompson (2003), reprinted with permission from AAAS. (Bottom panels) Monthly-mean values of December-January mean Z_{500} regressed onto standardized values of November column integrated ozone over the South Pole 1979–2017. (Left) Raw data and (right) detrended data. To more faithfully represent the long term behavior of the ozone time series, the detrending was performed by 1) removing the linear trend between 1979–1995; and 2) removing the long-term (1996–2017) mean from there onward. Note the resemblance of all the patterns to the SAM.

easterly phase at the 50 hPa level, the EP fluxes are inhibited from streaming equatorward, rendering the wintertime stratospheric polar vortex more susceptible to breakdowns, and thereby resulting in a weaker polar vortex for the winter as a whole.[19] A weaker polar vortex, in turn, favors the

negative polarity of the NAO/NAM, which is marked by the prevalence of negative surface air temperature anomalies over the high latitude continents, as shown in the right panel of Fig. 13.40. Extended episodes of negative NAO/NAM polarity are often punctuated by cold air outbreaks over Europe and eastern North America, which occur in association with Greenland blocking, as discussed in Section 13.4.

[19] Holton and Tan (1980). See also Section 10.5.

The Southern Hemisphere annular mode (SAM) offers another example of how preferred patterns of variability mediate the atmospheric response to external forcing: in this case the secular trend associated with the intensification of the so-called "ozone hole". The top panels in Fig. 13.41 contrast the observed and simulated 500 hPa height field, representative of the tropospheric circulation, averaged over the years 1995–2009, when the ozone losses were at their peak, minus the prior years 1979–1985, when they were substantially smaller. With the deepening of the ozone hole, heights dropped over Antarctica and the surrounding seas, while they rose in temperate latitudes. The patterns strongly resemble the SAM, the leading mode of variability of the Southern Hemisphere circulation (see Section 8.4). The pattern derived by regressing monthly Z_{500} fields upon column-integrated ozone over the South Pole (bottom left) exhibits a remarkably similar structure. The similarity is not merely a consequence of the trend toward lower ozone concentrations, because the regression pattern based on the detrended ozone time series (bottom right) is similar.

Both the examples in this section involve the annular modes. Other prominent teleconnection patterns are apparent in the extratropical response to variations in tropical heating, as documented in Chapters 17 and 18.

Exercises

13.12* Give examples from the literature illustrating how the external forcing of teleconnection patterns discussed in this chapter has been invoked to explain interannual and interdecadal climate variability.

13.13* Show examples of how teleconnection patterns modulate the frequency of occurrence of extreme weather events.

13.14* (a) Show examples of how seasonal mean circulation anomalies associated with teleconnection patterns leave a distinctive, spatially dependent imprint upon the biosphere. (b) Show examples of how this imprint renders the histories of the time varying indices of the teleconnection patterns visible in time series based on data for tree rings and other paleoclimate proxies.

Part VI The Tropical General Circulation

The tropical atmosphere encompasses the latitude belt equatorward of the subtropical anticyclones at the Earth's surface and the tropospheric jet streams at the tropopause level. As shown in Section 2.6.1, the meridional extent of the tropics decreases with increasing rotation rate. On Earth, it extends to about 30°N/S.

We will use the term *tropical motion systems* to refer to the long and still-growing list of phenomena encompassed by the tropical general circulation. The systematic documentation of the three-dimensional structure of the monsoons, the ITCZs, tropical cyclones, and easterly waves dates back to World War II. Satellite observations, beginning in the late 1960s with visible imagery, have provided a comprehensive and increasingly detailed and quantitative representation of these systems. Significant milestones along the way include the discoveries of equatorially trapped waves (1966), the Madden–Julian Oscillation (1972), and the El Niño-Southern Oscillation (ENSO) phenomenon (1980–82).

Part VI is made up of eight chapters. Chapter 14 describes and interprets the annual mean circulation. It is followed by a short chapter (Chapter 15) on the relationships between mesoscale convection, gravity waves, and geostrophically balanced tropical motion systems (Rossby waves and vortices). It explains some of the fundamental differences between the relative magnitudes of the terms in the governing equations that render motion systems in the tropical troposphere different from the systems discussed thus far in this book.

Chapter 16 documents the seasonality of the climatological mean tropical circulation (i.e., the response to seasonal cycle in the meridional gradient of insolation). The response involves the monsoons and variations in the strength and structure of the ITCZ/cold tongue complexes over the eastern Pacific and Atlantic. Chapter 17 describes and interprets year-to-year variability of the tropical circulation observed in association with the *El Niño/Southern Oscillation* (ENSO) cycle, a coupled atmosphere–ocean phenomenon focused in the tropical Pacific that affects the entire global atmosphere.

Variations in the tropical circulation with periods shorter than a season are mainly a reflection of the atmosphere's own internally generated variability, though coupling with the ocean may have some influence. Chapter 18 documents variations on timescales longer than a week but shorter than a season, by far the most important of which is the Madden–Julian Oscillation (MJO). The day-to-day variability described in Chapter 19 is associated with a number of different kinds of tropical waves, some equatorially trapped and others off-equatorial, mainly in the summer hemisphere. Midlatitude baroclinic waves dispersing into the tropics at upper tropospheric levels and interacting with the ITCZ are also discussed.

Not all tropical motion systems are wavelike. Chapter 20 documents the structure and discusses the dynamics and thermodynamics of warm core vortices, with emphasis on intense tropical cyclones. The final chapter (Chapter 21) is concerned with atmospheric variability on timescales of a day or less, both externally forced and internally generated. It offers a global perspective on the observational evidence relating to gravity and inertio-gravity waves presented in Sections 2.9, 10.2, and 15.2, and documents their signature in the mechanical energy spectrum.

14 The Annual Mean Circulation of the Tropics

with Ángel F. Adames

This chapter documents and offers a dynamical interpretation of the annual mean tropical circulation. It is made up of six sections. The first documents the patterns of rain rate, vertical velocity, and low cloud coverage. The second and third document and interpret the upper and lower tropospheric circulations in terms of equatorially trapped planetary waves introduced in Chapter 10 and relate them to the observed rain rate distribution. The fourth section shows how atmosphere–ocean coupling is instrumental in shaping the annual mean climatology of the western hemisphere of the tropics, prominent features of which are the equatorial could tongue–ITCZ complexes in the eastern Pacific and Atlantic. The fifth section examines the vertical structure of the overturning circulation in the equatorial plane, with emphasis on the shallow boundary layer features and the deeper equatorially trapped stationary waves driven by longitudinal contrasts in the SST. The final section addresses the question of what determines the rain rate distribution.

14.1 RAIN RATE, VERTICAL VELOCITY, AND LOW CLOUDS

Figure 14.1 shows annual mean maps of the variables considered in this subsection. The top panel shows tropical rain rate based on microwave imagery (see also Figs. 1.14 and 1.15). A disproportionate fraction of the rain falls in the "eastern hemisphere" (i.e., from the GM eastward to the Date Line), which is dominated by the Indo-Pacific warm pool. Most of the rainfall in the "western hemisphere" is associated with the Pacific and Atlantic ITCZs, which mark the northern boundaries of the equatorial dry zones emanating from the Southern Hemisphere subtropical subsidence zones (SSZs). The corresponding vertical velocity patterns at the 300 and 850 hPa levels, shown in the next two panels, both strongly resemble the rain rate pattern. The regions of ascent in the eastern hemisphere and particularly over the warm pool are more pronounced in ω_{300}, whereas the ITCZs are more pronounced in ω_{850}.

Figure 14.2 provides additional details on the structure of the vertical velocity field associated with tropical rain rate. The top panels show vertical profiles of upward mass flux and the bottom panels show their vertical derivatives which, through the continuity of mass ($-\partial\omega/\partial p = \nabla \cdot \mathbf{V}$),

determine the vertical profiles of horizontal divergence. The divergence field is instrumental in forcing the planetary-scale circulation. The wet and dry composite profiles are created by (spatially) regressing annual mean ω and its vertical derivative at each pressure level upon annual mean precipitation P at grid points within the tropical domain (30°S–30°N). The "wet" profiles are averages of all grid points with P greater than the tropical mean P, each weighted by P at that grid point, and the "dry" profiles are computed similarly for the grid points with P less than the tropical mean. Very similar generated profiles can be obtained by making composites for grid points with annual mean precipitation above and below some specified thresholds (not shown). The vertical mass flux profiles for the tropics as a whole (top left) exhibit broad, flat mid-tropospheric extrema indicative of ascent at wet grid points and subsidence at dry grid points. Most of the low-level convergence into the areas of heavy rainfall is in the boundary layer below 850 hPa and the upper level divergence is strongest at the 175 hPa level, well above the level of peak detrainment in deep convective clouds, which is closer to 250 hPa. In the absence of radiative heating in the anvils, the peak divergence probably would not be quite as high.[1]

The right-hand panels of Fig. 14.2 contrast the vertical velocity ω profiles for the eastern and western hemispheres of the tropics. The former, with its broad minimum centered near ~450 hPa, can be characterized as "top heavy" and the latter, with its minimum at ~850 hPa as "bottom heavy." In the top-heavy eastern hemisphere profile, the ascent rate increases with height in the 800–450 hPa layer, and hence the horizontal flow in that layer is convergent, whereas in the bottom-heavy western hemisphere profile the opposite conditions prevail. The boundary layer convergence is strong in both profiles but stronger in the bottom-heavy one. The top-heavy profile exhibits stronger upper tropospheric divergence, and it peaks at a higher level. The shallowness of the ITCZ-related overturning circulations in the western hemisphere is corroborated by the meridional cross section shown in Fig. 14.3, a partial zonal average extending from the Date Line to the coast of South America. The mass flux into the ITCZ from both sides is restricted to the boundary

---

[1] Hartmann et al. (1984).

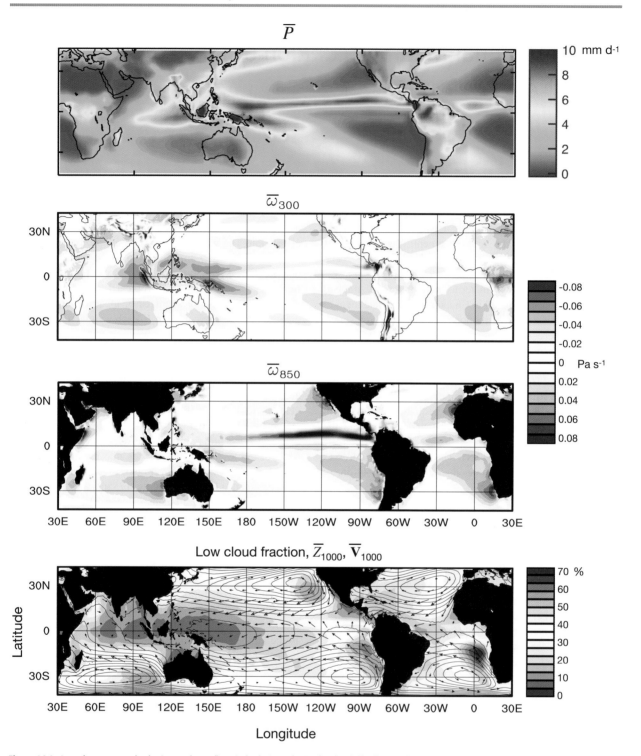

Figure 14.1 Annual mean maps: (top) rain rate; (second) vertical velocity at the 300 hPa level; (third) vertical velocity at the 850 hPa level; and (bottom) the percentage of the sky covered by low clouds, superposed on the 1000 hPa wind vectors and geopotential height (contour interval 10 m). Note that positive values of ω indicate descent.

layer. In the free atmosphere the flow is much weaker and in the opposite direction, so that the ITCZ is marked by divergence.[2]

The bottom panel of Fig. 14.1 shows the distribution of the areal coverage of low cloud decks that form near the top of the boundary layer. In contrast to the vertical velocity patterns, in which the blue-shaded regions of ascent reflect the areal coverage of deep convective clouds and the associated cirrus cloud shields emanating from their anvils, the low

[2] There exists an analogous feature in the equatorial Atlantic, but only during the boreal summer (Zhang et al., 2004).

Figure 14.2 (Top panels) Vertical profiles of annual mean vertical velocity ω in units of Pa s^{-1} and (bottom panels) horizontal divergence. These profiles are obtained by spatially regressing annual mean ω at each tropical grid point at each level upon annual mean P and inferring the divergence from the continuity equation. Composite profiles for the wet and dry grid points, indicated by the blue and red curves, respectively, are composites for all grid points wetter and drier than the tropical mean, P, with each grid point weighted by the departure of P from the tropical mean. (Left panels) Averages over the entire tropics (30°N–30°S). (Right panels) Profiles (wet grid points only) for the eastern hemisphere (Greenwich Meridian GM eastward to the Date Line, solid) and western hemisphere (Date Line eastward to the GM, dashed).

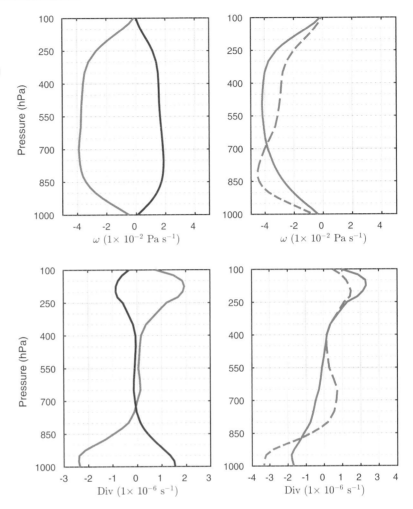

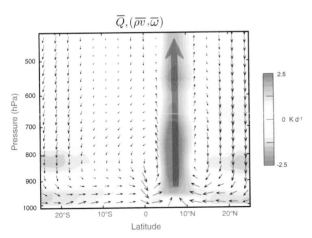

Figure 14.3 Meridional cross section showing annual mean diabatic heating rate $\overline{Q}$ (colored shading), and meridional and vertical mass flux $(\overline{\rho v}, \overline{\omega})$, zonally averaged from the Date Line to 80°W. The scaling of the mass flux vectors is arbitrary.

clouds are much more extensive in the gold-shaded regions of descent than in the rain belts. The inverse relation between the coverage of high and low clouds is reminiscent of the distributions in the *Tropic World* simulation in Section 2.4.

Exercise

14.1 Where do the copious amounts of water vapor that condense in the rain belts over the warm pool and in the ITCZs come from? Note the analogy with Tropic World in Exercise 2.12.

14.2 THE UPPER TROPOSPHERIC CIRCULATION

In the extratropical general circulation, the climatological mean distribution of precipitation cannot be understood without consideration of the transient variability associated with the storm tracks. In contrast, the relationship between the tropical circulation and the tropical rain rate climatology documented in the previous subsection can be understood, at least to first order, without consideration of the transients.

The wind and geopotential height fields at the 150 hPa level are shown in the top panel of Fig. 14.4 superimposed upon the 300 hPa vertical velocity field transcribed from Fig. 14.1. Note that extra height contours are included to bring out the subtle features in the equatorial belt. The height field exhibits an equatorially focused and equatorially

Figure 14.4 (Top) Annual mean vertical velocity ω at the 300 hPa level (colored shading), superimposed on 150 hPa wind vectors (scale at bottom right) and geopotental height (contour interval 100 m for the black contours and 20 m for the blue contours; the lowest of the blue contours corresponds to 14 200 m). (Middle panels) Geopotential and horizontal wind extending over a half cycle in (left) the equatorially trapped $n = 1$ Rossby wave and (right) the $n = -1$ Kelvin wave. (Bottom) The Gill (1980) solution for the response to an isolated heat source located on the equator, indicated by the shading, which consists of a superposition of the wave solutions, with a Rossby wave to the west of the heat source and a Kelvin wave to the east of it.

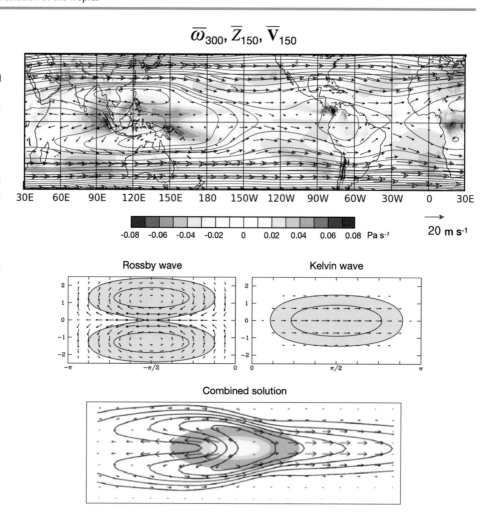

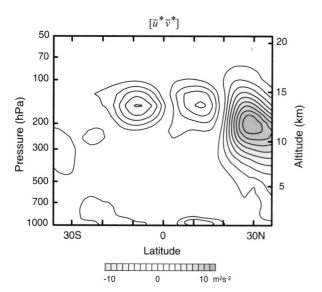

Figure 14.5 Annual mean northward transport of westerly momentum by the stationary waves as in Fig. 3.13, but with a contour interval 1.25 $m^2\ s^{-2}$; the zero contour is omitted.

symmetric wave pattern in which the strongest feature is the ridge extending across the Maritime Continent from about 90°E to 150°E and the troughs over the Atlantic and eastern Pacific. Weak ridges are also evident over South America and Africa. Along the equator easterly winds prevail over the warm pool, flanked by anticyclonic gyres, and westerlies prevail throughout the remainder of the tropics. The off-equatorial features tilt with latitude in the sense indicative of a meridional transport of westerly momentum from the outer tropics into the equatorial belt. This transport is clearly evident in the standing eddy transports shown in Fig. 3.13 and repeated with extra contours in Fig. 14.5. Were it not for the existence of cross-equatorial mean meridional circulations such as those associated with the monsoons, the convergence of the westerly momentum transport by the equatorial stationary waves would induce a strong equatorial superrotation.[3] The tilt of the gyres is also indicative of a poleward flux of wave activity out of the equatorial belt.

[3] Lee (1999).

To aid in the interpretation of the observed pattern, the analytic, steady state response to the outflow from an idealized equatorial mass source in a shallow water wave equation model linearized about a state of rest is also shown in Fig. 14.4. The mass source represents the divergence out of the rain areas at the cloud-top level in the bottom panels of Fig. 14.2. The middle panel shows the "building blocks" from which the solution is made up: an eastward propagating Kelvin wave and a westward propagating $n = 1$ equatorial Rossby wave, as described in Section 9.4. In the solution shown in the bottom panel, the Kelvin wave is dominant to the east of the mass source and the Rossby wave is dominant to the west of it. The Rossby wave is seen to be responsible for the anticyclonic gyres at the longitude of and to the west of the deep convection over the warm pool and the Kelvin wave is responsible for the equatorial ridge of high pressure waves accompanied by westerlies to the east of it. In the steady state solution the waves are damped as they propagate away from the mass source at the same rate as they are being forced by it.[4] We will show how this circulation develops in Section 15.3. Other wave-related processes that have a bearing on the stationary wave response to the deep cumulus convection over the Maritime Continent are (1) the modification of the linear response shown in Fig. 14.4 by the presence of the zonally averaged tropospheric jet streams[5] and (2) Rossby wave dispersion from the heat source by way of the extratropics,[6] both of which give rise to upper level subtropical troughs and equatorial westerlies to the east of the heating.

Exercise

14.2 Explain why most of the kinetic energy of the equatorial stationary waves resides in the zonal wind component.

14.3 THE LOWER TROPOSPHERIC CIRCULATION

The annual mean lower tropospheric circulation, as represented by the 850 hPa vertical velocity field transcribed from Fig. 14.1 and the 1000 hPa height and wind fields (oceans only), is shown in the top panel of Fig. 14.6. Although the divergence field that forces the horizontal wind field is nearly the mirror image of the upper tropospheric divergence field, the wind and height patterns are much different from their counterparts in the upper troposphere and not as amenable to a simple interpretation in terms of equatorially trapped planetary waves. One of the reasons the patterns are different

is because frictional drag is of first order importance in the momentum balance. The ITCZs are clearly evident, with the northeasterly and southeasterly trade winds feeding into the bands of strong ascent at 7°N in the Pacific and Atlantic sectors.

Other prominent features in the annual mean circulation of the tropical lower troposphere are the anticyclones over the subtropical eastern oceans, flanked on their eastern sides by subtropical subsidence zones (SSZs), marked by patches of orange in Figs. 1.18 and 14.6. The SSZs are collocated with equatorward low-level jets with divergent flow that curves westward and broadens to form the trade-wind belt. The SSZs are the source of much of the air in the equatorial dry zones. The relative vorticity of the air within the equatorward jets exhibits a strong cyclonic tendency due to the advection of planetary vorticity, which is balanced by the anticyclonic tendency due to the strong low-level divergence. In the heat balance, subsidence warming is balanced by cold advection and radiative cooling (Fig. 14.7). Most of the subsidence in the sinking branch of the Hadley cell is concentrated within these zones and most of the equatorward flow is concentrated within them and in the trade-wind belts immediately downstream of them. The five prominent subtropical anticyclones can be viewed as Rossby wave responses to these injections of mass into the subtropical boundary layer from above in these zones.

The SSZs are year-round features of the tropical general circulation. They can be simulated in a general circulation model with prescribed SST with and without mountains, as shown in the middle and bottom panels of Fig. 14.6. Hence, it can be concluded that orography and ocean dynamics are not the primary reason for their existence: they must be thermally induced. It is evident from Fig. 1.10 that in the tropics, the continents are warmer than the surrounding oceans at the top of the boundary layer (850 hPa) out to ~30°N/S latitude, even during winter. It will presently be shown that the relative warmth of the tropical continents is a shallow feature that is largely restricted to the layer below 700 hPa. Above that level the annual mean temperature field in the tropics is much more spatially uniform, for the reasons described in the next chapter. The land–sea thermal contrast in the T_{850} field can thus be interpreted as the signature of the steeper lower tropospheric lapse rates observed over land than over ocean – a consequence of the stronger sensible heat fluxes at the Earth's surface during the daytime hours and the limited supply of atmospheric water vapor available to support moist convection.

The sea level pressure pattern in Fig. 14.6 is, in large part, a hydrostatic response to the boundary forcing. with heat lows over the warm land and higher pressure over the cooler oceans. Why the subtropical highs are centered over the eastern oceans can be understood as follows. If the lower troposphere over the tropical oceans were uniformly warmer than over the continents they would be located over the central oceans, but cold advection in the equatorward flow on the eastern sides of the oceans, in combination with warm advection in the poleward flow on the western sides,

---

[4] This pattern is sometimes referred to as "the Gill solution" after Gill (1980).
[5] Monteiro et al. (2014)
[6] Sardeshmukh and Hoskins (1988)

Figure 14.6 (Top) Annual mean 850 hPa vertical velocity $\overline{\omega}$ superimposed on 1000 hPa wind vectors and geopotential height (contour interval 10 m) from ERA-I. Values of $\overline{Z}_{1000}$, labeled in terms of SLP values, range from ∼1025 Pa in the subtropical anticyclones to ∼1008 hPa in the equatorial trough over the warm pool. (Middle and bottom panels) Same as in the top panel, but from the Control (CTL) and No Mountain (NM) experiments described in Section 11.2. Contour interval 10 m, ranging from 60 to 210 m in the Control simulation, and from −130 to 20 m in the No Mountain simulation. The removal of mountains reduces the surface pressure everywhere due to conservation of atmospheric mass. Courtesy of Rachel H. White.

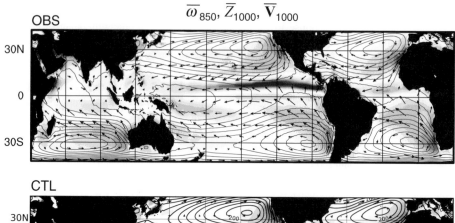

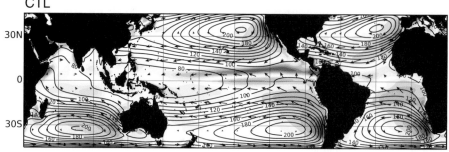

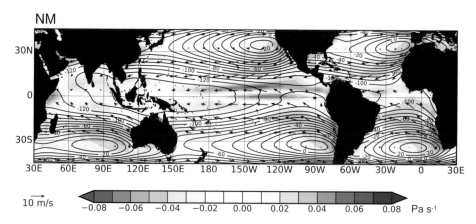

Figure 14.7 Annual mean 850 hPa horizontal temperature advection (colored shading) and wind (arrows). The longest vector is about 10 m s^{-1}.

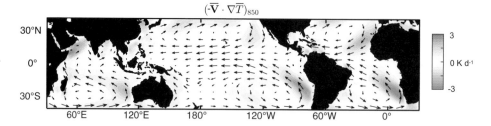

would displace the relatively cool subtropical highs eastward, enhancing the land–sea temperature contrasts on the eastern sides of the ocean, strengthening the cold advection in the SSZs. Through this feedback mechanism, one can envision the equatorward jets sharpening until their width approached the Rossby radius of deformation (∼1000 km), beyond which the flow is no longer geostrophic and Rossby wave dynamics is no longer applicable. Conservation of vorticity requires that the equatorward flow in the SSZs be strongly divergent: hence the strong subsidence.

The remarkable strength and resilience of the SSZs is due to the presence of positive feedbacks. First and foremost is a radiative feedback involving water vapor. In order for air parcels to sink in a stably stratified atmosphere, they must undergo diabatic cooling. Indeed, it is evident from Fig. 1.16 that the five SSZs coincide with regions of diabatic cooling in the 850–500 hPa layer. Hence, air columns within the SSZs are relatively dry (see Fig. 4.6) and thereby capable of emitting infrared radiation to space more efficiently than moister columns that are subject to a stronger greenhouse effect from the overlying water vapor.[7] They are dry because the air within them is subsiding. Conversely, the air is able to sink faster because it is dry. Another feedback involves the reflection of solar radiation by low clouds. The SSZs are marked by higher coverage of low clouds than the surrounding regions, as documented in the bottom panel of Fig. 14.1. The three strongest of the SSZs coincide with patches of negative cloud forcing in the bottom panel of Fig. 5.5. The low cloud decks in these regions shield the ocean surface from incoming solar radiation. The cooler the ocean surface, the more extensive the cloud coverage.

A third set of feedbacks involves the coupling with the underlying ocean. The surface wind field influences the underlying SST distribution, which, in turn, feeds back upon the atmospheric circulation. On the scale of ocean basins, the thermal contrasts between the cooler equatorward and warmer poleward atmospheric flow around the subtropical anticyclones is imprinted upon the ocean mixed layer by the energy fluxes at the air–sea interface. In addition, the surface winds induce currents (e.g., the warm Gulf Stream and Kuroshio on the western side and the cold Humboldt and California currents on the east side) that enhance the east–west thermal contrast. SST is colder on the eastern side of the ocean basins than on the western side, in accordance with these relationships, as can be verified by an inspection of Figs. 1.8 and 1.10. The equatorward surface winds on the eastern side of the tropical oceans also induce coastal upwelling, which is yet another cooling influence, albeit a more localized one.[8]

The distribution of SST and skin temperature over land exerts a strong influence upon atmosphere temperature averaged through the depth of the boundary layer, which feeds back hydrostatically upon the atmospheric flow: other things being equal, regions of higher SST and skin temperature tend to be marked by lower SLP and vice versa and lower SLP is more conducive to convergence. A qualitatively realistic representation of the boundary flow can be obtained using a one-layer, steady-state model forced only by the underlying temperature distribution.[9]

Exercises

14.3 Why is it that the 150 hPa wind vectors shown in Fig. 14.4 are parallel to the geopotential height contours at that level, whereas the surface wind shown in Fig. 14.6 tend to be directed down the height gradient?

14.4 Consider the zonal momentum balance that determines the zonal component of the surface wind along the equator, as depicted in Fig. 14.6. The leading terms are the frictional drag force F_x, the zonal pressure gradient force $-\partial \Phi / \partial x$ and the centrifugal-like force $-v \partial u / \partial y$ associated with the curvature of the cross-equatorial flow. Assume that the flow is steady state. Show that different two-way force balances prevail in longitudinal sectors with eastward- and westward-directed pressure gradients.

14.5 Why does the equatorward flow in the subtropical subsidence zones along the east coasts of the continents curve anticyclonically to form the trade winds, rather than cyclonically, as implied by the conservation of absolute vorticity in a two-dimensional horizontal flow?

14.6 Based on the decomposition in Exercise 4.9, what is the dominant term in the moisture transport in the subtropical subsidence zones versus farther to the west in the trade-wind belts? Explain your answer.

14.7 Why do low cloud decks tend to form in regions of strong subsidence?

14.8 Relate the SSZs to the cycling of air in the atmospheric heat engine depicted in the right panels of Fig. 2.8.

14.9 If the Earth were rotating backward, what would become of the ITCZ, the SSZs, and the equatorial cold tongues?

14.4 THE ITCZ/COLD TONGUE COMPLEXES

In this subsection, we elaborate further on the role of atmosphere–ocean interactions in shaping the annual mean climatology of the western hemisphere of the tropics, beginning with an important oceanic feature, the so-called *equatorial cold tongues* on the eastern sides of the Pacific and Atlantic basins. The Pacific cold tongue is discernible in the SST pattern in Figs. 1.6 and 1.8, whereas the Atlantic cold tongue is present only during summer and early autumn. Both cold tongues are marked by bands of net downward energy fluxes at the air–sea interface in Fig. 5.11. They are a consequence of the easterly climatological mean surface winds along the equator (an extension of the southeast trades, which flow across the equator into the ITCZ). The winds drag water westward with them, raising the sea level on the western sides of the basins and depressing it on the eastern sides. The westward drag force on the warm near-surface water deepens the thermocline on the western sides of the basins and raises it on the eastern sides, bringing cold, nutrient-rich water from below the thermocline close to the surface and into the sunlight, where photosynthesis can occur. The westward drag force also induces an ageostrophic flow (Ekman drift) of surface water in both Northern and Southern Hemispheres, giving rise to a narrow band of equatorial upwelling, which is visible in the ocean color imagery in Fig. 5.9. Because of the shallowness of the equatorial thermocline on the eastern side of the basins, the

---

[7] The dry air columns in these regions have been likened to radiator fins in the climate system (Pierrehumbert, 1995).

[8] The various dynamical processes that contribute to the prominence of the SSZs are discussed in greater detail in Rodwell and Hoskins (2001) and further elucidated in numerical experiments described in Miyasaka and Nakamura (2005).

[9] Lindzen and Nigam (1987).

Figure 14.8 (Top) 1000 hPa wind and divergence fields, contour interval 1×10^{-6} s^{-1}. (Middle) 1000 hPa wind vectors superimposed upon SST, contour interval 0.5°C. (Bottom) 1000 hPa wind vectors superimposed upon SLP, contour interval 0.5 hPa. The 25°C and 1016 hPa contours are bolded.

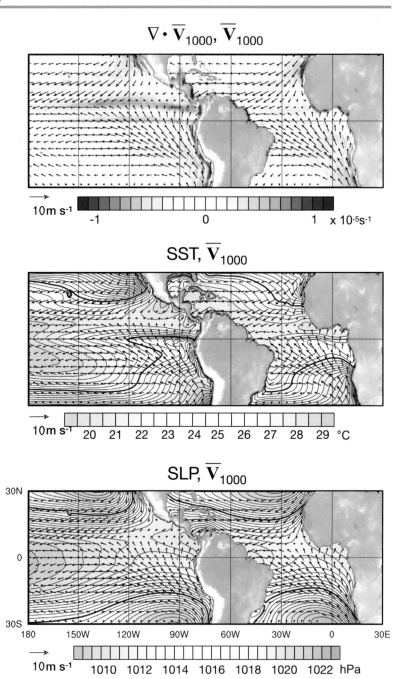

upwelled water is colder than it would be in the absence of the trade winds. The cold tongue weakens toward the west: SST rises by more than 3°C from 110°W to the Date Line.

Figure 14.8 shows $\mathbf{V}_{1000}$, $\nabla \cdot \mathbf{V}_{1000}$, SST and SLP fields over the western hemisphere of the tropics, which encompasses all the well-defined oceanic ITCZs and four of the five subtropical subsidence zones pointed out in the previous subsection. The SST gradients associated with the cold tongues, (middle panel) are by far the strongest observed anywhere in the tropics. Immediately to the north of the cold tongues are belts of relatively warm surface water that coincide with

the wind-driven, near surface, eastward North Equatorial Countercurrent (NECC) shown in the top panel of Fig. 16.3. The SST front that separates the cold tongue from the warmer water to the north is particularly sharp at the eastern end of the Pacific basin, where the cold tongue is centered ~0.5°S and the front at ~1–2°N in the annual mean.

The northward flow across the equator (Fig. 14.8, top panel) slows down as it passes over the cold tongue and speeds up again as it crosses the front, giving rise to a narrow band of divergence along the front itself. The northward wind stress drags the equatorial surface water northward,

SST

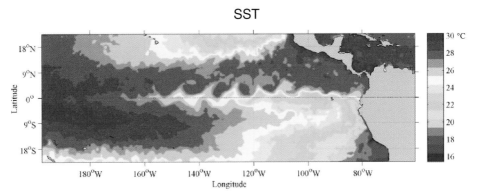

Figure for Exercise 14.13. Daily SST map in the central and eastern tropical Pacific on 25 June 2010. The cusp-shaped features centered along the equatorial SST front near 3°N are tropical instability waves. From Wang et al. (2017). © American Geophysical Union.

inducing upwelling to the south of the equator and downwelling to the north of, it giving rise to the observed equatorial asymmetry in the SST. The larger scale atmospheric asymmetries – the fact that the ITCZ is in the Northern Hemisphere and there is a strong northward cross-equatorial flow year-round – derive from the asymmetries in the land–sea geometry and the orography, as will be explained in Section 16.3.

The SLP field (Fig. 14.8, bottom panel) is smoother than the SST field (middle panel) but it exhibits many of the same features with reversed sign. For example, along the equator the SST and SLP profiles are almost mirror images of one another, reflecting the strong influence of SST upon boundary layer temperature and, in turn, the influence of the boundary layer temperature upon SLP by way of the hypsometric equation. Almost everywhere within the equatorial belt (10°N–10°S) the surface winds blow down the horizontal pressure gradient, reflecting the prevalence of a two-way balance between the pressure gradient force and frictional drag. For example, the northward cross-equatorial surface winds over the eastern oceans are driven by the cross-equatorial meridional pressure gradient, with higher SLP in the Southern Hemisphere, which in turn, is hydrostatically consistent with the lower Southern Hemisphere SST, which derives from the stronger trade winds. These examples serve to illustrate the pervasive atmosphere–ocean coupling observed in the deep tropics. It affects the annual mean climatology, it contributes to the seasonality discussed in the Chapter 16, and it plays a central role in ENSO physics, as discussed in Chapter 17.

Exercises

14.10 Explain the existence of the eastward North Equatorial Countercurrent (NECC), the westward South Equatorial Current (SEC), which is centered just north of the equator, just poleward of the SST front, and the SST front at 1–2° north of the equator.

14.11 Why does the northward flow across the equator slow down as it passes over the cold tongue and speed up again after it crosses to the warm side of the SST front?

14.12 (a) Explain how the northward wind stress across the equator in the eastern Pacific induces an upwelling/downwelling couplet, with upwelling to the south of the equator and downwelling to the north of it. (b) In what sense is this a positive feedback on the northward flow?

14.13 The westward tropical instability waves in the accompanying figure were first detected as sinusoidal variations in meridional current data from equatorial moorings with a period to about 20 days[10] and subsequently in satellite imagery. (a) What is the mechanism that gives rise to them? (b) Do they have any effect upon the atmosphere?

14.5 THE EQUATORIAL STATIONARY WAVES

The documentation of the annual mean tropical circulation in the previous subsections is based on wind, geopotential height, and vertical velocity data for just a few levels. Here we consider the vertical structure in greater detail, with emphasis on the fields of temperature and relative humidity, whose vertical structure is quite different from that of the variables that have been documented thus far. We will focus on the structure in the longitude–height plane averaged over the equatorial belt (10°N–10°S). This section reveals the vertical structure of the equatorial stationary waves.

The top panel of Fig. 14.9 shows the overturning circulation, indicated by vectors, superimposed upon the

[10] Halpern et al. (1988).

Figure 14.9 The equatorial stationary waves in the longitude–height plane, averages of annual mean fields averaged from 10°N to 10°S. Arrows indicate the sense of the circulation. The vertical velocity component is represented by the diabatic heating rate; (top) the eddy component $\overline{Q}^*$ and (bottom) the total $\overline{Q}$. Positive values, plotted as upward arrows, indicate ascent. The field indicated by the colored shading is (top) the eddy component of temperature $\overline{T}^*$ and (bottom) total relative humidity $\overline{RH}$.

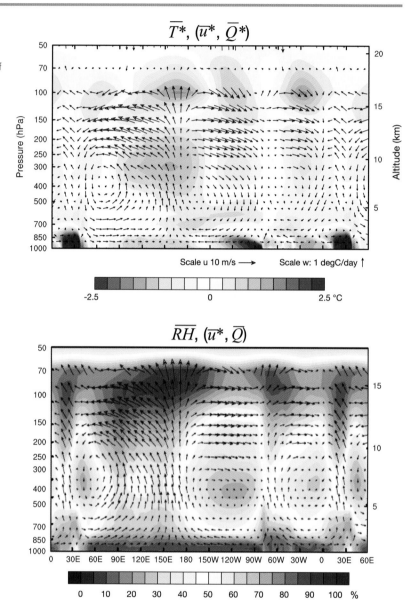

temperature perturbations. The vertical velocity is represented by the heating rate $\overline{Q}$ and is thus scaled by the inverse of the static stability. It should be emphasized that these are not closed circulation cells. They are the projection of a three-dimensional circulation onto the equatorial plane. In some regions, the convergence and divergence associated with the meridional wind component (not shown) makes an important contribution to the overturning circulation. This plotting convention reveals more clearly the extension of the overturning circulation into the lower stratosphere. The most striking feature in the section is the plume of upward pointing arrows over the western Pacific, extending from cloud base all the way up into the lower stratosphere. Weaker plumes of ascent (relative to the surrounding oceans) are discernible at the longitudes of Africa and South America. At tropospheric levels, these features can be interpreted as highly smoothed representations of the ascent in the updrafts in mesoscale convective systems.

The overturning circulation is strongest around the 300 hPa level. It is thermally direct, with rising of warm air over the warm pool and the tropical continents and sinking of cooler air over the eastern Pacific and Atlantic, at the longitudes of the dry zones. Above the 150 hPa level, the circulation is thermally indirect, with rising of cold air and sinking of warmer air. The planetary-scale plumes of cold, ascending air are not directly related to the mesoscale updrafts in the layer below. They are the signature of air parcels being lifted along gently sloping trajectories as they pass over these regions, forced not by the convective scale updrafts, but by the overturning planetary-scale circulation, which extends upward into the stably stratified layer above the convective cloud tops. At these high levels, the diabatic heating, indicated by the streams of upward-pointing arrows, is due to radiative damping of the wave-induced temperature perturbations.[11]

In the lower troposphere, the overturning circulation is more complicated because it is influenced not only by the planetary waves, but also by the boundary forcing, as explained in the previous two subsections. The features below the 700 hPa level are obviously related to the Pacific and Atlantic equatorial cold tongues, the Indo-Pacific warm pool, and the thermal signatures of Africa and South America. These shallow features account for much of the structure in the SLP field, as explained in the previous section. Here as well, the overturning circulation is thermally direct, with subsidence over the cold tongues and ascent over the warm pool and the continents. In the intermediate (850–450 hPa) layer, the temperature perturbations are weak, their structure is more complicated, and the energy conversion is small.

In the bottom panel of Fig. 14.9, the vertical arrows represent total vertical velocity; that is, the vertical component of the overturning circulation shown in the top panel plus the zonal mean vertical velocity at each level. In this representation, ascent is evident over the warm pool, over Africa and South America. The field of relative humidity (RH) is represented by colored shading. A strong correspondence between ω and RH is clearly evident. The air in the equatorial dry zones is remarkably dry, not only in the boundary layer, but throughout the entire depth of the troposphere. In fact, the strongest RH perturbations are observed in the mid-troposphere, where the vertical velocities in the equatorial planetary waves are strongest. The dryness of the mid-troposphere over the eastern tropical oceans tends to inhibit the occurrence of deep convection, even in the ITCZs, thus contributing to the shallower vertical velocity profile for the western hemisphere of the tropics in the top right panel of Fig. 14.2 The drier profile is also conducive to radiative cooling of the lower troposphere.

Exercises

14.14 (a) Based on simple energy balance considerations, explain why the boundary layer tends to be warmer over the tropical continents than over the tropical oceans in the annual mean and even during winter. (b) Why is the continental warmth more pronounced in temperature at the 850 hPa level (Fig. 1.10) than in skin temperature (Figs. 1.6 and 1.7)?

14.15 From an inspection of the longitudinal dependence of temperature and vertical velocity at the 100 hPa level in Fig. 14.9, it is evident that most of the upward flux of mass through the equatorial cold point takes place over the warm pool, where the temperature is as much as 1.5 K lower than in the zonal average. What are the implications for concentrations of stratospheric water vapor?

14.16 The overturning circulation in the equatorial plane in Fig. 14.9 is sometimes referred to as the "Walker Circulation" or "Walker cell" and spoken of as if it were analogous to the Hadley cell, except that it involves overturning in the equatorial plane instead of the meridional plane. What is misleading about that analogy?

14.6 WHAT DETERMINES THE ANNUAL MEAN RAIN RATE CLIMATOLOGY?

Deep convection can occur only in regions in which the air in the lower troposphere is sufficiently warm and moist to support it. There are various metrics for quantifying the warmth and humidity of an air column. In this section, we will compare how several of them relate to annual mean rain rate over the oceans.

The distribution of annual mean tropical rain rate most strongly resembles the distribution of *column saturation fraction* CSF shown in the second panel of Fig. 14.10. CSF is defined as the observed column-integrated water vapor W divided by the water vapor loading of a hypothetical column with the same temperature profile, but saturated at all levels. It is not dependent upon temperature: its spatial variability is determined by the atmospheric transport of water vapor, which is strongly related to the vertical velocity field and the irrotational component of the horizontal wind field.[12] In contrast, column-integrated water vapor W, whose climatological mean distribution is shown in the third panel of Fig. 14.10, is temperature dependent, as evidenced by its strong meridional gradient. It does not resemble the P field quite as strongly as CSF does, but it can be said that heavy (>5 mm d^{-1}) climatological mean rain rates are largely confined to the domain in which $W > 48$ mm.[13] The distribution of W, in turn, is strongly influenced by the underlying SST pattern (bottom panel). In the boundary layer, where variations in relative humidity are much smaller than those in the free atmosphere, specific humidity q is controlled by the distribution of SST (see Fig. 4.6). Heavy climatological mean tropical rain rates are largely confined to the domain in which SST > 27°C[14] and the same is true of intense convectively driven systems like tropical cyclones.[15]

The role of atmospheric variability in controlling the distribution of tropical rain rate comes into sharper focus in the instantaneous fields, a sample of which is shown in Fig. 14.11. While it is evident from the top panel that virtually all the high rain rates over the tropical oceans are restricted to regions with SST > 27 °C, that rough "rule of

---

[11] For example, as described in Fueglistaler et al. (2004).

[12] Bretherton et al. (2004).
[13] Mapes et al. (2018).
[14] Gadgil et al. (1984); Graham and Barnett (1987).
[15] Palmén (1956).

Figure 14.10 Annual mean (top) rain rate; (second) column saturation fraction, as defined in the text; (third) column-integrated water vapor $\overline{W}$; and (bottom) SST. The scales on the color bars are customized to illuminate the relationships between rain rate and the other variables.

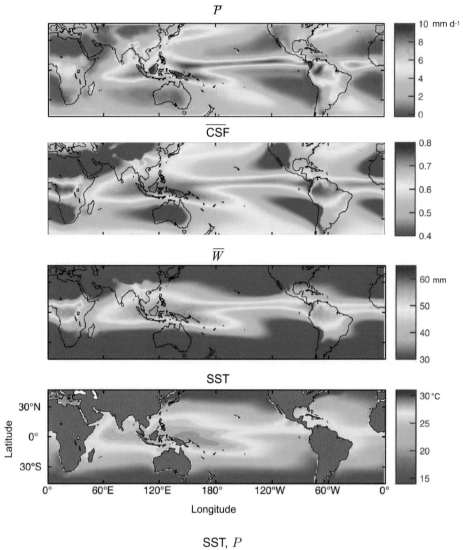

Figure 14.11 Instantaneous fields: Thin contours enclose regions with rain rates in excess of 20 mm d^{-1}, superimposed on (top) SST. The heavier reddish-black contour represents the 27°C isotherm. (Bottom) Column water vapor W, rendered in colored shading.

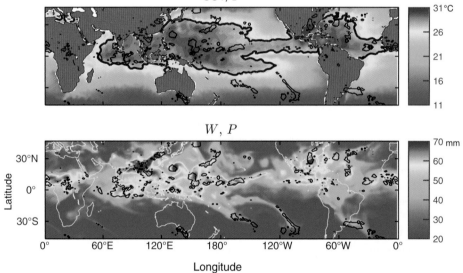

thumb" accounts for only a modest fraction of the space/time structure of P. The correspondence between instantaneous values of P and W, (nonlinear color scale, bottom panel) is much stronger. Here as well, virtually all the highest rain rates occur in association with values of W in excess of ~48 mm, indicated by the warm colors, in relatively slowly evolving filaments in which the air is moist enough to support the development of mesoscale convective systems. These moist patches develop in regions of ascent in tropical waves and they are continually being reshaped and drawn out into filaments by nonlinear advection, much like patches of strong cyclonic vorticity in extratropical latitudes.

15 Tropical Convection

with Ángel F. Adames

The governing equations for the tropical and extratropical general circulations differ in two respects: one relating to the relative importance of the terms in the horizontal equation of motion and the other to the terms in the thermodynamic energy equation. The extratropics are nearly in geostrophic balance. In the tropics, such a dramatic simplification of the horizontal equation of motion cannot be justified, but there exists a nearly two-way *weak temperature gradient* (WTG) balance between the vertical advection of dry static energy and the diabatic heating rates $-\omega s \approx Q$. These behavioral differences derive from the latitudinal dependence of the Coriolis parameter f due to the sphericity of the Earth. Because of the smallness of f, geostrophic (or thermal wind) balance is not generally applicable to large-scale tropical motions. In the absence of strong inertial stability $\eta = (f + \zeta)$, strong horizontal pressure and temperature gradients cannot be sustained. The ratio of the restoring forces for horizontal versus vertical displacements f^2/N^2 is much smaller in the tropics than in the extratropics, from which it follows that the horizontal wind field would be almost barotropic were it not for the release of latent heat in deep convection, which substantially reduces the effective static stability. Another distinction is the greater strength of the precipitation-related diabatic heating in the tropical troposphere that exists by virtue of the higher column water vapor W, as documented in the previous chapter.[1]

Under WTG balance, the horizontal temperature advection and the tendency term in the thermodynamic energy equation are sufficiently small that they can be neglected. The weak temperature gradients observed in large-scale tropical motion systems derive not from the small imbalances between $-\omega s$ and Q, but from the requirement that the temperature field remain in hydrostatic balance with the pressure (or geopotential) field as it evolves in accordance with the horizontal equations of motion. The horizontal divergence is determined by the vertical mass flux in moist convection. It follows that the humidity field needs to be taken into account in investigating the dynamics of tropical motion systems.[2]

This relatively short chapter explores the implications of WTG balance for interpreting the tropical (tropospheric) general circulation. Appendix D offers a formal justification for the WTG approximation based on consideration of

the relevant scaling, and it also provides a more complete explanation of how it affects the governing equations, taking into account the critical role of the moisture. The main text is divided into four sections. The first documents the extent to which the local, vertically integrated, annual mean energy balance is shaped by the release of latent heat and cloud-related radiative transfer. The second describes how downwelling gravity waves rapidly transfer the enthalpy imparted by diabatic heating in mesoscale convective updrafts to the large-scale environment, heating the tropics as a whole, thereby maintaining an environment that supports WTG-equilibrated, *convectively coupled waves* and *warm core vortices*. The third section describes how, in this environment, short-lived bursts of deep convection along the equator give rise to gravity waves and much more slowly evolving, WTG-equilibrated, equatorially trapped planetary waves referred to as "moisture modes". The final section discusses the self-aggregation of moist convection in a WTG-equilibrated environment.

Exercise

15.1 Relate the vertical derivative of dry static energy DSE to the lapse rate Γ in height coordinates.

15.1 THE LOCAL, VERTICALLY AVERAGED ENERGY BALANCE

The vertically integrated thermodynamic energy balance in Eq. (5.5) may be written as

$$\{Q\} - F_{SH} = LP + \{Q_R\}. \tag{15.1}$$

We will refer to $\{Q\} - F_{SH}$ as the *adjusted total heating*; that is, the net heating that takes place within the atmosphere. The radiative heating term $\{Q_R\}$ can be decomposed into a cloud-related component $\{Q_{rc}\}$ and a residual $\{Q_{res}\}$, which is presumably dominated by cooling due to the emission of infrared to space irrespective of cloud cover. Using these definitions, Eq. (15.1) can be rewritten as

$$\{Q\} - F_{SH} = LP + \{Q_{rc}\} + \{Q_{res}\}. \tag{15.2}$$

Figure 15.1 shows the adjusted total heating for the 1000–100 hPa layer, based on annual mean data (top panel), together with the cloud-related component $LP + \{Q_{rc}\}$ and

[1] Charney (1963).
[2] Sobel and Bretherton (2000); Sobel et al. (2001).

272

Figure 15.1 (Top) Vertically integrated (1000–100 hPa) annual mean diabatic heating rate minus the upward flux of sensible heat at the Earth's surface, $\{\overline{Q}\} - \overline{F}_{SH}$. (Middle) Vertically integrated cloud-related heating $L\overline{P} + \{\overline{Q}_{rc}\}$ obtained by spatially regressing $\{\overline{Q}\}$ upon $L\overline{P}$ as described in the next figure. (Bottom) The residual heating $\{\overline{Q}_{res}\}$, a proxy for clear sky radiative cooling.

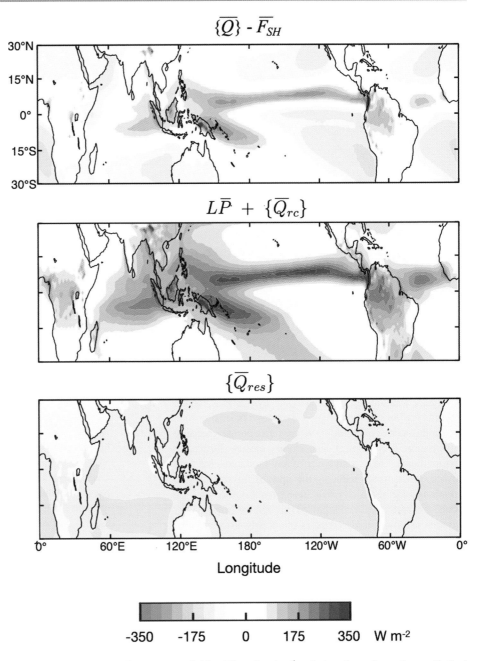

the residual radiative heating $\{Q_{res}\}$. Owing to the smallness of the sensible heat fluxes, especially over the oceans, the adjusted total heating bears a strong resemblance to the distribution of rain rate shown in Fig. 14.1, with an excess of heating over cooling in the rain belts and a deficit in the dry zones.

In the middle and bottom panels in Fig. 15.1, the annual mean $\{Q\} - F_{SH}$ is partitioned into a cloud-related component $LP + \{Q_{rc}\}$ and a residual identified with the clear sky radiative cooling. The cloud-related component of the radiative heating $\{Q_{rc}\}$ is assumed to be linearly proportional to the annual mean precipitation P. The constant of proportionality is derived by regressing grid point values of annual mean $\{Q - F_{SH}\}$ upon the field of P, as indicated in the scatter plot shown in the middle panel of Fig. 15.2. In performing the regression, spatial means are not removed from either

variable. The cloud-related (condensation plus radiative) heating estimated in this manner, shown in the middle panel of Fig. 15.1 is thus the LP field rescaled by the regression line in Fig. 15.2, which is the equivalent to $(LP + \{Q_{rc}\})/LP$. The residual $\{Q_{res}\}$, shown in the bottom panel of Fig. 15.1 is spatially uniform, consistent with the notion that it is dominated by clear sky radiative cooling.

The relationships involving the cloud-related heating come into sharper focus when the data are partitioned into ocean and land grid points, as in the left and right panels of Fig. 15.2. Over the ocean, the linear relationship explains over 98% of the variance of the adjusted total heating. That the regression line exhibits a slope of 1.18 indicates the net radiative heating in the shield made up of anvils of deep convective clouds acts as a positive feedback upon the organized convection, increasing the cloud-related heating

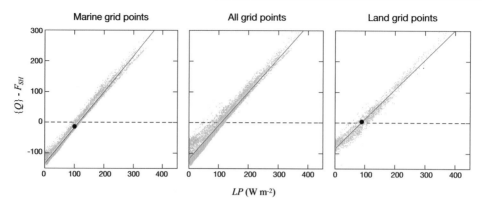

Figure 15.2 Scatter-plot *representations* of the annual mean $\{\overline{Q}\} - \overline{F}_{SH}$ shown in the top panel of the previous figure as a function of the rate of release of latent heat, *LP*. Each point in the plot represents an individual grid point within the tropical domain (30°N–30°S). The middle panel is based on all grid points. In the left and right panels the data are partitioned into land and ocean grid points as indicated: the circles indicate the centroids of the grid points. The red lines are least squares best fit regression lines based on data fin the respective domains.

by 18% compared to the heating attributable to condensation alone. Within the tropical ocean domain, the area averaged cloud-related heating nearly balances the residual heating $\{Q_{res}\}$, so there is little export of MSE.

The corresponding scatter plot for tropical land grid points, shown in the right panel of Fig. 15.2, is also indicative of a linear relationship between $\{Q - F_{SH}\}$ and P, and it accounts for 96% of the variance. The slope is smaller than that for the ocean grid points (0.96 vs. 1.18), perhaps because precipitation over tropical land occurs mostly during the daytime hours when cloud cover of any kind cools the land surface, substantially reducing the upward flux of infrared radiation, some of which would have been absorbed by high clouds and water vapor as it passed through the troposphere. Daytime cloud cover also reduces the downward shortwave radiation, some of which would have been absorbed in its passage through the atmosphere. As is the case over the oceans, on average the clear sky radiative cooling nearly balances the cloud-related heating.

To the extent that the WTG approximation is valid, vertical velocities in the tropics can be inferred from the heating rates. Formally, we can write for annual mean quantities

$$\omega \equiv \frac{dp}{dt} = \frac{Q - \mathbf{V} \cdot \nabla T}{s(p)}, \qquad (15.3)$$

where $s(p)$ is the static stability $\partial \mathrm{DSE}/\partial p$ as defined in Eq. (1.3). In the extratropics the temperature advection term dominates, whereas in the tropics the diabatic heating term dominates, and in the WTG approximation the temperature advection term is assumed to be negligible. In the annual mean climatology, the most serious violation of this assumption is neglecting the patches of cold advection in the subtropical subsidence zones, as shown in Fig. 14.7.

Exercises

15.2 The annual mean rain rate averaged over the tropics is on the order of 0.3 cm per day. Calculate the associated diabatic heating rate due to latent heat

release, expressed in W m^{-2}, and compare it with the heating rates estimated in this section.

15.3 As noted in Section 5.4.2, the MSE of the air in the upper branch of the Hadley cell is higher than that in the lower branch, rendering the tropical troposphere as a whole stable with respect to moist convection. Discuss the role of cloud-related radiative heating in maintaining this stable stratification.

15.4 Estimate and compare the typical magnitudes of the three terms in the thermodynamic energy Eq. (15.3) in the subtropical subsidence zones.

15.5 It is evident from the middle panel of Fig. 14.8 that in the annual mean, boundary layer air warms by more than 5 K as it flows northward along the eastern side of the Pacific and Atlantic from the Southern Hemisphere SSZs, across the equator, to the ITCZ. In this region the air sea fluxes shown in Fig. 5.11 are downward and thus act to cool the atmospheric boundary layer. How does the air warm?

15.2 MAINTENANCE OF WTG BALANCE

Throughout most of the tropics, the vertical profile of MSE exhibits a distinct minimum in the lower troposphere above the inversion at the top of the boundary layer. The air immediately below the inversion exhibits a lower potential temperature than the air above it, but it is much more moist and therefore possesses more MSE (e.g., as demonstrated in Section 5.4.2). Deep cumulus convection is inhibited by the presence of the inversion at the top of the boundary layer and by the dryness of the air in the layer immediately above it, which is entrained into any plumes that manage to penetrate through the inversion. The drier the entrained air, the more the entrainment dilutes the buoyancy of the plumes. Boundary layer convergence thickens the moist boundary layer, it weakens the temperature inversion at the top of it, and it raises the relative humidity of the overlying layer in the free atmosphere, all of which serve to reduce the convective inhibition and thus render the environment more conducive to convection. Boundary layer convergence, in turn, is sensitive to the underlying SST gradients and it varies in response to the passage of atmospheric waves. The updrafts in organized mesoscale convective systems suffer less loss of buoyancy due to the entrainment of dry air than isolated convective cells, and are thus able to ascend nearly all the way up to their equilibrium level in the upper troposphere; that is, the level at which the MSE that they carry with them from the boundary layer is comparable to that of the environmental air.[3] Cloud-tops in this so-called *penetrative convection* typically range as high as 150 hPa over the warm pool, and some extend much higher.[4]

If deep convection were uniformly distributed throughout the tropics, the spectrum of vertical velocity would thus be dominated by a mesoscale peak. In reality, the convection is also modulated by synoptic and mesoscale waves, with which it interacts. In contrast to the vertical velocity spectrum, which exhibits substantial variance on all scales, the spectra of kinetic energy, geopotential height, and temperature are dominated by synoptic and planetary-scale tropical motion systems. The spatial spectrum of temperature is much redder than that of vertical velocity because gravity and inertio-gravity waves excited by the convection are effective in smoothing the perturbations and because of the existence of planetary-scale, convectively coupled waves described in the next section, which develop in a WTG-equilibrated environment.

The gravity waves that smooth the fields inhabit the vast space between the mesoscale updrafts, in which the air is unsaturated and stably stratified. They are excited by the temperature contrasts between the warmer convective-scale updrafts and their cooler environment. As they disperse away from the updrafts, they spread the warmth to the large-scale environment and eventually throughout the entire tropics, homogenizing the temperature field. The smoothing effect of these waves is represented schematically in Fig. 5.3. The uniformity of the residual radiative cooling in Fig. 15.1 is evidence that the latent heat released in convective updrafts is dispersed throughout the tropics. Compelling evidence of the existence of gravity waves radiating out from convective plumes is afforded by satellite imagery of low cloud decks during volcanic eruptions like the one shown in Fig. 2.20. The expanding ring of clear sky is the signature of adiabatically induced subsidence warming, which spreads the warmth of the updraft to the surrounding air.

The expanding downwelling gravity wave in Fig. 2.20 can be simulated in a one-dimensional shallow water wave equation model in which a localized heat source at $x = 0$ is turned on at time $t = 0$. In the solution for an infinite one-dimensional domain shown in Fig. 15.3, patches of subsidence develop along the flanks of the heat source as soon as it is turned on and propagate outward at the

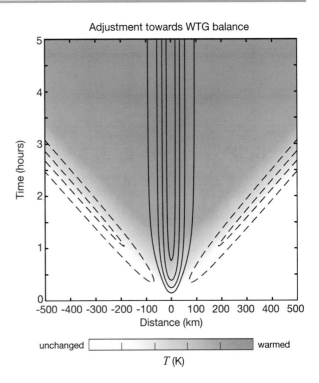

Adjustment towards WTG balance

unchanged   warmed
T (K)

Figure 15.3 Equatorial time–longitude (x, t) section showing the response to a localized heat source at $x = 0$ turned on at time $t = 0$ in the analytic solution of the shallow water wave equations in a one-dimensional space domain. Away from the heat source the flow is adiabatic. The colored shading indicates the temperature change and the contours represent the vertical velocity; solid contours ascent and dashed contours subsidence, zero contour omitted, scaling arbitrary. Based on Fig. 2 in Adames and Maloney (2021a).

phase speed of a gravity wave $c = \sqrt{gh_e}$, where h_e is the equivalent depth. To convert the perturbations in the height of the free surface in a shallow water wave equation model into temperature perturbations, it is necessary to prescribe a vertical structure. The assumed vertical scale affects the amplitude of the temperature perturbations, but not their shape or their tendency to propagate radially outward from the source with a fixed phase speed. As the downwelling wave passes a fixed grid point, the air warms adiabatically and thereby equilibrates (i.e., comes into WTG balance) with the heat source. Meanwhile, steady state ascent directly over the concentrated heat source produces adiabatic cooling that balances the diabatic heating. In this manner, the entire domain shown here eventually adjusts to the turning on of the heat source. If L is the size of the domain or the scale of a planetary wave in which the heat source is embedded, then WTG balance prevails on timescales $\tau \gg L/c$. By that time, the gravity waves that were responsible for the adjustment are gone and all that remains is the rectified warming that is spread uniformly throughout the adjusted domain. The temperature field derived from the corresponding two-dimensional simulation, shown in Fig. 15.4, assumes the form of an expanding ring analogous to the one observed when a stone is thrown into a pond. Though the heat source remains on after $t > 0$, the temperature at a fixed grid point

---

[3] Riehl and Malkus (1958).
[4] Highwood and Hoskins (1998).

Adjustment towards geostrophic balance

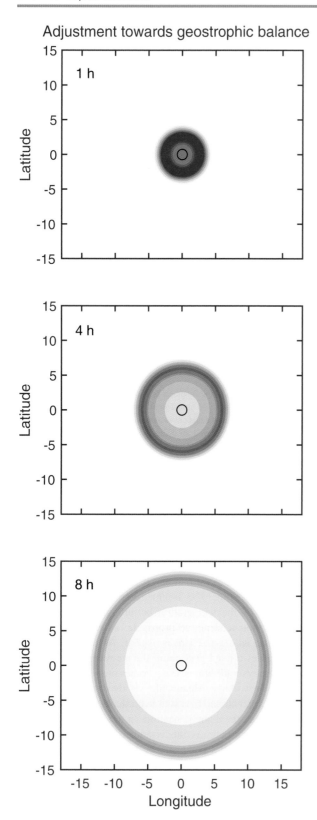

Figure 15.4 As in Fig. 15.3, but for a two-dimensional space domain, and showing only the temperature change due to a heat source at the origin, "turned on" at $t = 0$.

gradually drops after the passage of the wave, but it never quite reaches zero.

Figure 15.5 shows composite tropospheric temperature (thickness) fields based on vertical velocity pulses at individual equatorial grid points, highpass filtered in order to isolate the relatively weak inertio-gravity wave signature in the reanalysis. Why the composite is based on downwelling rather than upwelling events, is explained in the caption of Fig. 10.12 in Appendix F. A burst of convectively induced subsidence at the reference grid point gives rise to a warm disk that expands into the surrounding region behind a downwelling gravity wave. The process is analogous to the one in the numerical simulation in the previous figure, with two notable exceptions. (1) It is three-dimensional rather than two-dimensional: the waves are dispersing upward into the stratosphere as well as radially. (2) The burst of subsidence lasts for only an hour or two, the decorrelation time of the vertical velocity time series, whereas the heating in the numerical experiment was prescribed to be continuous after $t = 0$. Since the temperatures in the vicinity of the reference grid point do not remain elevated, what starts out as a warm disk evolves more rapidly into a warm ring.

The left and right panels of Fig. 15.5 are different in two respects. (1) Downwelling events correspond to negative extrema in $w_{300} + w_{700}$ in the left panel and $w_{300} - w_{700}$ in the right panel; and (2) temperature in the left panel is represented by 1000–200 hPa thickness, whereas in the right panel it is represented by 500–200 hPa thickness. The fields shown in the left panel thus correspond to relatively deep waves in which vertical velocity perturbations of the same polarity extend through the depth of the troposphere, whereas those in the right panel correspond to shallower waves in which the temperature perturbations exhibit a phase reversal in the mid-troposphere. Consistent with theoretical expectations, the deeper waves radiate outward from the reference grid point at about twice the rate of the shallower waves.[5]

15.3 CONVECTIVELY COUPLED WAVES

The vertical profiles of amplitude of the geopotential height and wind fields in convectively coupled waves resemble the profiles of divergence shown in the bottom left panel of Fig. 14.2, with maxima of opposing polarity in the boundary layer and in the upper troposphere centered around the 150 hPa level. In contrast to the high frequency, downwelling IG waves shown in Fig. 15.5, which assume the form of concentric circles radiating away from the reference grid point, the convectively coupled waves project strongly upon the modal structures of zonally propagating, equatorially trapped planetary waves.

Figure 15.6 shows the signature of convectively coupled waves in the 850 hPa geopotential height and wind fields in response to rain rate variations at reference grid points

[5] Holton and Hakim (2012).

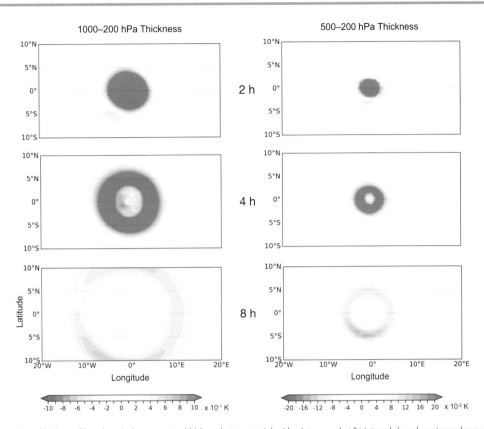

Figure 15.5 Composites of highpass filtered tropical temperature (thickness) at equatorial grid points spaced at 5° intervals based on strong downwelling events in the vertical velocity field. The patterns are shifted zonally to place the reference grid points at 0° longitude and then averaged to create a single set of composite maps, lagged as indicated with respect to the downwelling events. In the left panels the downwelling events correspond to negative extrema in $w_{300} + w_{700}$ and the temperature field is based on 1000–200 hPa thickness. In the right panels the downwelling events correspond to negative extrema in w_{300} minus w_{700} and the temperature field is based on 500–200 hPa thickness. The thickness fields are highpass filtered in order to remove the signature of the more energetic day-to-day variability, and the diurnal cycle and its higher harmonics are filtered out. For further specifics, see Appendix F. From Pahlavan et al. (2022). © American Meteorological Society. Used with permission. This preliminary version has been accepted for publication in the *Journal of Atmospheric Sciences* and may be fully cited. The final typeset copyedited article will replace the EOR when it is published.

along the equator arbitrarily referenced to the Date Line. In this case, the data are 5–100 d bandpass filtered. Removing the variability with periods longer than 100 d is equivalent to considering departures from the climatological mean, taking into account both the seasonal and ENSO-related variability. Removing the variability with periods shorter than 5 d filters out IG waves. Another distinction between Figs. 15.5 and 15.6 is that the former is a composite based on downwelling events at reference grid points, whereas the latter was created by lagged-linear regression, using the corresponding rain rate rather than vertical velocity as the reference time series. (Composites based on unsmoothed hourly rain rate data at reference grid points do not yield nearly as clear a gravity wave signature as composites based on downwelling events in the vertical velocity field.)

Kelvin and Rossby wave signatures reminiscent of the theoretical solutions shown in Fig. 14.4 are evident in all three panels of Fig. 15.6. By construction, the pattern at $t = 0$ (top panel) is characterized by strong convergence centered over the patch of enhanced rain rate. In contrast to the fields in the numerical simulations in which the forcing is "turned on" at $t = 0$, a planetary wave pattern is already present, with the easterly polarity of a Kelvin wave just to the east and a

suggestion of an equatorial Rossby wave over the convection itself. With the passage of time, the Kelvin wave propagates eastward and the Rossby wave propagates westward and becomes more distinct than it was at $t = 0$. At $t = 2$ d, the patch of enhanced rain rate assumes a swallowtail shape, with wingtips pointing westward, a characteristic feature of rain rate signatures, as will be discussed in Section 18.3. At $t = 4$ d it exhibits weak maxima in the cyclonic Rossby wave gyres and along the equator in the region of convergence $-\partial u/\partial x$ in the Kelvin wave. Patterns similar to those in Fig. 15.6 can be obtained from the analytic solutions for waves on an equatorial beta-plane in a shallow water wave equation model with weak frictional damping in which an equatorial heat source, a delta function in time, is applied at $t = 0$.[6]

The simultaneous regression patterns based on monthly mean data, shown in the bottom panel of Fig. 15.7, resemble the theoretical steady state solutions shown in Fig. 14.4 even more strongly. The corresponding patterns for the 150 hPa level, shown in the top panel, are quite different. The geopotential height patterns exhibit a quadrupole configuration centered on the reference grid point, with centers located

---

[6] See the right panel of Fig. 2 in Heckley and Gill (1984).

Figure 15.6 Geopotental height and horizontal wind fields at 850 hPa regressed upon rain rate averaged over 5° latitude × 5° longitude boxes centered on equatorial grid points spaced at 5° intervals. Latitude and longitude are relative to the center of the reference grid box, which is arbitrarily placed on the Date Line. The patterns were shifted zonally to place the reference grid points at the Date Line and then averaged to create a single set of composite maps, lagged as indicated with respect to rain rate at the reference grid points. These data have been bandpass filtered to retain fluctuations with periods between 5 and 100 d. Amplitudes are arbitrary.

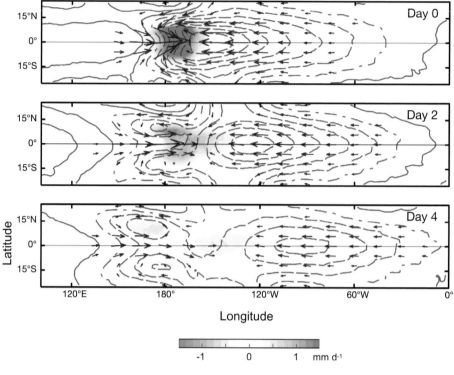

Figure 15.7 As in Fig. 15.6, but simultaneous regression patterns based on monthly mean data. (Top) The 150 hPa height and wind anomaly fields and (bottom) the corresponding 850 hPa fields. The longest vectors correspond to wind speeds of ∼5 m s^{-1} in the top panel and 2.5 m s^{-1} in the bottom panel. The contour interval is 2 m. Positive contours are solid, negative contours dashed, and the zero contour is bolded.

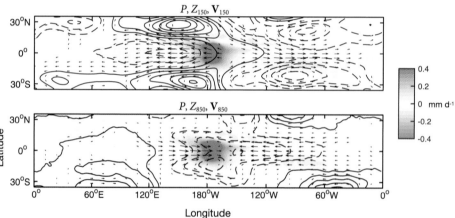

Figure 15.8 As in Fig. 15.7, but showing the velocity potential (heavier lines) and the velocity streamfunction (lighter lines). Contour interval 1 × 10^6 m^2 s^{-1} in the top panel and 0.5 × 10^6 m^2 s^{-1} in the bottom panel. Positive contours are solid, negative contours dashed, and the zero contour is bolded.

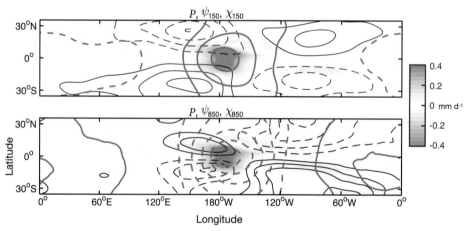

~28°N/S, much farther poleward than the anticyclonic Rossby wave gyres in the bottom panel. These features are quasi-stationary Rossby waves in the westerlies. The corresponding regression patterns for the irrotational and nondivergent wind components are shown in Fig. 15.8 together with the velocity potential and streamfunction fields. The irrotational flow is toward the reference grid point at the lower level and away from it at the upper level, while the irrotational component emphasizes the Rossby wave response to it. Owing to the beta-effect, the irrotational meridional flow induces a vorticity tendency at the longitude of the reference grid point: cyclonic in both hemispheres at the lower level and anticyclonic at the upper level. The former is balanced by frictional drag and the latter by zonal advection. Similar patterns emerge as the response to an equatorial mass source/sink in a shallow water wave equation model with a prescribed zonally symmetric background flow.[7]

Convectively coupled waves are more prominent over the oceans than over land, for several reasons. Cloud radiative feedback on the wave-related diabatic heating field is greater over the oceans than over land, as discussed in Section 15.1. Another possible contributing factor is that convectively coupled waves over and near land have to compete with the diurnal cycle for control of the available boundary layer moisture.

That the tropical heating and vertical velocity fields in the free troposphere tend to be in WTG balance allows for the existence of a special class of tropical motion systems known as *moisture modes* in which condensation heating plays a central role in the dynamics. In these systems, rain rate and the column-integrated water vapor are strongly coupled. They are so-named because their evolution can be inferred by considering the moisture budget under WTG balance, as shown in Appendix D.4. For example, whether a midlatitude Rossby wave propagates eastward or westward or remains stationary depends upon the potential vorticity tendency to the east and west of it due to the potential vorticity advection by the wave. In a similar manner, the direction and rate of propagation of a tropical moisture mode is determined by the pattern of wave-induced moisture advection.

In order for moisture modes to exist, the troposphere must be sufficiently humid to support convection that lasts long enough to allow the atmosphere to adjust to a WTG balance through the process described in Section 15.2. Thus, for them to be classified as moisture modes, their MSE perturbations must be determined mainly by the latent energy contribution; that is, $\delta(Lq) \gg \delta(\text{DSE})$. Scale analysis of the MSE budget reveals that this criterion is met when the area of enhanced rain rate propagates much more slowly than the gravity waves that spread the convectively induced warming to the surrounding regions and eventually to the tropics as a whole, and that requires that the phase speed of the waves be not much greater than ~5 m s^{-1}.[8] The vertical velocity perturbations in moisture modes are strongest in

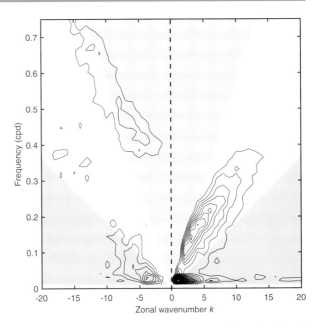

Figure 15.9 Two-sided wavenumber–frequency spectrum for satellite-observed brightness temperature, a proxy for rain rate, normalized using the procedure described in Section 10.2.4. The colored shading corresponds to ranges of phase speed; < 9 m s^{-1} (blue), which corresponds to moisture modes; > 28 m s^{-1} (yellow), in which the convective coupling has little effect on the dynamics; and intermediate values (unshaded), for which WTG balance may be applicable, but the latent heat term does not dominate the MSE perturbations. Adapted from Adames et al. (2019).

the mid-troposphere and the wind and geopotential height perturbations in the upper atmosphere tend to be out of phase with those in the lower troposphere.

Figure 15.9 shows a two-sided wavenumber–frequency spectrum for rain rate, as inferred from brightness temperature. We will be examining the features in this diagram in further detail in Chapters 18 and 19. The slowest of these waves, with phase speeds on the order of 5 m s^{-1}, qualify as moisture modes, whereas in the fastest waves, with phase speeds of >30 m s^{-1}, the release of latent heat has little influence upon the dynamics. It is notable that much of the variability falls in the intermediate range, in which the dynamics of the waves is influenced by, but not dominated by, moist thermodynamics. In simple models of wavelike moisture modes, the phase speed is largely determined by the horizontal and vertical moisture gradients, and it is influenced to a lesser degree by the surface latent heat flux.

The MJO, with its prominent rain rate signature and its slow phase speed, is widely regarded as an example of a moisture mode. We will examine its moisture budget in Section 18.3.

Exercises

15.6 In Fig. 15.8 the horizontal wind field is decomposed into an *irrotational* (or divergent) component, the gradient of a scalar *velocity potential* χ field, and a *nondivergent* (or rotational) component,

---
[7] Monteiro et al. (2014).
[8] Adames et al. (2019); Ahmed et al. (2021).

Figure 15.10 Samples of monthly mean rain rate (standardized by dividing it by its global mean value) as simulated in three different general circulation models run in an aquaplanet mode without rotation and with prescribed SST of (top) 295 K and (bottom) 305 K at all grid points. Reprinted from Bony et al. (2016).

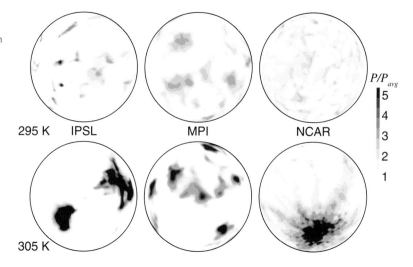

295 K   IPSL   MPI   NCAR

P/P_{avg}

5
4
3
2
1

the curl of a scalar *streamfunction* ψ field. Show how the χ and ψ fields are derived from the horizontal wind field.

15.7   In deriving the moisture budget Eq. (15.1) we made use of Eq. (D.25) in Appendix D. (a) Show that Eq. (D.25) can be written in the form

$$\frac{\partial Lq}{\partial t} = -\mathbf{V} \cdot \nabla Lq - \omega \frac{\partial Lq}{\partial p} - Q_c - \frac{\partial}{\partial p}\overline{\omega_* \text{MSE}_*}.$$

(b) Show that under conditions of WTG balance, the equation for the time rate of change of MSE is

$$\frac{\partial}{\partial t}\text{MSE} = -\mathbf{V} \cdot \nabla Lq - \omega \frac{\partial}{\partial p}\text{MSE} + Q_R - \frac{\partial}{\partial p}\overline{\omega_* \text{MSE}_*}.$$

and that it is essentially equivalent to Eq. (D.25).

15.4 SELF-AGGREGATION OF TROPICAL CONVECTION

It is evident from satellite imagery that at any instant in time, most of the deep convection in the tropics tends to be patchy, aggregated into discrete bands like the ITCZs, vortices, or irregularly shaped clusters in which the convective clouds are interconnected by large shields of cirrus clouds. The expanses between the aggregates are largely free of deep convective clouds. The role of convectively coupled waves in enhancing convection in some large-scale regions and suppressing it in others was touched on in the previous section and will be illustrated further in Chapters 18 and 19. In this section, we explore the role of *convective self-aggregation*,

which is observed to occur in numerical simulations run with spatially uniform solar forcing and without planetary rotation.

Convective self-aggregation can be interpreted as an instability of the radiative–convective equilibrium state (Section 2.1) involving interactions between and among moist convection, radiation, environmental moisture, and surface fluxes of latent and sensible heat. It occurs in tropical motion systems on scales ranging from a few hundred kilometers (as in cyclones and synoptic scale waves) to planetary (e.g., in "Tropic World" discussed in Section 2.4). It is not observed in dry models or in models without interactive radiation. Examples are shown in Fig. 15.10 – maps of rain rate averaged over a typical month in simulations performed with three different global models run with spatially uniform insolation and without rotation. Each model was run with prescribed SST of 295 K and 305 K. Aggregation occurs in all the runs. In all three models the degree of aggregation increases with SST.

Convective self-aggregation is simulated in experiments conducted with a wide variety of numerical models, some two-dimensional and others three-dimensional, some global and others regional, some with parameterized cumulus convection and others with sufficient resolution to be regarded as "convection permitting." Yet despite its pervasiveness, the mechanisms responsible for self-aggregation remain elusive because there are so many of them to be considered, and the behavior of the models is highly sensitive to their resolution and parameterization schemes.[9]

---

[9] Wing et al. (2017).

16 The Seasons in the Tropics

When plotted as partial zonal averages in Fig. 16.1, the seasonality of the zonal mean circulation in the eastern and western hemispheres of the tropics is quite different. In the eastern hemisphere (from the Greenwich Meridian eastward to the Date Line), the zonal mean circulation is dominated by the seasonally reversing Australasian monsoon, which is strong and nearly synchronous with the annual cycle in the meridional profile of insolation. In contrast, in the western hemisphere, the seasonality is not as pronounced and the annual cycle is lagged by about two months relative to the solstices. The seasonality in the rain rate climatology over the tropical continents is remarkably diverse, reflecting a variety of competing planetary-scale and regional influences.

In the first two sections of this chapter, we will document the contrasts between the tropical circulations in DJF and JJA, when the monsoonal influences are dominant, and between MAM and SON, in which the marine influences in the western hemisphere are more prominent. In the third section, we will document the existence of a weak semiannual cycle in equatorial rain rate – a response to the annual march of insolation. In the final section, we will offer some general remarks about the peculiarities of the seasonality of rainfall in different parts of the tropics, leaving it to the interested reader to explore the topic further.

16.1 DJF VERSUS JJA CIRCULATIONS

The annual mean tropical circulation considered in the previous chapter represents an average of stronger seasonally varying circulations. In particular, the seasonal climatologies reveal cross-equatorial flow and other aspects of the tropical circulation, which nearly cancel in the annual mean.

16.1.1 The Zonally Varying Flow

The dominant features of the DJF and JJA climatologies are the so-called *monsoon circulations*. The word *monsoon* stems from the Portuguese *monção* and originally referred to the seasonal reversal of the surface wind in south and east Asia; it is now more widely used with reference to seasonally reversing tropical motion systems over Australasia, Africa, and the Americas.

The contrasting JJA and DJF rain rate climatologies are shown in Fig. 16.2. A northward shift of the convection

from DJF to JJA is evident at all longitudes – much greater over and near the continents than over the oceanic ITCZs. Figure 16.3 documents the corresponding shift in the lower tropospheric circulation. The region of heavy rain rates over South Asia in JJA lies at the downstream end of a broad low-level air current that can be traced all the way back to the southeasterly trade winds in the Southern Hemisphere along the equatorward flank of the so-called "Mascarene high" in the SLP field. While flowing northward across the equator and northeastward across the Arabian Sea, air parcels pick up sensible heat and moisture from the underlying sea surface. Some of these warm, moist air parcels are lifted as they flow over the Ghats range along the west coast of India, some are funneled into the northern part of the Bay of Bengal and over the adjoining coastal plains and mountain ranges, and some of them continue eastward across Indochina toward the Philippines. Eventually most of this warm, moist boundary layer air ascends in deep convective clouds, in some of these areas yielding seasonal mean rainfall in excess of 500 mm per month. The heaviest rain rates, with values on the order of 5 m per month during the monsoon season, are observed along the lower slopes of the Himalayas to the north of the lowlands at the head of the Bay of Bengal. The remarkable mass of the Asian summer monsoon rainfall is reflected in the large summer and autumn stream flows in the Ganges, Bramaputra, Mekong, and other rivers that drain this region and in the freshness of the surface waters of the Bay of Bengal (Fig. 4.12).

The corresponding DJF patterns are weaker. Dry air in the northeasterly flow off East Asia crosses the equator and converges into a band of moderately heavy rain rate extending westward from the South Pacific Convergence Zone (SPCZ) across Southern Indian Ocean.

The oceanic ITCZs maintain their integrity and remain in the Northern Hemisphere year round. At most longitudes, they are slightly more pronounced, located farther northward, with stronger inflow from the south during JJA. This enhanced northward flow feeds into the North American monsoon in the Pacific sector and the West African monsoon in the Atlantic sector. The SPCZ is also present year round, but it is more extensive in the austral summer (DJF).

Four of the five subtropical subsiding equatorward jets and the associated subtropical anticyclones on the eastern sides of the respective ocean basins are year-round features. That they are slightly stronger during summer than during

Figure 16.1 Meridional profiles of zonally averaged climatological mean fields in (left) the eastern hemisphere (0° eastward to 180° longitude) and (right) the western hemisphere (180° eastward to 0°). The top panels show skin temperature and rain rate, contour interval 2 mm d^{-1} starting at 4 mm d^{-1}. In the bottom panels the colored shading indicates the meridional wind component at the 1000 hPa level. Courtesy of Ángel Adames.

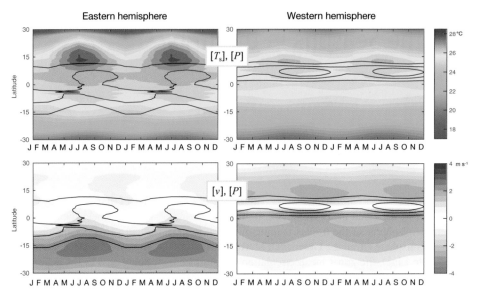

Figure 16.2 Climatological seasonal mean rain rate in mm d^{-1}: (top) DJF; (bottom) JJA. Courtesy of Katrina Virts.

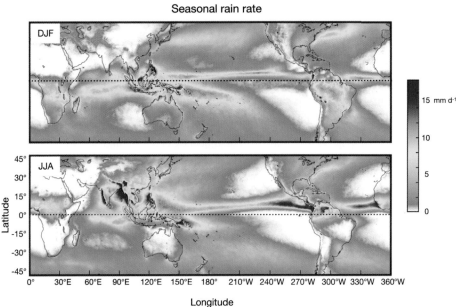

winter suggests that the thermal contrast between the warm continents and cold oceans plays a role in maintaining them.[1]

At the 150 hPa level (Fig. 16.4), divergent outflow from the tops of the plumes of ascending air in the Asian summer monsoon induces a planetary-scale Rossby wave response to the northwest, referred to as the *Tibetan anticyclone*. The so-called *tropical easterly jet* along its equatorward flank exhibits wind speeds in excess of 30 m s^{-1} as it crosses over the southern tip of India. Analogous, but much weaker upper tropospheric anticyclones are observed poleward of the regions of heavy summer monsoon rainfall over the Americas and Africa. The tropopause bulges upward over the Tibetan anticyclone, creating an island of tropospheric

air surrounded by a sea of stratospheric air.[2] A secondary feature in the JJA Z_{150} and $\mathbf{V}_{150}$ patterns is an anticyclone along 14°S in the Indian Ocean sector. In combination, the Northern Hemisphere (Tibetan) and Southern Hemisphere anticyclones resemble the equatorially trapped Rossby wave couplet shown in Fig. 15.6, but the axis of symmetry is to the north of the equator and the Northern Hemisphere feature is strongly accentuated. Whereas the Tibetan anticyclone is forced by the deep cumulus convection in the Indian summer monsoon, its Southern Hemisphere counterpart is induced by the conservation of vorticity in the southward cross-equatorial flow. In this manner, a heat source in one hemisphere induces a Rossby wave response in the other hemisphere.

---

[1] Monsoons play a role as well; see Rodwell and Hoskins (2001) and Wu et al. (2012).

[2] Park et al. (2009).

$$\overline{\omega}_{850}, \overline{Z}_{1000}, \overline{\mathbf{V}}_{1000}$$

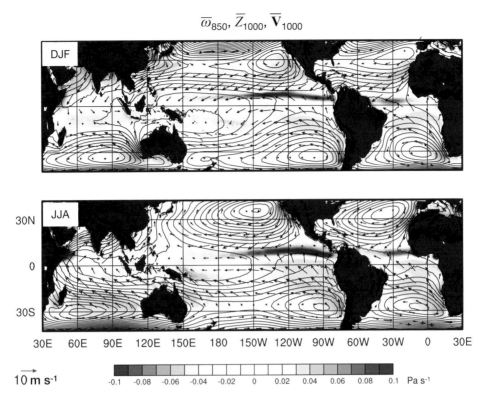

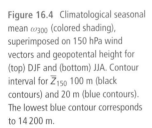

Figure 16.3 Climatological seasonal mean $\overline{\omega}_{850}$ (colored shading), superimposed on 1000 hPa wind vectors and geopotential height (contour interval 10 m): (top) DJF; (bottom) JJA. If values of $\overline{Z}_{1000}$ were converted to SLP, they would range from ~1026 hPa in the subtropical anticyclones to ~1008 hPa over the warm pool.

$$\overline{\omega}_{300}, \overline{Z}_{150}, \overline{\mathbf{V}}_{150}$$

Figure 16.4 Climatological seasonal mean ω_{300} (colored shading), superimposed on 150 hPa wind vectors and geopotental height for (top) DJF and (bottom) JJA. Contour interval for $\overline{Z}_{150}$ 100 m (black contours) and 20 m (blue contours). The lowest blue contour corresponds to 14 200 m.

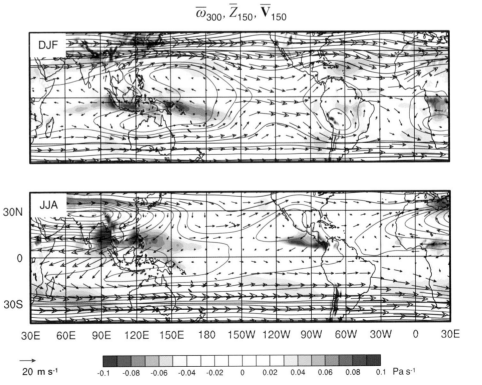

Figure 16.5 (Top and bottom panels) 150 hPa wind and relative vorticity fields 3 days before and 3 days after a flareup of convective activity over the northern Indian Ocean. Orange shading indicates positive values; that is, cyclonic (NH) and anticyclonic (SH). The most extreme values correspond to $\pm 7 \times 10^{-5}$ s^{-1}. (Middle panel) Outgoing longwave radiation (OLR) anomalies and 950 hPa wind on the day of the flareup. OLR ranges from 100 (blue) to 310 W m^{-2} (orange). The figures were constructed by regressing the daily fields upon a daily time series of OLR in a rectangular box coincident with the blue patch in the middle panel. The patterns resemble composites because total fields rather than anomaly fields are used in performing the regression. Adapted from Fig. 11 of Hoskins et al. (2020). © The Authors. CC BY 4.0.

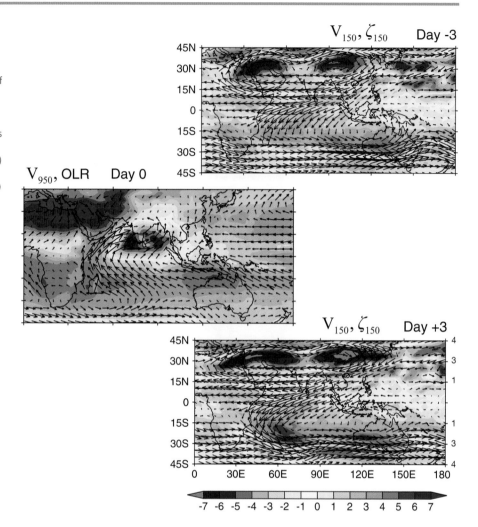

Further evidence of the cross-equatorial coupling in this feature is shown in Fig. 16.5 – composites of daily data 3 days before and 3 days after bursts of convective activity in the South Asian summer monsoon. The response of the circulation to the enhanced deep convection is characterized by a strengthening of the cross-equatorial flow and a strengthening of the downstream anticyclone in the southwest Indian Ocean. Much of the cross-equatorial flow from the Northern Hemisphere to the Southern Hemisphere in the upper branch of the Hadley cell during JJA takes place in association with such bursts of deep convection.

Despite the bias of the deep convection toward the Southern Hemisphere, the DJF flow pattern shown in Fig. 16.4 is nearly equatorially symmetric. Like the annual mean flow, it bears a strong resemblance to the idealized response to an equatorial heat source, except that the Rossby wave couplet is shifted toward the east of the heating rather than toward the west of it. The anticyclonic flow around the Rossby wave couplet in the western Pacific contributes to the strength of the climatological mean westerly jet stream along 30°N.

Figure 16.6 shows JJA minus DJF differences for a suite of variables. The SST/skin temperature differences (top panel) are dominated by the continents, which exhibit a

strong response to the annual cycle in insolation by virtue of the small heat capacity of land surfaces. The SLP (second panel) amplitudes are highest near the expansive high latitude Northern Hemisphere continents. (Land is masked out in order to deemphasize the shallow thermal signature.) Zonally averaged SLP is lower along 30°N and higher along 20–30°S in JJA than in DJF – in the sense as to drive a stronger northward cross-equatorial flow in JJA relative to DJF. The differences in column-integrated water vapor W (third panel) exhibit a more zonally symmetric structure, with higher values in the summer hemisphere, where SST is higher, and lower values in the winter hemisphere, with a nodal line along 7°N, the latitude of the ITCZs in the annual mean climatology. The dipole pattern, which is also apparent in the rain rate (bottom panel) is indicative of a northward shift of the ITCZ from DJF to JJA.

16.1.2 The Zonally Symmetric Flow

In contrast to the annual mean Hadley circulation, which exhibits a high degree of equatorial symmetry with only weak cross-equatorial mass transports, the JJA and DJF

JJA–DJF

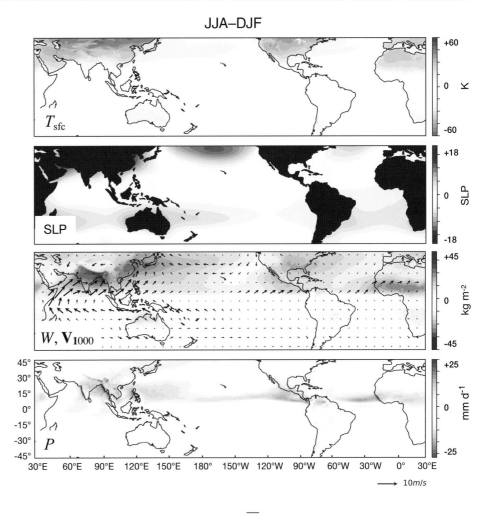

Figure 16.6 Climatological mean JJA minus DJF difference fields for SST/skin temperature T_{sfc}, SLP, column-integrated water vapor W, V_{1000} vectors and rain rate P, as indicated.

$\overline{Q}, \overline{\psi}$

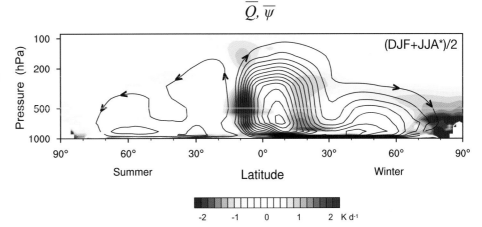

Figure 16.7 Climatological mean TEM meridional circulation and diabatic heating rate in the winter and summer hemispheres: an average of DJF and JJA* (i.e., JJA transposed left-right so that the North Pole is on the left and the South Pole on the right). Contour interval 16×10^9 kg s^{-1} (16 Sv). Courtesy of Ying Li.

circulation cells, shown in Figs. 1.27 (Eulerian) and 9.7 (TEM), straddle the equator, with rising branches in the summer hemisphere, sinking branches in the winter hemisphere, and flow from the summer hemisphere into the winter hemisphere in the upper branches centered ~150 hPa. They are represented as a single cell in Fig. 16.7, which (like Fig. 9.8) was constructed by transposing the JJA section left-to-right and averaging it with the DJF section. The ascending branch of the cell is within 15° of the equator in the summer hemisphere and the descending branch is between 15 and 30° in the winter hemisphere. The cell is thermally direct, transporting dry static energy from the summer hemisphere into the winter hemisphere. The moisture transport, which is in the same sense as the lower branches of the cell, is in the opposite direction, from the winter hemisphere into the summer hemisphere.

To interpret the cross-equatorial transport of westerly momentum by this cell requires consideration of the angular momentum balance. The Coriolis torque acting on the low-level flow from the winter hemisphere feeding into the belt of heaviest rain rates in the tropics of the summer hemisphere is balanced by frictional drag, but the balance in the return flow in the upper troposphere requires consideration of the eddy transports. The existence of positive inertial stability implies that $[m]$ must decrease with latitude going away from the equator in both directions (see Exercise 3.20). Hence, the presence of strong cross-equatorial flow requires the existence of an angular momentum source/sink couplet, a source to balance the westward Coriolis torque in the summer hemisphere and a sink to balance the eastward torque in the winter hemisphere. The configuration of $[u^*v^*]$ implied by such a couplet is a narrow maximum centered over the equator in the upper troposphere, coincident with the peak in $[v]$ but in the opposite direction of the mass flux. Indeed, it is apparent from Fig. 3.11 that there is near-equatorial peak in $[u^*v^*]$, which gives rise to a couplet in the eddy forcing $G = -(cos\,\phi)^{-2}\partial([u^*v^*]cos^2\phi)/\partial y$ (Fig. 3.14) almost exactly where it needs to be. This feature is also clearly evident in the pattern of EP fluxes shown in Fig. 9.8, where $[u^*v^*]$ is represented by the stream of arrows flowing across the equator from the summer hemisphere into the winter hemisphere at upper tropospheric levels. The correspondence is not fortuitous.

The cross-equatorial mass flux from the summer hemisphere to the winter hemisphere in the upper troposphere tends to be concentrated within the longitudinal sectors of the monsoons and much of it occurs during episodic meridional wind bursts associated with enhanced deep convection in the summer hemisphere like the ones shown in Fig. 16.5. Air parcels in these plumes emanating from the monsoon convection cross the equatorial belt $10°N$–$10°S$ in a matter of just a few days, during which time they can be viewed as conserving angular momentum, undergoing an a westward acceleration as they approach the equator and an eastward acceleration as they flow poleward in the winter hemisphere. While crossing through the equatorial belt, these air parcels thus exhibit a pronounced angular momentum deficit (i.e., negative values of u^*), resulting in an eddy transport $[u^*v^*]$ directed opposite to the mean meridional mass flux $[v]$. Hence, the balance requirement is satisfied simply because the cross-equatorial flow is concentrated in longitude and to some degree also in time. The mass flux and the momentum transports should not be viewed as separate entities with the transports "driving" the mean meridional circulation: they are both manifestations of the same (zonally and time-varying) circulation.

Figure 16.8 shows the corresponding pole-to-pole circulation cell averaged over JJA minus that averaged over DJF. The diabatic heating rates and ascent rates at the latitudes of the annual mean, the ITCZs, and in the rain belts of the boreal summer monsoons are enhanced in JJA relative to DJF and vice versa for the SPCZ and the rain belts of the austral summer monsoons. The belts of low-level convergence and divergence along the flanks of the cell are

associated with the dipole patterns in W and P in the bottom two panels of Fig. 16.6.

Exercises

16.1 Revisit the balance of forces in the surface wind field (Exercise 14.4), taking into account the strong cross-equatorial flow in the Asian summer monsoon.

16.2 Making use of the supplementary figures on the companion web page, document the seasonality of the Pacific and Atlantic equatorial cold tongue complexes.

16.3 Making use of the bottom panel in Fig. 16.4 and neglecting forcing by the transients, diagnose the zonal momentum budget in the entrance and exit regions of the jet over southern Asia in JJA.

16.4 The DJF V_{150} and Z_{150} fields are predominantly equatorially symmetric, despite the existence of the monsoon circulations. Explain.

16.5 Note the strong zonal Z_{1000} gradient and the strong southerly winds off the east coast of Africa during JJA in Fig. 16.3. In the geophysical fluid dynamics literature, this feature is referred to as the *Somali jet*, or *Findlater jet*, after its discoverer.[3] What is it that maintains the strong pressure gradient?

16.6 A prominent feature in Fig. 3.13 is cross-equatorial transport of westerly momentum from the winter hemisphere into the summer hemisphere by the equatorial stationary waves. Comparing with Fig. 1.25, we find that this feature is matched by a cross-equatorial mass flux in the opposite direction. (*a*) Show that the equatorial stationary wave configuration in Fig. 16.4 is consistent with a cross-equatorial transport of westerly momentum from the winter hemisphere into the summer hemisphere. (*b*) Describe how a balance between $\mathcal{G}$ and $f[v]$ can be achieved on the northern and southern flanks of these nearly coincident maxima in meridional mass flux and momentum transport. (c) Roughly what fraction of the angular momentum that the atmosphere extracts from the underlying surface in the subtropics and tropics of the winter hemisphere is transported poleward in the winter hemisphere and what fraction is transported across the equator into the summer hemisphere?

16.7* In Chapter 3, it was implicitly assumed that the cross-equatorial mass fluxes are negligible and that the balance requirement for the conservation of angular momentum is fulfilled independently by the Northern and Southern Hemisphere circulations. (a) Revisit these assumptions in light of the discussion in this section. (b) Suggest some ways of verifying that tropical convection in the summer hemisphere influences the general circulation of the winter hemisphere.

---

[3] Findlater (1969).

Figure 16.8 Climatological mean streamfunction for the JJA minus DJF TEM meridional circulations, contour interval 32 Sv, superimposed upon the diabatic heating rate $\overline{Q}$, indicated by colored shading. Courtesy of Ying Li.

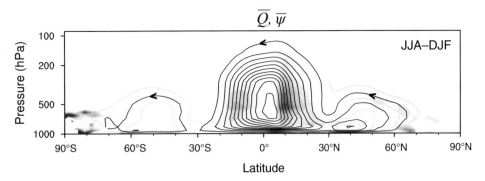

Figure 16.9 Climatological monthly mean meridional wind component v_{1000} on the 7.5°N latitude circle. Eastern hemisphere, from the Greenwich Meridian eastward to the Date Line (black line) and western hemisphere, from the Date Line eastward to the Greenwich Meridian (blue line). Courtesy of Ángel Adames.

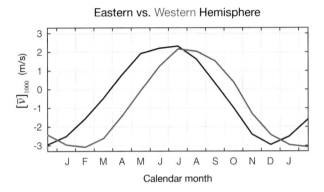

16.2 SON VERSUS MAM CIRCULATIONS

The seasonal variation in the zonally averaged mean meridional circulation [v] is nearly perfectly sinusoidal, with a peak in late July, lagged by about 5 weeks relative to the forcing because of the combined "thermal inertia" of the atmosphere and the ocean mixed layer.[4] It is evident from Fig. 16.9 that the phase is a month or two later in the western hemisphere than in the eastern hemisphere – a consequence of the higher ratio of land to ocean in the eastern hemisphere.

A "marine monsoon" resembling a north–south seesaw with the same nodal line at ~7°N, with greater emphasis on the Pacific and Atlantic ITCZs, dominates the MAM to SON difference patterns. Maps of MAM and SON SST and surface wind $\mathbf{V}_{1000}$ over the western hemisphere of the tropics are shown in Fig. 16.10. The boreal spring MAM is the season in which the ITCZ/cold tongue complexes come closest to being equatorially symmetric. The Pacific ITCZ is located just a few degrees of latitude north of the equator and there is often a much weaker secondary ITCZ at comparable latitudes of the Southern Hemisphere. The Atlantic ITCZ is diffuse and touches the equator on the western side of the

basin. The Pacific cold tongue is at its weakest and centered on the equator and there is no cold tongue on the eastern side of the Atlantic. It is as if the ITCZ were in the process of shifting across the equator into the Southern Hemisphere, but has not had sufficient time to complete the transition. During SON the northward flow associated with the observed Northern Hemisphere "oceanic monsoon" is a prominent feature of the circulation. The ITCZ is displaced as far north as it ever reaches and the equatorial Pacific cold tongue is at or near its coldest.[5]

SON–MAM difference fields analogous to those in Fig. 16.6 are shown in Fig. 16.11. Note that the scaling is different in the plots: the SON–MAM differences are not as strong as the JJA–DJF patterns, with the exception of the meridional wind component at the Earth's surface, which is stronger. Because skin temperatures over the continents are quite comparable in these two seasons, the SST differences are more clearly discernible and it is not necessary to mask out the land in the SLP panel. Because of the relatively long lag of the annual cycle in SST relative to that in insolation, SST tends to be higher in the autumn in the respective hemisphere. The equatorial cold tongue regions are cooler in SON, like most of the Southern Hemisphere oceans. [Recalling from Section 14.3 that it is the coolness of the oceans relative to the tropical continents that gives rise to the subtropical anticyclones, one might expect the anticyclones to be strongest during spring, when the oceans are coolest relative to the continents, and indeed, that is the case.] The line of demarcation between positive and negative SON minus MAM SLP differences corresponds almost perfectly to the latitude of the ITCZ in the annual mean climatology. The change in the SLP gradient is most pronounced at the latitude of the ITCZ, but it is in the sense as to strengthen the northward flow throughout the broader belt extending from the equator to ~12°N.

The MAM to SON change in W in Fig. 16.11 resembles the SST change, but it is amplified in the warmest regions (i.e., along the ITCZs) by virtue of the nonlinearity inherent in the Clausius–Clapeyron equation. The change in rain rate P also exhibits a north–south dipole pattern. It is much

---

[4] For example, see Dima and Wallace (2003).

[5] Mitchell and Wallace (1992), Figs. 3 and 4.

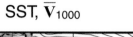

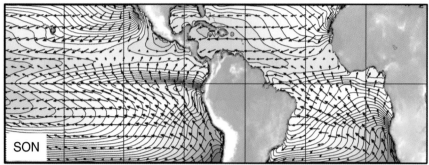

Figure 16.10 Climatological mean $\mathbf{V}_{1000}$, in m s^{-1}, scale at bottom, superimposed upon sea surface temperature. (Top) SON; (bottom) MAM.

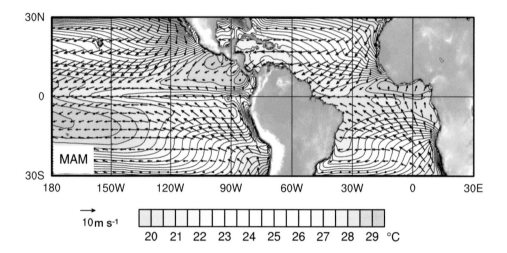

narrower in meridional extent than the dipoles in the other variables shown in this figure because the ITCZ retains its sharpness as it shifts northward and southward with the seasons, in response to the thermal forcing. It is remarkable that the maximum in v_{1000} and the node in W extend across South America and Southeast Asia and there is even a hint of them in the Indian Ocean sector.

Over the eastern Pacific and Atlantic sectors, where the ITCZ is north of the equator, the strengthening of the northward cross-equatorial flow enhances the upward fluxes of latent and sensible heat through the ocean surface, thereby reinforcing the coldness of the cold tongues. The air flowing into the ITCZs from the south is thus cooler in SON than in MAM, reinforcing the thermal contrast, and thus serving as a positive feedback that helps to prolong the Northern Hemisphere "marine summer monsoon" at nearly full strength into October.

That the existence of positive feedbacks increases the amplitude and the lag of the atmospheric response to the prescribed periodic forcing can be demonstrated by considering the governing equation for a harmonic oscillator of the form

$$c\frac{dX}{dt} = F_o \cos\omega t - \gamma X, \qquad (16.1)$$

where X is some measure of the amplitude of the departures from the annual mean state. In this specific example, c is a measure of the heat capacity of the ocean, F_o is a measure of the insolation forcing at the top of the atmosphere, ω is 2π radians per year, t is time in units of years measured, for example, relative to the Northern Hemisphere summer solstice, and γ is the rate of damping, mainly due to the emission of infrared radiation to space. The solution of Eq. (16.1) is of the form $X_o \cos\omega(t+\phi)$ where $X_o^2 = F_o^2((c\omega)^2 + \gamma^2)^{-1}$ and $\tan^{-1}\phi = -c\omega/\gamma$. Hence, as the effective damping rate γ decreases in response to the addition of a positive feedback, the lag of the response, relative to the phase of the forcing, increases. If the feedback were strong enough to almost completely cancel the damping, the forcing and the response would be in quadrature.

Consistent with the northward shift in the belts of heavy rain rates, the boundary between the Northern Hemisphere and Southern Hemisphere Hadley cells shifts northward from MAM to SON. The difference pattern shown in Fig. 16.12 is dominated by a circulation cell centered on the latitude of the ITCZ, with ascent to the north and descent to the south. The difference cell is much narrower than its JJA–DJF counterpart shown in Fig. 16.8.

Figure 16.11 Climatological mean SON minus MAM difference fields for SST/skin temperature T_{sfc}, SLP, column-integrated water vapor W, V_{1000} vectors and rain rate P, as indicated.

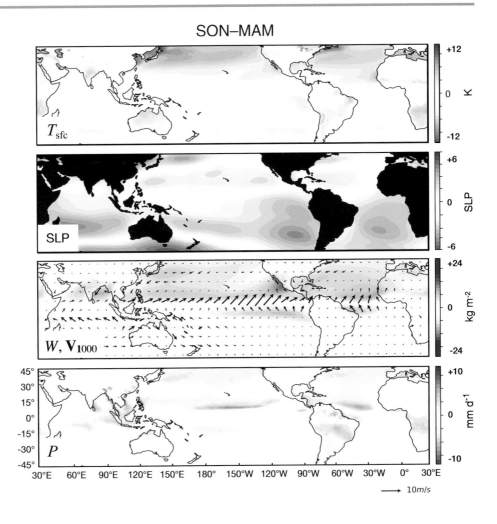

SON–MAM

16.3 WHY IS THE ITCZ IN THE NORTHERN HEMISPHERE?

The same processes and feedbacks that give rise to the annual cycle in the ITCZ–cold tongue complexes are instrumental in explaining why the primary ITCZs reside in the Northern Hemisphere not only in the annual mean, but year round.

In the early general circulation literature, the trade winds from the Northern and Southern Hemisphere were viewed as converging into the doldrums, a mythical rain belt, marked by light winds, centered over the equator. The discovery that the trade winds converge, not at the equator but on average at ~7°N was a major paradigm shift. The appearance of a secondary ITCZ in the eastern Pacific during FMAM was regarded as a curiosity when it was first noticed in the observations, but it took on greater significance after double ITCZs separated by a colder equatorial belt emerged in many simulations with coupled atmosphere–ocean models, even when run in an aquaplanet mode.

The features of interest stand out clearly in the fields of annual mean SST and rain rate shown in Fig. 14.10. If the SST field is prescribed, the rain rate climatology can be faithfully simulated in models, as evidenced by comparing the top two panels of Fig. 14.6. But simulating both the rain rate and SST fields in a free running, coupled atmosphere–ocean model has proven to be a much more formidable challenge because of the complexity of the feedbacks relating to its strength and position. The models have tended to exaggerate the strength of the equatorial Pacific cold tongue and the secondary ITCZ in the Southern Hemisphere, such that the overall configuration is more equatorially symmetric than observed. These biases vary substantially

Figure 16.12 Climatological mean TEM meridional circulations superimposed upon diabatic heating rate $\overline{Q}$: (top) MAM, (middle) SON and (bottom) SON minus MAM. Contour interval 16 Sv, zero contour omitted. Courtesy of Ying Li.

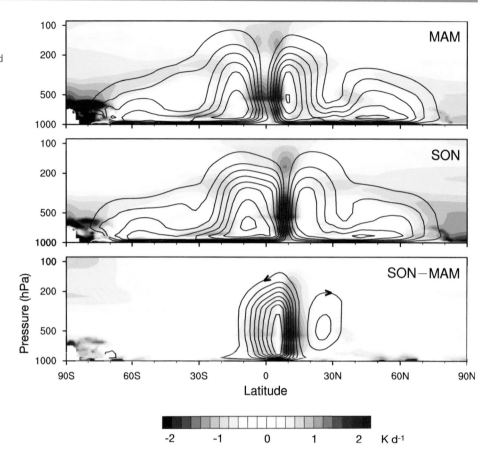

from model to model, and even the same model run with different convective parameterization schemes. There has been very little progress in reducing these biases across three generations of climate models spanning the past twenty years.

The unresolved modeling issues notwithstanding, it is clear that the observed bias of the ITCZ toward the Northern Hemisphere can only be due to two factors:

- Africa and the Americas extend farther westward in the Northern Hemisphere than in the Southern Hemisphere, forcing monsoon circulations in JJA that have no counterparts in DJF. Once established, these circulations, with northward low-level flow across the equator, tend to persist into SON because of positive wind-evaporation and cloud feedbacks.
- The Andes protrude into the strong westerly flow beneath the jet stream at subtropical latitudes, deflecting some of it equatorward and downward into the subtropical subsidence zone (SSZ, see Section 14.3) on their upwind side along the coast of Chile. The direct dynamical response to this orographic enhancement of the SSZ, augmented by wind-evaporation and cloud feedbacks, renders the entire Southeast Trades stronger than the Northeast Trades year-round.[6]

[6] Takahashi and Battisti (2007).

> **Exercise**
>
> 16.9* Review the history and current status of efforts to simulate the ITCZ in coupled atmosphere–ocean models, in which SST is simulated, rather than prescribed.

16.4 THE EQUATORIAL SEMIANNUAL CYCLE

Over the equator, insolation is ~8% stronger at the times of the equinoxes than at the times of the solstices. In response to this forcing, rain rate in the equatorial belt (mainly in the continental sectors) exhibits a weak semiannual cycle with maxima about a month after the equinoxes. This feature of the annual march is represented by averages for the post-equinoctial months April and October minus the average for the post-solsticial months January and July shown in Fig. 16.13. Rain rate is indeed heavier in the equatorial belt toward the end of the equinoctial seasons. The semiannual cycle is also evident in diabatic heating rate and the mean meridional circulations, shown in the top panel. The patterns are centered slightly northward of the equator.

Figure 16.13 Amplitude of the climatological mean semiannual cycle as represented by [(Apr + Oct) - (Jan + Jul)] /4 fields. (Top) TEM meridional circulation (contour interval 4×10^9 kg s^{-1} (4 Sv), zero contour omitted), superimposed upon diabatic heating rate Q; (bottom) rain rate. Courtesy of Ying Li.

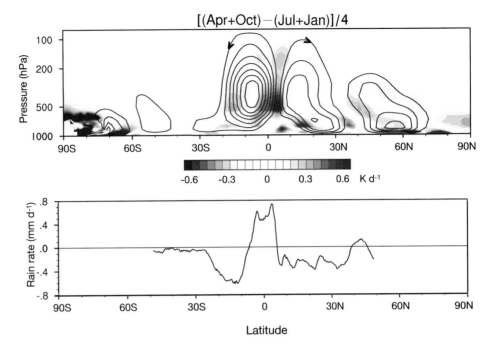

16.5 ABRUPT SEASONAL TRANSITIONS

Away from the bounding continents over much of the Pacific and Atlantic Oceans, precipitation is concentrated in the ITCZs, which undergo a gentle latitudinal migration in response to the seasonal cycle in meridional gradient in insolation, from near the equator in March to ~10°N in September. In contrast, in certain longitudinal sectors, such as in the Indian Ocean and over South Africa and Central America, summer monsoon-related precipitation exhibits a rather abrupt onset around the time of the summer solstice and it extends poleward as far as ~20°N/S. The dynamics that govern the location of the ITCZ and the attendant meridional overturning circulations are fundamentally different from those that govern the monsoons. The distinctions between them are exemplified by Fig. 16.1 in which the left panels for the eastern hemisphere are dominated by the monsoons and the right panels for the eastern hemisphere are dominated by the ITCZs, and by the schematics shown in Fig. 16.14.[7]

The ITCZs tend to be situated over the belts of highest SST and they are associated with symmetric instability in the near-surface flow driven by the gradients in SST.[8] The poleward extent of the ITCZ appears to be limited to within 12° of the equator by the symmetric stability of the boundary layer flow associated with the increase in the absolute value of planetary vorticity with latitude. The strength of the ITCZ precipitation and meridional overturning circulation is primarily governed by the amplitude of the eddy transport of westerly momentum out of the deep tropics associated with midlatitude baroclinic eddies. The extrema in the latitude

of the ITCZ are lagged by almost 3 months relative to the annual cycle in insolation because the large heat capacity of the near-surface layer of the ocean that participates in the seasonal cycle.

In contrast, precipitation in the monsoons is located in the subtropics (~20° latitude) and is nearly co-located with the maximum in near-surface MSE. The overturning circulation is relatively unconstrained by restoring inertial forces and it is shielded from the extratropical transient eddies by the anticyclone that develops in response to the divergence at the cloud-top level. Hence, the strength of the precipitation and meridional circulation in these systems is largely determined by the amplitude of the near-surface MSE maximum.

The dynamics that govern the ITCZ and the monsoons have been illuminated using atmospheric general circulation models coupled to a slab ocean, the so-called aquaplanet, forced by the observed seasonal cycle in insolation.[9] For slabs with depths on the order of 50 m or so, precipitation is concentrated in an ITCZ that experiences seasonal migrations about the equator, reaching as far as ~12° latitude three months after the summer solstice. But for a sufficiently shallow mixed layer depth (e.g., 1 m), as summer approaches the meridional gradient in MSE becomes large enough and the location of the MSE maximum shifts far enough poleward that the organized convection develops much farther poleward, near 20° latitude. The onset of this monsoon-like entity occurs suddenly around the time of the solstice.[10] The development process involves feedbacks associated with the low-level advection of MSE and the shielding of the

[7] See the review by Geen et al. (2020) and references therein.
[8] Tomas and Webster (1997).

[9] Bordoni and Schneider (2008).
[10] "Continental monsoons" can even be found on aquaplanets with realistic mixed layer depths when the heat convergence in the subtropics by the wind-driven ocean currents is included in the model.

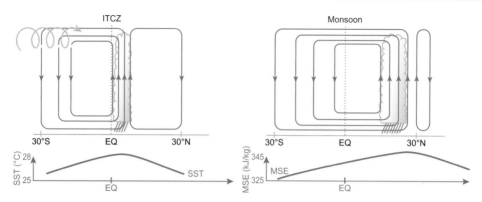

Figure 16.14 Schematic illustration of the circulation and precipitation in the ITCZ and "continental" monsoon regimes. The gray cloud denotes clouds and precipitation, red contours denote the streamfunction. (Left) In the ITCZ regime, convergence and precipitation are located near the Equator, and approximately co-located with the peak SST, with Hadley cells that are predominantly driven by meridional eddy momentum transports associated with baroclinic eddies, as indicated by the helical arrow. (Right) In the monsoon regime, the convergence zone is located farther from the Equator, with the mid-tropospheric zero contour of the streamfunction aligned with the near-surface MSE maximum and highest rain rates just equatorward of it. Equatorward of the latitude of maximum rainfall in the summer hemisphere, the Hadley cell is only weakly influenced by the eddies. It extends across the Equator and is nearly angular momentum conserving. In both the ITCZ and monsoon regimes, the summer Hadley cell is much weaker, if present at all. Adapted from Geen et al. (2020). © American Geophysical Union.

overturning circulation by the onset of upper-level easterlies. There exists a critical depth (heat capacity) on the aqua-planet that marks the transition from a slow seasonal ITCZ migration, between roughly 10°S and 10°N, to a monsoon circulation with precipitation centered at ∼20° latitude in the summer hemisphere that begins abruptly in early summer and terminates abruptly in early fall. The shallow mixed layer depths that support the monsoons[11] on an aquaplanet (and their abrupt onsets) have a heat capacity that is similar to that associated with the layer of soil that participates in the seasonal cycle, and are thus a good analog for the dynamics of the Indian, American, and African summer monsoons.

Over most of India, May is the hottest month of the calendar year. Apart from an occasional pre-monsoonal thunderstorm, hot, dry conditions prevail until the onset of the summer monsoon, which normally takes place in early June. The change of seasons usually takes the form of discrete event in which rain and a surge of cooler southwesterly winds spreads northward across India over an interval of about a week. This apparent abrupt "regime shift" can be viewed as a composite of two elements: the seasonal transition in the rain rate climatology from the dry winter monsoon to the wet summer monsoon, and the passage of the wet phase of the intraseasonal oscillation that will be described in Chapter 18. The timing of the intraseasonal oscillation varies from year to year, so the onset of the rainy season is smeared out over a month or so in the climatology. The onset of the "North American monsoon" is equally abrupt in the climatology: over parts of Arizona, June is the driest month of the year and July is the wettest.

Not all distinctive features in the seasonality of the tropical circulation are related to land–sea thermal contrasts. For example, at latitudes near 30–35°N, Japan and the coastal plain of China experience a period of heavy spring rain rate, which migrates northward from April to June. When it first forms in late spring, this feature is a low-level convergence zone induced by the flow around the Tibetan Plateau at the jet stream level. Downstream of the Plateau, in the boundary layer, a warm, moist, southerly, tropical air stream converges with a cooler, much drier northerly air stream that has been deflected northward around the Plateau. As the tropospheric jet stream shifts to the north of the Plateau, the orographically induced wave weakens and the convergence zone (locally known as the Baiu or Mei-yu front) shifts northward and eventually disappears. This front-like patch of enhanced rain rate is distinct from the monsoon, which affects tropical and subtropical latitudes (e.g., see Fig. 1.15) and the online animation and peaks in midsummer.[12] Like the sudden onset of the monsoon in some regions, it shows up clearly in animations for individual years like the sample shown online than in monthly mean climatologies because of its short timescale and because of year-to-year variability in its timing.

The rain rate climatology at individual tropical and sub-tropical stations exhibits a variety of "annual marches" in response to abrupt monsoon onsets, spring rains associated with the Mei-yu or Baiu front over east Asia, late summer and autumn maxima attributable to tropical cyclones, and other regional features. A few examples are shown in the graphic that accompanies Exercise 16.12*.

Exercises

16.10 Describe the changes in the North American summer monsoon and the circulation of adjoining regions that typically takes place during the first week of July.

16.11* Show how daily mean rain rate climatologies can be used to explore the monsoon-related abrupt seasonal transitions.

---

[11] Geen et al. (2020).

[12] Chiang et al. (2020).

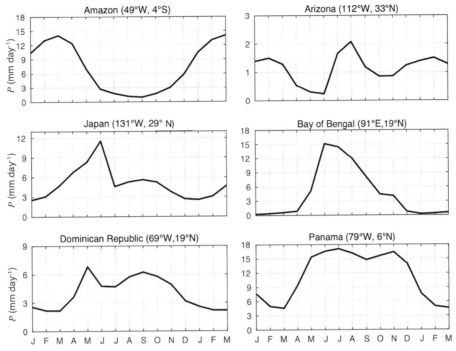

Figure for Exercise 16.12 Climatological monthly averaged rain rates at selected tropical locations.

16.12* The accompanying figure shows rain rate climatologies for a selection of grid points with interesting features. For one or more of them, describe and interpret the associated circulation features.

16.13 The accompanying figure shows the monthly mean climatology of rain rate averaged over the equatorial belt 8°N–8°S and the sum of the contributions of the annual and semiannual cycles. Explain the features in this figure based on the material presented in this section.

16.14 When daily rain rate data are composited on a time axis relative to each year's summer monsoon onset date rather than calendar date, they exhibit a very strong shift to rainy conditions over the course of a week or so, centered close to the onset date, followed by a relaxation back toward drier conditions over the next week or two. Explain.

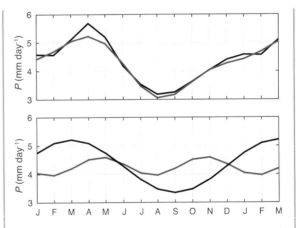

Figure for Exercise 16.13. (Top) Total (black curve) and the sum of the first two harmonics (i.e., the annual and semiannual cycles; red curve). (Bottom) The first harmonic (black curve); and the second harmonic (red curve).

17 El Niño–Southern Oscillation

with Ángel F. Adames and Xianyao Chen

In this chapter, we consider the leading mode of year-to-year climate variability, the *El Niño–Southern Oscillation* phenomenon, widely referred to as ENSO. *El Niño* connotes the episodic weakening of the equatorial Pacific SST cold tongue.[1] *Southern Oscillation* refers to a "seesaw" in sea-level pressure (SLP) between the eastern and western ends of the tropical Pacific Ocean.[2] The *Southern Oscillation index*, a measure of the strength of the east-to-west SLP gradient across the central Pacific, tends to be below normal during El Niño and above normal during La Niña (a strengthening of the cold tongue).[3] ENSO refers to an irregular cycling back and forth between (warm) El Niño and (cold) La Niña conditions.

El Niño events typically last 9 to 12 months and are followed the next year by La Niña events. They are spaced 3 to 7 years apart. In Fig. 17.1, the time history of ENSO from 1950 onward is documented in terms of five standardized indices:

- An index based on SST anomalies: departures from the seasonally varying climatological mean SST averaged over the rectangular box extending from $90°$W to the Date Line and from $6°$N to $6°$S) minus the concurrent global mean SST.[4] Episodes of positive anomalies, shaded in red, correspond to El Niño events. For example, strong El Niño events occurred in 1982, 1997, and 2015. We will use this SST index as a reference time series for creating regression maps and cross sections shown in some of the figures in this chapter.

- An indicator of the state of the Southern Oscillation (SO), here defined as the difference between SLP averaged over two broad regions of the tropical ($20°$N–$20°$S) oceans: the eastern Pacific from the Date Line to the coast of South America minus the remainder of the tropics, including the continental sectors. During El Niño events the index (inverted in Fig. 17.1) is negative. Since SLP is normally

higher on the eastern side of the Pacific basin than on the western side, negative values of the SO correspond to a weakening of the west-to-east gradient that drives the equatorial trade winds.

- An index of the strength of the equatorial trade winds in the central Pacific: the 1000 hPa zonal wind component averaged over $150°$E–$150°$W, $6°$N–$6°$S.

- The difference between sea level at San Francisco, USA, and Perth Australia, based on tide-gauge records. The former mirrors the sea level perturbations on the eastern side of the Pacific basin and the latter the perturbations on the western side.

- An index of rain rate in the equatorial central Pacific: rain rate averaged over $150°$E–$150°$W, $6°$N–$6°$S.

The relationships between the time series shown in Fig. 17.1 can be summarized and related to the climatology as follows. SLP is always higher and sea level is always lower on the eastern side of the Pacific basin than on the western side. The zonal SLP gradient drives the westward component of the southeast trades, which extend across the equator from the Southern Hemisphere in the core of the equatorial dry zone. During El Niño events the SLP and sea level gradients flatten, the trade winds in the central Pacific relax, and the Southern Hemisphere dry zone shrinks. The remarkable strength of coupling between the different variables represented in Fig. 17.1 is evidenced by temporal correlations, as documented in Table 17.1. Over and around the tropical Pacific the amplitudes of these year-to-year variations are as large as the seasonal variations described in the previous chapter.

This chapter is organized as follows. The first section explains (i) the feedback mechanism that renders the coupled atmosphere–ocean system over the tropical Pacific subject to large nonseasonal variability, (ii) the switching mechanism that gives rise to the oscillatory behavior, and (iii) why the cycling back and forth between El Niño and La Niña is irregular. The second section describes and interprets the three-dimensional structure of the atmospheric ENSO signature, and the third section the related zonally averaged structure in the $[u]$ and $[T]$ fields. The final section describes the global, stationary planetary wave response to the geographical redistribution of tropical rain rate.

---

[1] Wyrtki (1975).
[2] Walker and Bliss (1932).
[3] Bjerknes (1966, 1969), Rasmusson and Carpenter (1982).
[4] Referred to as the equatorial Pacific "cold tongue index CTI" (Deser and Wallace, 1987). For further specifics concerning the time series used in constructing the figures shown in this chapter, see the online supplementary information.

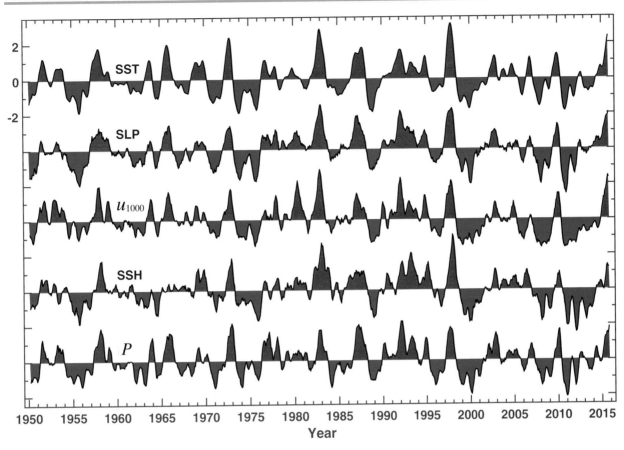

Figure 17.1 Time series of a suite of climate variables that exhibit a pronounced ENSO signature. Top to bottom: equatorial Pacific SST anomalies averaged over the region $180°$–$90°$W, $6°$N–$6°$S, from which the global mean SST has been subtracted, month by month; the Southern Oscillation (SO), represented by tropical ($20°$S–$20°$N) SLP averaged from the Date Line to the South American coast minus SLP over the remainder of the tropical oceans; 1000 hPa zonal wind u_{1000} averaged over the equatorial central Pacific ($150°$E–$150°$W, $6°$S–$6°$N); sea surface height SSH, the difference between sea level based on tide-gauge records at San Francisco and a station near Perth on the west coast of, Australia; and rain rate P in the central equatorial Pacific ($150°$E–$150°$W, $6°$S–$6°$N), a composite of station records prior to 1979 and outgoing longwave radiation from 1979 onward. The index of the SO is inverted. All the time series have been standardized by subtracting out their means and dividing each by its own standard deviation.

Table 17.1 *Correlation matrix for indices shown in Fig. 17.1 based on monthly mean data (1950–2015).*

	SST	SLP	u-wind	SSH	P
SST	1.00				
SLP	0.90	1.00			
u-wind	0.84	0.86	1.00		
SSH (tide gauge)	0.79	0.80	0.75	1.00	
P	0.87	0.89	0.84	0.71	1.00

17.1 THE PHYSICS OF ENSO

In this section, we briefly summarize the physics of ENSO, as understood on the basis of theory and from the analysis of observations and climate models.[5] ENSO is intrinsically

due to the coupling between the atmosphere and ocean in the tropical Pacific. Without the coupling, ocean models forced by the annual cycle of wind stress and surface fluxes do not produce El Niño events, and atmospheric models forced by the annual cycle of SST do not exhibit a Southern Oscillation. As such, ENSO is a true *mode* of natural variability in climate. Indeed, linear stability analysis[6] of the equatorial Pacific atmosphere–ocean system yields a leading (damped) eigenmode that is ENSO-like with a period of roughly 3 to 4 years. The dynamics of the ENSO mode involves the interplay between processes described in the next two subsections: the so-called "Bjerknes feedback" which renders the coupled atmosphere–ocean system highly sensitive to small perturbations from its equilibrium state and the delayed dynamic adjustment of the tropical Pacific Ocean by the equatorial Kelvin waves and the equatorial Rossby waves that mediate the thermocline depth and upper ocean currents.

[5] See Battisti et al. (2019) for a review of the dynamics of ENSO and other coupled atmosphere–ocean phenomenon.

[6] Neelin et al. (1998); Thompson and Battisti (2000, 2001).

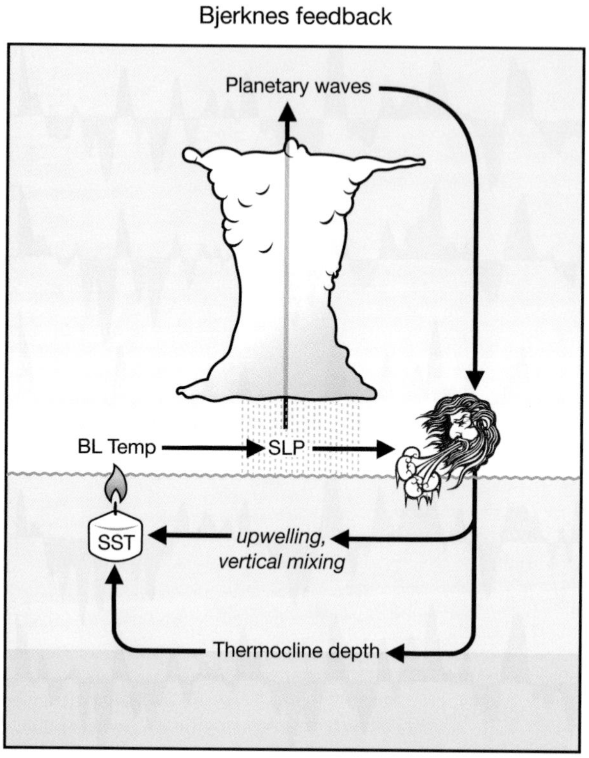

Bjerknes feedback

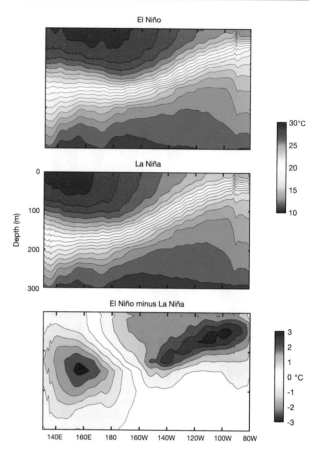

Figure 17.2 Schematic showing the chain of causality that gives rise to the Bjerknes feedback mechanism, which is responsible for the large coupled atmosphere–ocean variability in the equatorial Pacific basin. The zonal wind stress forces the ocean and the resulting sea surface temperature anomalies feed back upon the wind field by way of the horizontal temperature gradient in the atmospheric boundary layer. See text for a more complete explanation.

Figure 17.3 Subsurface ocean temperature in the equatorial plane during strong El Niño and La Niña events and the temperature difference (El Niño minus La Niña). El Niño and La Niña are defined as months in which the ENSO SST index (the top time series in Fig. 17.1) is more than 1.5 standard deviations above zero (El Niño) and below zero (La Niña).

17.1.1 Atmosphere–Ocean Coupling: The Bjerknes Feedback

The climatological-mean equatorial trade winds induce an east–west slope in sea level and thermocline depth along the equator: from west to east, the sea level slopes downward and the thermocline slopes upward. As a result, along the equator ocean heat content is greater toward the western side of the basin than toward the eastern side. The climatological mean trade winds also drive a poleward Ekman drift of the surface waters in both hemispheres, which induces upwelling and vertical mixing along the equator. Since the thermocline is close to the surface in the eastern equatorial Pacific, upwelling there can bring cold, nutrient-rich water up to the surface. These processes are instrumental in maintaining the equatorial cold tongue in SST, as discussed in Section 14.4.

The equatorial trade winds, in turn, are mainly driven by the west-to-east SLP gradient – the hydrostatic expression of the boundary layer temperature gradient along the equator, which mirrors the underlying SST gradient. The stronger the cold tongue, the stronger the SST gradient, the stronger the SLP gradient, and the stronger the westward component of the trade winds. Stronger trade winds, in turn, favor a stronger cold tongue. This circular chain of causality is referred to as the *Bjerknes feedback*, summarized schematically in Fig. 17.2.

With the onset of El Niño, the equatorial trade winds in the central Pacific weaken, reducing the time-mean rate of upwelling as well as the subsurface mixing that brings colder, denser water up to the surface. In the eastern Pacific, the thermocline deepens in response to an eastward surge of warm water in an equatorial Kelvin wave issued from the weakened trade winds in the central Pacific. The flattening of the equatorial thermocline is clearly discernible in the equatorial cross sections shown in Fig. 17.3. With the deepening of the thermocline on the eastern side of the basin, the entrainment of cold water into the surface layer is further reduced, resulting in additional warming. The cold tongue accordingly weakens, and with it, the west-to-east SLP gradient that drives the trade winds along the equator. The same physics plays out to amplify an incipient La Niña event.

17.1.2 Delayed Ocean Feedbacks

The Bjerknes feedback mechanism accounts for the strong coupling between SST and surface wind in ENSO, but it does not explain the cycling back and forth between El

El Niño minus La Niña

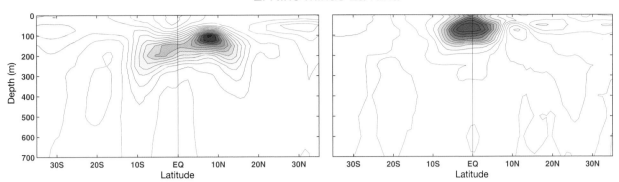

Figure 17.4 As in Fig. 17.3, but meridional cross sections of the subsurface ocean temperature difference (El Niño minus La Niña) in the Pacific basin, zonally averaged over the sector extending from the Date Line to (left) 130°E and (right) to 80°W. Contour interval 0.1°C.

Niño and La Niña. To understand why El Niño events do not last forever, it is necessary to consider how the thermal structure of the upper ocean in the equatorial belt is perturbed by the ENSO-related zonal surface wind anomalies. The equatorial temperature changes, shown in the bottom panel of Fig. 17.3, are largest just above the thermocline. In the eastern Pacific, where the thermocline is relatively shallow, the warming extends all the way up to the surface, but in the western Pacific, where the thermocline is much deeper, the subsurface cooling does not appreciably affect SST.

It is instructive to consider the subsurface ocean temperature changes, not only in the equatorial plane, but also in the meridional plane, as revealed by the cross sections shown in Fig. 17.4. That the warming on the eastern side of the basin is narrowly focused on the equator confirms that it is a Kelvin wave signature extending eastward from the wind forcing in the central Pacific, analogous to the thermally forced Kelvin wave signature in Fig. 14.4. In contrast, the cooling in the western Pacific is much broader, with off-equatorial maxima along 6°N and 6°S suggestive of a westward propagating equatorial Rossby wave generated by anomalous cyclonic shear of the wind stress along the flanks of the ENSO-related westerly wind stress anomalies near the Date Line.

The switching mechanism in ENSO can be understood as follows. The El Niño-related westerly wind stress anomaly in the central Pacific generates oceanic Rossby waves. These waves have little or no direct influence upon SST, but they shallow the thermocline off the equator as they propagate to the western boundary of the basin, as shown schematically in the top two panels of Fig. 17.5. Reflection of these waves off the islands that form the western boundary gives rise to an upwelling Kelvin wave that propagates eastward into the central and eastern Pacific, shallowing the thermocline and reversing the warming trend in the central and eastern Pacific. The Kelvin waves eventually restores the west-to-east thermocline slope in the central Pacific, thereby terminating the El Niño event. The transit time for a Rossby wave to go from the central Pacific to the western boundary and return in the form of a Kelvin wave is ~6 months. Hence, the Kelvin waves continues to shallow the thermocline in

the central and eastern Pacific for many months after the termination of the El Niño event, producing anomalous cooling of the surface characteristic of La Niña conditions, which are likewise amplified by the Bjerknes feedback. In this manner, the same wind forcing that triggers the onset of an El Niño event also generates the off-equatorial Rossby wave that will lead to its demise ~9 months later, and to the onset of La Niña.[7]

The idealized ENSO cycle, including its switching mechanism, can be represented mathematically by a set of equations referred to as the "delayed oscillator,"[8] which has also been used to model the behavior of populations of predators and their prey. On the basis of this model, one would expect the ENSO phenomenon, as described above, to be oscillatory. The period of the ENSO cycle is determined by a competition between two processes: the strength of the Bjerknes feedback, which governs the timescale of SST change associated with a change in the local wind stress in the central-eastern Pacific (~6 months), and the timescale associated with the basin-wide dynamical ocean adjustment – that is, the time it takes for Rossby waves to go from the central Pacific to the western boundary and return to the central Pacific in the form of a Kelvin wave, which is also about 6 months. The period of the ENSO cycle is much greater than 2 × 6 months because the oceanic dynamical response is delayed relative to the feedback processes that give rise to the initial SST and wind stress anomalies.

17.1.3 Irregularity and Nonlinearity of ENSO

Irregularities in the period of the ENSO cycle are mainly due to two factors: the existence of a strong climatological seasonal cycle in the tropical Pacific, and stochastic forcing of the linear ENSO mode by uncoupled atmospheric variability. The Bjerknes feedback requires the presence of a basic state cold tongue. Because the strength of the cold tongue exhibits a strong annual cycle, so does the strength of the Bjerknes

---

[7] Battisti (1988).
[8] Schopf and Suarez (1988); Battisti and Hirst (1989).

Figure 17.5 The switching mechanism in ENSO. The top panel depicts El Niño conditions with westerly wind anomalies in the central Pacific and elevated SST at the eastern end of the basin. The cyclonic shear on the flanks of the anomalous westerlies gives rise to a pair of westward propagating, upwelling Rossby waves, indicated by the blue shading. In the middle panel, these waves are shown reflecting off the western side of the basin and returning as an upwelling Kelvin wave that propagates eastward along the equator. In the bottom panel, the Kelvin wave is approaching the eastern side of the basin, facilitating the reemergence of the equatorial cold tongue. The SLP pattern adjusts to the changing SST pattern nearly instantaneously and the westerly wind anomalies weaken and eventually reverse, marking the onset of the La Niña phase of the ENSO cycle.

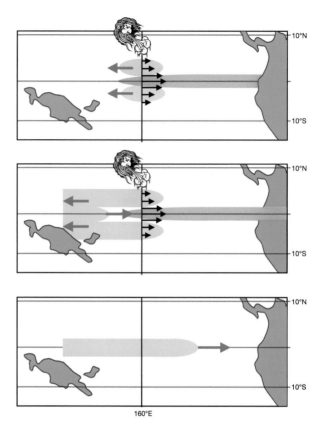

feedback. Linear stability analysis of the coupled tropical Pacific atmosphere–ocean system performed assuming variations about the seasonally varying, climatological basic state shows how the existence of the climatological annual cycle renders the ENSO mode (a Floquet mode) irregular. Consistent with the time series of the ENSO SST index shown in Fig. 17.1, the ENSO mode exhibits a broad spectral peak on the interannual timescale, with warm phases (El Niño events) that peak toward the end of the calendar year.[9] The boreal spring, when the climatological mean equatorial cold tongue is weakest, is the time of greatest uncertainty as to what the coupled system is going to do for the next year. In contrast, the boreal autumn or the next midwinter is the time when the ENSO-related anomalies are strongest and most persistent from one month to the next. Most El Niño events end abruptly in late boreal winter or spring, as documented in Fig. 17.6.

Irregularity in the ENSO mode is also due to stochastic atmospheric variability that is linearly independent of the ENSO mode. In particular, random variability of the wintertime jet exit region over the North Pacific of the kind described in Chapter 13 appears to have influenced the timing of most of the El Niño and La Niña events in the modern record.[10] Stochastic forcing of the linear ENSO

mode also contributes to the "diversity" of spatial patterns of individual El Niño and La Niña events. Other factors that may contribute to the irregularity of ENSO include the Madden–Julian Oscillation, and stochastic wind forcing that depends on the state of ENSO, and the nonlinearity in the ENSO cycle itself.

On average, warm episodes tend to be stronger than cold episodes, though it should be acknowledged that much of the difference is attributable to the high amplitude warm episodes (El Niño events) that peaked in 1997, 1982, 1972, and 2015, which are evident in the time series of the various ENSO indices shown in Fig. 17.1. More robust evidence of nonlinearity is the asymmetry between the duration of warm and cold events. Most El Niño events exhibit a single peak late in the calendar year and last for less than a year, whereas most La Niña events exhibit multiple late-year peaks spaced about a year apart.[11] These and other nonlinear characteristics of the time series of the SST index can be explained as the response to a nonlinearity in the relationship between anomalies in SST and thermocline depth.[12]

The complexity of ENSO-related variability in space and time is a topic of ongoing research, but it will not be explored

[9] Thompson and Battisti (2000, 2001).
[10] Vimont et al. (2003), updated in Battisti et al. (2019); Chang et al. (2007).

[11] Okumura and Deser (2010).
[12] Lübbecke and McPhaden (2017); Battisti et al. (2019). Cane et al. (1986) explicitly included this nonlinearity in the first model to realistically simulate ENSO-like variability.

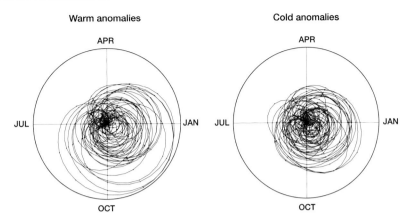

Figure 17.6 Three-month running mean of the absolute value of the amplitude of a SST index of ENSO, plotted parametrically as a function of time calendar (month) in polar coordinates such that the 12 months of each calendar year are arrayed in a counterclockwise loop encircling the origin. Each data point in the series is plotted once: positive values in the left panel and negative values in the right panel. Successive points in time are connected by straight line segments. The circular frames of the panels represent amplitudes of 2.5°C. The ENSO SST index used to create this figure and the figures from here on in this chapter is referred to as "Niño 3.4," defined as SST averaged over the box (5°N–5°S, 120–170°W). The name derives from indices defined by Rasmusson and Carpenter (1982). Regression patterns based on the two different ENSO SST indices used in this chapter are virtually identical as documented in the solution to Exercise 17.1(g).

any further in this chapter. From here onward, we will focus on the linear behavior, as inferred from regression analysis of the observations based on the SST index shown in Fig. 17.1 and a closely related SST index (Niño 3.4) described in the caption of Fig. 17.6.

Exercises

17.1   The time series of monthly values of the ENSO SST index in Fig. 17.1 is posted on the companion web page. There are two versions: an unsmoothed one and a 5-month, centered, running mean smoothed one. By construction, the time-means are zero. (a) Calculate the standard deviation, the moment coefficient of skewness. (b) Choose your own definition of an El Niño event and count the number of events in the record and list the calendar years in which they occurred both chronologically and in order of amplitude. (c) Compare your list with one or more of the lists in the literature. (d) Based on your own definitions, estimate the durations of the El Niño and La Niña events separately, and for each of them, list the calendar months in which they started, peaked, and ended. (e) Calculate the lag-correlation function out to 5-months for the whole time series and separately for each calendar month: that is, J to F, J to M, J to A, J to M, J to J, etc.; then F to M, F to A, F to M, F to J and F to J; etc. Note the seasonality of the "memory" and relate it to

Fig. 17.6 and to entries returned from an internet search of "ENSO spring predictability barrier." (f) Using your definition of an El Niño event with sign reversed to define a La Niña event, repeat (b), (c) and (d) and compare your results for El Niño and La Niña events.[13] (g) Compare the ENSO indices: the cold tongue index (CTI) and "Niño 3.4" index defined in the caption of Fig. 17.6.

17.2   Would ENSO-like variability exist (a) in the absence of SST anomalies and (b) in the absence of anomalous currents and thermocline displacements (i.e. in a "slab ocean" model)?

17.3*   How would ENSO, as simulated in a fully coupled atmosphere–ocean model without a climatological mean annual cycle, be different from the observed?

17.2 SPATIAL STRUCTURE

Figure 17.7 contrasts wind and height fields at the 150 hPa level and the vertical velocity field at the 400 hPa level during strong El Niño and La Niña events. The La Niña pattern (top panel) looks like an exaggeration of the climatology, with ascent centered over the Maritime Continent flanked by anticyclonic Rossby gyres. Along the equator the highest

---

[13] To compare your results to what others have found, see Okumura and Deser (2010).

$$\omega, \mathbf{V}, Z$$

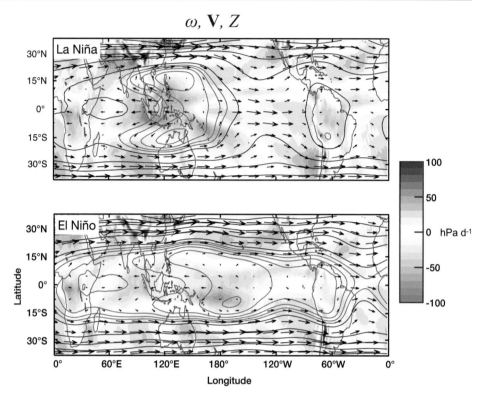

Figure 17.7 Contrasting 400 hPa vertical velocity (colored shading) and 150 hPa geopotential height and wind patterns during strong La Niña and El Niño; that is, months during which the Niño 3.4 SST index defined in the previous figure caption is more than 1.5 standard deviations above or below zero. Contour succession for 150 hPa height: (. . . , 14 100, 14 200, 14 210, 14 220, . . .) m; the first black contour at the separation between the black and blue contours corresponds to 14 210 m. The longest vectors are ~20 m s^{-1}. Adapted from Adames and Wallace (2017). © American Meteorological Society. Used with permission.

heights are just to the east of the enhanced convection. In both the Z and u fields, the amplitude of the equatorial stationary waves is enhanced relative to the climatological mean. In contrast, in the El Niño composite (bottom panel) the stationary waves are anomalously weak except in the Atlantic sector.

The corresponding composites for 1000 hPa height and wind are shown in Fig. 17.8. In the La Niña composite, the equatorial trade winds extend almost all the way westward to New Guinea. The ITCZ is well defined and displaced slightly to the north of its climatological mean position. The equatorial dry zone is pronounced, with strong subsidence along the equator toward its western end. Tropical islands in this subsidence belt receive little, if any rain during La Niña. During El Niño, the equatorial easterlies and the dry zone retreat into the eastern Pacific, with warm pool convection encroaching from the west and the ITCZ convection from the north. This strong dependence of rain rate upon SST is reflected in the strong linear correlation between rain rate over the central equatorial Pacific and the SST index in Fig. 17.1 and Table 17.1. In contrast, the Maritime Continent and tropical islands along the northern limit of the ITCZ experience drought.

Figure 17.9 shows a suite of spatial patterns constructed by linearly regressing selected fields upon the standardized SST index shown in Fig. 17.1. The pattern of sea level (bottom panel) is consistent with the vertical cross sections shown in Figs. 17.3 and 17.4. The anomalously high sea level on the eastern side of the Pacific is narrowly focused on the equator and extends all the way to the coast of the Americas and northward and southward along the coast. The negative sea level anomalies in the west are wider, with

off-equatorial maxima suggestive of equatorially trapped Rossby waves. That the positive anomalies extend poleward as a coastally trapped Kelvin wave as far as San Francisco and the negative anomalies extend westward and poleward along the west coast of Australia explains the strong linear correlation between the ENSO SST-based index and the index based on tide-gauge records at Perth and San Francisco (Fig. 17.1; Table 17.1).

The SST pattern (second panel from bottom) bears a strong resemblance to the leading EOF of the global SST anomaly field (i.e., departures from the seasonally varying climatological mean at each grid point).[14] This pattern is so prominent that it is clearly recognizable even in instantaneous SST anomaly maps during El Niño events used in real-time monitoring of ENSO. From the Date Line eastward, it strongly resembles the underlying pattern of sea level. Individual El Niño events vary with respect to the strength and shape of the region of positive SST anomalies in the eastern equatorial Pacific. Some events tend to be more basin-wide (i.e., extending from the South American coast to beyond the Date Line), while others tend to be more focused in the central Pacific. The basin-wide events tend to be stronger.[15]

The SLP regression pattern (center panel Fig. 17.9) is dominated by the east-west, planetary-scale Southern Oscillation dipole, but it also exhibits a patch of anomalies over the equatorial cold tongue, of negative polarity during El Niño. The anomalous low pressure is the hydrostatic expression of the anomalous warmth of the boundary layer above the weakened cold tongue, analogous to a "heat low."

[14] Weare et al. (1976); Zhang et al. (1997); Chen and Wallace (2015).
[15] Takahashi et al. (2011).

$$\omega, \mathbf{V}, Z$$

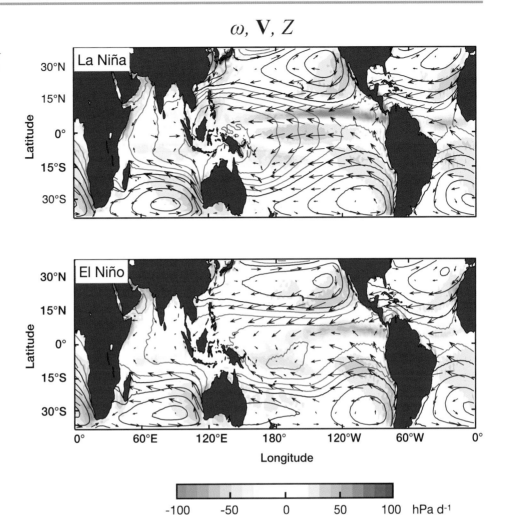

Figure 17.8 As in Fig. 17.7 but for ω_{850}, $\mathbf{V}_{1000}$, and Z_{1000}. The longest vectors are ~10 m s^{-1}. Contour interval 10 m.

The SLP anomalies are attended by a band of southward (down-gradient) surface wind anomalies extending along the northern flank of the cold tongue. In the central Pacific, the meridional wind component associated with the Southern Oscillation SLP signature also makes a substantial contribution to the low-level convergence along the equator, which determines the rain rate signature of ENSO.

The distribution of ENSO-related rain rate anomalies shown in the second panel from top of Fig. 17.9 is generally consistent with the distribution of boundary layer and surface wind convergence, but it does not closely match the SST field (second panel from bottom). For example, the ENSO-related SST anomalies are of almost uniform amplitude from the Date Line eastward to the coast of South America, whereas the rain rate anomalies are strongest near and slightly to the west of the Date Line.

The pattern of ENSO-related Z_{150} anomalies is shown in the top panel of Fig. 17.9, together with the corresponding 400 hPa vertical velocity anomalies, which (assuming WTG balance) can also be interpreted as an indicator of the thermal forcing due to the release of latent heat. The height pattern exhibits anticyclonic gyres straddling the equator, centered at 150°W, somewhat to the east of the strongest convection. This pattern is quite different from the linear response to an equatorial heat source for a resting basic state, shown in Fig. 14.4, which is characterized by a Kelvin wave response to the east of the heat source with westerly wind anomalies, and anticyclonic Rossby wave gyres to the west of it. The mismatch between the patterns may be due to the fact that the basic state flow (Fig. 14.4) is characterized by equatorial westerlies to the east of the ENSO-related positive rain rate anomalies and easterlies to the west of them – conditions that might affect the propagation of Rossby waves away from the heat source.

Much of the popular literature about ENSO gives the impression that the anomalous divergence associated with the rain rate anomalies over the central and eastern equatorial Pacific is due to the weakening of the climatological mean overturning circulation in the equatorial plane; that is, the so-called "Walker circulation." It is evident from Fig. 17.10 that the anomalous boundary layer convergence and upper tropospheric divergence over this region during El Niño is not exclusively due to changes in the zonal wind component; anomalous confluence/diffluence of the meridional wind component also makes a substantial contribution.

The three-dimensional structure of the ENSO-related anomalous circulation, shown in Fig. 17.11, is made up of the elements described in the foregoing discussion. The

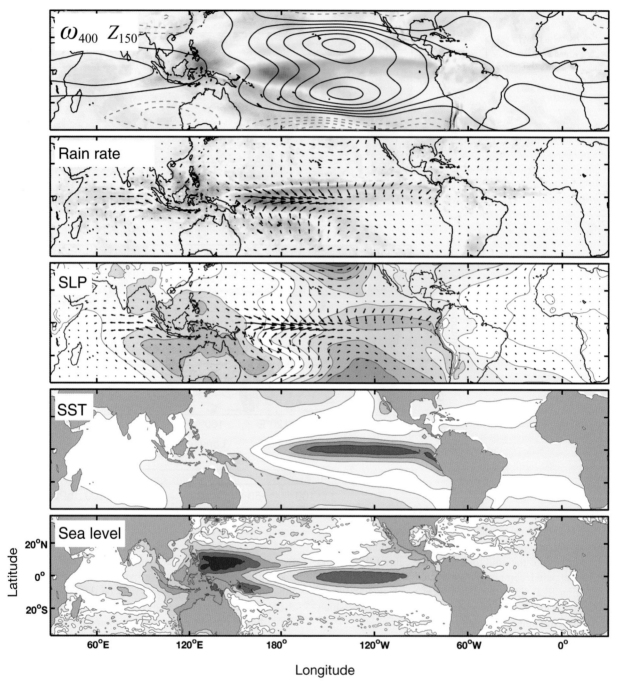

Figure 17.9 The signature of ENSO as represented by regressing various fields upon the time series of the ENSO SST index shown in the top panel of Fig. 17.1. Vectors are 1000 hPa surface winds; the longest arrow corresponds to 2 m s^{-1}. Green vs. brown shading correspond to enhanced vs. suppressed rain rate and ascent versus descent. Orange shading indicates positive SST, sea level and SLP anomalies and blue shading negative ones. Contour intervals for Z_{150} 50 m, SLP 0.2 hPa, SST 0.2°C, and sea level 2 cm. For Z_{150} positive contours solid, negative contours dashed, the zero contour is omitted. For SLP, SST, and sea level.

overturning circulation that determines the rain rate signature is the superposition of a shallow, mainly meridional circulation associated with the southward displacement and strengthening of the ITCZ and a deeper, planetary-wave circulation to which the zonal overturning in the equatorial plane and meridional overturning both contribute. The zonal wind stress associated with the planetary wave signature

contributes to the Bjerknes feedback depicted in Fig. 17.2 and to the generation of the oceanic Kelvin and Rossby waves in Fig. 17.5.

The ENSO-related rain rate anomalies bear a close relationship to the anomalies in the distribution of *exp*(SST), as shown in Fig. 17.12. Hence, in ENSO, as in the mean climatology, SST influences the rain rate distribution mainly

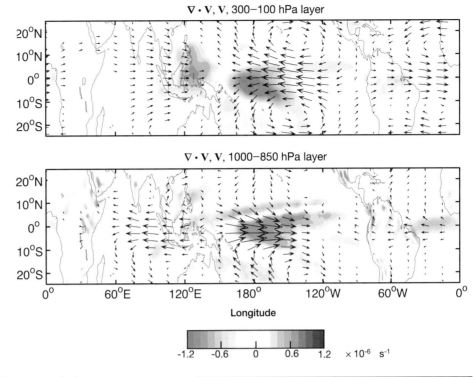

Figure 17.10 Wind and divergence regressed upon the ENSO SST index Niño 3.4, defined in the caption of Fig. 17.6. (Top) averaged from 300 to 100 hPa and (bottom) averaged from the surface to 850 hPa. The longest arrow corresponds to ~4 m s^{-1} (top) and ~1 m s^{-1} (bottom).

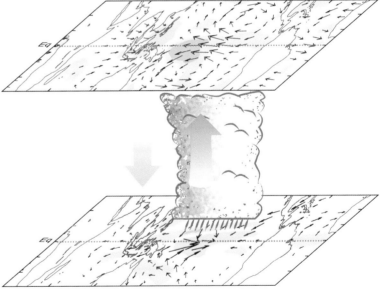

Figure 17.11 A schematic version of the regression fields shown in the previous figure. From Adames and Wallace (2017). Shaded vertical arrows highlight the most prominent features in the vertical velocity field. © American Meteorological Society. Used with permission.

by way of shaping the pattern of column-integrated water vapor W. The distributions P and W are similar in the vicinity of the cold tongue, but elsewhere in the tropics the predominantly negative rain rate anomalies observed during El Niño events do not appear to be attributable to local SST anomalies: rather, they appear to be part of the atmospheric planetary-wave response to the anomalous diabatic heating over the central Pacific.

The close correspondence between the patterns of ENSO-related anomalies in rain rate, boundary layer (1000–850 hPa) convergence and upper tropospheric (100–300 hPa) divergence in Fig. 17.10 shows that to first order, the vertical profiles of diabatic heating rate and horizontal divergence

in ENSO are indicative of the response to a heat source centered in the middle troposphere. But there is more to these patterns than a single modal structure that can be represented as separable functions of (x,y) and p. ENSO-related 300 and 850 hPa vertical velocity fields are compared in Fig. 17.13. The former exhibits an east–west dipole pattern, indicative of an eastward displacement of the heaviest rain rate from the Maritime Continent toward the central Pacific. In contrast, the latter exhibits a north–south dipole, indicative of a southward displacement of the ITCZ. That the positive pole is dominant in both dipoles is indicative of an overall increase in equatorial rain rate during El Niño at the expense of off-equatorial rain rate.

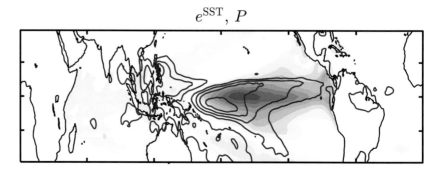

$$e^{\mathrm{SST}}, P$$

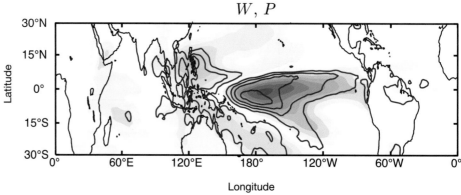

$$W, P$$

Figure 17.12 Demonstration that (1) ENSO-related rain rate anomalies resemble anomalies in column water vapor W, and (2) over the tropical Pacific, the W anomalies are mainly a thermodynamic response to variations in saturation vapor pressure at the air–sea interface due to the presence of ENSO-related SST anomalies. The contoured field in both panels is rain rate P (contour interval 0.75 mm d^{-1}). The bottom panel shows the column water vapor W. Inspired by the Clausius–Clapeyron equation, the top panel shows the antilog of the ENSO-related SST anomalies, scaled to match the observed W field in the bottom panel as well as possible.

Figure 17.13 Vertical velocities (top) ω_{300} and (bottom) ω_{850} regressed upon the ENSO SST index Niño 3.4.

$$\omega_{300}$$

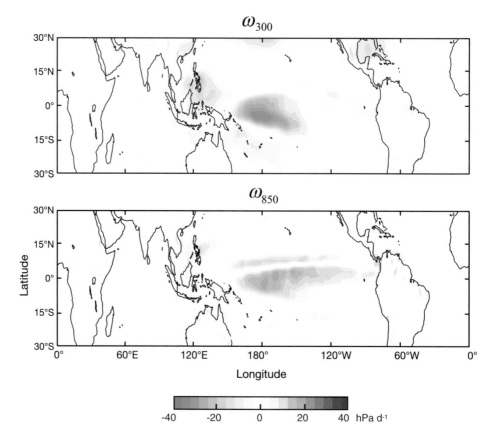

$$\omega_{850}$$

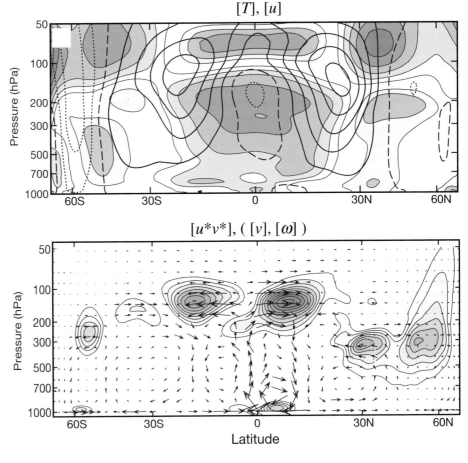

Figure 17.14 ENSO-related zonally symmetric anomalies as inferred from linearly regressing the various fields upon the ENSO SST index Niño 3.4. (Top) Temperature (colored shading, contour interval 0.1 K); zonal wind (black contours at intervals of 0.5 m s^{-1}, zero contour dashed, dotted contour negative). (Bottom) The anomalous mean meridional circulation (arrows: longest vertical components 5 hPa d^{-1}; longest meridional components 0.2 m s^{-1}) and the anomalous northward eddy transport of westerly momentum (contours at intervals of 0.5 m^2 s^{-2}, zero contour omitted; colored shading for anomalies >1 m^2 s^{-2}, orange northward, blue southward). From Adames and Wallace (2017). © American Meteorological Society. Used with permission.

Exercise

17.4 (a) Explain how the sharp zonal gradient of climatological mean rain rate at the eastern edge of the Indo-Pacific warm pool (Fig. 1.14) induces skewness in ENSO-related variations in time series of rain rate. (b) How is it reflected in distinctions between the spatial patterns of rain rate anomalies over the equatorial central Pacific observed during El Niño and La Niña?[16]

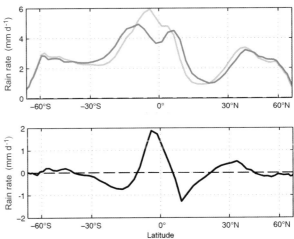

Figure 17.15 (Top) Zonally averaged rain rate observed during El Niño (gold) and La Niña (blue). (Bottom) The difference, El Niño minus La Niña. From Adames and Wallace (2017). © American Meteorological Society. Used with permission.

17.3 THE ZONALLY SYMMETRIC COMPONENT

The zonally symmetric component of the ENSO signature, shown in the top panel of Fig. 17.14 is characterized by anomalous warmth of the tropics and by an intensification and equatorward displacement of the tropospheric jet streams in both hemispheres, indicative of a narrowing of the tropics. The anomalous tropical warmth is sufficient to significantly elevate global mean temperature.[17] The ENSO-related wind and temperature anomalies are in thermal wind balance: the anomalous warmth of the tropical upper tropo-

sphere and coolness of the lower stratosphere is required to balance the observed wind anomalies and vice versa. Hence, to explain the patterns of anomalies in both fields requires consideration of both the thermal forcing and the dynamical forcing of the zonally symmetric flow.

The thermal forcing can be inferred from the observed changes in the zonally symmetric component of tropical rain rate in the ENSO cycle shown in Fig. 17.15. Integrated over

[16] See Hoerling et al. (2001).
[17] Newell and Weare (1976); Navato et al. (1981); Yulaeva et al. (1994).

$$\rho\,[u]\cos^2\phi$$

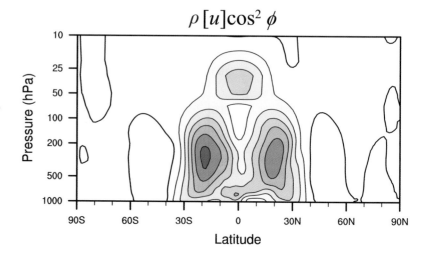

Figure 17.16 Fields regressed upon the time series of standardized atmospheric angular momentum (AAM) anomalies in Fig. 3.5: (top) meridional cross section of $\rho[u]cos^2\phi$ (contour interval 0.08 kg m^{-2} s^{-1}; solid contours are positive and the zero contour is bold); (bottom) SLP (colored shading) and V_{1000} (vectors, scale to right of color bar).

$$\text{SLP, } \mathbf{V}_{1000}$$

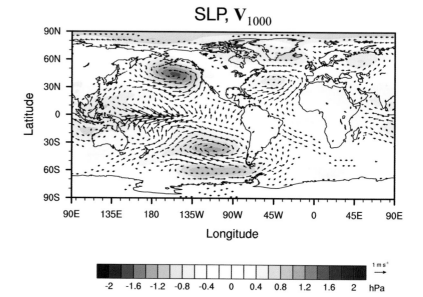

the tropics, there is not much difference in rain rate between El Niño and La Niña. However, the equatorial rain belt is narrower and more intense and the outer tropics are drier during El Niño than during La Niña. Consistent with these changes in rain rate, the Hadley cell is intensified during El Niño (bottom panel, Fig. 17.14), but only in the inner tropics (15°N–15°S).

The ENSO-related dynamical forcing is characterized by anomalous poleward transports of westerly momentum centered ~15°N/S, shown in the bottom panel of Fig. 17.14. This feature is a consequence of the anomalous weakness of the equatorial stationary waves observed during El Niño (Fig. 17.7),[18] which results in a reduced poleward EP flux. A reduced EP flux, in turn, is indicative of a reduced equatorward transport of westerly momentum, which is equivalent to an anomalous poleward transport of westerly momentum relative to the climatological mean. In analogy

with the annual climatological mean poleward transports of westerly momentum, this enhanced poleward transport drives an anomalously strong Hadley-like circulation on its equatorward flank and a thermally indirect (Ferrel-like) circulation cell on its poleward flank. Hence, it contributes to the maintenance of the observed configuration of anomalous mean meridional circulations shown in the bottom panel of Fig. 17.14. These anomalous momentum transports also force zonal wind anomalies. Poleward of 15°N/S, where the transports converge, they induce an anomalous eastward acceleration and equatorward of 15°N/S, where they diverge, they induce an anomalous westward acceleration.

Based on the evidence presented here, it appears that the anomalous warmth of the tropical troposphere during El Niño could be dynamically forced, a response to the anomalously weak equatorial stationary waves and consequently to anomalously strong westerlies at the 150 hPa level centered 20–25°N/S. Thermal wind balance requires that $-\partial[T]/\partial y$ be anomalously strong in the layer below. To meet

[18] See also Grise and Thompson (2012).

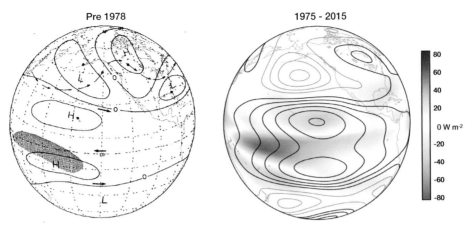

Figure 17.17 ENSO teleconnections to the extratropics. (Left) Schematic for upper tropospheric geopotential height during the boreal winter season that was created in 1980 based on data ending in 1978. Shading indicates the region of enhanced equatorial rain rate anomalies during El Niño as inferred from OLR data. From Horel and Wallace (1981). (Right) Pattern of 150 hPa height regressed on the ENSO SST index shown in the top panel of Fig. 17.1 based on an independent period of record: November–March data from 1979 through 2015. Contour interval 10 m. Green shading indicates enhanced and brown suppressed rain rates as inferred from OLR anomalies.

this condition, $[T]$ must be anomalously strong throughout the tropics. Anomalous thermal forcing, for example, due to enhanced latent and sensible heat fluxes over the equatorial cold tongue, might contribute to the anomalous global warmth (i.e., positive $<T>$) during El Niño, but it cannot explain why the warmth, shown in the top panel of Fig. 17.14, is largely confined to the tropics.

Because of the prevalence of westerly wind anomalies in the pattern in the top panel of Fig. 17.14, atmospheric angular momentum (AAM) tends to be anomalously high during El Niño and low during La Niña.[19] In Fig. 17.16 it is shown that regressing $[u]$ on the time series of AAM (Fig. 3.5, bottom panel) yields a pattern similar to the pattern in the top panel of Fig. 17.14, based on the ENSO SST index, and regressing the SLP and V_{1000} fields upon it also yields ENSO-like patterns. Evidently, ENSO is the leading contributor to variations in AAM on the interannual timescale.

Exercises

17.5 Are the ENSO-related $[u]$ and $[T]$ anomalies shown in the top panel of Fig. 17.14 (qualitatively) in thermal wind balance?

17.6 Making use of the balance requirement for global AAM with reference to the bottom panel of Fig. 17.16, explain how the ENSO-related AAM anomalies are maintained against frictional drag during El Niño and La Niña events.

17.4 ATMOSPHERIC TELECONNECTIONS

With the availability of an extended archive of gridded, global upper air analyses starting in the late 1970s it soon

became evident that the upper tropospheric signature of ENSO extends into the extratropics. The schematic in the left panel of Fig. 17.17 showing the global pattern of 500 hPa height anomalies during an El Niño in boreal winter was created in 1980. The wave train oriented along a great circle extending across the Pacific–North American sector has been observed in nearly every El Niño event since that time, as evidenced by the regression map shown in the right panel of Fig. 17.17 and the signatures of the strong El Niños in 1982, 1997, and 2015 shown in Fig. 17.18. The wave train has also been replicated in numerous simulations with general circulation models forced with observed and idealized tropical SST anomalies.

Anomalous divergence above the region of enhanced tropical convection gives rise to a patch of anomalous anticyclonic vorticity along its poleward flanks. The associated pattern in the streamfunction and geopotential height fields disperses poleward and thence eastward to create a stationary planetary-scale Rossby wave train.[20] A wave train with a similar horizontal scale and orientation is generated in a shallow water wave equation model with an equivalent depth of ~1.5 km, in which the vertical structure is nearly barotropic, as shown in Fig. 17.19. (Similar results are obtained if the divergence is set equal to zero, as was done in creating Fig. 11.14.)

Figure 17.20 shows the eddy components of the baroclinic and barotropic circulation patterns observed during the warm phase of ENSO. The former is represented by geopotential height anomalies at the 150 hPa level, which corresponds to the strongest outflow index convection (e.g., see Fig. 14.2) and the latter corresponds to the level of minimum horizontal divergence in large-scale extratropical motion systems, where the barotropic component of the flow is most prominent. The baroclinic signature consists mainly of equatorially trapped Kelvin and Rossby waves, while

[19] Dickey et al. (1994); Rosen et al. (1984).

[20] Hoskins and Karoly (1981).

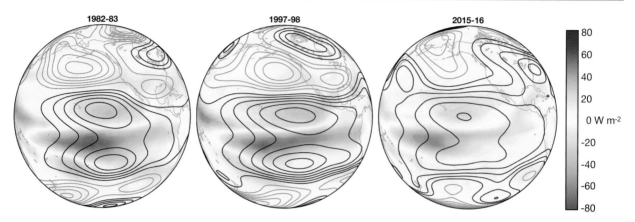

Figure 17.18 ENSO teleconnections to the extratropics during three strong El Niño events as indicated by November–February 150 hPa height anomalies. Contour interval 10 m; black positive; gray negative. Shading indicates the pattern of rain rates anomalies during El Niño as inferred from OLR data. Green shading indicates enhanced and brown suppressed rain rate.

Figure 17.19 Response to a circular heat source centered on the equator at $0°$ longitude in a shallow water wave equation model with a westerly background flow consisting of pure superrotation with a velocity of $10\ \mathrm{m\ s^{-1}}$. The equivalent depth is ~1.5 km, large enough to ensure that the waves are able to disperse globally and not just in a tropical waveguide. The contours represent geopotential, with high values along the equator. From Lim and Chang (1983). Published (1983) by the American Meteorological Society.

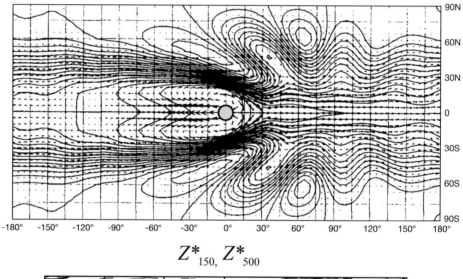

$$Z^*_{150},\ Z^*_{500}$$

Figure 17.20 Eddy components of the global baroclinic and barotropic ENSO signatures, where the term *baroclinic* denotes of reverse polarity between the tropical upper and lower troposphere and *barotropic* denotes of the same polarity. Here the barotropic component is represented by the Z^*_{150} field (colored shading) and the barotropic component by the Z^*_{500} field (contours at 3 m intervals, solid positive, dashed negative, the zero contour omitted). The patterns are obtained by regressing the respective fields upon the ENSO SST index Niño 3.4. The polarities of the patterns are thus indicative of the sign of the anomalies observed during El Niño. For justification for representing the baroclinic and barotropic components of the global circulationin this manner, see Adames and Wallace (2017).

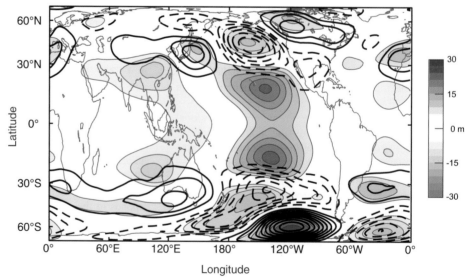

$$Z_{150}, \, d(\text{SST}) \, / \, dt$$

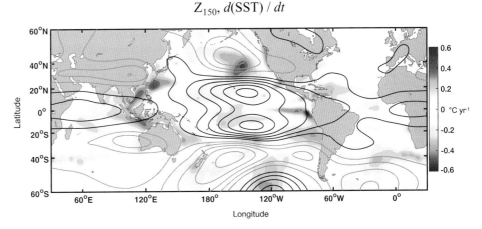

Figure 17.21 Contours indicate the global ENSO signature in the 150 hPa height field, the same as in the top panel of Fig. 17.9 but for the boreal winter months DJF only (positive contours black, negative contours gray, zero contour omitted; contour interval 50 m). Colored shading indicates the concurrent SST tendency, estimated by subtracting the SON SST of El Niño years from the MAM SST of the subsequent year.

the barotropic component consists mainly of extratropical wave trains dispersing along a great circle, as in the shallow water wave equation simulation shown in the previous figure. (The almost perfect alignment of the extratropical features in the two fields confirms that the extratropical response is dominated by the barotropic component.)

The extratropical response to a tropical heat source is mediated by the background flow and, in particular, by the configuration of the subtropical jet streams. When the nonlinear terms are included, the forcing can be expressed as the advection of vorticity by the divergent component of the flow. The apparent Rossby wave source can be quite different from the linear $f\nabla \cdot \mathbf{V}$ term in the barotropic vorticity equation. In particular, an equatorial region of divergence situated in a region of easterly winds can give rise to a Rossby wave source farther poleward, within the subtropical westerlies, from which it can excite a strong extratropical response.[21] If the background flow is zonally varying, the longitude of the Rossby wave source may be different from that of the heat source. In an equatorially symmetric background flow, the extratropical response tends to be equatorially symmetric, even when the heat source is biased toward one hemisphere.[22] Theory also explains why the extratropical response to ENSO-related tropical heating anomalies tends to be stronger in the winter hemisphere, where the westerlies are stronger.

Another factor that could affect the extratropical response is the excitation of preferred modes of variability of the extratropical circulation, in particular, the PNA pattern shown in the left panel of Fig. 13.13. If this process were dominant in shaping the extratropical signature of ENSO, one might expect patterns formed by regressing upper tropospheric geopotential height upon ENSO indices to closely resemble the PNA pattern, but this is not the case. For example, the northernmost center of action in the ENSO response shown in Fig. 17.17 encompasses

most of Canada, whereas the corresponding center in the PNA pattern is restricted to western Canada. The response does appear to be affected by feedbacks involving the high frequency transients, which cause the centers of action in the extratropical wave train to become more zonally elongated, giving rise to a zonally averaged extratropical response with anomalous warmth over midlatitudes that has sometimes persisted for a year or longer beyond the demise of the El Niño event believed to have initiated it.[23]

By modulating cloudiness and surface air temperature and wind, ENSO-related extratropical circulation anomalies imprint a signature upon the mixed layer in the other tropical oceans and in the extratropical Pacific, as shown in Fig. 17.21. In effect, the atmospheric planetary wave response conveys the ENSO signature to the world ocean.[24] The workings of this so-called "atmospheric bridge" have been investigated in numerical experiments with coupled atmosphere–ocean models in which the globe is partitioned into a region in which the SST is prescribed to match the historical record of SST, and into regions in which ocean is represented by a slab with a prescribed heat capacity, which responds passively to anomalous energy fluxes at the air–sea interface. Prescribing the SST history in the tropics as a whole elucidates the contribution of tropical ocean to the low-frequency variability of the extratropical circulation, while prescribing the SST history in the tropical Pacific only serves to highlight the impact of ENSO in particular on the global climate system. The atmospheric bridge accounts for the warming of the Indian Ocean during the later stages of El Niño that often persists into the subsequent boreal summer monsoon season and for the warmth of the subtropical North Atlantic, which persists through the boreal spring. During El Niño, anomalous tropical convection forces trains of Rossby waves that propagate poleward to the extratropical North and South Pacific, resulting in higher surface wind speeds,

---

[21] Sardeshmukh and Hoskins (1988).
[22] Sardeshmukh and Hoskins (1988).

[23] Lau et al. (2005, 2006).
[24] Alexander (1992); Lau and Nath (1996).

negative surface air temperature anomalies, and enhanced upward energy fluxes at the air–sea interface. These processes are responsible for the negative SST anomalies over much of the extratropical North and South Pacific in Fig. 17.21.

Exercises

17.7* (a) Summarize and interpret the global climatic impacts of ENSO, and go into greater depth for a selected region of your choice. (b) Give an example of how ENSO prediction a season or more in advance can serve to mitigate the adverse impacts of ENSO.

17.8 (a) In Fig. 17.21, why is cooling observed over the equatorial Pacific cold tongue region? (b) Explain and interpret the features in the SST tendency field in the Indo-Pacific warm pool region and in the extratropical oceans.

17.9* Describe the impacts of ENSO upon the terrestrial and marine biosphere and on the carbon cycle.

17.10* Most of the documentation of the spatio-temporal structure of ENSO in this chapter is based on linear regression analysis. In reality, no two El Niño (or La Niña) events are exactly alike. Discuss the literature relating to subtypes of El Niño events.

17.11* ENSO is clearly the dominant mode of interannual variability involving the interactions between the tropical oceans and the global atmosphere, but other modes have been identified. The most notable are the so-called (a) *Pacific* and *Atlantic meridional modes* and (b) the *Indian Ocean dipole*. Review the literature relating to these modes, including papers that question their reality.

17.12 Are there precursors of El Niño events?

17.13 Does El Niño exist apart from ENSO?

18 Intraseasonal Variability of the Tropical General Circulation

with Ángel F. Adames

The last two chapters were devoted to the seasonal cycle in the tropical general circulation and to ENSO-related *interannual variability*. In this chapter, we consider the variability on the *intraseasonal* timescale, defined here as fluctuations with periods ranging from 20 to 90 days (or frequencies ranging from 1 to 5 cycles per season). In this frequency range, there exists a phenomenon of singular importance, the Madden–Julian Oscillation (MJO),[1] a moisture mode characterized by eastward propagating fluctuations in equatorial rain rate and a distinctive planetary wave signature with teleconnections to extratropical latitudes. Like ENSO, the MJO modulates atmospheric angular momentum, and the amplitude of the equatorial stationary waves.

This chapter is organized as follows. The first section presents observational evidence of the existence of the MJO based on data for rain rate, velocity potential, and upper tropospheric winds, analyzed in several different ways. The second describes the three-dimensional structure of the MJO in greater detail as it evolves through a cycle, with emphasis on the wind and geopotential height fields. The third section examines the moisture balance, which plays a key role in the maintenance and eastward phase propagation of the MJO, a characteristic of moisture modes.

The middle sections of the chapter are mainly concerned with the impacts of the MJO upon the global general circulation. Section 18.4 documents its distinctive signature in temperature, moisture, and cloudiness in the vicinity of the tropical cold point. Section 18.5 shows how the MJO modulates the zonal flow at the tropospheric jet stream level, leading to variations in atmospheric angular momentum. Section 18.6 shows how MJO-related rain rate anomalies force teleconnection patterns that affect the extratropical circulation, particularly during the winter season.

Section 18.7 documents the seasonality of the MJO, with emphasis on the boreal summer JJA, when its structure is quite different from that observed during the remainder of the year – less equatorially symmetric and more focused on the South Asian monsoon region. The final two sections briefly describe other planetary-scale tropical phenomena that occupy the same range of the frequency spectrum as the MJO: the so-called *Pacific–Japan pattern* and equatorially trapped, convectively coupled Rossby waves.

---

[1] Madden and Julian (1971, 1972); Regarding its discovery, see Li et al. (2018).

18.1 EVIDENCE FOR THE EXISTENCE OF THE MJO

At any given time, tropical convection is organized on a spectrum of space scales ranging up to what have come to be referred to as *superclusters*, with dimensions of 1000 km or even larger. The configurations of these features change erratically from day to day, but in time-lapse animations some of the larger clusters can be followed for a week or longer as they migrate across the Indo-Pacific warm pool from west to east. Figure 18.1 shows a time–longitude section of 20–90 d bandpass filtered OLR, a proxy for rain rate, and 150 hPa velocity potential χ_{150} anomalies averaged over the equatorial belt (15°N–15°S). Patches of enhanced and suppressed convection are discernible in OLR, migrating eastward across the warm pool at a rate of roughly 4 degrees of longitude per day ($\sim$5 m s^{-1}), slow enough to be classified as a moisture mode. These features are mirrored in the χ_{150} field, in which their signature is smoother and more wavelike. In contrast to the OLR anomalies, which are largely restricted to the warm pool sector, some of the χ_{150} extrema can be tracked as they propagate all the way around the globe. Averaged over the whole tropical belt, their eastward phase speed is on the order of 10 m s^{-1}.

To document the horizontal structure and evolution of the MJO, as represented by these same variables, a series of lagged-regression patterns based on χ_{150} at the reference grid point (0°, 120°E) is shown in the left column of Fig. 18.2. The reference time series is the 90 d highpass filtered fields in order to ensure that the regression patterns are not unduly influenced by sampling variability associated with the annual cycle or by ENSO. The eastward propagating patches of enhanced and suppressed convection in the time–longitude section (Fig. 18.1) are clearly evident in this sequence. That the patterns for lags of $+10$ d and -10 d are almost identical, but of opposing polarity suggests that the preferred period of the MJO-related perturbations is around 40 d. The lagged-regression maps for this reference time series are thus of the form

$$A(x,t) = A_1(x)\cos \omega t + A_2(x)\sin \omega t \qquad (18.1)$$

where $\omega = 2\pi/40$ d, A_1 is the simultaneous regression pattern, and A_2 is the lagged-regression pattern 10 d later. Baroclinic waves and many other phenomena that result

OLR, χ_{150}

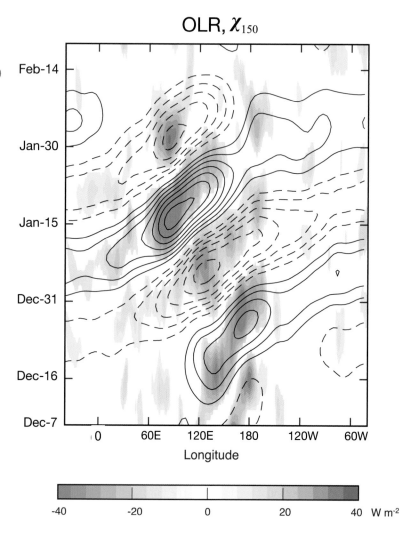

Figure 18.1 Time–longitude section of OLR (colored shading) and global 150 hPa velocity potential χ_{150} anomalies averaged over the equatorial belt (15°N–15°S) from December 1992 to mid February 1993. Contour interval 3×10^6 m^{2}s^{-2}. Blue shading indicates enhanced rain rates.

from flow instabilities are well described by Eq. (18.1) and normal mode solutions derived from linear stability analysis assume a similar form.

To what extent are these lagged-regression maps for the reference grid point (0°, 120°E) typical of those for tropical grid points in general, and to what extent are they indicative of the atmosphere's leading mode of tropical variability on the intraseasonal timescale? To address these questions, EOF analysis is performed upon the global field of daily, 90 d highpass filtered velocity potential. Since the MJO-related perturbations are strongest and out of phase with one another at the 150 and 850 hPa levels, the analysis is based on $\Delta\chi$, defined as $\chi_{150} - \chi_{850}$. The leading EOF, shown in the middle panel of the right column of Fig. 18.2, is remarkably similar to the simultaneous regression map that appears directly across from it in the figure. Moreover, EOF 2, shown in the top and bottom panels with opposing signs, bears a strong resemblance to the +10 and −10 d lagged correlation maps for the reference grid point in the left column. Hence, EOF 1 resembles A_1 in Eq. (18.1), EOF 2 resembles A_2, and their associated principal component (PC) time series project positively upon cosine and sine

functions with the same period. The resemblance between the lagged-regression maps and the EOFs confirms that the lagged-regression maps for the reference grid point (0°, 120°E) are indeed typical of those for tropical grid points, and that the EOFs are not artifacts of linear algebra.

The left panel of Fig. 18.3 shows the autocorrelation functions for PC1 and PC2 and the corresponding cross-correlation function. The implied quasi-periodicity in the cross-correlation function of ∼40 d is consistent with Eq. (18.1) and with the fact that PC2 appears with reversed polarity in the top and bottom panels in the right column of Fig. 18.2. Consistent with the orthogonality constraint that applies to EOFs, PC1 and PC2 are uncorrelated with one another at zero lag. PC1 exhibits its positive and negative extrema while PC2 is near zero and changing signs and vice versa: PC2 is linearly proportional to the time derivative of PC1 and vice versa. Time series with these properties are said to be *in quadrature* with one another. Another characteristic of such "paired modes" is that they account for comparable fractions of the total variance to within the range of uncertainty of the sampling variability. They are the embodiment of Eq. (18.1).

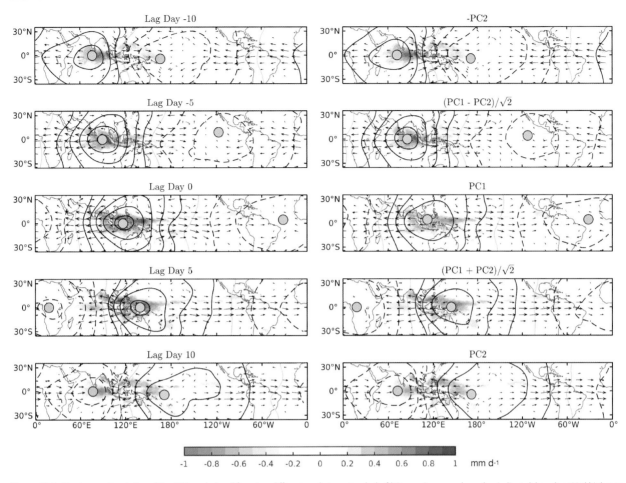

Figure 18.2 Structure and evolution of the MJO as deduced from two different analysis protocols. (Left) Regression maps, lagged as indicated, based on 90 d highpass filtered daily data. The reference time series is standardized χ_{150} at $0°$, $120°$ E, over the Maritime Continent. Colored shading represents rain rate. The vectors indicate $\mathbf{V}_{150}$. The contours indicate χ_{150}, contour interval 10^6 m^2 s^{-1}, solid contours denote positive values of χ_{150}, consistent with divergence. In each panel, the primary positive center in the χ_{150} field is indicated by a circle: red for divergence and blue for convergence. (Right) The same fields regressed upon linear combinations of PC1 and PC2 of the standardized 90 d highpass filtered daily $\Delta\chi = \chi_{150} - \chi_{850}$. The five panels in both columns are spaced 1/8 MJO cycle apart so the sequence spans 1/2 cycle. The longest arrows correspond to horizontal velocities of $\sim$3 m s^{-1}.

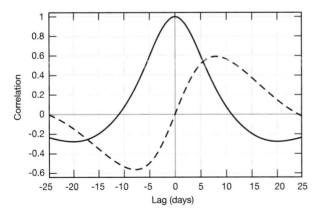

Figure 18.3 Autocorrelation function of the leading principal component (PC1, solid) and the cross-correlation function between PC1 and PC2 dashed) of the global $\Delta\chi$ field. Positive lags indicate PC1 leading PC2, so the pronounced positive peak at positive lags is indicative of eastward phase propagation.

The second and fourth maps in the right column of Fig. 18.2, created by regressing the same fields upon the

standardized sum and difference of PC1 and PC2, closely resemble the $+5$ and -5 d lagged-regression maps across from them in the left column. It follows that PC1 and PC2 comprise a two-dimensional phase space in which cycles of the MJO assume the form of counterclockwise loops. During intervals when the MJO is active, the state of the atmosphere, as represented in this phase space, executes circular orbits around the origin with a period of $\sim$40 d, as shown in the right panel of Fig. 18.4. During inactive intervals, it wanders around aimlessly close to the origin.

The leading pair of PCs of the $\Delta\chi$ field is one of several indices that have been used to characterize time variations in the MJO. The most widely used MJO index is based on EOF analysis of the combined near-equatorial, meridionally averaged 850 hPa zonal wind, 200 hPa zonal wind, and OLR fields.[2] The two leading PCs are referred to as *real time multivariate indices* RMM1 and RMM2 – real time because they can be constructed without recourse to time filtering.

[2] Wheeler and Hendon (2004).

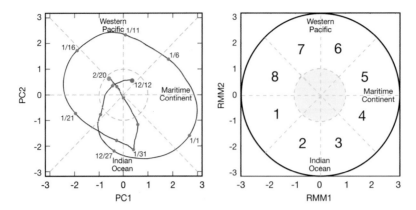

Figure 18.4 Representations of the MJO in a two-dimensional phase space, (left) defined by the leading principal components of the global χ_{150} field and (right) by the RMM1 and RMM2 indices as defined in Section 18.1. The trajectory shown in the left panel corresponds to a segment of the time–longitude section shown in Fig. 18.1, as indicated by the calendar dates. The right panel describes a widely used protocol for compositing calendar dates by MJO Category or Phase. The place names indicate the regions of enhanced rain rate corresponding to the various phases. The two phase spaces are nearly equivalent.

Apart from being lagged by about 2 days, the time series of RMM1 and RMM2 are very similar to PC1 and PC2 of $\Delta\chi$. They are used in constructing several of the figures shown later in this chapter. Another widely used MJO index based on the OLR field alone is the so-called OLR MJO index (OMI).[3] Analyses based on OMI place relatively greater emphasis on the patterns over the warm pool region, whereas those based on $\Delta\chi$ place greater emphasis on the far field response.[4]

Spectra of OLR, $\Delta\chi$, and u_{150}, shown in Fig. 18.5, exhibit prominent peaks at eastward frequencies around one cycle per 40 d, consistent with the reversal of the patterns in Fig. 18.2 over a time span of 20 d. Such peaks are decisively statistically significant, even in the absence of an *a priori* hypothesis that they should exist.[5] The detection of such a peak was, in fact, the basis for the discovery of the MJO.

Exercises

18.1 Describe what regression maps analogous to the ones in Fig. 18.2 would look like for the other half of the MJO cycle.

18.2 What do the time derivative fields for the variables in Fig. 18.2 look like at the time of maximum PC2?

18.3 EOFs 1 and 2 explain roughly comparable fractions of the variance of the $\Delta\chi$ field (29% vs. 26%). How would the sequence of patterns in the right panel of Fig. 18.2 be different if EOF1 were much stronger than EOF2?

18.4 Why are two time-varying indices needed to characterize the time-varying state of the MJO, whereas a single index was used in the previous chapter to characterize the time-varying state of ENSO?

18.5 Explain some of the distinctions between the spectra of the variables shown in Fig. 18.5. (a) Why is the variance of the velocity potential χ_{150} so much more concentrated at the low zonal wavenumber than that of OLR? (b) Why does the spectrum for Z_{150} exhibit a prominent spectral peak at zonal wavenumber $k = -1$?

18.6 The spectra of OLR, χ_{150}, and u_{150} shown in Fig. 18.5 exhibit a secondary peak for westward propagating planetary waves that mirrors the 40 d MJO-related peak in the eastward propagating waves. How should this feature be interpreted?

18.2 MJO-RELATED WIND, TEMPERATURE, AND GEOPOTENTIAL HEIGHT FIELDS

Using the indices based on $\Delta\chi$ defined in the previous section, we extend the description of the MJO to include other variables. In constructing some of the figures in this section and in the next section, rather than showing conditions during a discrete phase of the MJO, we average maps or longitude–height sections over a sequence of adjacent MJO phases extending over 1/4 cycle, zonally shifting each one so that all the maxima in $\Delta\chi$ come into alignment. The resulting patterns tend to be slightly smoother and simpler than the patterns in the regression maps for discrete phases of the MJO because the weak imprint of the land–sea geometry is smeared out by the longitudinal shifting. We will retain the land–sea geometry in these plots for the purpose of orientation, but the reader should bear in mind that the patterns are intentionally blurred by the zonal shifting.

Figure 18.6 shows regression maps of 150 hPa zonal wind u and geopotential height Z based on PC1 of $\Delta\chi$. As in the middle panels of Fig. 18.2, the enhanced convection, indicated here by the red circle, is centered over the Maritime Continent and it is flanked by u_{150} anomalies, positive to the east of it over the western Pacific, and negative to the

---

[3] Kiladis et al. (2014).
[4] See the Appendix of Adames and Wallace (2014b).
[5] See the tutorial in Madden and Julian (1971).

Figure 18.5 Two-sided wavenumber–frequency spectra for equatorially symmetric features based on data averaged from 15°S–15°N. Unlike the spectra shown in Section 10.2, these spectra are not divided by a smoothed reference spectrum. Gradations in shading are arbitrary. The spike at $k = -1$ in the top right panel corresponds to the external Rossby mode. Courtesy of George Kiladis.

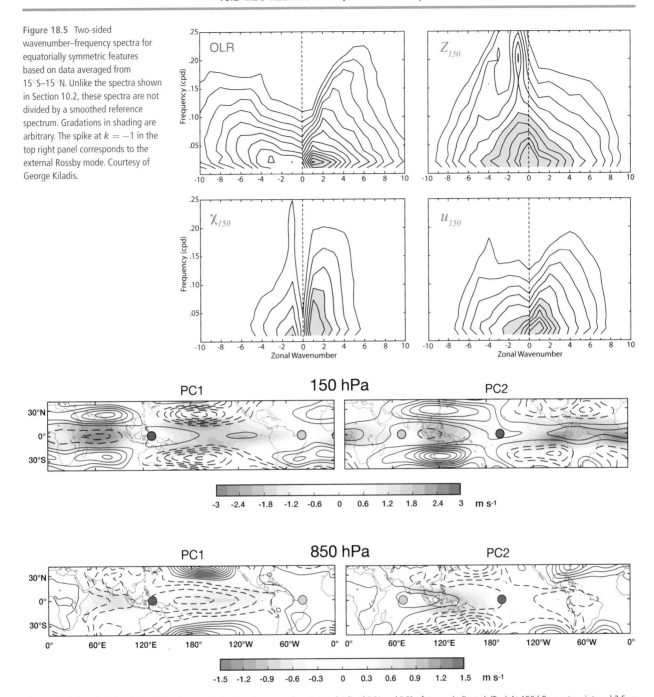

Figure 18.6 Zonal wind u (colored shading) and Z (contours) regressed upon standardized PC1 and PC2 of $\Delta\chi$ as indicated. (Top) At 150 hPa, contour interval 2.5 m; (bottom) at 850 hPa, contour interval 1.5 m. The small red and green circles indicate centers of regions of enhanced and suppressed convection corresponding to extrema in $\Delta\chi$: red enhanced and blue suppressed.

west of it over the eastern Indian Ocean. The Z_{150} anomalies are centered at the same longitudes as the u_{150} anomalies but their meridional profile is quite different. The strongest anomalies are centered not along the equator (as is the case for u_{150}) but along 28°N/S. We will refer to these features as the *flanking Rossby waves*. In the corresponding map based on PC2, the u_{150} and Z_{850} fields bear a similar relation to the enhanced convection, but all the patterns are shifted ~60° of longitude eastward relative to their counterparts in PC1.

Over the Maritime Continent the strongest MJO-related u_{150} anomalies are observed at the times of peak PC2.

At the 850 hPa level (Fig. 18.6, bottom panel), the pattern for PC1 resembles the response to an isolated equatorial heat source in a shallow water wave equation model shown in Fig. 14.4 but with reversed polarity, which is made up of equatorially trapped $n = -1$ Kelvin and $n = 1$ Rossby modes. The patch of negative Z_{850} anomalies along the equator to the east of the heating is the Kelvin wave signature

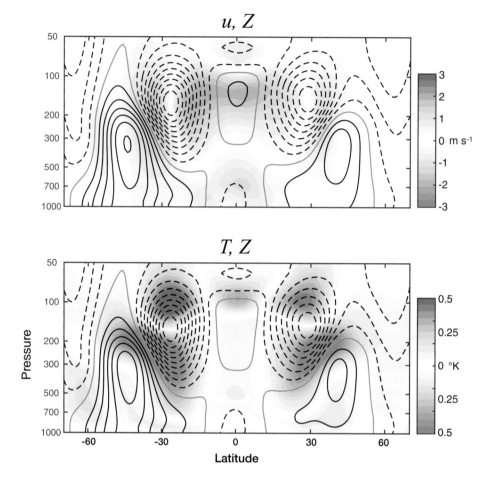

Figure 18.7 Meridional cross sections of zonal wind, geopotential height, and temperature over the Maritime Continent averaged over 120–160°E as inferred by regressing these fields upon standardized -PC2 of $\triangle \chi$, when rain rate is enhanced over the Indian Ocean and suppressed over the western Pacific. Contour interval for Z in both panels 2 m. Colored shading denotes u (top) and T (bottom). Adapted from Adames and Wallace (2014a). © American Meteorological Society. Used with permission.

and the off-equatorial negative anomalies to the west of it is the cyclonic Rossby wave gyre signature. The pattern for PC2 also matches the theoretical solution fairly well over the Indian Ocean sector, where suppressed convection is attended by positive Z_{850} anomalies, but the pattern over the Pacific sector is weaker because SST in the equatorial belt isn't warm enough to support deep convection, and it is also disrupted by the presence of the Himalayas and the Andes.

The corresponding 150 hPa zonal wind pattern shown in the top panel of 18.6 is similar in shape but of opposing polarity. The zonal wind anomalies at this level are flanked by geopotential height anomalies centered along ~28°N/S, which assume a quadrupole configuration, with lows poleward of the westerly wind anomalies and highs poleward of the easterly anomalies. These flanking Rossby waves are not present in the Matsuno or the Gill solutions for the planetary wave response to equatorial mass or heat sources because in those treatments, the background flow is a state of rest.[6] They emerge when a realistic basic state for the upper troposphere, with westerly jets centered at 30°N/S, is included.[7]

In contrast to the rain rate signature of the MJO, which is largely restricted to the Indo-Pacific warm pool, the u and Z anomalies propagate all the way around the equatorial belt. It is evident from Fig. 18.1 that the perturbations tend to lengthen and propagate more rapidly as they exit the warm pool and pass over the cooler sector of the tropics, where they are decoupled from the convection. This explains why the superclusters migrate eastward across the warm pool only about half as fast (5 m s^{-1}) as the phase speed of the MJO estimated by dividing the circumference of the Earth by its ~40 d period (~10 m s^{-1}).

Now let us consider the structure of the MJO in the meridional plane in greater detail. The top panel of Fig. 18.7 shows meridional cross sections at the longitude of the Maritime Continent (120°E) at the time of the minimum in PC2, when the convection is enhanced over the Indian Ocean and suppressed over the western Pacific and strong westerly wind anomalies prevail at upper tropospheric levels over the Maritime Continent. The Z perturbations in the flanking Rossby waves along 28°N/S are in their negative polarity, bracketing the broad belt of upper tropospheric westerly wind anomalies centered on the equator. The other notable tropical feature in the zonal wind field is the patch of equatorial lower tropospheric easterly anomalies. The corresponding MJO-related temperature section shown in

[6] Matsuno (1966); Gill (1980).
[7] Monteiro et al. (2014).

Figure 18.8 MJO-related fields derived by regressing variables upon standardized PC1, whose peak corresponds to the time when the highest rain rates are observed over the Maritime Continent. The fields are smoothed by averaging over a quarter cycle of the MJO. The figure shows equatorial (10°N–10°S) vertical cross sections of temperature, contoured, and the corresponding zonal and vertical mass transports $(\rho u, \rho w)$ indicated by vectors, the same in all three panels, superimposed upon (top) $\rho \nabla \cdot \mathbf{V}$, (middle) $\rho \partial u/\partial x$ and (bottom) $\rho \partial v/\partial y$, all indicated by colored shading. The color bar applies to all three panels. The longest zonal mass flux vector corresponds to 0.6 kg m^{-2} s^{-1}, and the longest vertical mass flux vector to 7.5 × 10^{-4} kg m^{-2} s^{-1}. The contour interval for temperature is 0.05 K; positive contours are solid, negative contours are dashed, and the zero contour is omitted. See Adames and Wallace (2014b). © American Meteorological Society. Used with permission.

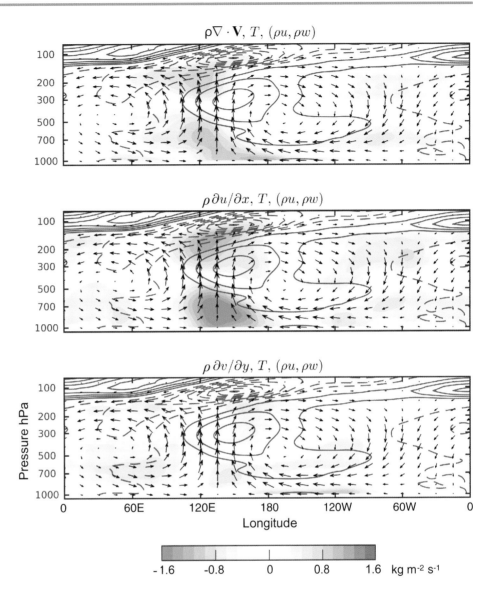

the bottom panel of Fig. 18.7 is dominated by the flanking Rossby wave signature. It also exhibits equatorial anomalies, positive in the 700–150 hPa layer and negative at the level of the equatorial cold point just above 100 hPa.

The MJO-related temperature perturbations indicated by the contours in Fig. 18.8 are weak, with peak amplitudes ranging up to only ∼1°C. The MJO-related overturning circulation in the equatorial plane is thermally direct: that is, the region of enhanced convection is anomalously warm. The temperature anomalies are strongest near the 300 hPa level, with the warmest air slightly to the east of the strongest convection. The lower tropospheric temperature field is barely influenced at all by the MJO.

The corresponding pattern of horizontal divergence, indicated by the colored shading in Fig. 18.8, slopes westward with height. As required for the continuity of mass, the region of planetary-scale ascent lies above a region of low-level convergence. The belt of low-level convergence extends almost all the way to the eastern end of the section and deep-

ens toward the west. West of 120°E the layer of convergence becomes detached from the Earth's surface and the boundary layer flow becomes divergent. Vestiges of the convergent layer extend westward to 60°E at mid-tropospheric levels. That the planetary-scale MJO-related planetary-scale updraft slopes upward toward the west is primarily due to the diffluence of the meridional wind component (i.e., $\partial v/\partial y$ in the expression for horizontal divergence) in the equatorial belt, shown in the bottom panel of Fig. 18.8. The frictional drag force acting on the easterly (Kelvin wave-related) flow contributes to the boundary layer convergence. Similar considerations apply to the shallow layer of diffluence in the westerly flow on the western side of the rain area.

18.3 THE MJO-RELATED MOISTURE BUDGET

As noted above, by virtue of its slow phase speed while passing over the warm pool, the MJO qualifies as a moisture

Figure 18.9 Distribution of
vertically integrated (1000–100 hPa)
vertical velocity $\{\omega\}$, indicated by
colored shading in all three panels, at
the time of maxima in PC1 of $\triangle\chi$,
superimposed upon (top)
column-integrated water vapor W
(contour interval 0.4 mm); (middle)
rain rate P (contour interval
0.25 mm d^{-1}); and (bottom) $\partial W/\partial t$
(contour interval 0.25 mm d^{-1}).

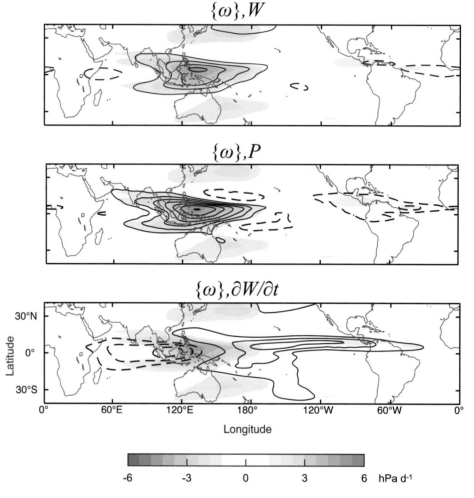

mode, in which WTG balance prevails and the release of
latent heat in moist convection exerts a profound influence
upon the dynamics. The smallness of the MJO-related temperature perturbations in Fig. 18.8 and the tight coupling
between the rain rate, velocity potential, and horizontal wind
fields is consistent with this interpretation. Consideration of
the moisture balance should thus be helpful in understanding
why the phase propagation in the MJO is eastward.

The MJO signature in column-integrated water vapor W
defines the envelope in which deep convection can occur. It
is mainly determined, not by the underlying SST pattern,
as in the seasonally varying climatology and in ENSO,
but by water vapor transports associated with the MJO-
related, planetary-scale circulation and precipitation fields.
The strong correspondence between MJO-related column-
integrated water vapor W and the fields of rain rate P
and vertically averaged vertical velocity $\{\omega\}$ is shown in
the top and middle panels of Fig. 18.9. The well-defined
pattern of $\partial W/\partial t$, with extensive, zonally elongated patches
of moistening to the east of the region of enhanced rain rate
and drying to the west of it (bottom panel), accounts for the
eastward phase propagation of the MJO. That the patch of
drying along the equator protrudes into the rain area from
the west while moistening is still occurring along 10°N/S
accounts for the distinctive swallowtail shape of the patch of

enhanced rain rate in the MJO. (See also Fig. 15.7 and the
accompanying discussion).

It is shown in Appendix D that under WTG balance, the
equation of the conservation of water vapor, expressed in
terms of the tendency of latent heat, can be written in the
form

$$L\frac{\partial W}{\partial t} = -\{\mathbf{V}\cdot\nabla Lq\} - \left\{\omega_c\frac{\partial}{\partial p}\mathrm{MSE}\right\} - \left\{(\omega_{rc}+\omega_{res})\frac{\partial Lq}{\partial p}\right\}$$
$$+ F_{LH} + F_{SH}, \tag{18.2}$$

where q is specific humidity, MSE is moist static energy,
F_{SH} and F_{LH} are the surface fluxes of sensible and latent
energy, and L is the latent heat of condensation and ω_{rc}
and ω_{res} are the cloud-related and residual radiative heating,
respectively. In order to produce a positive tendency in W,
the water vapor transport $-\{Tr(q)\}$ needs to be in excess
of what is required to account for the observed rain rate P.
The excess transport is given by the sum of the three terms
in brackets. The vertical advection is partitioned into three
terms one involving ω_c, which balances the condensation and
convective heating rate Q_c (see Appendix D) and the other
two terms involving vertical velocity components that balance the radiative heating rate. The two remaining terms in
Eq. (18.2) are the boundary fluxes of latent and sensible heat.

Figure 18.10 Water vapor balance in the MJO at the time of peak PC1 of $\triangle \chi$, as inferred from regression analysis. In all panels the distribution of column water vapor W is indicated by contours at increments of 0.25 mm and the various terms in the budget equation (18.2) divided by L are indicated by colored shading. The top panel shows evaporation in units of mm d^{-1} and the second shows tendencies resulting from the water vapor transport over and above that required to supply the moisture required to account for the MJO-related variations in rain rate. The three remaining panels show the contributions of horizontal advection of moisture, the vertical advection of moist static energy by the component of the vertical velocity that balances the condensation and convective heating, and the vertical advection of moisture by the component of the vertical velocity that balances the cloud-related diabatic heating, as explained in Appendix D.4.

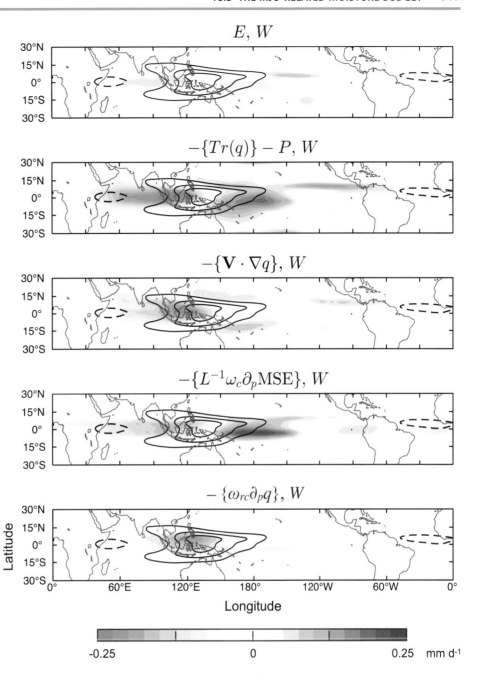

Figure 18.10 shows the leading terms that appear on the right-hand side of Eq. (18.2) regressed upon PC1. They are divided by L so they can be compared with the column water vapor tendency $\partial W / \partial t$ shown in the bottom panel of the previous figure. Evaporation $E = F_{LH}/L$, shown in the top panel, is enhanced to the west of the region of enhanced W and P and suppressed to the east of it. If the moisture budget of the MJO were dominated by this term, the rain area would propagate westward rather than eastward. The sensible heat flux term F_{SH}/L (not shown) is smaller than $E = F_{LH}/L$ over the tropical oceans by about a factor of five. The vertical advection term involving the descent that balances the clear sky radiative cooling $\omega_{res}\,\partial q/\partial p$ is likewise negligible in the moisture balance of the MJO. The second panel in Fig. 18.10

shows the excess transport over and above that required to account for the MJO-related variations in rain rate. It exhibits a robust pattern, much stronger than the one in evaporation, with moistening to the east of the rain area and drying to the west of it, and thus accounts for the observed eastward propagation.

The remaining panels of Fig. 18.10 show the contributions from each of the three circulation-related terms in Eq. (18.2). (The term involving ω_{res} in Eq. (18.2) is not included because it is much more spatially uniform that the other terms and thus has much less influence upon the circulation.) The horizontal advection term promotes drying to the west of the region of enhanced rain rate. In this region the low-level westerly wind anomalies carry drier air

Figure 18.11 (Top) Isometric showing relative humidity (colored shading), the overturning mass circulation (vectors), and specific humidity (contour interval 0.025 g kg^{-1}). The gray horizontal plate depicts the Z_{1000} anomalies and the white arrows depict the boundary layer flow. M and D correspond to the regions in which the moisture tendency exhibits positive and negative extrema, respectively. (Bottom) Meridional cross sections showing relative humidity (shading), the meridional mass circulation (arrows), and the zonal mass flux (contour interval 0.15 kg m^{-2} s^{-1}). All fields represent averages over the half-cycle of the MJO in which PC1 of $\Delta\chi$ is positive, shifted in longitude so that the centroid of the rain areas coincide, and only the equatorially symmetric component is shown. Adapted from Adames and Wallace (2015). © American Meteorological Society. Used with permission.

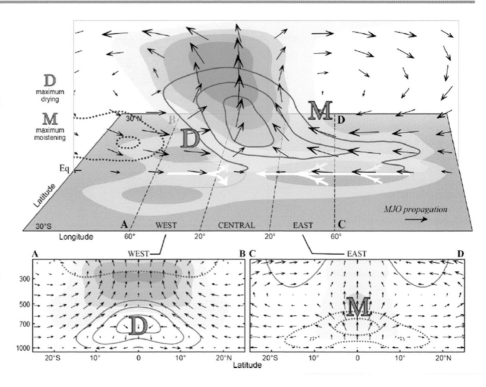

eastward while the meridional winds on the western flanks of the cyclonic Rossby wave gyres carry drier subtropical air equatorward. Upon close inspection, there is also a hint of off-equatorial moistening by the poleward flow to the east of the region of enhanced rain rate. A quarter cycle earlier, the Rossby wave anticyclones over the Maritime Continent advect moisture westward into the Indian Ocean, promoting the development and eastward propagation of deep convection (not shown). The vertical advection of MSE by ω_c is responsible for most of the moistening to the east of the rain area. It is dominated by relatively shallow convection driven by boundary layer mass convergence associated with the low-level MJO-related easterly wind anomalies along the equator. The vertical advection of moisture by the cloud-related, radiatively forced velocity field ω_{rc} (i.e., the ascent that balances the radiative heating in the anvils of the deep convective clouds), shown in the bottom panel, moistens the air directly over the region of enhanced rain rate. Although it is the smallest of the three largest transport terms in Eq. (18.2) in terms of spatially averaged r.m.s. amplitude and it does not contribute to the eastward phase propagation of the MJO, it is instrumental in maintaining moisture modes. Eliminating it in numerical simulations results in a substantial weakening or complete elimination of the MJO.[8]

The dynamical processes that contribute to the transport can be inferred from the three-dimensional representation of the MJO-related moisture, wind, and vertical velocity fields shown in Fig. 18.11. The isometric in the top panel shows the overturning circulation in the equatorial plane feeding moist, boundary layer air into the rain area from the east and lifting

it, deepening the moist layer toward the west. Within the rain area itself, a plume of humid air extends all the way up to the cold point tropopause. This moist, westward flowing air stream originates in the Kelvin wave response to the diabatic heating. Underneath this sloping, planetary-scale stream of ascending moist air, a narrow tongue of drier air, centered on the equator, tunnels into the rain area from the west.

The meridional cross sections in the bottom panels of Fig. 18.11 show the associated meridional overturning circulations. As noted above, the low-level easterly flow to the east of the rain area (right panel) is confluent. The air tunneling into the rain area from the west is dry because it originates in the subtropics and is advected equatorward in the low-level equatorward flow on the western flanks of the Rossby wave gyres to the west of the rain area. Its dryness is reinforced by subsidence within the low-level westerly air stream. The pervasive "swallowtail shape" of the patch of enhanced rain rate over the Maritime Continent in Figs. 18.2 and 18.9 derives from this narrow tongue of dry air. That it is also discernible in the middle panel of Fig 15.6 indicates that it is a robust feature of the planetary wave signature that develops in response to a patch of enhanced equatorial rain rate. The upward mass flux in the sloping updraft, in turn, drives the planetary-scale circulation that feeds it.

It is notable that although most of the MJO-related spectral peak lies within the blue-shaded moisture mode domain in Fig. 15.7, the zonal wavenumber $k = 1$ component falls within the unshaded domain. Zonal wavenumber 1 is strongly influenced by the circumglobal propagation of the MJO. As noted in Section 18.2, its rate of eastward phase propagation is substantially higher over the segment of the tropics extending from the central Pacific eastward to

[8] Jiang et al. (2020); Adames and Maloney (2021b).

Figure 18.12 MJO-related anomalies at the 100 hPa level at the time of peak RMM2, when the convection is enhanced over the western Pacific and suppressed over the Indian Ocean. The contours in all five panels correspond to 100 hPa temperature anomalies (solid contours positive, dotted contours negative; contour interval 0.1 °C). Colored shading refers to the variables indicated in the titles. "Clouds" refers to the fractional coverage by high, thin clouds sensed by the CALIPSO satellite. Adapted from Figs. 5, 7, and 9 of Virts and Wallace (2014). © American Meteorological Society. Used with permission.

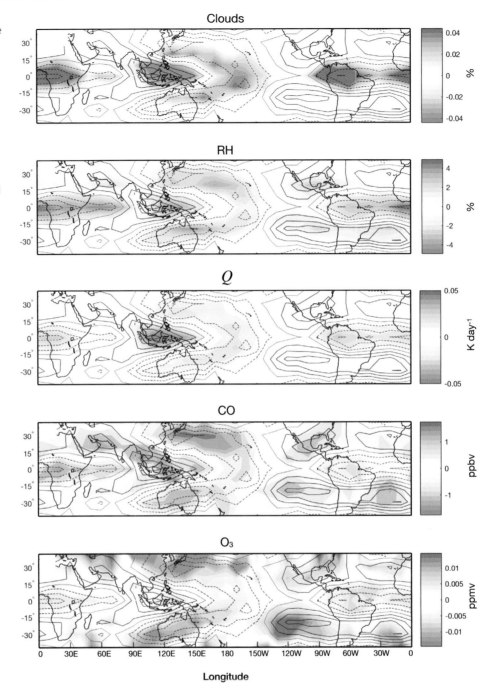

Africa than the 5 ms⁻¹ value observed over the warm pool. In this drier, cooler segment of the tropics the effective static stability is higher, resulting in a stronger restoring force for vertically displaced air parcels and consequently a higher gravity and Kelvin wave phase speed.

Exercise

18.7 (a) Would MJO-like variability exist on an aqua-planet? (b) Would it exist in the absence of wind-induced SST anomalies?

18.4 MJO-RELATED PERTURBATIONS NEAR THE TROPOPAUSE

By lifting and depressing the tropical tropopause layer (TTL) in the vicinity of the equatorial cold point, the MJO perturbs the coverage of high, thin cirrus clouds, relative humidity, and the concentrations of chemical tracers, such as CO. Figure 18.12 shows the signature of the MJO in these variables superimposed upon 100 hPa temperature. At these levels, the MJO-related temperature perturbations are induced adiabatically by the vertical velocity field and damped by radiative cooling (i.e., $Q' \approx -\alpha T'$, where the primes (')

refer to the MJO-related anomalies and α is a positive constant) so temperature can be viewed as a surrogate for diabatic heating rate, which corresponds to the vertical velocity in isentropic coordinates. Colder, ascending air tends to be cloudier, more humid, richer in CO and other tropospheric tracers, and lower in ozone than warmer, subsiding air. At the 100 hPa level the MJO-related overturning circulation is thermally indirect, with rising of colder air and sinking of warmer air, driven from below by the mechanical work term $-\omega\Phi$ (Fig. 6.4) in the same manner

as the climatological mean equatorial stationary waves at these levels.

It is evident from the equatorial longitude–height section shown in Fig. 18.13 that the MJO-related temperature perturbations above the 150 hPa level slope eastward with height. That they are eastward propagating implies that they are downward propagating, like those in Kelvin waves. However, because they are convectively coupled rather than freely propagating, their period is much longer than that of the free Kelvin wave mode documented in Section 10.2.

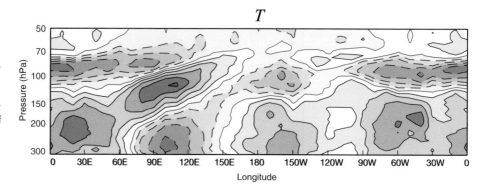

Figure 18.13 MJO-related temperature perturbations in the equatorial belt (5°N–5°S) at the time of peak RMM2, when convection is enhanced in the western Pacific sector relative to time mean. Contour interval 0.1°C. Adapted from Fig. 4 of Virts and Wallace (2014), courtesy of Katrina Virts. © American Meteorological Society. Used with permission.

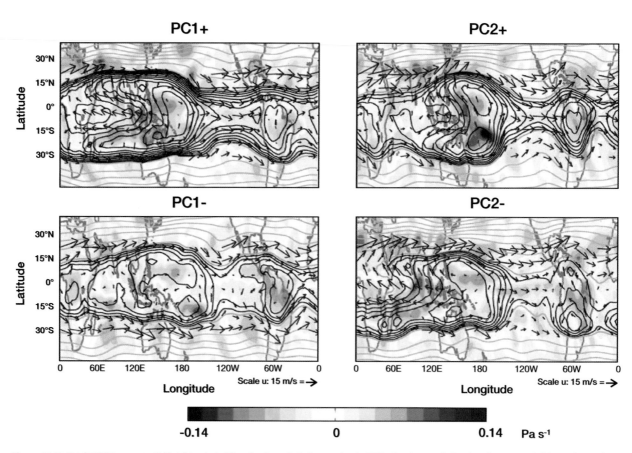

Figure 18.14 Total 150 hPa geopotential height and wind (i.e., the climatological mean plus the MJO-related anomalies) and total ω_{300} sampled during the quadrants of the MJO cycle centered at the times of maxima and minima of PC1 and PC2, as indicated, and only on the days when $\sqrt{(PC1^2 + PC2^2)} > 1$. Contour interval 100 m for the black contours and 20 m for the blue contours; the lowest of the blue contours corresponds to 14 200 m. The analysis in this figure is based on the RMM indices in Wheeler and Hendon (2004). From Fig. 6.2 of Dima (2005).

Figure 18.15 Time–latitude section showing MJO-related variations in $[u^*v^*]$ (colored shading) and $[u]$ (contours) at the 150 hPa level as they evolve through a full MJO cycle. Contour interval for $[u]$ 0.25 m s^{-1}. The small red and blue circles denote the times of maximum and minimum AAM, respectively, and the white arrows indicate the direction of the momentum transport.

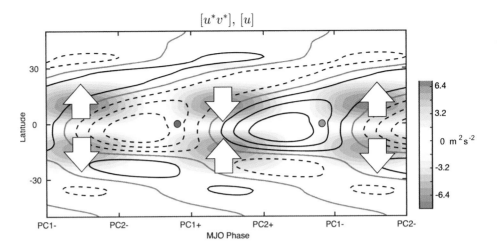

18.5 ZONALLY SYMMETRIC MJO-RELATED PERTURBATIONS

Like ENSO, the MJO modulates the amplitude and structure of the equatorial planetary waves in the total wind field. Around the time of the maximum of PC1 in the $\Delta\chi$ field, when the region of enhanced convection is propagating eastward across the Maritime Continent, the MJO-related equatorial zonal wind anomalies come into alignment with the u^* perturbations in the climatological mean planetary waves and the stationary waves in the total wind field become anomalously strong. The strengthening of the planetary waves is particularly noticeable in the upper troposphere around the 150 hPa level, as shown in the top left panel of Fig. 18.14. It is in this phase of the MJO that the EP flux out of (and the eddy transport of westerly momentum into) the equatorial belt by the total wind field (i.e., climatological mean stationary waves plus MJO-related perturbations) are at their strongest. The anomalous equatorward momentum transports converge within the equatorial belt, inducing an eastward acceleration, as shown in Fig. 18.15. The newly formed patch of anomalous equatorial westerly wind anomalies is flanked by easterly anomalies, the residual from the previous half MJO cycle. What causes the successive easterly and westerly belts to propagate poleward with time is not entirely clear.

The eddy-induced perturbations in the $[u]$ field induce zonally symmetric MJO-related perturbations in the other fields by way of the mean meridional circulations, as discussed in Chapter 7. Figure 18.16 shows the mean meridional circulations, together with the temperature perturbations that they induce. To validate the mean meridional circulation fields in the reanalysis, observed fields of remotely sensed sub-visible cirrus clouds and specific humidity are shown in the other panels. The $[T]$ field oscillates in concert with the $[u]$ field so as to be in thermal wind balance with it. In the "westerly phase" of the MJO, the upper troposphere is anomalously warm and dry, owing to the induced subsidence, and the lower stratosphere is anomalously cold and moist, owing to the induced ascent. The mean meridional circulation is thermally indirect, inducing

the $[T]$ anomalies at levels above 150 hPa and maintaining them in the presence of radiative damping at the expense of the zonal kinetic energy associated with the $[u]$ anomalies.

The MJO also modulates total atmospheric angular momentum (AAM). Starting around the time of peak PC1, when the region of enhanced convection is propagating eastward across the Maritime Continent, the surface winds over the tropical Pacific exhibit an anomalous easterly component, as shown in Figs. 18.6 and 18.8. The resulting frictional torque at the Earth's surface induces a positive AAM tendency. The associated negative SLP anomalies extend all the way to the Andes (same figure) inducing a pressure torque in the same sense. AAM reaches a peak a quarter cycle later, when the torques reverse sign.[9]

Exercises

18.8 Considering the strength of the poleward flow in the upper branch of the annual mean Hadley cells, is it conceivable that the poleward propagation of successive belts of westerly and easterly wind anomalies in Fig. 18.15 could simply be due to meridional advection?

18.9 How is the zonally symmetric MJO-related variability reflected in the spectra shown in Fig. 18.5?

18.10* Summarize the current status of research on the influence of the QBO upon the MJO.

18.6 MJO TELECONNECTIONS TO HIGHER LATITUDES

The global extent of the MJO is most clearly evident at upper tropospheric levels. Figure 18.17 shows the r.m.s. amplitude of the Z^* perturbations averaged over the 300–100 hPa layer. The flanking Rossby waves along 28°N/S are discernible over the warm pool. Poleward of this latitude there are patches of high amplitude variability suggestive of wave trains emanating from the tropics and dispersing along great

---

[9] Weickmann et al. (1997).

Figure 18.16 Correlation coefficients between zonally averaged values of various fields, as indicated, and the MJO index RMM2, which peaks 2 d before PC2 of $\triangle\chi$. RMM2 peaks when convection is enhanced in the western Pacific and suppressed over the Indian Ocean. Zonally averaged temperature anomalies, as inferred from regression of $[T]$ upon the same index, are shown in every panel (contour interval 0.1 K, positive contours solid, negative contours dashed, zero contour bolded). In the top right panel $[v]$ is indicated by the horizontal component of the arrows, and diabatic heating rate by the vertical component, positive upward. The largest magnitude of the component vectors are $r = 0.38$ for $[v]$ and $r = 0.52$ for $[Q]$. The bottom left panel shows cloud fraction, while the bottom right panel shows specific humidity. Adapted from Virts and Wallace (2014). © American Meteorological Society. Used with permission.

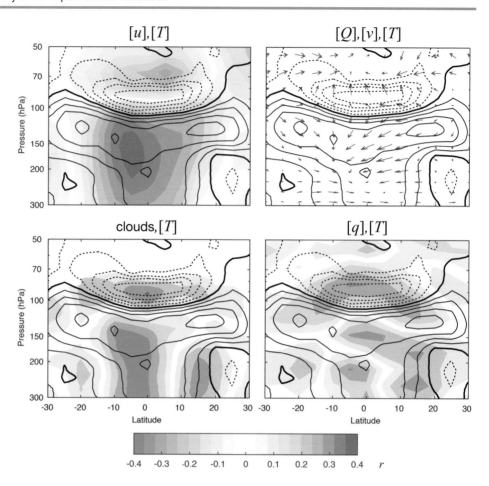

Figure 18.17 Root mean squared (r.m.s.) amplitude of the 100–300 hPa layer-mean Z^*, averaged over the MJO cycle, based on year-round data, superimposed upon the annual climatological-mean 100–300 hPa u (contour interval 10 m s^{-1}; zero contour bolded, negative contour dashed). Adapted from Fig. 11 of Adames and Wallace (2014a). © American Meteorological Society. Used with permission.

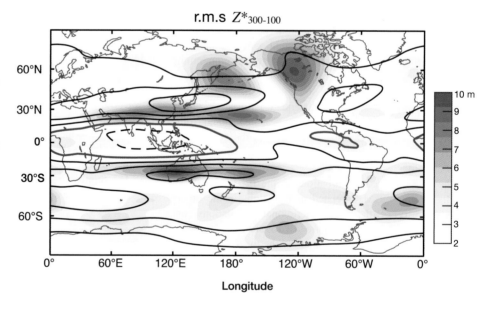

circles. This impression is confirmed by the top panels of Fig. 18.18, which show Z^* averaged over the same layer regressed upon PC1 and PC2 of $\triangle\chi$. The similarity between corresponding patterns for upper and lower levels (compare top and bottom panels in each column), reflects the equivalent barotropic structure of low-frequency geopotential height perturbations at extratropical latitudes, in contrast to the baroclinic structure of tropical motion systems. The dispersion of the MJO planetary-scale Rossby wave signal into the extratropics occurs mainly at upper tropospheric levels, but the circulation is perturbed throughout the depth of the troposphere.

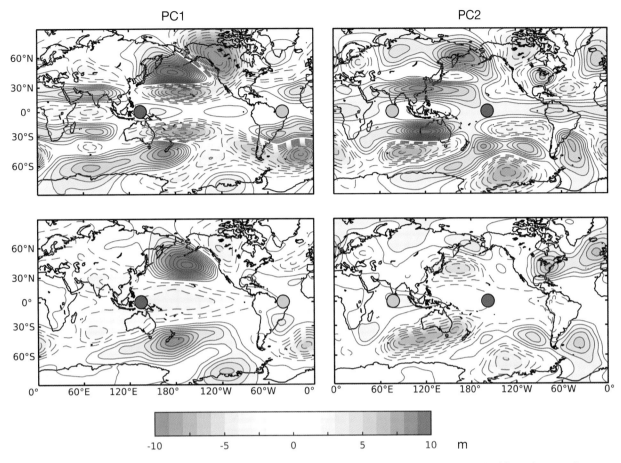

PC1   PC2

-10   -5   0   5   10   m

Figure 18.18 Geopotential height anomalies regressed upon (left) PC1 and (right) PC2 of $\triangle\chi$. (Top) 100–300 hPa layer-averaged and (bottom) 1000-500 hPa layer-averaged. The red and green circles indicate the regions of enhanced and suppressed rain rate, respectively, as in Fig. 18.2. Contour interval 1 m. Adapted from Figs. 7 and 8 of Adames and Wallace (2014a). © American Meteorological Society. Used with permission.

Figure 18.19 Latitude–calendar month sections of MJO-related r.m.s. amplitude in the sector 60°E to the Date Line. (Top) u_{150} (colored shading) and Z_{150} (bold contour 4 m, contours increasing by 2 m). (Bottom) u_{850} (colored shading) and OLR (bold contour 2 W m^{-2}, contours increasing at intervals of 1 W m^{-2} thereafter).

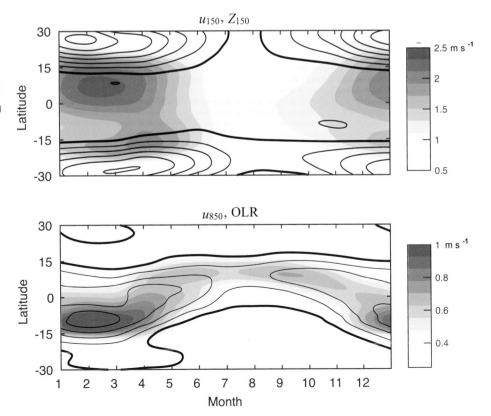

Figure 18.20 Rain rate (colored shading), V_{850} (vectors), and χ_{850} (contours) regressed on linear combinations of PC1 and PC2 of $\Delta\chi$ corresponding to the migration of the MJO's equatorial rain area from west to east across the Indo-Pacific warm pool during the boreal South Asia rainy season JJAS. The sequence spans an interval of about 20 days (1/2 cycle). The longest arrows are $\sim$2 m s^{-1}. Contour interval 10^6 m^2 s^{-2}. Colored circles as defined in the caption of Fig. 18.2.

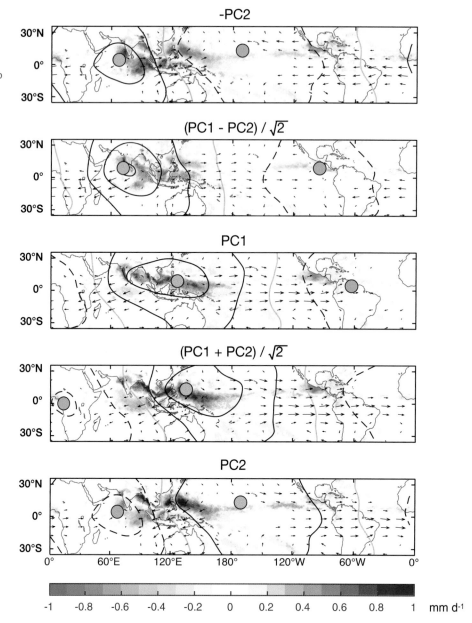

The MJO-related patterns account for only a modest fraction of the variance of the extratropical circulation, but the stronger features in them are statistically significant. The extratropical wave train evolves through the course of the MJO-cycle. A PNA-like pattern that seems to emanate from the flanking Rossby wave to the east of the Maritime Continent dominates the Northern Hemisphere teleconnections at the time of the PC1 extrema. Through its influence on the extratropical circulation, the MJO modulates the frequency of wintertime blocking events over both the North Pacific and the North Atlantic.[10]

The dispersion of the MJO-related wave train from the tropics into higher latitudes is dependent upon the seasonally varying background flow. In the Northern Hemisphere a robust extratropical response to the MJO is observed only during the winter season, when the climatological mean westerlies are strong, and in the Southern Hemisphere it is also stronger during winter than during summer.[11]

18.7 SEASONALITY OF THE MJO

All the figures in the previous sections are based on data for all 12 calendar months. They are representative of the MJO structure, which is much the same throughout most of the year. The notable exception is JJAS, when the circulation over the warm pool is distorted by the South Asian monsoon.

The seasonality of the MJO is revealed by latitude–calendar month sections of r.m.s. amplitude shown in Fig. 18.19. In the u_{150} and Z_{150} fields (top panel) it is

---

[10] Henderson et al. (2016).

[11] See Fig. 4 of Adames et al. (2016).

Figure 18.21 (Top) The Pacific–Japan (PJ) pattern, defined by regressing various fields during JJAS upon the standardized time series of OLR averaged over the rectangular box (10°–20°N, 120°–130°E). (Top) OLR (colored shading), $\mathbf{V}_{850}$ (vectors), and 850 hPa height (contour interval 2 m). (Bottom) The PJ-related u_{850} (colored shading), $\mathbf{V}_{850}$ (repeated), and the climatological mean $\overline{u}_{850}$ (contour interval 2.5 m s^{-1}, zero contour not shown). The longest vector corresponds to ∼3 m s^{-1}.

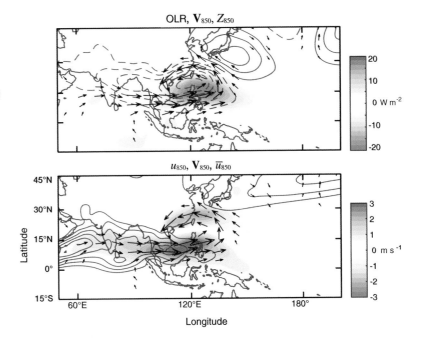

OLR, $\mathbf{V}_{850}$, $\mathbf{Z}_{850}$

u_{850}, $\mathbf{V}_{850}$, $\overline{u}_{850}$

manifested mainly in the amplitude of the perturbations. In the 8 months of the year in which they are present, the flanking Rossby waves are always centered near 28°N/S. They are entirely absent during the monsoon season and the zonal wind perturbations equatorward of them are very weak as well. In contrast, the seasonality in the OLR and u_{850} hPa fields (bottom panel) is manifested mainly in the northward and southward migration of the amplitude maxima between ∼12°N and 12°S, lagging the annual cycle in insolation by 2–3 months.

The seasonality of the OLR and low-level zonal wind perturbations is related to the equatorial asymmetry of the swallowtail pattern whose "wing" (rain band) extends into the summer hemisphere and is accentuated relative to the one that extends into the winter hemisphere. During JJAS, the southern wing virtually disappears and the MJO's rain area is dominated by the WNW–ESE-sloping northern wing. As the wing propagates eastward across a north–south "picket line" of stations or grid points, the rain band propagates northward across it as a diffuse cyclonic shear zone preceded by easterly wind anomalies and followed by westerly anomalies, as shown in Fig. 18.20.[12]

18.8 OTHER PATTERNS OF INTRASEASONAL VARIABILITY IN JJA

During JJAS, there exists another convectively coupled phenomenon that varies on the intraseasonal timescale: the so-called *Pacific–Japan* (PJ) pattern.[13] A suitable index for characterizing its variability is OLR in the rectangular box (10°–20°N, 120°–130°E, centered just to the east of the

Philippines. The OLR field regressed upon this PJ index is shown in the top panel of Fig. 18.21, together with the regression pattern for the 850 hPa wind field. The enhanced convection observed in association with positive values of the index is accompanied by cyclonic low-level circulation anomalies centered to the north of it, with anomalous westerlies across the Philippines, which can be characterized as an extension of the climatological mean low-level jet associated with the South Asian summer monsoon (bottom panel). The PJ-related extensions and retractions of this jet resemble the extensions and retractions of the upper tropospheric, subtropical jet that give rise to the PNA pattern, as described in Section 13.1.2. Just as the PNA pattern feeds on the kinetic energy of the wintertime subtropical jet stream, the PJ pattern extracts kinetic energy from the low-level monsoon jet. Whether this is the primary reason for its existence is not firmly established. In any case, the PJ pattern is no less important than the MJO in modulating summer monsoon rainfall over Southeast Asia on the intraseasonal timescale.

The MJO and the PJ pattern are not perfectly orthogonal in the OLR field and as a result, they are incorporated into a single, paired EOF mode, often referred to as the *monsoon intra-seasonal oscillation* (MISO), when the domain of the analysis is expansive enough to include both phenomena. Northward propagating, so-called "active" and "break" periods of the summer monsoon over India have been attributed to the northward propagating rain bands associated with the MISO. Figure 18.22 shows a sequence of patterns spanning a half cycle associated with the MISO (middle column) decomposed into contributions from the MJO (left column) and the PJ pattern (right column). The PJ pattern is represented by a single spatial pattern that oscillates in place, whereas the MJO and MISO involve a pair of patterns that appear in temporal quadrature with one another.

---

[12] Hartmann and Michelsen (1989).
[13] Nitta (1987); Kosaka and Nakamura (2006).

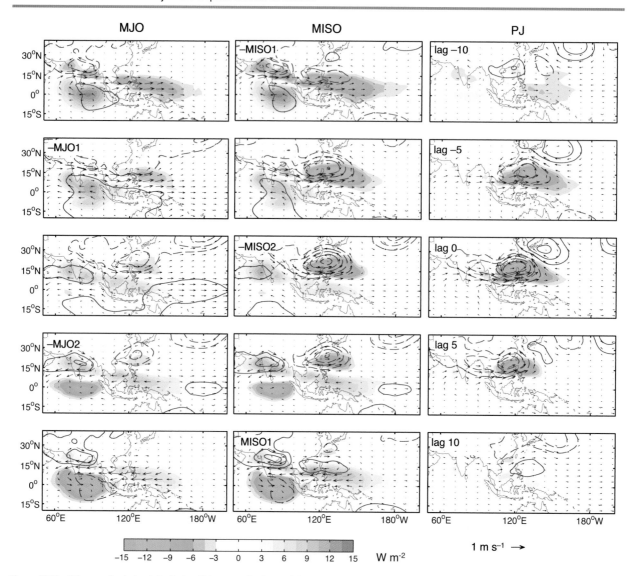

Figure 18.22 OLR anomalies (colored shading) and V_{850} anomalies (vectors) during the South Asian summer monsoon rainy season JJAS at approximately 5 d intervals (top to bottom) spanning a half cycle of three different tropical intraseasonal oscillations. (Left), the MJO as represented by the OMI, a pair of standardized OLR indices (Kiladis et al., 2014). (Middle) the monsoon intraseasonal oscillation (MISO) as represented by standardized PC1 and PC2 of 20–96 filtered OLR within the domain (22.5°S–40°N, 57.5°E–180°E). (Right) The Pacific–Japan (PJ) pattern, as defined by standardized OLR anomalies averaged over the domain (10°–20°N, 120°–130°E). The phases of the three oscillations are adjusted to synchronize them as much as possible.

The northward propagation of the rain bands in the MISO is seen to be entirely attributable to the eastward propagating, tilted rainband of the MJO. When each of these three oscillations is defined by its own index, ∼85% of the variance of the combined MISO indices can be explained by a least-squares best fit based on the two MJO indices and the single PJ index.

18.9 EQUATORIALLY TRAPPED ROSSBY WAVES

The only member of the family of convectively coupled, equatorially trapped planetary waves that varies on the intraseasonal timescale is the Rossby wave (Fig. 14.4, top panel). The other family members vary on shorter timescales,

as will be discussed in Section 19.1. The Rossby wave is an integral part of the MJO signature, but they exist apart from it. When the MJO-related variability is regressed out of the equatorial geopotential, wind, and rain rate fields, an appreciable fraction of the residual variability projects upon Rossby waves with westward phase speeds on the order of 4-5 m s^{-1}, slow enough to be classified as a moisture mode. Its westward propagating signal is most clearly discernible in time–longitude sections and EOFs over the central and western Pacific sector, particularly to the south of the equator, as shown in Fig. 18.23. The associated vertical velocity perturbations modulate the rain rate in the SPCZ and the equatorial zonal wind perturbations force an oceanic Kelvin wave response that perturbs the thermocline depth.

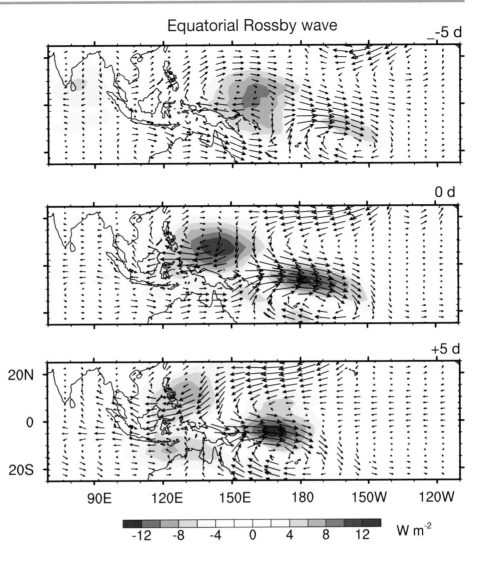

Figure 18.23 Lagged-regression maps depicting an equatorially trapped Rossby (ER) mode, based on extended EOF (EEOF) analysis (see Appendix C) of OLR anomalies over the Pacific sector (140°E–110°W, 10°N–10°S) for the calendar months November–April. Low OLR (blue shading) is indicative of enhanced rain rate. The vector field represents the corresponding horizontal wind perturbations at the 850 hPa level. Scaling is arbitrary. From Gonzalez and Jiang (2019). © American Geophysical Union.

It is instructive to compare the equatorially trapped Rossby (ER) patterns shown in Fig. 18.23 with the MJO in Fig. 18.8, bearing in mind the inverse relationship between OLR and rain rate and that the MJO pattern reflects the structure of waves propagating across the entire width of the warm pool, whereas the ER pattern is limited to the central and western Pacific, on the eastern edge of the warm pool. The MJO and ER patterns are similar in some respects, but the former includes a Kelvin wave component, which is responsible for a large fraction of the moistening to the east of the region of enhanced rain rate over the Western Pacific. In the absence of a strong Kelvin wave contribution, the zonal moisture gradient, which is particularly strong over the central Pacific, with drier air toward the east, comes into play. Rain rate and column water vapor are anomalously high in the westerly phase (Fig. 18.23), when moist air is being advected eastward, amplifying or at least sustaining the waves. For reasons that are not so easily explained, the moistening extends somewhat to the west of the westerlies, resulting in westward phase propagation, as observed. The intraseasonal variability over the Pacific sector tends to be dominated by ER waves during La Niña when the east-to-west moisture gradient, which favors westward phase propagation, is anomalously strong, and by the MJO during El Niño, when it is weak or nonexistent.[14]

Exercises

18.11 How are MJO-related patches of enhanced convection affected by the land–sea geometry of the Marine Continent?

[14] Gonzalez and Jiang (2019).

19 Day-to-Day Variability of the Tropical Circulation
with George Kiladis

Tropical weather systems with timescales shorter than a few weeks can be divided into three broad categories: equatorially trapped waves, off-equatorial waves, and tropical vortices. The equatorially trapped waves are similar in horizontal structure to the free stratospheric Kelvin, MRG, and IG wave modes described in Section 10.2, but they are convectively coupled; that is, they exhibit distinctive rain rate signatures revealing the existence of overturning circulations. The other property that marks them as convectively coupled is that their phase speeds are slower than those of the free modes. Over most of the tropics they are westward propagating because the background or "steering flow" is from east to west and because of the β-effect. In some sectors, the waves are able to amplify by drawing upon the baroclinicity and/or moisture gradients of the background field, while in others, they appear to be "coasting", advected by background flow. Extratropical baroclinic waves sometimes intrude into the tropics, giving rise to "cold core" cyclonic vortices in the upper troposphere. These three categories are not mutually exclusive. Systems that originate as equatorially trapped waves may become detached from the equatorial waveguide and vice versa, and a cyclonic vortex that forms in a wave trough may take on a life of its own, separate from its parent wave train.

ENSO and the MJO both perturb the planetary-scale environment in which these higher frequency tropical weather systems develop and thereby modulate their frequency of occurrence and their amplitudes on intraseasonal and interannual timescales. Just as the heat and momentum transports by baroclinic waves feed back upon the mean flow in which they are embedded, the horizontal transport and vertical fluxes of moisture, heat, and momentum by these three families of tropical weather systems play a role in shaping the climatological mean flow, as well as the statistics of its interannual and intraseasonal variability. The first two sections of this chapter discuss the various species of equatorially trapped waves and off-equatorial waves. Disturbances that develop in the troughs of upper level baroclinic waves originating in midlatitudes and dispersing into the tropics are discussed in the third section.

19.1 EQUATORIALLY TRAPPED WAVES

The reader was introduced to equatorially trapped stratospheric planetary waves back in Section 10.2. The phase speeds of these modes are in close agreement with the theoretical dispersion relationships ((Fig. 10.8) and their horizontal structures (Figs. 10.10 and 10.11) closely resemble the solutions of the linearized shallow water wave equation model on an equatorial beta-plane, and they exhibit distinctive signatures in space–time spectra of dynamical fields (Fig. 10.9). The stratospheric waves are "free" waves that can develop in response to a spectrum of tropospheric diabatic heating perturbations that are not necessarily organized in any particular way.[1] These same linear solutions are relevant to the interpretation of convectively coupled, equatorially trapped waves in the troposphere.[2]

Figures 19.1 and 19.2 show time–longitude sections of rain rate along the equator (2.5°S–7.5°N), as inferred from satellite imagery; details are given in Appendix F. The first overlaps with the section shown in Fig. 18.1. The same two MJO-related rain areas are evident, but in this high resolution version they are embroidered with finer scale features. The section shown in Fig. 19.2, which is based on data for a different year, spans a longer time interval and extends all the way around the equatorial belt. Both sections are crisscrossed by a myriad of mesoscale- and synoptic-scale disturbances, some eastward and others westward propagating. These relatively narrow bands pass through the broad, eastward propagating envelopes that mark the wet phase of the MJO, modulating the rain rate on the day-to-day timescale. That the eastward and westward propagating bands frequently pass through one another without any apparent interaction is an indication that they are well represented by linear dynamics. The eastward propagating bands tend to be spaced about 5 days apart, whereas the westward propagating bands exhibit a wider range of scales, including mesoscale features with recurrence times around 1 and 2 days. The westward bands spaced two days apart are especially prominent near 120°E in Fig. 19.1 during mid-December and near 20°E. There are also diurnal pulses of convection, some of them near the coastlines and mountain ranges of the Maritime Continent.

Figure 19.3 provides a statistical description of these features: two-sided wavenumber–frequency spectra of rain rate as inferred from outgoing infrared radiation (OLR) produced using the procedure described in Section 10.2.4,

---

[1] Holton (1972); Salby and Garcia (1987).
[2] Webster (2020).

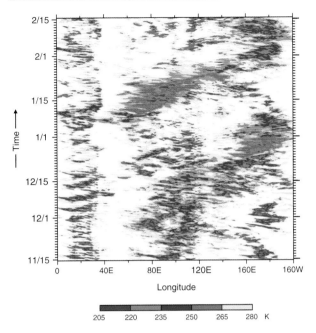

Figure 19.1 Time–longitude section of satellite-sensed brightness temperature (blue shading at 15K increments), a proxy for rain rate, averaged over the eastern hemisphere of the equatorial belt (2.5°S–7.5°N), from November 15, 1992 to February 15, 1993.

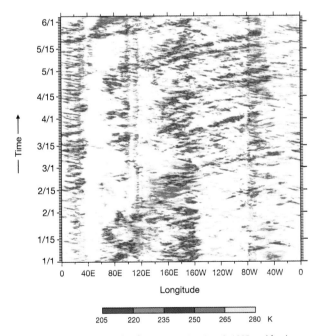

Figure 19.2 As in Fig. 19.1, but from January 1 to June 3, 1987, and for the entire latitude circle.

in which the raw spectrum is normalized by dividing it by an empirically derived, "red noise" background spectrum. The "signal strength" data in the top left panel are the same as in Fig. 15.9. The two-sided spectra are decomposed into equatorially symmetric and antisymmetric components as in Figs. 10.9 and 18.5. The sections in the left panels are for the entire year, whereas those in the right panels are for

June–August. In the symmetric spectra shown in the top panels, the more prominent features appear to be associated with Kelvin and equatorial Rossby waves and the higher frequency meridional mode $n = 1$, westward-propagating, inertio-gravity waves. In the antisymmetric spectra shown in the bottom panels, the $n = 2$ westward propagating inertio-gravity wave and a continuum of power spanning the range of the $n = 0$ mixed Rossby–gravity (MRG) wave are prominent features. The band of enhanced variance radiating outward from the origin on the right-hand side of the symmetric spectra corresponds to a 15 m s^{-1} phase speed, consistent with the slope of the eastward propagating bands of enhanced convection in Figs. 19.1 and 19.2 and substantially slower than the stratospheric Kelvin waves. This is the signature of the convectively coupled Kelvin wave. The MJO is clearly discernible in the equatorially symmetric spectra. Other features of interest will be identified later on.

The similarity between the features in the wind and rain rate spectra and the strong coherence between the wind and rain rate time series[3] testifies to the strong coupling between convection and dynamics in the equatorial belt. The space–time distribution of rain rate within the inner tropics on the day-to-day timescale is largely controlled by these equatorially trapped modes. The precise physical processes that determine the phase speed of the coupled waves is still an area of active research, but it is becoming increasingly clear that the influence of moist thermodynamics upon the dynamics is greatest for the waves with the slowest phase speeds, as discussed in Section 15.3. The latent heating due to convection tends to cancel the adiabatic cooling in regions of ascent, reducing the effective static stability relative to what it would be in a dry atmosphere.[4] Reduced static stability, in turn, results in a slower phase speed of the coupled waves relative to their dry (free mode) counterparts.

In the following subsections, we will examine the structure of the various equatorially trapped planetary waves by filtering the data so as to isolate the various space–time spectral features in Fig. 19.3, and then projecting (linear regressing) the various kinematic and thermodynamic fields onto the appropriate reference time series. Equatorial Rossby waves, which vary on the intraseasonal timescale, were considered in Section 18.9. Here we consider higher frequency Kelvin waves, MRG waves, and IG waves.

19.1.1 Kelvin Waves

Convectively coupled Kelvin waves are the most prominent of the equatorially trapped modes. As can be inferred from the time–longitude sections shown in Figs. 19.1 and 19.2, they are most active over the Indian and especially the Pacific Ocean sectors and occur both within and outside of the broader MJO-related rain bands, modulating daily rain rates. To isolate them, we perform EOF analysis on

[3] Wheeler et al. (2000).
[4] Kiladis et al. (2009).

Figure 19.3 Two-sided wavenumber–frequency spectra of OLR (left panels) based on data for all seasons; (right panels) for JJA only. The spectra are averages of spectra computed latitude by latitude from 15°S to 15°N. The top panels show spectra for the equatorially symmetric component and the bottom panels for the equatorially antisymmetric component. The contours indicate the ratio of the power in the computed spectrum to that in a smoothed red noise background spectrum. Contour interval 0.1, contours and shading begin at 1.1, above which the signal is statistically significant at the 95% level. Dispersion curves are shown for the $n = -1$ Kelvin wave, the $n = 1$ equatorial Rossby wave, the $n = 1$ and $n = 2$ inertio-gravity waves, and the $n = 0$ MRG wave, plotted for an equivalent depth of 25 m. The boxes represent regions of the wave–number-frequency domain assigned to westward propagating IG (WIG) waves (Section 19.1.3) and off-equatorial waves (TD, Section 19.2.1).

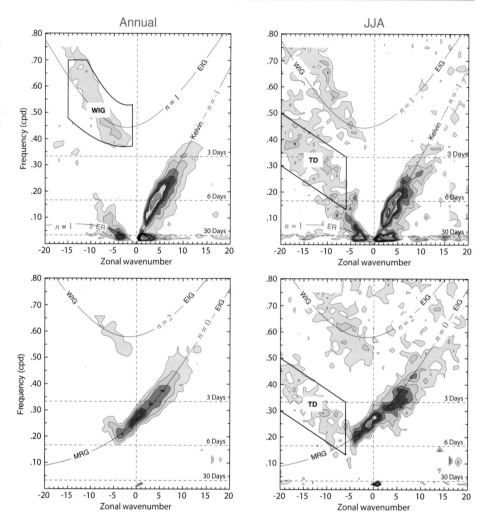

tropical (20°N–20°S), 2–25 day, eastward-filtered OLR for the months JJA, when they are most prominent. The two leading EOFs account for comparable fractions of the variance and their PC time series vary in quadrature with one another, from which it can be inferred that they represent a propagating signal. The PCs exhibit spectral peaks with periods ~5 d, consistent with the spectral peak corresponding to the Kelvin wave in Fig. 19.3. Hence, the time history of the Kelvin wave activity over the Pacific sector should be well-represented by these PCs and their preferred spatial structure well-represented by the corresponding EOFs and other maps and vertical cross sections generated by regressing the various fields upon the PC time series.

Figure 19.4 shows OLR and 850 hPa geopotential height and wind anomalies regressed upon PC1 for lags of -3, 0, and +2 days. The wind and height fields are dominated by the equatorially symmetric component. The u and Z anomalies occur in phase, as in the canonical Kelvin wave structure shown in the top panel of Fig. 10.10. The convective signal is less like the canonical one: the OLR anomalies start out centered on the equator over the western Pacific, but as the waves propagate eastward it migrates into the region of

the ITCZ to the north of the equator.[5] The OLR signature of the enhanced convection occurs in phase with the lower tropospheric convergence, $-\partial u/\partial x$. Another departure from the canonical structure is the blocking of the geopotential signal by the Andes (bottom panel), spreading it poleward in both directions in orographically trapped Kelvin waves. The Kelvin wave signal eventually does seep across the Andes barrier, as does the convective signal, while maintaining the same general relationship between their convective and kinematic fields.

The average eastward phase speed of convectively coupled Kelvin waves, as inferred from Fig. 19.4, is ~15 m/s, consistent with the gravity wave speed depicted by the middle sloping line in Fig. 19.3. It lies within the unshaded region of Fig. 15.9 in which the wave dynamics are influenced by coupling with convection, but not to the extent that the moisture variations dominate the moist static energy; that is, they are convectively coupled Kelvin waves but they do not qualify as moisture modes. As noted in Section 10.2, pure Kelvin waves are nondispersive: their phase speed (ω/k) and

[5] Straub and Kiladis (2002).

Figure 19.4 Structure of the convectively coupled $n = -1$ Kelvin waves at the 850 hPa level, constructed by regressing fields upon the leading PC of eastward 2–25 day filtered OLR for JJA between $20°$N and $20°$S. The figure shows OLR (shading, blue negative, indicative of enhanced rain rate), wind (vectors), and Z perturbations (contours), lagged as indicated relative to the PC1 time series. The longest vectors correspond to wind speeds of 4 m s^{-1}. Contour interval 10 m, negative contours dashed, the zero contour omitted.

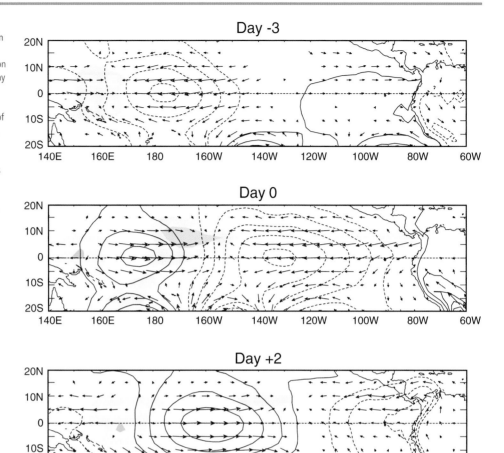

group velocity $(\partial\omega/\partial k)$ are the same, so they hold together as their wave fronts propagate over long distances, as seen in the sections shown in Figs. 19.1 and 19.2.

The vertical structure of the convectively coupled Kelvin wave, documented in Fig. 19.5, bears a strong resemblance to the MJO. As in Fig. 19.4, the OLR minimum, centered at $\sim170°$W, lies in the region of low-level convergence to the east of the westerly wind maximum at the 850 hPa level, where $\partial u/\partial x < 0$. The zonal wind and the corresponding convergence field tilt westward with height so that the strongest convergence at the Earth's surface occurs slightly to the east of the deepest convection, as in the MJO (Fig. 18.8). The perturbations in specific humidity shown in the bottom panel of Fig. 19.5 also tilt westward with height like those in the MJO. The tunneling of drier boundary layer air on the western side of the region of enhanced rain rate is not as pronounced as in the MJO because the advection of drier air around the western flanks of cyclonic Rossby gyres is absent. However, the evolution of the convection as the wave passes by is analogous to that in the MJO, with shallow cells giving way to deeper cells and finally to top-heavy, stratiform cells, as implied by upward progressioin of the moist layers from right to left in the bottom panel of Fig. 19.5. As in the MJO, above the 250 hPa level

the waves tilt eastward with height, with zonal wind and geopotential height in phase and temperature leading them by a quarter cycle (Fig. 19.5, top, middle), as in the Kelvin waves shown in Fig. 10.10. The upward dispersing signal of the forced waves is discernible all the way up to the stratopause (not shown). The vertical velocity, indicated by the arrows in the third panel of Fig. 19.5, varies in-phase with rain rate, consistent with the structure of convectively coupled, equatorially trapped waves.

19.1.2 Mixed Rossby–Gravity Waves

Like Kelvin waves, MRG waves are commonly observed in both the stratosphere, where they are free modes (Section 10.2), and in the troposphere, where they are convectively coupled. They are most prominent in the western and central Pacific, in the range of longitudes in which the equatorial dry zone is flanked by the ITCZ to the north and the South Pacific Convergence Zone (SPCZ) to the south.

Motivated by the strong MRG peak on the westward side of the antisymmetric spectrum shown in the bottom panels of Fig. 19.3, EOF analysis is performed on $20°$S–$20°$N, 2–10 day westward-filtered OLR. The MRG wave in the Pacific

Figure 19.5 Vertical structure of convectively coupled Kelvin waves in the equatorial plane. The top panel shows the longitudinal profile of OLR along 7.5°N. The three panels below it show zonal wind u, temperature T, and specific humidity q perturbations, as indicated, on Day 0 at the time of the peak in PC1, which corresponds to Day 0 in the previous figure. Colored shading indicates u, T, and specific humidity q perturbations, gold positive and blue negative. Contour intervals 0.2 m s^{-1} for u, 0.05 K for T, and 0.4 g kg^{-1} for q. The zero contours are omitted. The heavy arrows in the third panel highlight the regions of ascent and subsidence.

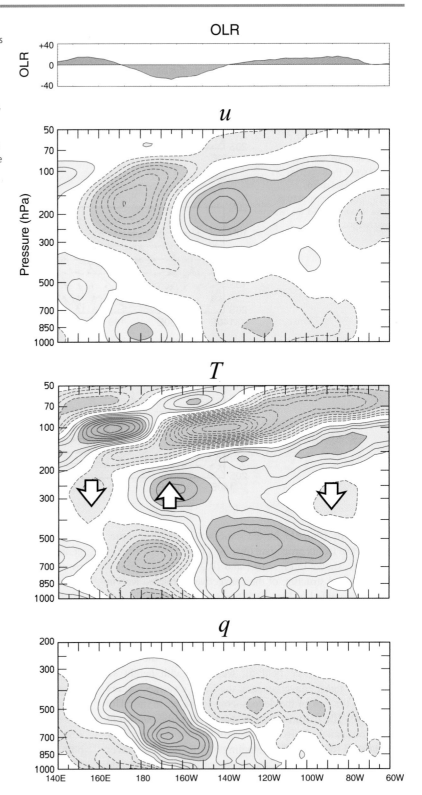

sector emerges as the leading set of paired modes. Figure 19.6 shows the fields derived by regressing OLR and 850 hPa wind onto the PC1 time series. The wind perturbations in the waves in the central Pacific assume the form of westward propagating gyres centered on the equator, like their stratospheric, free mode counterparts (Fig. 10.11). However, the convectively coupled waves track west-northwestward as they move over the western Pacific warm pool and eventually cease to be equatorially trapped MRG waves. As in the canonical MRG divergence pattern, the OLR signal over

Day -1.5

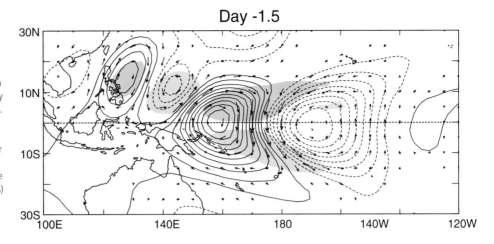

Figure 19.6 Horizontal structure of convectively coupled $n = 0$ mixed Rossby–gravity (MRG) wave constructed by regressing fields upon the leading PC of westward 2–10 day filtered OLR between 20°N and 20°S. The OLR perturbations are indicated by colored shading: blue denotes negative values and the threshold for shading corresponds to the 95% significance level. Also shown are the corresponding 850 hPa wind (vectors) and streamfunction (contours) perturbations (top) 1.5 d before and (bottom) at the time of the peak in PC1 as indicated. The longest arrows correspond to wind perturbations of 2 m s^{-1}. Contour interval 1×10^5 m^2 s^{-1}, negative contours are dashed, and the zero contour is omitted.

Day 0

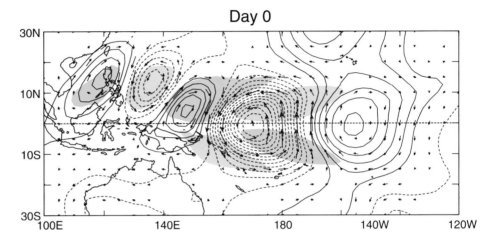

the central Pacific is antisymmetric about the equator; that is, when convection is enhanced in the ITCZ it tends to be suppressed in the SPCZ at the same longitude and vice versa. Except in this transition zone in the western Pacific, the zonal wavelength is ~8500 km and the westward phase speed ~25 m s^{-1}, comparable to that of the stratospheric waves, whose propagation characteristics are governed by dry dynamics. Hence, despite the prominent signature of the MRG wave in the rain rate spectrum (Fig. 19.3, bottom panels), moist thermodynamics has relatively little impact in slowing its phase speed. The same is true of convectively coupled IG waves discussed in the next subsection.

Like the OLR field, the lower tropospheric divergence and temperature fields exhibit odd symmetry about the equator and are out of phase: regions of poleward flow are relatively warm and the flow within them is convergent, favoring enhanced convection (low OLR) and vice versa. A robust feature of the observed waves that is not present in the linear solutions is the poleward tilt of the features in the OLR field with longitude and the associated tilts in the geopotential height and wind fields, which give rise to a poleward transport of westerly momentum. As the waves propagate westward into the warm pool region, their northern centers become progressively more dominant and

the waves in the OLR and Z fields come into alignment, with enhanced rain rate in the cyclonic gyres.

In equatorial longitude–height sections along the equator shown in Fig. 19.7, the MRG waves in the central Pacific exhibit a pronounced eastward phase tilt with height from ~700 hPa up to around the 150 hPa level and a reversed tilt above that level, opposite to that in the eastward propagating Kelvin wave in Fig. 19.5. The stratospheric tilt is in the same sense as the free MRG waves at the 50 hPa level (Fig. 10.11). Hence, at a fixed location, the waves exhibit upward phase propagation at tropospheric levels and downward phase propagation at stratospheric levels. As the waves propagate west-northwestward into the warm pool, the vertical tilt of their axes virtually disappears, and the waves assume a vertical structure characterized by gyres of opposing polarity in the lower troposphere and in the upper troposphere, the upper one peaking around the 150 hPa level.

19.1.3 Inertio-Gravity Waves

In time–longitude sections at tropical latitudes, the convective activity is frequently modulated by westward propagating convective disturbances with a phase speed ~20 m s^{-1}

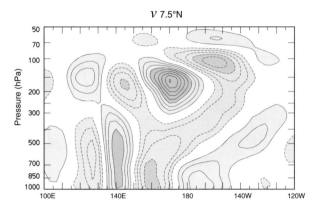

Figure 19.7 Vertical cross section of meridional wind v at 7.5°N in convectively coupled $n = 0$ mixed Rossby–gravity (MRG) waves, constructed by regressing these fields upon the leading PC of westward 2–10 day filtered OLR between (20°N–20°S). Solid contours and gold shading indicate northward flow. Contour interval 0.2 m s^{-1}.

and a period of ∼2 days, some of which are visible in Fig. 19.1. The associated patches of enhanced rain rate, which tend to be spaced about 4000 km apart, are often swallowtail shaped, as in the MJO but facing in the opposite direction, with narrow wings that assume the form of squall lines. They are discernible in the wet phases of the MJO and other low frequency planetary waves. These disturbances correspond to the westward propagating inertio-gravity waves in the spectra shown in the top left panel of Fig. 19.3. When the OLR data are filtered so as to retain the zonal wavenumbers and frequencies in the box labeled "WIG" in Fig. 19.3 and wind and the 200 hPa height and wind fields are regressed upon the time series of the leading PC, the patterns that emerge, shown in the middle and bottom panels of Fig. 19.8, closely resemble the idealized structure of the $n = 1$ westward propagating inertio-gravity wave shown in the top panel. The u, Z, and divergence (a proxy for rain rate) fields exhibit even symmetry about the equator, with u and Z out-of-phase and divergence in quadrature with them. Along the equator the zonal wind perturbations are westward in the highs and eastward in the lows (the opposite of the phase relationship in eastward propagating Kelvin waves) and off the equator they are in the opposite direction (relative to the wave in the geopotential height field) as those in MRG waves.[6] The axes of the waves over the western Pacific exhibit a northwest–southeast tilt; that is, the waves are propagating toward the west-southwest rather than directly westward, perhaps in response to the equatorial asymmetries in the rain rate climatology. Westward propagating IG waves are also observed over North Africa[7] and the Amazon.[8]

What makes the WIG waves so much more prominent than their eastward propagating counterparts with compa-

rable zonal wavenumbers and frequencies is not entirely understood. Numerical simulations of IG waves propagating in a variety of background flows suggest that WIGs are favored even in an atmosphere with a resting basic state, but their prominence is accentuated by the presence of easterly low-level vertical wind shear $\partial u / \partial z < 0$, as is observed over the western Pacific sector.[9]

The apparent preference for a 2-d period is suggestive of a tendency for phase locking with the diurnal cycle in heating, as described by Chen and Houze[10]:

> *Because the life-cycle of large convective system can take up to a day, they leave the boundary layer filled with air of lower moist-static energy and a cloud canopy that partially shades the ocean surface from the sunlight the following day. So the day after a major large convective system, the surface conditions do not favor another round of convection; therefore, convection occurs in neighboring regions unaffected by the previous convective systems* [the finite time required for the boundary-layer to recover] *leads to a tendency for the large systems to occur every other day at a given location. This 2-day* [rhythm] *appears to phase-lock with westward-propagating equatorial inertio-gravity waves of similar frequency.*

This mechanism does not account for the remarkable longitude/time continuity of the rapid westward propagating cloud bands in Figs. 19.1 and 19.2, but it might contribute to the tendency for them to be spaced two days apart from time to time. If the phase locking were pervasive, one might expect westward propagating waves to exhibit a prominent two-day spectral peak analogous to the one for the diurnal tide. That this is not the case is evidence that the westward propagating IG wave exists independently of the contribution of the diurnal forcing and that the phase locking with the diurnal cycle is ephemeral.

19.1.4 Evolution of the Convection in Equatorially Trapped Waves

Analysis of the fields of vertical velocity, temperature, and relative humidity suggest that as the area of enhanced rain rate in the MJO approaches a station, it is first manifested as an increase in shallow convection, which deepens over the course of a few days, and subsequently (after the heaviest rain rates have passed to the east of the station) becomes top heavy, with most of the rain falling from stratiform cloud anvils rather than from the bases of convective updrafts. This inference is based on numerous ground-based observations of cloud types and on radar imagery. Convectively coupled Kelvin waves behave in a similar manner and the same is true of the other equatorially trapped waves as well.[11] This remarkable level of "self-similarity" cannot be explained in terms of the life cycle of individual mesoscale convective

---

[6] See also Takayabu (1994); Haertel and Kiladis (2004); Kiladis et al. (2009).
[7] Tulich and Kiladis (2012).
[8] Mayta and Adames (2021).

[9] Tulich and Kiladis (2012).
[10] Chen and Houze (1997)
[11] Kiladis et al. (2009).

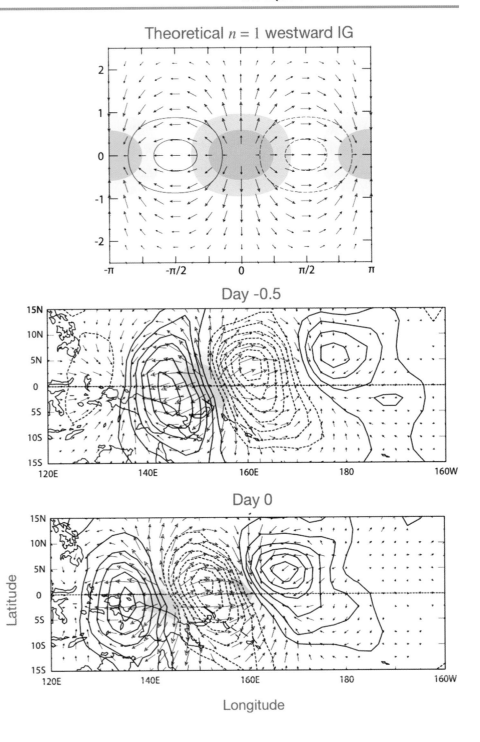

Figure 19.8 (Top) Theoretically derived structure of the $n = 1$ westward inertio-gravity wave (WIG). Colored shading represents divergence (blue positive, gold negative), arrows represent the wind field, and contours indicate the corresponding perturbations in geopotential. (Bottom panels) Observed WIG waves. Perturbations in OLR (shading, blue negative), 200 hPa wind vectors, and Z (contours) associated with the leading EOF of the $n = 1$ WIG-filtered OLR in the domain 20°S–20°N. Contour interval 2 m, with negative contours dashed. The longest vectors correspond to around 2 m s^{-1}.

systems, because the timescales on which they evolve are entirely different.

At any given instant in time, a convectively coupled wave can be viewed as being made up of an ensemble of mesoscale convective systems, each at a different stage of its own life cycle. That the convection in the waves evolves in the same manner that it evolves in mesoscale convective systems implies that the populations of cloud types spend a disproportionate fraction of their lifetimes in the phase of their life cycle that is favored by the large-

scale environment at the time and place that they develop, which in turn, is controlled by the convectively coupled, large-scale waves. The early stages of the convective life cycle tend to last longer in the phase of the convectively coupled that precedes the heaviest rain rates, and the late stages tend to be emphasized in the phase that follows the rain rate maximum.[12]

---

[12] Mapes et al. (2006).

Transient eddy kinetic energy at 850 hPa JJA

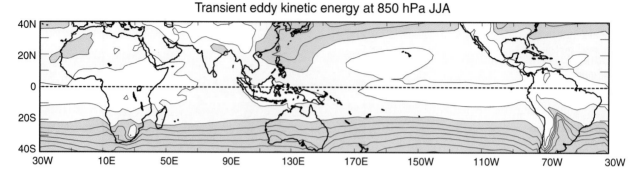

Figure 19.9 2–30 d bandpass filtered transient kinetic energy at the 850 hPa level in JJA. The contours surrounding the unshaded regions correspond to 8 m²s⁻².

Exercises

19.1 Identify three longitudinal sectors in Figs. 19.1 and 19.2 that exhibit diagonal stripes at roughly the diurnal frequency and relate them to the underlying continent–ocean geometry in the equatorial belt (e.g., making use of the base map in Fig. 1.9). Offer a possible explanation of why the diurnal variability is especially prominent in these particular sectors.

19.2 As the Kelvin wave shown in Fig. 19.4 propagates eastward across the Pacific on Day 0 and Day +2, the associated perturbations in rain rate tend to be focused on the ITCZ, rather than along the equator, where the wind perturbations are strongest. Explain why this is the case.

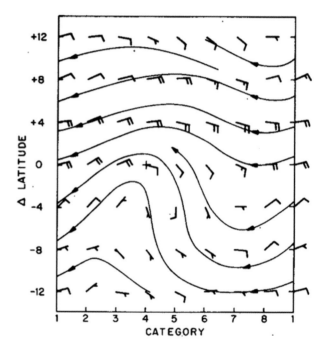

Figure 19.10 Composite of an African easterly wave based on analysis of station data, showing the "inverted trough" structure. Streamlines and wind (in m s⁻¹) at 700 hPa. The "categories" correspond to days relative to the passage of the wave troughs over the stations. Since the waves are zonally propagating and have a wavelength of ∼ 3000 km, they can also be interpreted as increments of longitude about 3° in width. The analyzed meridional profiles for the individual categories are shifted in latitude so that the 700 hPa vorticity maximum in Category 4 coincides with 0° latitude. From Fig. 3(c) of Reed et al. (1977). Published (1977) by the American Meteorological Society.

19.2 OFF-EQUATORIAL WAVES

Despite the existence of equatorially trapped planetary waves and the MJO, the climatology of kinetic energy of 30-day highpass-filtered transients at the 850 hPa level, shown in Fig. 19.9, is largely devoid of equatorial features. Within the tropics, K_E tends to be higher in the summer hemisphere. In the boreal summer JJA the most active regions are the Sahel, longitudes around the Philippines, and the Ganges Valley of northern India. The transients in all these regions are associated with off-equatorial waves that assume a variety of forms, ranging from modified versions of the equatorially trapped waves described in the previous section to highly nonlinear structures with tight cyclonic vortices that sometimes develop into intense cyclones. Many of the off-equatorial waves exhibit distinctive characteristics that reflect the peculiar dynamic and thermodynamic environments of the regions in which they develop.

19.2.1 Easterly Waves

The trade wind regions of the subtropical North Pacific and Atlantic are host to off-equatorial, westward propagating disturbances. So-called *easterly waves* were first identified over the Caribbean in the 1930s. They are manifested on Northern Hemisphere synoptic charts as westward moving "inverted troughs" in the northeasterly trades (i.e., upside down compared to the troughs in the westerlies in the pressure or streamfunction fields), whose passage over a fixed point or slowly moving ship is typically associated with a short interval of rainy weather accompanied by a windshift from northeasterlies to southeasterlies as the trough goes by.

The representation of easterly waves over Africa shown in Fig. 19.10 was constructed by binning the time sequence of 700 hPa wind data at individual stations in accordance with

Figure 19.11 Time–longitude sections of (left) rain rate, as inferred from unfiltered satellite-sensed brightness temperature, and (right) the meridional wind component at the 700 hPa level averaged over 5°N–15°N, from July 1 to September 30, 1996.

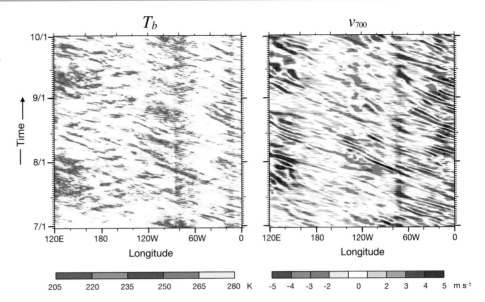

eight phases of the waves (Categories 1–8). The longitudinal span of a full cycle corresponds to roughly 30 degrees of longitude. The highest rain rates occur in categories 2 and 3. The sharpness of the features is reminiscent of case studies of strong easterly waves in the early literature.

Figure 19.11 shows time–longitude sections of rain rate and v_{700} averaged from 5°N to 15°N, extending from West Africa westward across the Atlantic and Pacific. Many of the more prominent westward propagating features originate over Africa along the right side of these sections. They propagate westward at a rate of 8–12 m s^{-1}, much more slowly than the roughly 2-day pulses associated with the convectively coupled inertio-gravity waves discussed in the previous section. Easterly waves that originate over West Africa in Fig. 19.11 can be followed as they propagate across the Atlantic, Central America, and into the Pacific.

The signature of easterly waves shows up clearly in the two-sided spectra of brightness temperature and OLR shown in Fig. 19.3. Since they occur almost exclusively to the north of the equator during northern summer, their power is more or less evenly split between the symmetric and antisymmetric JJA spectra, as a band of enhanced power within a band of westward propagating waves starting at frequencies of ∼0.20 cycles per day, starting at zonal wavenumber $k = 6$ and extending toward higher frequencies and wavenumbers. This distinctive feature inspired the filtering protocol defined by the trapezoidal domain, labeled "TD" in recognition of the fact that the inverted troughs of easterly waves sometimes develop closed cyclonic circulations that may intensify and become tropical depressions (TD) in the pressure field, as will be described in Section 20.1.[13]

The left panels of Fig. 19.12 show the three-dimensional structure of easterly waves in the Pacific sector in late summer JAS, created by regressing various fields upon the leading (paired) PC of TD-filtered OLR in the domain 20°N–20°S, 120°E–120°W. The top panel is a vertical cross section of v along 7.5°N and the patterns in the bottom panel are for the 850 hPa level. The zonal wavelength of the waves decreases from about 80 degrees of longitude over the eastern Pacific to about 30 degrees at the western end of the section, an effect ascribed to "wave accumulation" due to weakening of the trade winds as one proceeds westward across the western Pacific.[14] Rain rates in these synoptic scale waves tend to be enhanced in regions of anomalous cyclonic flow in the lower troposphere.

The waves undergo a transition from eastward-tilting with height over the eastern Pacific to an equivalent barotropic structure over the western Pacific analogous to those in the MRG waves shown in Fig. 19.7. Easterly waves propagate westward because they are embedded in the easterly flow equatorward of the subtropical anticyclones. The westward propagation is slightly augmented by the beta-effect. Over the eastern and central Pacific, where convection tends to be focused within the ITCZ, the waves are constrained to propagate westward, but over the western North Pacific, where the steering flow curves northward, around the western side of the climatological mean anticyclone, the waves propagate west-northwestward. The poleward component of the phase propagation is enhanced by the transverse component of the beta-effect, referred to as the "beta drift."[15]

Like other tropical waves, easterly waves derive energy from convective-coupling, but in addition, waves passing over West Africa feed on the strong low-level thermal contrast between the hot desert to the north and the cooler equatorial ocean to the south. The baroclinicity is reflected in the strong easterly vertical wind shear associated with the so-called "African easterly jet" centered along ∼10°N at the 600 hPa level. Figure 19.12 (right panels) shows the structure

[13] Takayabu and Nitta (1993); Wheeler and Kiladis (1999); Dunkerton et al. (2009).

[14] Sobel and Bretherton (1999); Webster (2020).
[15] Boos et al. (2014).

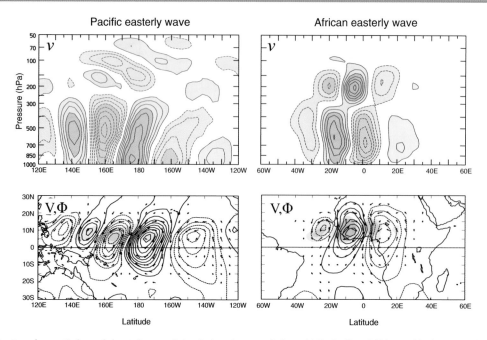

Figure 19.12 Structure of convectively coupled easterly waves during the boreal summer. (Left panels) The Pacific and (right panels) African easterly waves as inferred from regression analysis of fields upon PC1 of JAS TD-filtered OLR in their respective domains. (Top panels) Vertical cross sections of the meridional wind component v along 7.5°N. Contour interval (left) 0.2 m s^{-1} and (right) 0.4 m s^{-1}. Solid contours and gold shading indicate northward flow. The zero contour is omitted. (Bottom panels) the corresponding horizontal structure in the OLR field is indicated by colored shading: blue denotes negative values and the threshold for shading corresponds to the 95% significance level. Also shown are the corresponding perturbations in 850 hPa wind (vectors) and streamfunction (contours) at the time of peak PC1. The longest arrows correspond to 5 m s^{-1}. Contour interval 1×10^5 m^2 s^{-1}, negative contours are dashed, and the zero contour is omitted.

of easterly waves in this sector, referred to as *African easterly waves*, obtained as in the left panels but for the domain 20°S–20°N, 60°W–60°E. In the layer below 700 hPa, where the baroclinicity is concentrated, the waves tilt eastward with height. Above the African easterly jet (~600 hPa) the waves tilt (albeit weakly) in the opposite sense. The eastward tilt of the wave axes with height is indicative of an equatorward, down-gradient heat transport, indicative of a conversion from zonal to eddy available potential energy ($C_A > 0$). That the anomalies in vertical velocity (as inferred from OLR) and the meridional wind component occur out of phase with one another implies that hot, formerly Saharan air is rising while cooler marine air is sinking, converting eddy available potential energy into eddy kinetic energy (i.e., $C_E > 0$.) In these respects, African easterly waves resemble upside-down extratropical baroclinic waves.

The baroclinic energy conversion is not the whole story. It is evident from the bottom right panel of Fig. 19.12 that the perturbations in the streamfunction field exhibit meridional tilts, positive (i.e., eastward with increasing latitude) to the south of the African easterly jet axis and negative to the north of it. These tilts are indicative of a down-gradient transport of westerly momentum into the easterly jet, implying a barotropic energy conversion from zonal to eddy kinetic energy (i.e., $C_K > 0$). It follows that both C_A and C_K are energy sources for the easterly waves in this sector.[16]

[16] Kiladis et al. (2006).

As the African easterly waves propagate westward over the Atlantic, the enhanced convection comes into alignment with the troughs of the waves at the Earth's surface. This phase relationship is indicative of frictionally induced convergence in regions of cyclonic vorticity and vice versa. The wave troughs are warm relative to their surroundings and the waves are characterized by the rising of warm air and the sinking of cool air, but here the mechanical energy derives not from the north–south heating contrast, but from the zonal gradients in the release of latent heat, as is typical of convectively coupled waves.

19.2.2 Submonthly (12–20 day) Monsoon Gyres

Over monsoon regions such as southern Asia, northern Australia, and Central and South America, the convective variability on the intraseasonal timescale is dominated by the MJO, but another spectral peak in rain rate is evident with a period in the 12–20 day range. The disturbances that give rise to this variability assume the form of convectively coupled off-equatorial Rossby gyres. Figure 19.13 shows an example of such a disturbance that is prominent over southeast Asia during northern summer, as defined by the leading EOF of JJA 12–20 day filtered OLR in the entire tropical belt, from 20°S–20°N. Similar patterns can be obtained by regressing fields upon OLR averaged over the rectangular box (10–20°N, 120–130°E), the same region that was used as an index of the Pacific–Japan (PJ) pattern discussed in Section 18.8 (not shown). As in the PJ pattern, enhanced convection

Figure 19.13 Structure of convectively coupled submonthly monsoon gyres at the 850 hPa level, as inferred from regression analysis of fields upon PC1 of JJA 12–20 day filtered OLR in the entire tropical belt (20°N–20°S). Blue shading indicates negative OLR perturbations indicative of enhanced convection. The longest vectors correspond to wind perturbations of 5 m s^{-1}. Contour interval for geopotential 20 m^2 s^{-2}. Negative contours are dashed. The zero contour is omitted.

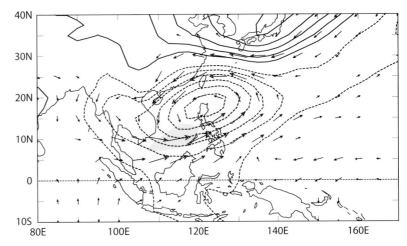

is observed within enhanced low-level westerly anomalies, with a cyclonic circulation to the north. This gyre is smaller in zonal scale than the PJ pattern, so it modulates the strength of low-level monsoon westerly jet exit region more locally over the Philippines, and it includes a weak circulation in the opposite sense over India. In contrast to the PJ pattern, which is a standing oscillation with geographically fixed nodes and antinodes, the 12–20 day monsoon gyres propagate westward at about 5 m s^{-1}, reaching the Bay of Bengal in about 5 days and northern India several days after that (not shown). These slow moving "Ganges valley cyclones" modulate the monsoon rains and they are partially responsible for the local peak in 850 hPa eddy kinetic energy in Fig. 19.9.

Exercise

19.3 The time-longitude section of brightness temperature, a proxy for rain rate, shown in the left panel of Fig. 19.11, exhibits both eastward and westward propagating waves, whereas the corresponding section for the meridional wind component at 700 hPa, shown in the right panel, is dominated by westward propagating waves. Explain why the two sections are so different.

19.3 EXTRATROPICAL INFLUENCES

We have already seen an example of how the influences of extratropical disturbances can extend deep into the tropics. The trailing cold fronts of extratropical cyclones can generate shallow but strong pressure surges that propagate equatorward along the eastern side of the major mountain ranges, as depicted in Fig. 12.15, triggering patches of enhanced tropical convection that are capable of forcing equatorially trapped waves. In this section, we will show how waves originating within the extratropical storm track may disperse into the tropics and interact with tropical motion systems. The example shown here is for the Pacific ITCZ

during the boreal winter. Another example is the flareups in monsoon-related rainfall over northern Australia observed in association with the passage of baroclinic waves in the Southern Hemisphere storm track to the south of Australia.[17]

Figure 19.14 shows one-point regression maps for OLR, 200 hPa streamfunction and winds for a grid point near Hawaii. The reference time series is relative vorticity (the Laplacian of streamfunction), highpass filtered with a cutoff at a period of ~25 days to eliminate the influence of the MJO. The pattern is dominated by synoptic scale waves with their axes tilted southwest to northeast, dispersing southeastward from the storm track in the western Pacific. Their shape and orientation are reminiscent of the later stages of the LC1 life cycle in Fig. 8.9 (bottom left). The associated pattern in OLR outlines an elliptical patch of enhanced rain rate along the ITCZ, situated in the region of cyclonic relative vorticity advection to the east of the advancing negative streamfunction (cyclonic vorticity) anomalies. The wave axes tilt westward with height as in baroclinic waves in the extratropics (not shown).[18] The southwest-to-northeast (SW–NE) tilt of the wave axes in Fig. 19.14 is indicative of a poleward transport of westerly momentum and it also reflects the southeastward group velocity of the Rossby wave train. The lagged-regression map for 2 days later reveals that the waves in the streamfunction field are propagating southeastward with a phase speed of ~7 m s^{-1}. The patch of enhanced rain rate along the ITCZ is keeping pace with them and by this time it too has developed a SW–NE orientation. Meanwhile, the positive upstream center over the central western Pacific has weakened and a new negative center has developed south of the equator along 110°W.

While the eastward propagating patch of enhanced rain rate in Fig. 19.14 is largely restricted to the ITCZ (not shown), at later lags the associated signature in OLR and other indicators of cloud-top temperature extend northeastward into the subtropics within the zone of southwesterly flow and cyclonic relative vorticity advection. In individual

---

[17] Narsey et al. (2017).
[18] Kiladis (1998).

Figure 19.14 One-point regression maps for OLR (bold contour) and for 200 hPa wind (vectors) and streamfunction (contour interval $1 \times 10^5 \, \text{m}^2\text{s}^{-1}$). The reference time series is DJF 30-day highpass filtered 200 hPa relative vorticity at 20°N, 160°W with sign reversed. The bold contour corresponds to the 95% significance level for negative OLR (enhanced rain rate). (Top) simultaneous; (bottom) lagged by 2 days.

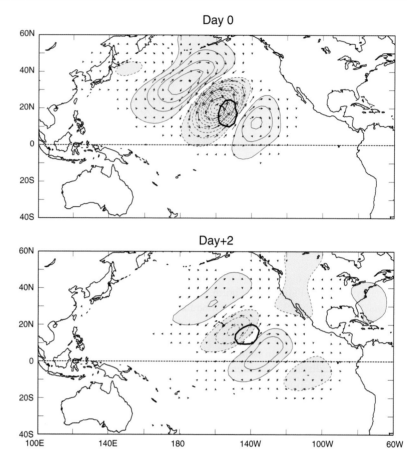

Figure 19.15 Enhanced OLR satellite image for 00 UT, 3 December, 2019. Upper tropospheric outflow from a zone of enhanced convection along the ITCZ is feeding into the rain bands in advance of a cold front associated with an extratropical cyclone off the coast of California. NOAA GOES-17 satellite image from December 3, 2019.

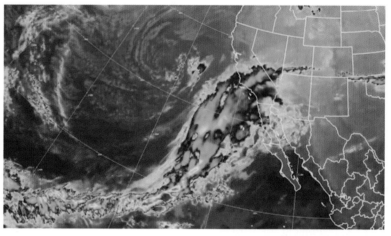

cases, this feature often assumes the form of a discrete rain band emanating from the ITCZ and merging with the cold frontal rain band of the baroclinic wave, as in the examples shown in Figs. 4.4 and 19.15. The cold frontal zone of the LC1 baroclinic wave is thus energized by its encounter with the ITCZ.[19]

In the upper troposphere, where westerlies extend equatorward into the outer tropics, baroclinic waves originating

in the extratropics disperse equatorward until they encounter their critical latitude, where the climatological mean westerly wind speed is equal to their zonal phase speed, as described schematically (for stationary waves) in Fig. 11.15. During the boreal winter and spring, westerlies extend all the way across the equator into the Southern Hemisphere in the eastern Pacific sector and, just marginally, in a patch over the eastern Atlantic. Within these wave guides or "ducts", the waves are able to penetrate deeper into the tropics and the more slowly moving ones are able to disperse all the

[19] Kiladis (1998); Huaman et al. (2020).

Figure 19.16 200 hPa DJF extended *E* vectors (arrows) and zonal wind *u* (contours) (top) climatology, (middle) El Niño and (bottom) La Niña years. The *E* vectors are based on 25 d highpass filtered data. Contour interval 10 m s^{-1}, the zero contour is bolded and there is one dashed negative contour. Blue shading indicates easterlies; yellow $u > 20$ m s^{-1}.

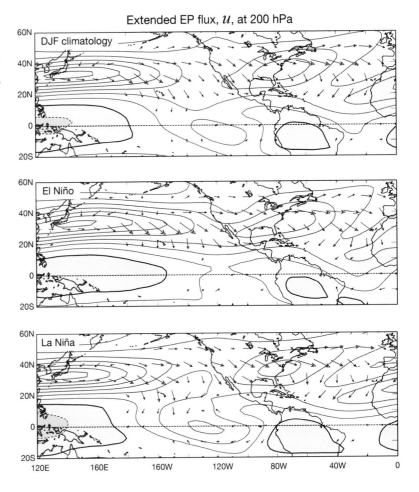

Extended EP flux, u, at 200 hPa

way across the tropics into the Southern Hemisphere. The barotropic component of the extended EP flux ***E*** (Eq. 12.18) is helpful in diagnosing such zonally varying features.

The top panel of Fig. 19.16 shows the climatological mean 25 day highpass filtered *E* vectors at the 200 hPa level in DJF. The *E* vectors to the east of the storm tracks curve equatorward and they penetrate deepest into the tropics over the eastern Pacific and eastern Atlantic westerly waveguides before they encounter their critical latitude. The equatorward orientation of the *E* vectors is consistent with the southwest-to-northeast tilt of the eddies in Fig. 19.14. The westerlies in the equatorial duct in the Pacific sector tend to be stronger during La Niña winters than during El Niño winters and the penetration of Rossby waves, as indicated by the southward extent of the *E* vectors, is accordingly deeper, as shown in the bottom panel of Fig. 19.16.[20]

---
[20] Webster and Holton (1982).

LC1 baroclinic waves and other extratropical disturbances whose upper tropospheric wave patterns extend into the subtropics usually assume the form of extended upper tropospheric troughs in the geopotential height field attended by cyclonic shear lines. Cold core cyclonic vortices at the jet stream level are manifested as "cutoff lows" in the geopotential height field. As a wave approaches its critical latitude, the trailing end of its trough may break off to form a cold core cyclonic vortex that slowly drifts westward or southwestward over the course of a few days. Depending upon the large-scale environment, it may remain decoupled from the low-level flow or it may induce a surface low directly beneath or to the east of it. A deep, convectively coupled upper tropospheric cold core cyclone may produce significant rainfall and, albeit rarely, may become involved in the formation of a tighter, more intense, longer-lived, warm core tropical cyclone like the ones discussed in the next chapter.

20 Warm Core Tropical Vortices

with George Kiladis

Warm core tropical vortices are distinctly different from any of the motion systems considered in previous chapters. In the literature they are referred to as tropical depressions, tropical storms, or tropical cyclones, in order of increasing intensity. Tropical cyclones (TCs) are also known by local names such as *typhoon* and *hurricane*. They are smaller in scale, more axisymmetric, longer lasting, and potentially more intense than the extratropical cyclones embedded in baroclinic waves and their formation and development mechanisms are entirely different. They exhibit a different vertical structure than the cold core cyclonic vortices touched upon in Section 19.3. Despite the large temperature perturbations in their warm cores, their azimuthally averaged vertical velocity and diabatic heating rates are nearly in WTG balance and the release of latent heat plays a central role in their dynamics, as in the so-called "moisture modes" discussed in Section 15.3. Collectively, these systems play a significant role in the global atmospheric general circulation, and quite possibly in the oceanic thermohaline circulation as well.[1] They form over the tropical oceans but may move over land and are thus an important facet of the seasonally varying rain rate (and wind) climatology of a large area of the tropics and subtropics.

This chapter is made up of five sections. The first describes the formation and development mechanisms that give rise to intense tropical cyclones. The second describes their observed structure. The third discusses tropical cyclone dynamics and thermodynamics in greater detail. The fourth section describes and interprets the global distribution of tropical cyclone tracks and the fifth describes how their frequency of occurrence in various parts of the tropical oceans varies seasonally and is modulated by ENSO and by the MJO.

20.1 FORMATION AND DEVELOPMENT MECHANISMS

If the background flow is convergent over an extended time interval, rings of tagged air parcels will contract and eventually form a cyclonic vortex, even in the absence of any preexisting relative vorticity. But most TCs develop within patches of cyclonic vorticity embedded within preexisting waves. They first become identifiable when the streamlines become sufficiently distorted to form closed cyclonic circulations in which air parcels become sequestered. The distortions that lead to such cut-off lows tend to be greatest in the vicinity of "critical latitudes," where air parcels are moving through the waves very slowly. An embryonic vortex or "pouch," as they have been called, will expand in area if the rotation within it amplifies and it may eventually become separated from the wave in which it formed.[2] Favored sites for the development of these closed cyclonic gyres are the troughs of easterly waves and the cyclonic shear zones along the flanks of equatorial westerly wind maxima that are observed, that is present during the passage of the westerly phase of an equatorial Kelvin or Rossby wave. Another favored location is to the west of the region of enhanced rain rate in the MJO. Since the process that leads to the formation of closed contours depends on the total flow, tropical cyclone formation is favored in regions of climatological mean low-level cyclonic vorticity.

The processes that lead to the intensification of TCs have been investigated in numerous theoretical studies and numerical simulations. The picture that emerges is not as tidy or as well agreed upon as in the case of baroclinic waves, but there exists a broad consensus with regard to the essential ingredients, which can be understood in terms of conceptual ideas to which the reader has been introduced in Chapters 2 and 7 of this book. In fact, there are many parallels between the flow in a slowly evolving tropical cyclone, as viewed in a cylindrical coordinate concentric with it, and the zonally symmetric component of the global general circulation, as viewed in a spherical coordinate system centered on the Earth's axis of rotation. Both can be viewed as examples of circular vortices whose relatively slow toroidal circulations maintain thermal wind balance between the nondivergent component of the horizontal wind field (i.e., the zonal wind component in the general circulation and the azimuthal wind component in a TC) and the temperature field.[3]

Because the centripetal acceleration is of first order importance, gradient, rather than geostrophic balance applies;

---

[1] Emanuel and Nolan (2004), Emanuel (2018).

[2] Dunkerton et al. (2009).
[3] The seminal paper applying these principles to tropical cyclones is Emanuel (1986). For a comprehensive review of the tropical cyclone literature, see Houze (2010) and Emanuel (2018).

344

$$-\frac{R}{p}\frac{\partial T}{\partial r} = \frac{\partial}{\partial p}\left(fu + \frac{u^2}{r}\right),\tag{20.1}$$

where u is the azimuthal wind speed and r is the distance from the storm center. In analogy with Eq. (7.15), we can define a state vector

$$\mathbf{S} = \delta m\mathbf{j} + \delta\theta\mathbf{k},\tag{20.2}$$

where $\mathbf{j}$ and $\mathbf{k}$ are unit vectors in the radial and vertical directions, and δm and $\delta\theta$ are the displacements of angular momentum and potential temperature surfaces from a reference, state-of-rest grid in which the m surfaces are vertical, with angular momentum m increasing outward and the θ surfaces are horizontal, with θ increasing upward. If the contour intervals for m and θ are constant, the areas of individual (m,θ) grid boxes are inversely proportional to potential vorticity PV, as in the "membrane thickness" analogy in Section 7.3. In fact, if, in that section, we had written the equations in spherical coordinates rather than in Cartesian coordinates, we could have defined $\mathbf{S}$ in terms of δm and $\delta\theta$ there as well, and the equations would have been more exact. Just as the m surfaces bulge inward toward the Earth's axis of rotation in westerly jet streams in the global general circulation, in a strong cyclonic circulation they bulge inward toward the axis of rotation of a TC. The isentropes are pinched together radially inward of the strongest cyclonic circulation, just as they are pinched together poleward of a westerly jet. It follows that the intensification of TCs is marked by a buildup of PV in the interior core region, where the areas of the (m,θ) grid squares are very small.

In analogy with Eq. (7.20) we can define the toroidal (alternatively *transverse* or *secondary*) circulation $\mathbf{\Psi} = v\mathbf{j} + \omega\mathbf{k}$, where v is the azimuthally averaged radial wind component, positive outward, and ω is the corresponding vertical component in pressure coordinates, and v and ω satisfy a two-dimensional continuity equation

$$\frac{1}{r}\frac{\partial}{\partial r}vr + \frac{\partial\omega}{\partial p} = 0.\tag{20.3}$$

In analogy with Eq. (7.21), we can define the forcing vector

$$\mathbf{\Gamma} = \frac{\mathcal{F}}{\partial m/\partial r}\mathbf{j} + \frac{\mathcal{Q}}{\partial\theta/\partial p}\mathbf{k}\tag{20.4}$$

that drives the transverse circulation. The essential elements in the forcing vector in a TC are the diabatic heating $\mathcal{Q}$, which is dominated by the release of latent heat, mainly in the interior of the vortex, and the frictional drag $\mathcal{F}$ on the intense cyclonic circulation, mainly in the boundary layer. As in the global general circulation, the eddy forcing also plays a significant role, as explained in Section 20.3.1.

Exercises

20.1 (a) Based on the tropical cyclone climatology, show that the planetary vorticity f plays an essential role in their formation. (b) Describe the mechanism by which planetary vorticity is converted to the relative vorticity associated with the rotating cyclone.

20.2 Why do virtually all intense tropical vortices rotate cyclonically?

20.2 OBSERVED STRUCTURE

The top left panel of Fig. 20.1 shows the azimuthally averaged azimuthal wind component u based on a composite consisting of 100 intense TCs, whose mean central SLP is 945 hPa. The circulation is cyclonic, except in the upper troposphere outward of radius $r \sim 400$ km. Wind speeds of up to 50 m s^{-1} are observed at $r \sim 100$ km. In the upper troposphere, it weakens with height and reverses sign in the outer part of the vortex, so that the flow at the 125 hPa level consists of a cyclonic "core" surrounded by an anticyclonic ring. At all levels, angular momentum $m = ur + fr^2/2$, indicated by the blue contours in the bottom right panel, increases monotonically outward. In the lower troposphere its radial gradient $\partial m/\partial r$ is enhanced inward of the radius of maximum wind ($r \sim 100$ km) and it is weaker in the much wider ring of anticyclonic radial wind shear outward of that radius. In the upper troposphere, the m contours bulge radially outward due to the anticyclonic rotation.

The temperature cross section, shown in the top right panel of Fig. 20.1, exhibits a distinct warm core, strongest at the 300 hPa level, at which the air near the center of the cyclone is ~ 12 K warmer than in a typical tropical sounding. The warm core coincides with the layer of anticyclonic vertical wind shear in the upper part of the vortex; that is, the layer in which the term in parentheses in Eq. (20.1) decreases with height. The potential temperature surfaces, shown in the bottom right panel of Fig. 20.1, are displaced downward at the center of the vortex, especially in the upper troposphere where the temperature departures are largest.

The transverse circulation in the composite, shown in the bottom left panel of Fig. 20.1, is dominated by boundary layer inflow,[4] and a deep updraft, strongest at $r \sim 100$ km that slopes slightly outward with increasing height, with an outflow layer centered at around 125 hPa. Inflow extends all the way up to ~ 300 hPa, albeit much weaker than that in the boundary layer.

From $r = 400$ km outward to well beyond 1000 km, the azimuthally integrated (i.e., circumference-weighted) boundary layer inflow (taken here as the layer below the 850 hPa level) is almost independent of radius and approximately equal to 2×10^{10} kg s^{-1}. From $r = 400$ km inward to $r = 150$ km it decreases by about 30%, presumably because of low-level convergence into the TC's outer rain bands. If the ascent rate in the innermost 150 km of the composite TC were spatially uniform, the vertical velocity at the 500 hPa level would be ~ 30 cm s^{-1}. At the innermost grid points ($r \lesssim 25$ km), the composite exhibits weak subsidence, with vertical velocities of up to 10 cm s^{-1} at the 200 hPa level. Indeed, one of the hallmarks of intense TCs is existence of an *eye*, surrounded

---

[4] The extraordinary shallowness of the inflow layer in TCs is substantiated by the analysis of Zhang et al. (2010).

Composite TC

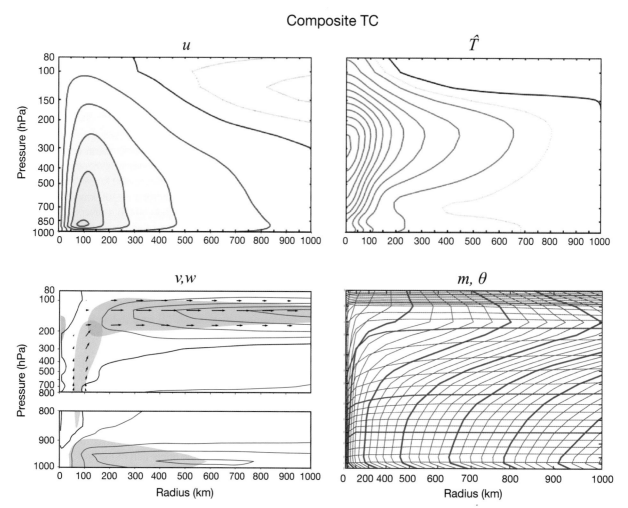

Figure 20.1 Radial cross sections of azimuthally averaged fields in a composite of 100 intense tropical cyclones, as represented in the ERA5 reanalyses, with 25 km horizontal resolution. (Top left) The azimuthal wind field u. Contour interval 10 m s^{-1}. The black contour is the zero line; solid contours indicate a cyclonic and dashed contours an anticyclonic circulation. (Top right) temperature $\widehat{T}$, the departure relative to the mean temperature of the $r = 900$–1100 km ring. Contour interval 1K, dotted contour 0.5K. The black contour is the zero line and the area of light blue shading is colder than the reference temperature at the same level. (Bottom left) The transverse circulation: inflow > 10 m s^{-1} indicated by gold shading; ascent > 0.3 m s^{-1} by blue shading; outflow > 10 m s^{-1} by green shading, and subsidence > 10 cm s^{-1} by purple shading. Contours indicate the radial mass flux per meter in the vertical: blue denotes outflow at rates of 2.5, 5, 7.5, and 10 $\times 10^6$ kg m^{-1} s^{-1} and red denotes inflow at rates of 20, 15, 10, and 2.5 $\times 10^6$ kg m^{-1} s^{-1}. The black contour corresponds to the zero line. (Bottom right) angular momentum m, increasing with radius, and potential temperature θ, increasing upward. Contours of m at intervals of 1×10^5 m^2 s^{-1} starting from the left, with every fifth contour bolded; contours of θ at intervals 4 K, with bold contours every 20 K starting at 320 K. In order to make the widths of all the squares equal for the reference state, a resting atmosphere with constant f, r^2 rather than r is used as the radial coordinate. The area enclosed by pairs of adjacent m and θ contours is thus inversely proportional to potential vorticity. Courtesy of Yumin Moon.

by an outward-sloping updraft called the *eye wall cloud*. The eye is typically cloud-free above the boundary layer.

The azimuthal flow can also be represented in terms of the distribution of angular momentum, as shown in the bottom right panel of Fig. 20.1. Air parcels lose angular momentum as they spiral inward toward the center of the TC due to frictional drag on the cyclonic circulation. The drag on the weakly inflowing air above the boundary layer is due to vertical mixing by deep convective-scale clouds – sometimes referred to as "cumulus friction."

As air parcels ascend moist adiabatically in the interior of the TC, they cross θ surfaces toward higher values as a consequence of the release of latent heat. As they rise, the air parcels begin to spiral outward, closely paralleling the θ sur-

faces. In the upper branch of the transverse circulation, the flow is nearly adiabatic and frictionless. Angular momentum m increases radially outward as the azimuthal asymmetries increase and air parcels that were formerly external to the TC circulation become entrained into it, bringing their relatively higher angular momentum.

The composite fields shown in Fig. 20.1 are based on gridded ERA5 reanalyses data with 25 km horizontal resolution. To document the structure of intense TCs in the innermost 100 km, Fig. 20.2 shows the corresponding cross sections for Hurricane Patricia,[5] which was monitored

[5] Rogers et al. (2017).

Hurricane Patricia

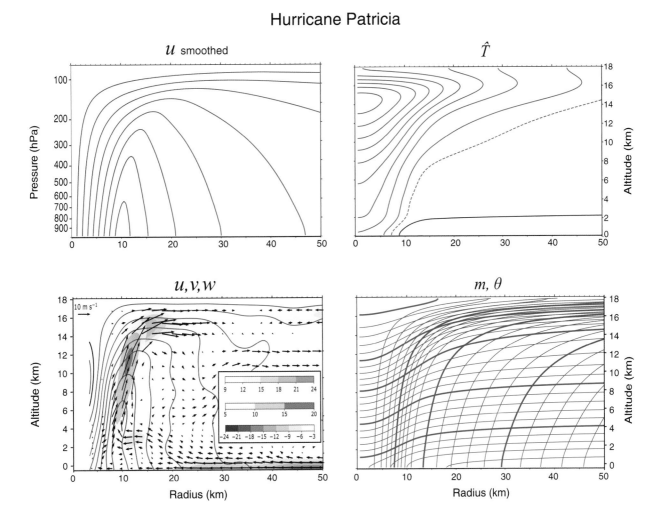

Figure 20.2 Radial cross sections of azimuthally averaged fields in Hurricane Patricia at 1200 UT 23 Oct. 2015. (Top left) The smoothed azimuthal wind u, contour interval 10 m s^{-1}, contours ranging from 10 to 80 m s^{-1}. (Top right) Temperature $\hat{T}$, the departure from the tropical standard atmosphere. The solid red contours are at 2 K intervals; the black contour corresponds to zero; the dashed red contour to +1 K. (Bottom left) The transverse circulation indicated by colored shading (see inset for key, with radial inflow in orange, updraft in blue, and radial outflow in green) and by the vectors; the scale at the top left is applicable to both radial and vertical velocities. The red contours indicate the unsmoothed azimuthal circulation, contour interval 10 m s^{-1}. (Bottom right) Angular momentum m estimated from the smoothed azimuthal wind field shown in the top left panel and the corresponding potential temperature θ field inferred from the thermal wind equation. Contours as in Fig. 20.1. In this figure the radius scale is linear. Courtesy of Jonathan Martinez, Michael Bell, and Yumin Moon.

with reconnaissance aircraft equipped with Doppler radar. The temperature field was inferred from the azimuthal wind field assuming thermal (gradient) wind balance Eq. (20.1). It is evident that the structure of the fields in Patricia was qualitatively similar to those in the composite, but the peak winds were almost twice as strong (80+ vs. 50 m s^{-1}), the temperature departures and SLP depression were about twice as large, and the updraft and strongest azimuthal winds were observed at $r \sim 10$ km, compared to $r \sim 100$ km in the composite. The existence of an upper tropospheric "warm core" reflects the weakening of the entire cyclonic circulation with increasing height, as in the composite TC, but the baroclinicity is concentrated within the innermost 50 km. In contrast to the composite TC, the warm cores of intense TCs like Patricia extend almost all the way down to the ocean surface within the "eye." This remarkably tight feature is related to the outward tilt of the eye wall cloud and the associated cyclonic shear zone, which gives rise to anticyclonic vertical

wind shear and strong baroclinicity concentrated at the edge of the eye.

Owing to the stronger radial gradients of azimuthal wind in Patricia than in the TC composite, the m contours are more concentrated in the cyclonic shear zone on the inner flank of the strongest winds, the θ surfaces are more depressed over the center of the TC, and the static stability $-\partial\theta/\partial p$ in the lower and middle troposphere is stronger. The stronger gradients in m and θ reflect the high concentration of potential vorticity in the eyes of individual intense TCs.

In Patricia, much of the ascent in the transverse circulation was concentrated in a well-defined eye wall cloud, a ring with a radius of ~ 10 km and a width of only a few km. At the 500 hPa level the ascent rate averaged over the storm's innermost 13 km was ~ 5 m s^{-1}, more than an order of magnitude stronger than the ascent rate in the innermost 150 km of the TC composite in Fig. 20.1, but owing to the much smaller area, the *integrated* mass flux is estimated to be 1.5×10^9 kg, roughly an order of magnitude smaller. It is not possible to make a direct "apples with apples" comparison of the radial profiles of the upward mass fluxes in Patricia and the composite because features on the scale of Patricia's eye wall cloud are not resolved in the composite and the aircraft-borne radar data in Patricia do not extend radially outward beyond $r \sim 75$ km. If the ascent in the transverse circulation was dominated by the ascent in the eye wall cloud, the integrated mass flux in Patricia would be at least as large as that in the composite. That it is an order of magnitude smaller indicates that most of the upward mass flux in the TCs included in the composite must have occurred in mesoscale updrafts in the rain bands within the innermost 150 km rather than in the eye wall cloud. Mesoscale updrafts do not extend as high as the eye wall cloud, but they tend to increase in intensity with increasing radius, so even though they cover only a very small fraction of the area of the interior of the TC, they could still account for most of the upward mass flux in the transverse circulation. This interpretation draws support from the fact that (1) if the domain of the integration for Patricia is extended outward from 13 to 35 km, the vertical mass flux doubles and if it is extended outward to 75 km it doubles again, and (2) if the domain of the integration for the composite is extended outward from 150 to 400 km, the vertical mass flux doubles.

Exercises

20.3 From what radius would a ring of air parcels, initially at rest at 10° latitude, need to contract while conserving angular momentum in order to acquire an azimuthal wind speed of 40 m s^{-1} when it reaches a radius of 100 km?

20.4 A *Rankine vortex* consisting of an inner disk of radius r_o in which the horizontal flow is in solid-body rotation and an outer ring in which u varies in inverse proportion to r bears some intriguing similarities to the axisymmetric flow in tropical cyclones. (a) Prove that the relative vorticity ζ is 2ω in the inner disk and zero in the outer ring. (b) What happens in the limit as $r_o \to 0$? (c) Can this mathematical treatment be extended to the radial flow in a TC? (d) How is the azimuthal flow in a TC different from that in a Rankine vortex?

20.5 At what radius in the composite TC are the centripetal acceleration and the Coriolis force relating to the azimuthal flow of comparable magnitude? Assume a latitude of 15°.

20.3 FURTHER INTERPRETATION

This section offers further interpretation of the observed structure of TCs, drawing upon relationships discussed in previous chapters.

20.3.1 Dynamical Considerations

The essential features in Figs. 20.1 and 20.2 can be simulated in axisymmetric models of TCs:[6]

- the "warm core,"
- the deep, balanced azimuthal circulation,
- the strong updraft in the cyclonic shear zone on the inner flank of the strongest winds,
- the radially outward tilt of the shear zone and the updraft with increasing height,
- the shallowness of the inflow layer,
- the tendency for the m and θ surfaces to parallel one another at upper tropospheric levels outside the eye wall.

As TCs intensify, the m surfaces are pulled inward across the radius of maximum winds, and packed into the cyclonic shear zone radially inward of it, while the θ surfaces are pulled downward in the interior of the vortex, creating the warm core, which is strongest around the 300 hPa level. With the tightening of both the radial gradient of m and the vertical gradient of θ, PV builds up in the 1000–300 hPa layer radially inward of the radius of maximum winds. If, as in Section 7.3, we think of the m and θ contours in the radial plane as forming a mesh painted onto an imaginary membrane, whose thickness is proportional to the PV, the tightening of the gradients implies a shrinking of the (m,θ) grid boxes and hence, a thickening of the membrane.

The deformations of the m and θ surfaces also alter the shape of the (m,θ) grid boxes. Following an air parcel in the transverse circulation ascending and flowing radially outward in the eye wall cloud, the boxes evolve from rectangular to rhomboidal and become increasingly elongated. As the m and θ surfaces come into alignment, the restoring force experienced by air parcels moving diagonally across the long axis of the grid boxes weakens and less and less forcing is required to maintain the circulation. In the limit, the restoring force disappears entirely as air parcels exit freely from the TC following the mutually parallel m and θ surfaces. This limit corresponds to the threshold for what is referred to as *conditional symmetric instability* (CSI). Whereas gravitational instability prevents θ from increasing with pressure and inertial instability prevents m from increasing radially inward, CSI prevents either from occurring, thereby maintaining the integrity of the (m,θ) grid.

In the TC literature, moist isentropes labeled with values of equivalent potential temperature or saturation equivalent potential temperature θ_{es} are often used in lieu of θ surfaces in radial cross sections. In the eye wall clouds of intense TCs like Patricia, the transverse circulation is channelized between mutually parallel m and θ_{es} surfaces from the time it leaves the boundary layer to the time it exits in the outflow.

In an axisymmetric model, the only forcings are diabatic heating and frictional drag, as described in Eq. (20.4). In nature there exists an additional forcing associated with the

---

[6] See, for example, Emanuel (1986).

departures from axial symmetry because TCs are embedded in a background flow with vertical and horizontal shear as well as moisture gradients. The spiral rain bands in TCs distort the flow in a manner analogous to the tilted ridges and troughs of baroclinic waves. Their orientation is such that they transport angular momentum radially inward, while producing an outward flux of wave activity. The convergence of this "eddy transport" near the radius of maximum winds results in a cyclonic forcing that promotes the strengthening of the vortex. The eddy forcing is appreciable and evidence based both on observations and numerical experiments suggests that it plays an important role in the development of TCs, particularly in their early stages.[7] On the other hand, it is notable that tropical cyclone development can be simulated in axisymmetric models.

20.3.2 Thermodynamic Considerations

Like other convectively coupled motion systems, TCs develop in regions of elevated SST – 27°C is often cited as a threshold – and high column water vapor W. In soundings 500 km or farther away from the center of the composite TC, the middle troposphere is relatively dry (RH ~50%) and equivalent potential temperature exhibits a distinct mid-tropospheric minimum, typical of tropical soundings. Mid-tropospheric RH and θ_e both increase radially inward from $r \sim 500$ km, as mid-tropospheric air is moistened by the deep convection in the outer rain bands. At the radius of maximum azimuthal winds, RH reaches values in excess of 95% and θ_e is virtually invariant with height – conditions more conducive to uninhibited moist adiabatic ascent in a neutrally stable environment than to deep convection in a conditionally or convectively unstable environment. Another indication of the absence of instability in the interior of TCs is the lower frequency of occurrence of lightning flashes in the eye wall cloud than in the spiral rainbands in the outer part of the storm and the fact that flash rates are higher in weaker TCs than in stronger ones.[8] It follows that conventional convective instability cannot be responsible for the deepening and maintenance of intense TCs. Rather, they are fueled by the extraordinarily strong fluxes of latent and sensible heat from the underlying ocean surface.

The near surface air cools by about 2 K as it spirals inward toward the center of the storm[9] due to adiabatic expansion as it crosses the isobars toward lower pressure. As it expands while acquiring heat from the underlying ocean surface, its potential temperature rises. The inflow air then ascends moist adiabatically in the eye wall cloud, cooling as it expands, but not nearly as much as it would have, were it not for the release of the latent heat that it picked up while spiraling

inward in the boundary layer. As the air trajectories come into alignment with the potential temperature surfaces in the outflow layer, the heating rate drops off. By this time the air is so cold that virtually all the water vapor that was in it has condensed out.

The air that detrains from the top of the eye wall cloud should be viewed as exiting the TC, rather than returning to the boundary layer and passing through the same set of states again and again, as in the schematics in Figs. 2.4 and 2.7 or in the idealized Carnot cycle. It has spent only about a day in the transverse circulation in the TC, but it will take a week or longer for it to reject the latent and sensible heat that it acquired while in the boundary layer by emitting infrared radiation, enabling it to descend from higher toward lower potential temperature surfaces. By the time it reaches the boundary layer again the TC will most likely be far away or nonexistent. In this sense the "working fluid" in the TC "heat engine" is analogous to the air that is taken into an internal combustion engine, heated, allowed to expand, and emitted as exhaust. TC exhaust is colder than the intake air because the energy that it acquired while passing through the transverse circulation resides in the form of potential energy rather than enthalpy, but that has no bearing on the work done. TCs and internal combustion engines are examples of so-called "open cycle" engines, whereas in the discussion in Section 2.3 the general circulation is treated as a "closed cycle."

The rate of kinetic energy generation by a TC is the product of the kinetic energy generated per kilogram of air processed in its transverse circulation times the mass of air per unit time that is being processed. The kinetic energy generation per unit mass is equal to the work done on the air by the radial pressure gradient as it spirals inward toward the eye wall cloud, which is equal to $-\delta p/\rho$, or the drop in geopotential $-\delta \Phi$. In the TC composite, the pressure drop δp from the standard atmosphere (1013 hPa) to the center of the TC (945 hPa) is 6800 Pa, so the potential energy released and the kinetic energy generated is ~5000 J kg^{-1}. As an estimate of the rate at which air is being processed in the transverse circulation of the TC, we use the upward radial mass flux through the 500 hPa surface integrated over the innermost 150 km of the TC, 1.3×10^{10} kg s^{-1}. Based on these estimates, the rate of kinetic energy generation in the composite TC is almost 10^{14} W, 40 times the world electrical power consumption and equivalent to a few percent of the output of the entire global general circulation.

The thermal efficiency of the composite TC is equal to the KE generation per unit mass of air passing through the transverse circulation divided by the quantity of latent and sensible heat added in the boundary layer; that is, $\delta\Phi/(c_p\delta T + L\delta q)$. As noted above, air spiraling inward in the boundary layer expands while acquiring sensible heat from the ocean surface. Its pressure drops by 68 hPa while its potential temperature θ rises by ~6° C. The input of sensible heat is thus $c_p(T/\theta)\delta\theta \sim 6000$ J kg^{-1}. If the Bowen ratio is 0.2, then the total (latent plus sensible) heat added is 6 times this amount and the thermal efficiency is

[7] Pfeffer and Challa (1981). The techniques that have been developed for diagnosing the global general circulation are being applied to diagnose the asymmetries on TCs; for example, see Guo and Kuang (2019); Nolan et al. (2007).
[8] Xu et al. (2017).
[9] Kowaleski and Evans (2015), Fig. 1.

Figure 20.3 Climatological mean tropical cyclone tracks. From Knapp et al. (2010, 2018).

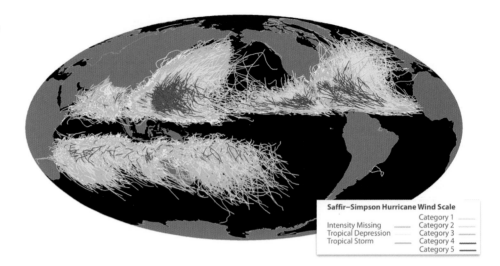

Saffir–Simpson Hurricane Wind Scale

	Category 1
Intensity Missing	Category 2
Tropical Depression	Category 3
Tropical Storm	Category 4
	Category 5

$\sim 5000/36,000 \sim 14\%$, over an order of magnitude higher than in the global general circulation.[10]

Exercises

20.6 (a) Estimate the latent and sensible heat fluxes in an intense tropical cyclone. (b) Compare your estimate with those over the western boundary currents during wintertime.

20.7 Do tropical cyclones have any impact on SST?

20.8* Do tropical cyclones influence the atmospheric general circulation by mediating the poleward heat transport by the oceanic thermohaline circulation?

20.4 TROPICAL CYCLONE GENESIS REGIONS AND TRACKS

Almost all TCs develop poleward of $10°$N/S, where the Coriolis force is effective in "spinning up" the azimuthal circulation during the early stages of development. As noted above, another necessary condition is that SST be high enough so that the condensation of water vapor is a strong heat source – widely used threshold values are $26°C$ and $27°C$. A third requirement is an absence of strong vertical wind shear that would disrupt their axial symmetry and vertical alignment and prevent deep convection from moistening the mid-troposphere in the interior of the vortex.[11] These criteria are instrumental in shaping the genesis regions of intense TCs shown in Fig. 20.3.

TCs are carried along with the vertically averaged "steering flow" in which they are embedded. The motion relative to the steering flow due to the β effect and the β drift is only on the order of 1–2 m s^{-1}. Initially, most TCs drift westward with the trade winds, often intensifying as they progress

along their tracks, especially if they pass over a patch of high SST. Eventually, most of the tracks curve poleward. Over the western North Pacific and western North Atlantic, the TC tracks pass around the western flanks of the climatological mean subtropical anticyclones. In the coastal regions of East Asia and other regions that lie along their major tracks, TCs account for an appreciable fraction of the annual mean rainfall.[12] Eventually, TCs lose their identity as they enter the westerlies and begin to interact with extratropical waves. If the timing is propitious, the potential vorticity and moisture that they carry with them can result in a dramatic intensification of the extratropical system that they merge with, along with substantial downstream impacts due to Rossby wave dispersion within the extratropical storms tracks.[13]

20.5 MODULATION OF TROPICAL CYCLONE ACTIVITY

The seasonality of tropical cyclone activity and the manner in which it is modulated by equatorial waves, the MJO, and ENSO are of interest in their own right and they serve to illustrate the environmental controls on the genesis and intensification of TCs. Here we use the term "TC" in reference to all warm core vortices sufficiently intense to be classified as tropical storms; that is, having sustained winds of at least 17 m s^{-1} at some time during their lifetimes.

20.5.1 Seasonality

TC genesis is inherently a warm season phenomenon, around the time of the highest SST. That the peak is a month or two later in the Atlantic/Caribbean than in the Pacific off the west coast of Central America is consistent with the SST climatology. Table 20.1 shows the distribution of the number of TC days, as a function of calendar month for the four ocean basins considered in this subsection. Over the North

---

[10] For alternative treatments of thermal efficiency, see Hack and Schubert (1986) and Nolan et al. (2007).

[11] Gray (1968).

[12] Khouakhi et al. (2017).

[13] Keller et al. (2019).

Table 20.1 *Tropical cyclone (TC) frequency over four different ocean basins, as indicated. The leftmost column shows the average number of "tropical cyclone days" per year (i.e., the sum of the 365 calendar dates with each one weighted by the average number of TCs in existence on that date) and the remaining columns list the percentage contributions of each calendar month to the total frequency. A TC is defined as a warm core vortex strong enough to produce sustained wind speeds of at least 17 m s^{-1} at the ocean surface at some time during its lifetime. The boundaries of the ocean basins are as in Figs. 20.4 and 20.5. Courtesy of G. Neljon Emlaw.*

	TC days	J	F	M	A	M	J	J	A	S	O	N	D
Indian	67	2	0	1	2	11	11	6	8	8	22	20	8
W. Pacific	263	2	1	2	3	4	7	15	21	18	14	8	5
E. Pacific	183	1	0	0	0	5	10	21	24	22	13	3	0
Atlantic	123	0	0	0	1	2	7	12	25	31	15	5	1

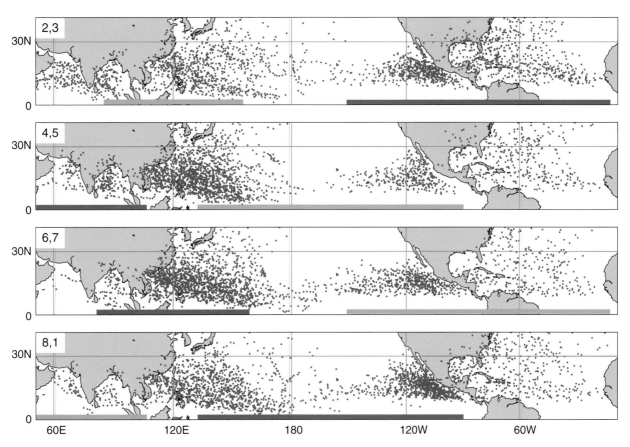

Figure 20.4 Composites of the daily positions of warm core vortices sufficiently intense to be classified as Tropical Storms at some time during their lifetimes (i.e., having sustained wind speeds of at least 17 m s^{-1}). The panels correspond to four sequential phases of the MJO, defined in accordance with the eight categories for RMM1 and RMM2 in Fig. 18.4. The colored bars along the baseline of the panels indicate the longitudes with anomalous MJO-related vertical wind shear; red denotes a weakening of the climatological mean westerly vertical wind shear, which is conducive to cyclogenesis, and blue for a strengthening of it, as determined from the 150 and 850 hPa zonal wind patterns shown in Fig. 18.6. Courtesy of G. Neljon Emlaw.

Atlantic and Pacific Oceans TC frequencies tend to peak in the warm season. Over South Asia and especially in the Bay of Bengal and the Arabian Sea, there are two TC seasons: one pre-monsoon and the other post-monsoon. The likely cause of the suppression of TC genesis during the peak of the monsoon season is the strong vertical wind shear between the low-level westerlies and the upper tropospheric easterlies that develops over this area in association with the Tibetan anticyclone.

20.5.2 Influence of the MJO

The MJO modulates TC activity by providing a favorable environment locally by enhancing the low-level cyclonic vorticity and convergence and reducing the vertical wind shear. Figure 20.4 shows a measure of TC activity during four different phases of the MJO life cycle. Coherent patches of enhanced and suppressed TC activity can be followed as they progress eastward from the Indian Ocean, across

El Niño

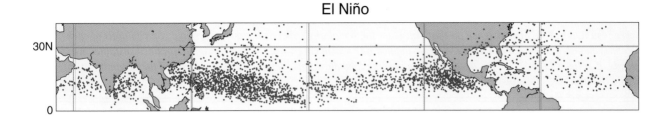

La Niña

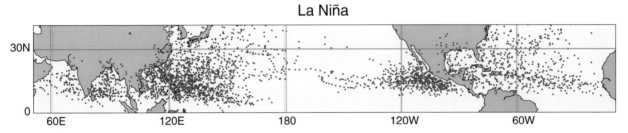

Figure 20.5 Composites of the daily position of TCs as in Fig. 20.4 but for (top) El Niño years and (bottom) La Niña years, classified on the basis of the ENSO SST index Niño 3.4, where values > 0.7°C are classified as El Niño and < −0.7°C as La Niña. Courtesy of G. Neljon Emlaw.

the Pacific, and into the Atlantic sector. This behavior can be understood as follows. Poleward of 10°N/S, where most TCs develop, the vertical wind shear $\partial u/\partial z$ is westerly almost everywhere. Anomalous easterly wind shear is thus conducive to TC development and intensification because it reduces the ambient vertical wind shear. MJO-related zonal wind anomalies tend to be out of phase at the 850 and 150 hPa levels and hence modulate the vertical shear. In Section 18.2, it was shown that unlike the MJO-related rain rate anomalies, which are largely confined to the longitudinal sector of the Indo-Pacific warm pool, the zonal wind anomalies propagate all the way around the equatorial belt and hence can potentially influence TC activity throughout the tropics. In each of the panels in Fig. 20.4, the corresponding anomalies in vertical wind shear, transcribed from Fig. 18.6, are indicated by the colored bars along the baseline. The correspondence between enhanced (suppressed) TC activity and easterly (westerly) shear is quite good and it is even more impressive in the phase-by-phase composites shown on the companion web page.

It is apparent from Fig. 18.20, that the MJO also perturbs the horizontal wind shear at lower tropospheric levels. The enhanced cyclonic shear over the eastern Pacific in the phase of the MJO cycle when rain rate is lowest over the Maritime Continent contributes to the enhancement of TC activity at

that time.[14] The modulation of TC activity by the MJO might be even stronger were it not for the fact that the planetary-scale MJO signature is weakest during the boreal summer monsoon season JJAS, the time of highest TC activity over the Pacific and Atlantic sectors.

20.5.3 Influence of El Niño and La Niña

Figure 20.5 shows composites of TC activity for contrasting polarities of the ENSO cycle. Relative to La Niña, TC activity during El Niño is enhanced over the central Pacific, where the vertical wind shear $\partial u/\partial z$ between the upper and lower troposphere (Figs. 17.7 and 17.8) is weaker, and suppressed over the South China Sea, the Atlantic, and the Caribbean, where it is stronger. Elevated SST over the equatorial central Pacific during El Niño might also contribute to the higher TC frequency.[15]

Exercise

20.9* Do tropical cyclones make an appreciable contribution to annual mean rainfall?

---

[14] Maloney and Hartmann (2001).
[15] Camargo and Sobel (2005).

21 Diurnal and Higher Frequency Variability of the Global Circulation

In motion systems with timescales ranging from hours up to and including the diurnal cycle, gravity and inertio-gravity waves are dominant. The influence of the Earth's rotation is discernible, but geostrophic balance does not prevail and Rossby wave propagation and dispersion do not play a dominant role in the dynamics.

The reader has encountered gravity and inertio-gravity waves in several chapters of this book, as summarized in Table 21.1. It was explained in Section 10.2.3 how they come in two forms: external waves analogous to those that propagate along the free surface of a liquid and internal waves analogous to those that perturb the density field in a stratified fluid with a rigid lid. External waves propagate strictly horizontally, whereas internal waves propagate both horizontally and vertically. Waves excited by tropical convection disperse upward into the stratosphere. The external waves propagate too fast to enable them to interact significantly with other elements of the tropospheric flow, but their divergence signature is strong enough to exert a discernible influence on rain rate, especially in the tropics.[1]

The first three sections of this chapter describe three different kinds of high frequency variability in which gravity wave dynamics plays an important role:

- diurnal variations in rain rate,
- gravitationally and thermally driven atmospheric tides,
- external inertio-gravity waves excited mainly by tropical heating.

The phenomena discussed in all three of these sections are global in scale but the impacts are most clearly discernible in the tropics.

The fourth (overview) section summarizes the properties of gravity waves discussed in various sections of this book. It shows examples of convectively and orographically induced gravity waves resolved by the vertical velocity field in the ERA5 Reanalysis and it documents their seasonally varying geographical distribution of r.m.s. amplitude. The fifth and final section shows and discusses the contribution of gravity waves to the global mechanical energy spectrum, to which the reader has been introduced in Section 6.8.

21.1 DIURNAL VARIATIONS

Over coastal and mountainous regions of the tropics, diurnal variations account for as much of the rain rate variability as the day-to-day variations associated with migrating weather systems. Judging from the intricacy of the patterns in Fig. 21.1, it is evident that diurnally varying atmospheric motions range from tens to tens of thousands of km in horizontal scale.

Figure 21.1 compares the variance of daily mean rain rate with the r.m.s. amplitude of the mean diurnal cycle in rain rate, sampled at 3-hourly intervals, a measure of the prominence of the diurnal cycle and its higher harmonics. The patterns are similar in some respects and both bear some resemblance to the annual mean rain rate pattern shown in the top panel of Fig. 14.1. But upon close inspection, it is evident that the diurnal variability tends to be more prominent over land, especially near coastlines. These distinctions between the patterns of day-to-day and diurnal variability reflect the different mechanisms that give rise to the convection. In contrast to the day-to-day variability, which is driven mainly by spatial gradients in the moisture field, the diurnal variability is driven mainly by daytime diabatic heating in the boundary layer over land, which gives rise to large values of convective available potential energy (CAPE) around the middle of the day. Convection over land is accordingly more vigorous than over the oceans, with higher rain rates[2] and a higher frequency of occurrence of lightning, as shown in Fig. 21.2. Near coastlines and over mountain ranges sea breezes and mountain-valley winds set the timing of the convection. For example, Fig. 21.3 shows the frequency of lightning over parts of the Maritime Continent evolves from mid-day (top) when the convective cells are concentrated over land, close to the coastline, to evening (middle), by which time they have propagated into the interior of the islands, to late night and early morning (bottom) when they are situated over the waters just offshore. Gravity and IG wave propagation in the direction normal to the coastline is clearly evident in the accompanying animation on the companion web page. The phase speed of these (internal) gravity waves, which is determined by their vertical structure and the ambient stratification, is typically on the order of $15 \, \mathrm{m \, s^{-1}}$.[3]

---

[1] King et al. (2015).

[2] Venugopal and Wallace (2016).
[3] Mapes et al. (2003); Ruppert et al. (2020).

Table 21.1 *Sections of prior chapters in which gravity and inertio-gravity waves are mentioned.*

Section	Topic
1.5	Gravity waves are one of three classes of waves that emerge in the solutions of the Navier–Stokes equations.
2.9	Gravity waves serve as "messengers" capable of transmitting dynamical "signals" over long distances through the atmosphere.
3.1	Orographically induced, vertically propagating gravity waves extract momentum from the flow in the upper atmosphere, where they break and deposit it in the boundary layer, where they are generated by pressure gradients across mountain ranges.
10.2	Gravity waves are solutions of the shallow water wave equations. There exists a family of equatorially trapped planetary waves comprising Kelvin waves, inertio-gravity waves, and mixed Rossby–gravity waves. These waves interact with the zonal mean zonal flow to produce the equatorial stratospheric QBO.
11.2	Terrain-related "surface roughness" associated with mountains and hills with horizontal wavelengths shorter than the model grid spacing gives rise to gravity waves that act as a brake on the upper-level westerlies in the Northern Hemisphere winter circulation.
15.3	Subsidence warming associated with gravity and IG waves is instrumental in homogenizing the temperature field in the tropical troposphere and keeping it in WTG balance.
19	Convectively coupled equatorially trapped planetary waves, including IG waves, are clearly discernible in the tropical troposphere.

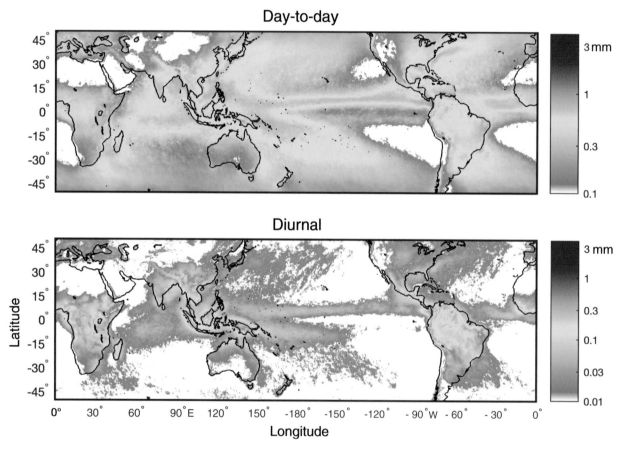

Figure 21.1 Annual mean r.m.s. variances of (top) 24-hour running mean rain rate and (bottom) mean rain rate at the eight observing times 00, 03, 06 ⋯ UT. Note that the lower bound of the color scale in the bottom panel is an order of magnitude lower than that in the top panel. Courtesy of Katrina Virts.

If the diurnal cycle and its higher harmonics assumed the form of discrete spectral lines, its entire signature would be geographically anchored, and fully captured by binning the data by time of day and averaging, for example, as was done in producing Fig. 21.3. However, in reality they are spectral bands, with enhanced variance at neighboring frequencies.

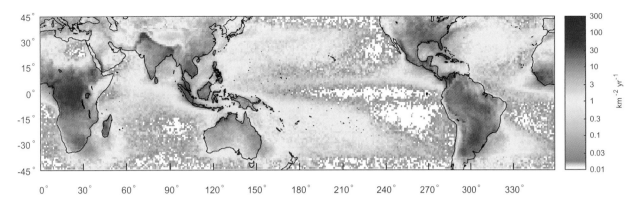

Figure 21.2 Annual mean frequency of occurrence of lightning, in strokes per square kilometer per year. Courtesy of Katrina Virts.

Annual mean lightning frequency

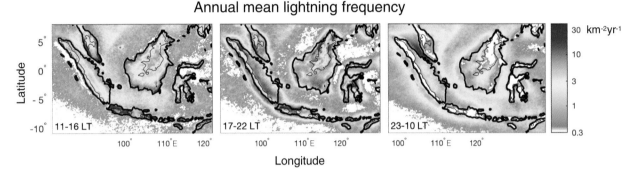

Figure 21.3 Annual mean frequency of occurrence of lightning over the Martime Continent during the afternoon, evening, and late night and morning hours as indicated. LT refers to local time. The units are in flashes km^{-2} yr^{-1}. An hour-by-hour animation is available as a supplement to this figure on the companion web page. Courtesy of Katrina Virts.

It follows that amplitude and phase of the response to the diurnal periodicity in insolation varies slightly from cycle to cycle, as reflected in the time–longitude sections shown in Figs. 19.1 and 19.2, and is consequently not fully geographically anchored and not so easily removed from time series.

Exercise

21.1 The diurnal variations discussed in this section are the response to day–night differences in diabatic heating. Under what conditions can diurnal variations in frictional drag induce a dynamical response?

21.2 ATMOSPHERIC TIDES

The diurnal variability described in the previous section is mainly boundary forced. Much of it derives from the fact that land surfaces heat up more in response to the insolation during the daylight hours and cool off more at night than the sea surface does, imparting a land–sea contrast to the "skin temperature," which is distributed through the boundary layer by the sensible heat fluxes and through the depth of the troposphere by condensation heating in deep convection. In addition to this boundary-forced component, there exists a radiatively forced component that derives from the diurnal cycle in insolation and its higher harmonics. The term *atmospheric tide* refers to this radiatively forced component (the so-called "thermal tide," abbreviated by S_1, S_2, etc., to denote the diurnal cycle and its higher harmonics), but it may also include the "lunar tide" $L_1, L_2 \ldots$ induced by the combined gravitational fields of the Sun and Moon, of which the second harmonic L_2 is the strongest. The insolation that drives the thermal tide varies almost sinusoidally with time during the daylight hours but is zero during the night and hence projects strongly upon both S_1 and S_2. The radiative heating is dominated by the absorption of ultraviolet radiation in the ozone layer and by the absorption of visible and near infrared radiation by water vapor and clouds in the troposphere.

Classical tidal theory is largely based on Laplace's tidal equations – linear solutions of the shallow water wave equations (10.2)–(10.4) in a spherical domain. The radiative forcing is prescribed in terms of a separable function of latitude, longitude, and time of day at each level (i.e., as if the Earth were an aquaplanet). The meridional dependence of tidal amplitude is represented in terms of *Hough functions*, which play a role analogous to the Hermite polynomials used to represent equatorially trapped waves.

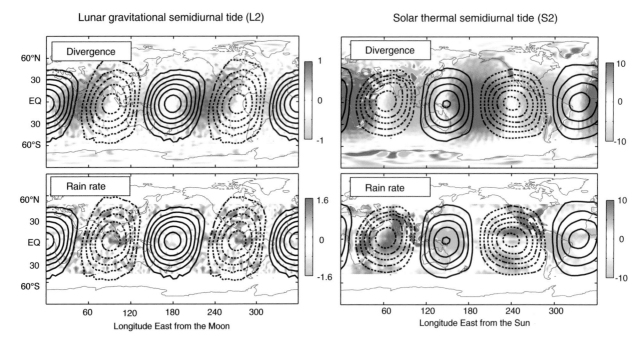

Figure 21.4 Semidiurnal cycles in the Moon's gravitational tide (L$_2$ left panels) and the solar thermal time (S$_2$, right panels), computed using four times daily data. Contours represent mass-weighted vertically averaged geopotential height in the 1000–100 hPa layer, departures from the zonal mean at each latitude. Contour interval 0.1 m for L$_2$ and 2 m for S$_2$; the zero contours are omitted and negative departures are indicated by dashed contours. The coastal geometry is applicable to the times when the Moon and Sun are directly over the Date Line and the Greenwich Meridian. The colored shading in the top panels indicate the mass-weighted horizontal divergence vertically averaged over the same layer in units of 10^{-8} s^{-1}. The shading in the bottom panels indicates corresponding departures from the zonal mean rain rate in units of μm h^{-1}. From Kohyama and Wallace (2016). © The Authors.

Atmospheric tides are well represented in fields derived from state-of-the-art reanalyses. The semidiurnal tides S$_2$ and L$_2$ in the geopotential height Z field exhibit a zonal wavenumber 2 structure centered over the equator, as shown in Fig. 21.4. Horizontal divergence varies in quadrature with geopotential, the convergence preceding the maxima in Z. Expansion of air parcels in advance of the Z minimum results in adiabatic cooling, which raises the relative humidity and thereby increases the rain rate, as explained in Section 14.6. The vertical structure of S$_2$ and L$_2$ is barotropic in the troposphere, where S$_2$ is about 20 times as strong as L$_2$, consistent with tidal theory.[4] Above the 100 hPa level L$_2$ amplifies much more rapidly with height than S$_2$. At the stratopause (1 hPa), S$_2$ is still stronger than L$_2$, but only by a factor of ~3. The classical (i.e., radiatively driven) diurnal tide S$_1$ is overshadowed by the boundary forced component described in the previous section. Considered in isolation, the radiatively-forced component of S$_1$ is stronger than S$_2$ in the troposphere and it amplifies with height much more rapidly than S$_2$ in the stratosphere.

The tidal signal is manifested in the two-sided spectrum of SLP shown in Fig. 21.5 as a set of very thin horizontal bands centered on the diurnal period and its higher harmonics. Because of the geographically fixed boundary-forced component, peaks are observed on both sides of the spectrum, rather than only on the westward propagating side.

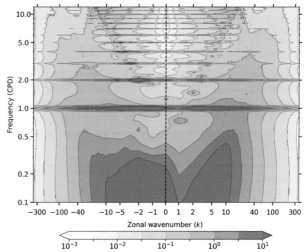

Figure 21.5 Two-sided wavenumber–frequency spectrum of tropical (30°N–30°S) SLP computed latitude-by-latitude and then meridionally averaged. From Pahlavan et al. (2022). © American Meteorological Society. Used with permission. This preliminary version has been accepted for publication in the *Journal of Atmospheric Sciences* and may be fully cited. The final typeset copyedited article will replace the EOR when it is published.

Exercises

21.2 Brier and Bradley (1964) showed observational evidence of a lunar synodic cycle in precipitation over the United States, with peak values a few days after the new and full moon. Could the existence of such a cycle conceivably be explained in terms of the tidal oscillations described in this section?

---

[4] Chapman and Lindzen (2012).

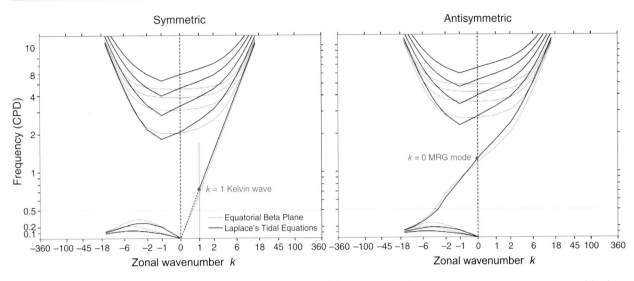

Figure 21.6 Dispersion curves (i.e., frequency ω as a function of zonal wavenumber k) for zonally propagating waves in a shallow water wave equation model with an equivalent depth of 10 km. The black curves are solutions of Laplace's tidal equations for waves in a spherical domain, calculated using the Kyoto University model, and provided by Takatoshi Sakazaki. The gray curves for waves on an equatorial beta-plane are analogous to ones shown in Fig. 10.14, but in this figure the equivalent depth is much larger. As in Fig. 10.14, both wavenumber and frequency scales are logarithmic except for the former in the range $k = -1$ to $+1$ and the latter in the range $0 < \omega < 0.5$ cpd, in which the scales are linear. Courtesy of Hamid A. Pahlavan

21.3 EXTERNAL MODES

The two-sided spectra shown in Chapter 10 relate to internal, equatorially trapped waves in the tropical stratosphere. Here we consider external waves extending from pole-to-pole and through the entire depth of the atmosphere. The reader was introduced to these waves in Section 2.9. Dispersion curves for them derived from an eigenanalysis of Laplace's tidal equations for the atmosphere at rest are shown in Fig. 21.6. There exists a one-to-one correspondence between the global external modes and the modes on an equatorial beta-plane shown in Figs. 10.8 and 10.14. The global modes extend from pole-to-pole, assuming the form of Hough functions, as in classical tidal theory, rather than Hermite polynomials. An important distinction is they do not decay with increasing latitude like the equatorially trapped modes do. The Kelvin wave has a dual identity as a gravity wave and as an acoustic wave in which the motion is purely horizontal, and is sometimes referred to as a *Lamb wave*.[5] While the curve for it extends all the way down to the origin in the figure, it should be recognized that the gravest Kelvin wave mode corresponds to $k = +1$, indicated by the small red circle: the dotted portion of the curve to the left of $k = 1$ exists in theory only. There is no equatorially symmetric solution for $k = 0$, as explained in Exercise 21.3.

Despite the one-to-one correspondence between them, there are some noticeable differences between the respective global and equatorially trapped modes. Compared to the equatorially trapped modes, the global modes asymptote much more slowly to the gravity wave solutions, lending greater importance to the distinction between gravity waves and the corresponding IG waves. Whereas the frequencies of

eastward and westward propagating equatorially trapped IG waves of a given zonal wavenumber k are almost identical, the westward propagating global waves exhibit somewhat lower frequencies than their eastward propagating counterparts. The differences range up to nearly 50% for the lowest zonal wavenumber.

The external modes are barely discernible in the two-sided spectra for 50 hPa wind components and temperature, shown in Chapter 10. They are more prominent in the spectra for geopotential height, but even in these, they account for only a small fraction of the variance. They show up more clearly as isolated peaks in normalized coherence spectra between SLP and 200 hPa height shown in Fig. 21.7. (The normalized cospectrum is analogous to a frequency dependent correlation coefficient. It is obtained by dividing the cospectrum in each frequency band by the square root of the product of the powers of the two variables in that band.) The spectral peaks associated with the external modes are situated along the dispersion curves (repeated from previous figure) at $k = 1, 2, 3\ldots$. The assumed value of 10 km for the equivalent depth h_e yields the best fit between theory and observations.[6]

The observed normalized cospectra fit the theoretical dispersion diagram rather well. The Kelvin (a.k.a. Lamb) wave shows up clearly in the left panel, extending upward and to the right, connecting a series of spectral peaks with periods of $33, 16.5, 11 \cdots$ hours, corresponding to zonal wavenumbers $k = 1, 2, 3 \cdots$ respectively. Peaks along the dispersion curve for the $n = 0$ (MRG) wave are evident in the right panel: the ones for positive zonal wavenumbers correspond to the gravest of the eastward propagating, equa-

[5] Vallis (2017).

[6] Sakazaki and Hamilton (2020) provide a comprehensive review of the literature relating to observational evidence of external modes and a more detailed analysis of the spectral peaks that appear in Fig. 21.7.

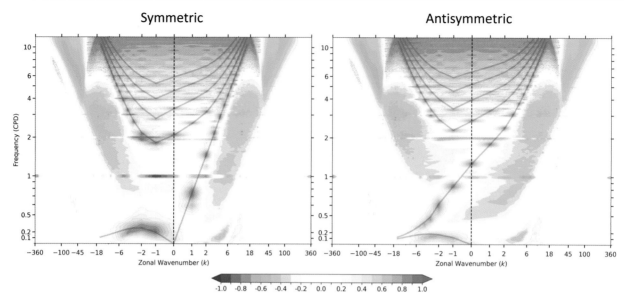

Figure 21.7 Two-sided wavenumber–frequency normalized cospectrum between the 200 hPa geopotential height and SLP for the tropics (30°N–30°S) computed latitude-by-latitude and then meridionally averaged and partitioned into equatorially symmetric and antisymmetric components as indicated. The larger the absolute value of the normalized cospectrum, the stronger the linear relationship between the geopotential height perturbations at the two levels. Solar and lunar tides are partially filtered out. The dispersion curves, provided by Takatoshi Sakazaki (Kyoto University), are derived from Laplace's tidal equations for a resting global atmosphere with an equivalent depth of 10 km. The wavenumber and frequency scales are logarithmic except for the former in the range $k = -1$ to $+1$ and the latter in the range $0 < \omega < 0.5$ cpd, in which the scales are linear. Courtesy of Hamid A. Pahlavan.

torially antisymmetric IG waves and those for negative zonal wavenumbers correspond to the gravest of the westward propagating, equatorially antisymmetric Rossby waves. For the very low zonal wavenumbers, the frequency of the $n = 0$ wave is intermediate between that of Rossby and IG waves and clearly distinguishable from both of them. The geopotential height patterns for respective pairs $k = \pm 1$, $\pm 2 \cdots$ are identical, but the wind fields are entirely different.[7] The westward propagating MRG waves exhibit a stronger wind signature than the eastward propagating ones by virtue of their much stronger nondivergent component. The zonally symmetric $k = 0$ peak involves an exchange of mass between Northern and Southern Hemispheres, as elaborated on in Exercise 21.3b. "Families" of peaks for other IG modes are also evident. The counterpart of the equatorially symmetric $n = 1$ equatorially trapped mode forms the left side of the V-shaped configuration that asymptotes to the westward propagating gravity wave at high frequencies. The $n = 2$ mode is also discernible. That coherences are generally higher inside and above the V-shaped configuration than outside and below it presumably reflects the absence of vertically propagating internal gravity waves, which detract from the coherence between geopotential height perturbations at different levels.

At frequencies < 0.5 cycles per day the cospectra for both symmetric and antisymmetric modes, shown in Fig. 21.7, are dominated by Rossby waves, both westward and eastward propagating. The westward propagating waves are the external modes discussed above. The westward $k = -1$ through $k = 4$ external Rossby modes have been shown to modulate tropical rain rate[8] but they propagate

westward much too rapidly for them to be convectively coupled. The frequency range of these signals in Fig. 21.7 does not extend down into the range of the convectively coupled equatorial Rossby waves discussed in Section 18.9. The eastward propagating $k = 6$ to 10 signals reflect the impact of synoptic scale waves impinging from extratropical latitudes, which occurs primarily in regions of low latitude westerlies, as discussed in Section 19.3. As in the spectra of 500 hPa Z shown in Fig. 13.30, these extratropical Rossby waves are advected eastward by the westerly background flow in accordance with Eqs. (11.1) and (11.2).

Exercises

21.3 (a) Show that a $k = 0$ mode with a meridional structure like that of a Kelvin wave could not exist because it would violate mass continuity. (b) Show how the $k = 0$, $n = 0$ mode, which lies on the dispersion curve for the MRG wave, (qualitatively) satisfies the shallow water wave equations (10.2)–(10.4). (c) Are there other zonally symmetric external modes?

21.4 The normal modes that are responsible for the isolated spectral peaks in the coherence spectrum shown in Fig. 21.7 are responsible for only a minute fraction of the variance of the tropical geopotential height field. Why do they show up so clearly?

21.5 Rossby waves with zonal wavenumbers $k > 3$ do not exhibit clearly discernible peaks in the coherence spectra shown in Fig. 21.7. Explain

---

[7] Pole-to-pole meridional profiles of geopotential for each mode are shown in Fig. 2 of Sakazaki and Hamilton (2020).

[8] King et al. (2015); Sakazaki (2021).

21.4 GRAVITY WAVES: AN OVERVIEW

Gravity and IG waves are an integral part of the atmospheric general circulation. In Chapter 2, we showed observational evidence of external gravity waves radiating out from volcanic eruptions and propagating all the way to the antipodal point and back. In Chapter 3, we argued that orographically induced waves with scales too small to be resolved, even in todays high resolution models, exert a strong drag force on the zonal winds at atmospheric levels extending from the troposphere upward into the mesosphere. In Chapter 5, we argued that they are instrumental in dispersing the heat released in mesoscale updrafts into the broader scale tropical environment. In Chapter 10, we showed the shallow water wave equations, which are the foundation for our theoretical understanding of gravity waves. In the context of this framework, we drew a distinction between external

modes (horizontally propagating waves with an equivalent barotropic vertical structure) and internal (vertically propagating) modes and we documented the existence of a family of equatorially trapped, internal, planetary-scale modes in the equatorial stratosphere. In Chapter 15, we showed observational evidence of the existence of gravity waves in the tropical troposphere that are instrumental in enforcing WTG balance in the tropics. For further specifics with regard to the sections and diagrams in previous chapters that relate to gravity waves, see Table 21.2.

On a day-to-day basis, the signature of internal gravity waves is clearly evident without recourse to filtering in the ERA5 Reanalysis lower stratospheric vertical velocity field shown in Fig. 21.8. The waves in this still image appear to be emanating out of organized convection in the troposphere, and this impression is confirmed by viewing sequences of snapshots at hourly intervals (not shown). These are just

Table 21.2 *Prominent phase speeds of inertio-gravity waves observed in the Earth's atmosphere as inferred from two-sided spectra of various fields, as indicated.*

Phase speed (equiv. depth)	Atmospheric variables	Figures	Type
310 m s^{-1} (10 km)	SLP; coherence between Z_{200} and SLP	2.19, 21.7	external
50 m s^{-1} (255 m)	stratospheric T	10.13	internal
	stratospheric u, v, T, Z	10.15, 10.16	
	tropospheric T	15.5	
23 m s^{-1} (54 m)	stratospheric u, v, T, Z	10.15, 10.16	internal
	tropospheric T	15.5	
15 m s^{-1} (23 m)	rain rate	19.3	convectively coupled

Figure 21.8 The top panel shows weather disturbances over the tropical western Pacific at 0200 UTC August 7, 2015, rendered in visible satellite imagery. The bottom panel shows a nearly concurrent instantaneous map of the 50 hPa vertical velocity field. The black contours indicate the 250 K brightness temperature. From Pahlavan et al. (2022). © American Meteorological Society. Used with permission. This preliminary version has been accepted for publication in the *Journal of Atmospheric Sciences* and may be fully cited. The final typeset copyedited article will replace the EOR when it is published.

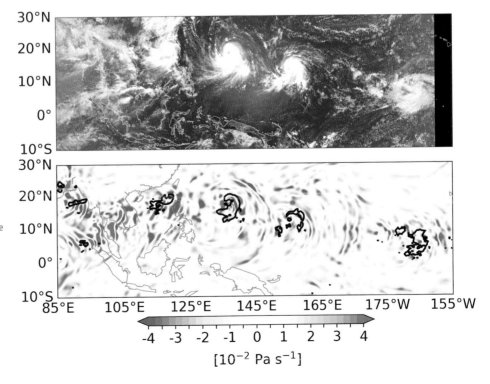

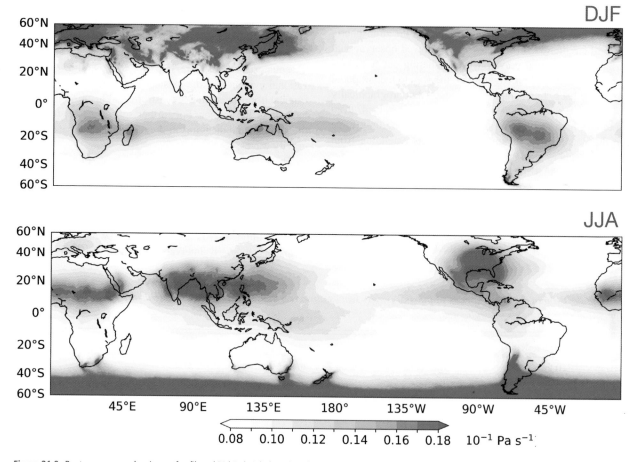

Figure 21.9 Root mean squared variance of unfiltered 50 hPa height based on data at 1-hour intervals. (Top) DJF; (bottom) JJA. From Pahlavan et al. (2022). © American Meteorological Society. Used with permission. This preliminary version has been accepted for publication in the *Journal of Atmospheric Sciences* and may be fully cited. The final typeset copyedited article will replace the EOR when it is published.

the resolved waves. Satellite imagery for the water vapor channel reveals a much richer spectrum of internal gravity waves, as shown in the solution for Exercise (21.7).

The climatology of the r.m.s. amplitude of these resolved gravity waves, as expressed in the temporal r.m.s. variance of the (unfiltered) 50 hPa vertical velocity field, is shown in Fig. 21.9. Consistent with the instantaneous distribution shown in the previous figure, the variance in tropical latitudes tends to be higher in the regions of heavy rain rate, which are biased toward the summer hemisphere. Variance is also enhanced over the ITCZ, but not as much as over the continental rain areas, where the deep convection is more vigorous.

As noted in Chapter 3, orographically induced gravity waves also play an important role in the general circulation. The climatological mean DJF vertical velocity field shown in Fig. 21.10 bears the imprint of the wrinkles in the Earth's crust on scales ranging from planetary down to the smallest resolvable scales in the ERA5 Reanalysis. The fine structure in this pattern, which becomes more apparent when the image is magnified, is largely a gravity wave signature. Not only does it exert a drag force on the upper atmosphere: it also mediates microclimates in regions of rough terrain.

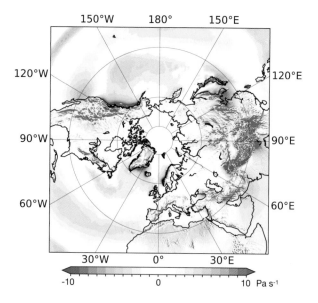

Figure 21.10 DJF climatological mean 500 hPa vertical velocity, the same as in the left panel of Fig. 11.3, but based on the ERA5 Reanalysis. In order to fully appreciate this figure, it is necessary to do in on specific regions and relate them to local topographic features. Courtesy of Hamid A. Pahlavan.

An orographic signature is also evident in Fig. 21.9, most clearly over extratropical latitudes of the winter hemisphere.[9]

21.5 GRAVITY WAVES IN THE GLOBAL MECHANICAL ENERGY SPECTRUM

It is instructive to partition the global mechanical energy spectrum shown in Fig. 6.9 into the contributions from "fast" and "slow: that is, higher frequency IG waves in which air parcels experience large accelerations, and lower frequency Rossby waves, in which the horizontal wind and temperature fields are in thermal wind balance and the Doppler-shifted phase speeds are much slower than those in IG waves. The flow in IG waves is fully three-dimensional and therefore its kinetic, potential, and mechanical energy are theoretically predicted to decrease in inverse proportion to wavenumber k to the 5/3 power.[10] In contrast, in Rossby waves, the flow is constrained by the Earth's rotation to be quasi-two-dimensional; that is, at extratropical latitudes the vertical velocity is nearly an order of magnitude smaller than in IG waves of comparable amplitude in the horizontal wind field.[11] This constraint renders the energy cascade from larger toward smaller scales less efficient, resulting in a steeper drop-off of energy with increasing zonal wavenumber. Turbulence theory predicts that the spectrum of such quasi-two-dimensional turbulence should follow a k^{-3} power law, as opposed to a $k^{-5/3}$ power law that prevails in fully three-dimensional flow, and that at wavelengths longer than the forcing, energy should be transferred from higher toward lower wavenumbers.[12]

With these theoretical predictions in mind, let us re-examine the mechanical energy spectrum shown in Fig. 21.11, the same one as in Fig. 6.9, but here partitioned into IG and Rossby wave contributions by performing a modal decomposition upon the wind, temperature, and geopotential height fields. It is evident that at scales smaller than baroclinic waves, the IG wave spectrum follows the $-5/3$ power law and the Rossby wave spectrum follows the -3 power law, in agreement with theory. Rossby waves are dominant at the larger scales, but because of the different slopes of the spectra, IG waves become increasingly prominent at the higher zonal wavenumbers. In the ERA Interim reanalyses IG waves make an appreciable contribution to the mechanical energy spectrum beyond $k \sim 12$, and beyond $k \sim 50$ they account for most of the energy. Integrated over the entire spectrum, Rossby waves account for slightly over 90% of the mechanical energy and IG waves account for the remainder. Turbulence theory also explains the buildup

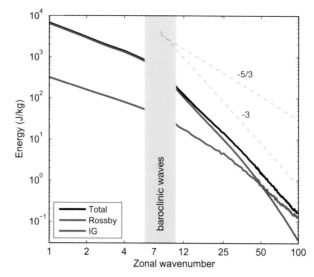

Figure 21.11 Mechanical energy ($A + K$) spectra from ERA Interim reanalysis plotted on a log–log scale, partitioned into Rossby waves and inertio-gravity waves by projecting data onto the theoretically-derived normal modes. Adapted from Žagar et al. (2017). © American Meteorological Society. Used with permission. For an analogous plot based on ERA5 data, see Stephan and Mariaccia (2021).

of energy in planetary-scale Rossby waves during the later stages of the "spin up" of general circulation models (Fig. 6.8) and the concentration of mechanical energy at lowest wavenumbers and the lowest frequencies in extratropical spectra (Fig. 13.30).

Observations of wind and temperature at discrete points are influenced by both Rossby waves and IG waves. The numerical weather prediction models in use in the 1970s and 1980s were incapable of resolving, much less predicting all but the largest scale IG waves. The goal of the early data assimilation schemes was to eliminate them from the analysis so that the more slowly evolving Rossby modes would be more accurately represented in the initial conditions for the numerical integrations. Today's state-of the-art operational numerical weather prediction models include the "fast" IG modes as well, and operational centers are experimenting with nonhydrostatic models that represent low frequency acoustic waves as well as Rossby and IG waves.

Exercises

21.6 Given that gravity waves are so prominent in vertical velocity at the 50 hPa level in Fig. 21.8, why is there no trace of them in the time–height sections for 50 hPa temperature and meridional wind shown in Figs. 10.6 and 10.7?

21.7* Show examples of internal gravity waves as revealed by satellite imagery.

---

[9] See also Podglajen et al. (2020).
[10] Kolmogorov (1962).
[11] See Chapter 11 of Vallis (2017) for a discussion of energy scaling.
[12] Kraichnan (1959); Charney (1971).

Appendix A Space and Time Averaging Operations

A.1 TIME AND ZONAL AVERAGING

The description and interpretation of the general circulation rely heavily on space and time means of the variables themselves and of their variances and covariances. For a two-dimensional field, the space domain of the analysis might extend over the entire Earth, it might be restricted to the extratropics of one hemisphere, or it might be even more strongly restricted. The interval used for the time averaging should be long enough to comprise a sufficient number of samples of the phenomena of interest to ensure that the results are statistically significant. Just how long it must be is dependent upon the application. In this book, the emphasis is on hemispheric and global domains and the reader can be assured that the sampling intervals are long enough that statistical significance is not at issue.

A.1.1 Basic Definitions

We will be concerned with terms of the form $[\overline{xy}]$, where x and y represent any scalar meteorological parameters (e.g., u, v, T, Φ etc.). The overbar and the brackets define the time and zonal averaging operators,

$$\bar{x} = \frac{1}{T} \int_0^T x \, dt \tag{A.1}$$

$$[x] = \frac{1}{2\pi} \oint x \, d\lambda, \tag{A.2}$$

where t is an extended time averaging interval such as used for calculating climatologies, and λ is longitude. Departures from a time average

$$x' = x - \bar{x} \tag{A.3}$$

will be referred to as transients and departures from a zonal average

$$x^* = x - [x] \tag{A.4}$$

will be referred to as eddies. By definition

$$\overline{x'} = 0 \tag{A.5}$$

and

$$[x^*] = 0. \tag{A.6}$$

Double averaging operators are redundant,

$$\bar{\bar{x}} = \bar{x} \tag{A.7}$$

and

$$[[x]] = [x]. \tag{A.8}$$

The time and zonal averaging operators are commutative,

$$\overline{[x]} = [\bar{x}]. \tag{A.9}$$

In order to visualize these identities, one can imagine x as representing a two-dimensional "observation matrix", x_{ij} where the index i refers to the longitude of a particular observation and j to its order in the time sequence. The terms in Eq. (A.9) represent averages over all the elements in the matrix. The order of the averaging (i.e., whether by rows first or by columns first) is irrelevant.

A.1.2 Decomposition of Product Terms

Time averages of products can be expanded in the form

$$\overline{xy} = \overline{(\bar{x} + x')(\bar{y} + y')}$$
$$= \bar{x}\,\bar{y} + \overline{x'y'} + \overline{x'\bar{y}} + \overline{\bar{x}y'}.$$

Since $\overline{x'\bar{y}} = 0$ and $\overline{y'\bar{x}} = 0$ and the double averaging operator on the first term is redundant, it follows that

$$\overline{xy} = \bar{x}\,\bar{y} + \overline{x'y'}, \tag{A.10}$$

where the first term on the right-hand side can be identified with the time mean and the second term with the transients. In a similar manner

$$[xy] = [x][y] + [x^*y^*], \tag{A.11}$$

where the first term involves zonal means and the second term involves eddies. The zonal mean of Eq. (A.10)

$$[\overline{xy}] = [\bar{x}\,\bar{y}] + \left[\overline{x'y'}\right]$$

can be expanded by writing $\bar{x} = [\bar{x}] + \bar{x}^*$ and $\bar{y} = [\bar{y}] + \bar{y}^*$ in the first term on the right-hand side. After simplifying the resulting equation, we obtain

$$[\overline{xy}] = [\bar{x}][\bar{y}] + [\bar{x}^*\,\bar{y}^*] + \left[\overline{x'\,y'}\right]. \tag{A.12}$$

In this formulation,[1] the first term is identified with the zonally averaged, time-mean circulation and resolves features such as the climatological-mean Hadley cell. The second term is identified with eddies that are steady in time and resolves features such as the monsoons. These features have come to be referred to as "stationary waves" although

---

[1] First derived by Priestly (1949).

362

the term "steady eddies" would be more consistent with the nomenclature used here. The final term gives the total contribution of the transients.

In an analogous manner, Eq. (A.11) can be time averaged and the first term on the right-hand side can be expanded in terms of time means and transients, which yields

$$\overline{[xy]} = [\bar{x}][\bar{y}] + \overline{[x]'[y]'} + \overline{[x^*y^*]}. \tag{A.13}$$

In this formulation,[2] the first term on the right-hand side is identical to the corresponding term in Eq. (A.12). The second term is the contribution from the transient zonally symmetric circulations and the third term is the total contribution from the eddies. Equation (A.13) can be derived directly from Eq. (A.12) simply by replacing all zonal averages and departures from zonal averages by time averages and departures from time averages, or alternatively, Eq. (A.12) could be derived from Eq. (A.13) by the reverse procedure.

The relationship between Eqs. (A.12) and (1.13) can be clarified by expanding $[\overline{xy}]$ as follows. Following Lorenz (1953) we begin by writing

$$x = [\bar{x}] + \bar{x}^* + [x]' + x'^*$$

and

$$y = [\bar{y}] + \bar{y}^* + [y]' + y'^*,$$

where the first term on the right-hand side refers to the time-mean, zonal-mean contribution, the second term to the zonally varying, time-mean contribution, the third term to the time-varying, zonal-mean contribution and the fourth term to the *transient eddy* contribution. After taking the product and averaging over time and longitude, we obtain

$$\overline{[xy]} = [\bar{x}][\bar{y}] + \overline{[x]'[y]'} + [\bar{x}^*\bar{y}^*] + \left[\overline{x'^*y'^*}\right]. \tag{A.14}$$

The first three terms on the right-hand side of Eq. (A.14) also appear in Eq. (A.12) and/or Eq. (A.13). The remaining term is the contribution of the transient eddies. By combining these three equations in various ways we obtain the identities

$$\left[\overline{x'y'}\right] = \overline{[x]'[y]'} + \left[\overline{x'^*y'^*}\right] \tag{A.15}$$

and

$$\overline{[x^*y^*]} = [\bar{x}^*\bar{y}^*] + \left[\overline{x'^*y'^*}\right], \tag{A.16}$$

which refer, respectively, to the contributions from all the transients (both zonal mean and eddies) and all the eddies (both time and transient).

Table A.1 shows convenient way of summarizing these relationships.

A.1.3 Statistical Interpretation

With the exception of the $[\bar{x}]\ [\bar{y}]$ term, the terms in the above expansions involve covariances (i.e., averages of products). For example, $\overline{[x'y']}$ is the temporal covariance between x and y and $[\bar{x}^*\ \bar{y}^*]$ is the longitudinal covariance

Table A.1 *Summary of the interrelationships between the space–time covariance terms. defined in this section.*

	Zonally symmetric	Eddy	Sum
Steady	$[\bar{x}]\,[\bar{y}]$	$[\bar{x}^*\,\bar{y}^*]$	$[\bar{x}\,\bar{y}]$
Transient	$\overline{[x]'[y]'}$	$\left[\overline{x'^*y'^*}\right]$	$[\overline{x'y'}]$
Sum	$\overline{[x][y]}$	$[\overline{x^*y^*}]$	$[\overline{xy}]$

between $\bar{x}$ and $\bar{y}$. The covariance between x and y can be expressed as the product of the correlation coefficient r between x and y and the product of the standard deviations of x and y and similarly for spatial or temporal averages of x and y. For example, the temporal covariance is given by

$$\overline{x'y'} = r(x,y)\sqrt{\overline{x'^2}\,\overline{y'^2}} \tag{A.17}$$

and the longitudinal covariance between $\bar{x}$ and $\bar{y}$ is given by

$$\left[\bar{x}^*\bar{y}^*\right] = r(\bar{x}^*,\bar{y}^*)\sqrt{\bar{x}^{*2}\,\bar{y}^{*2}}. \tag{A.18}$$

The standard deviations are the same as the r.m.s. amplitudes, which can often be roughly estimated on the basis of inspection of time series or maps. The correlation coefficient r ranges from -1 to $+1$: a high absolute value of r implies a strong linear relationship between the variables. The square of the correlation coefficient r^2 is the fraction of the variance of one variable that can be explained based on a knowledge of the other variable, assuming that the relationship is linear. The correlation coefficient r is also the slope of the least squares best fit regression line on a scatter plot of standardized values of the two variables.

In assessing the statistical significance of covariance statistics the Student's t-test is applied to the correlation coefficient using the Fisher transform. In choosing the appropriate number of statistical degrees of freedom to use in the formula, care must be taken to correct for the autocorrelation inherent in the time series of the variables.

A.1.4 Physical Interpretation

The variables x and y used in this section could represent any atmospheric scalar field, but we will be using them most often in reference to transports of quantities such as angular momentum, energy, and water vapor by the three-dimensional wind field, evaluated at fixed grid points. In this case, we can think of x as the concentration of the quantity that is being transported (e.g., angular momentum per unit mass of air, m), and y as the velocity component in the direction of the flux or transport. For example, if the poleward transport of angular momentum is of interest, then y would represent the meridional velocity component v. and the counterpart of the eddy covariance term $[x^*y^*]$ would be $[m^*v^*]$, the poleward eddy transport of angular momentum.

A.2 GLOBAL AVERAGING

In the foregoing section, spatial averaging was synonymous with zonal averaging. In principle, these operations can be performed for any prescribed spatial domain, one dimensional or two dimensional and the relationships derived in the previous section are generally applicable. For some purposes it is useful to consider global averages, denoted in this book by angle brackets $\langle x \rangle$ and departures from the global average, denoted by $\widehat{x}$,

$$x = \langle x \rangle + \widehat{x}. \tag{A.19}$$

Global averages are pole-to-pole, cosine-weighted averages of zonally averaged quantities. Hence,

$$\langle x \rangle = \langle [x] \rangle. \tag{A.20}$$

The [ ] operator, though redundant, is helpful in the sense that it serves as a recipe for computing global averages. Another useful identity is

$$x = \langle x \rangle + [\widehat{x}] + x^*, \tag{A.21}$$

and, owing to the mutual spatial orthogonality of these three components in a global domain, the variance of x can be partitioned in the same way,

$$x^2 = \langle x \rangle^2 + [\widehat{x}]^2 + x^{*2}. \tag{A.22}$$

A.3 VERTICAL AVERAGING

Fields of energy-related variables and concentrations of trace constituents are sometimes vertically integrated or averaged over the mass of a column extending from the Earth's surface to the "top of the atmosphere" or over the depth of a discrete layer such as the troposphere or the boundary layer. The vertical integration is most straightforward in pressure coordinates: when geometric height or the log of pressure is used as the vertical coordinate, the field variable needs to be weighted by density, ρ. To denote the mass-weighted, vertically integrated field variable x, we will use curly brackets $\{x\}$. An example is column-integrated water vapor W, which is often analyzed in combination with two-dimensional fields such as precipitation P and evaporation E.

Appendix B The Zonal Momentum Balance

The equation that governs the local time rate of change of zonal wind can be written in pressure coordinates in the form

$$\frac{\partial u}{\partial t} = -u\frac{\partial u}{\partial x} - v\frac{\partial u}{\partial y} - \omega\frac{\partial u}{\partial p} + \frac{uv\,tan\phi}{R_E} - \frac{\partial \Phi}{\partial x} + fv + \mathcal{F}_x$$

(B.1)

where $x = R_E\,cos\phi\,\lambda$ is the zonal coordinate, $y = R_E\,cos\phi$ is the meridional coordinate, λ is longitude, ϕ is latitude, and R_E is the radius of the Earth.[1] The terms uw/R_E and $2\Omega\omega cos\phi$ have been neglected because they are at least two orders of magnitude smaller than the corresponding terms in v. The advective terms in Eq. (B.1) can be rewritten in the form

$$-u\frac{\partial u}{\partial x} - v\frac{\partial u}{\partial y} - \omega\frac{\partial u}{\partial p}$$

$$= -\frac{\partial}{\partial x}u^2 - \frac{1}{cos\phi}\frac{\partial}{\partial y}uv\,cos\phi - \frac{\partial}{\partial p}u\omega$$

$$+ u\left(\frac{\partial u}{\partial x} + \frac{1}{cos\phi}\frac{\partial}{\partial y}v\,cos\phi + \frac{\partial\omega}{\partial p}\right)$$

where the term in parentheses vanishes because of the continuity of mass. Substituting back into Eq. (B.1) and making use of the identity

$$\frac{1}{cos^2\phi}\frac{\partial}{\partial y}\left(uv\,cos^2\phi\right) = \frac{1}{cos\phi}\frac{\partial}{\partial y}uv\,cos\phi - \frac{uv\,tan\phi}{R_E}$$

we obtain

$$\frac{\partial u}{\partial t} = -\frac{\partial}{\partial x}u^2 - \frac{\partial}{cos^2\phi}\frac{\partial}{\partial y}uv\,cos^2\phi$$

$$- \frac{\partial}{\partial p}u\omega - \frac{\partial \Phi}{\partial x} + fv + \mathcal{F}_x.$$

(B.2)

---
[1] Holton and Hakim (2012) p. 41.

In the zonal average, the terms $\partial u^2/\partial x$ and $\partial\phi/\partial x$ drop out because of the identity

$$[\partial\,(\,)\,/\partial x] = \frac{1}{R_E}\oint\frac{\partial\,(\,)}{\partial x}dx = (\,)\bigg|_0^L = 0.$$

(B.3)

Expanding the $[uv]$ and $[u\omega]$, making use of Eq. (A.11) yields

$$\frac{\partial[u]}{\partial t} = -\frac{1}{cos^2\phi}\frac{\partial}{\partial y}[u][v]\,cos^2\phi - \frac{\partial}{\partial p}[u][\omega] + f[v] + [\mathcal{F}_x]$$

$$- \frac{1}{cos^2\phi}\frac{\partial}{\partial y}[u^*v^*]\,cos^2\phi - \frac{\partial}{\partial p}[u^*\omega^*]$$

(B.4)

Writing the mean meridional motion terms in the form

$$\frac{1}{cos^2\phi}\frac{\partial}{\partial y}[u][v]\,cos^2\phi = \frac{[u]}{cos\phi}\frac{\partial}{\partial y}[v]cos\phi + \frac{[v]}{cos\phi}\frac{\partial}{\partial y}[u]cos\phi$$

and

$$\frac{\partial}{\partial p}[u][\omega] = [u]\frac{\partial[\omega]}{\partial p} + [\omega]\frac{\partial[u]}{\partial p},$$

and making use of the zonally averaged continuity equation in spherical coordinates

$$\frac{1}{cos\phi}\frac{\partial}{\partial y}[v]cos\phi + \frac{\partial[\omega]}{\partial p} = 0$$

(B.5)

we obtain, after some minor rearranging,

$$\frac{\partial[u]}{\partial t} = [v]\left(f - \frac{1}{cos\phi}\frac{\partial}{\partial y}[u]cos\phi\right) - [\omega]\frac{\partial[u]}{\partial p}$$

$$- \frac{1}{cos^2\phi}\frac{\partial}{\partial y}[u^*v^*]cos^2\phi - \frac{\partial}{\partial p}[u^*\omega^*] + [\mathcal{F}_x].$$

(B.6)

Appendix C Methods of Identifying Teleconnection Patterns

This appendix presents a brief description of the statistical analysis techniques that have been used to identify teleconnection patterns, with emphasis on the ones used in this book. All involve operations in the data matrix $\mathbf{X}$ for the variable x, assumed to be represented on a grid, whose columns (j) represent maps at various times and whose rows (i) are time series at fixed grid points. For the applications in this text, x represents departures from the seasonally varying climatology.

C.1 THE CORRELATION (OR COVARIANCE) MATRIX

For each grid point (i), it is possible to generate a *one-point (temporal) correlation vector or map* R_i, consisting of the correlation coefficient r_{ij} between the temporal variations in x_i, referred to as the "reference" grid point and the temporal variations in x at all grid points x_j. The *correlation matrix* $\mathbf{R}$, whose elements are r_{ij} is the ensemble of one-point correlation maps for all grid points. Its diagonal elements are all equal to 1 and $r_{ij} = r_{ji}$. In a similar fashion, for each grid point it is possible to generate a one-point (temporal) *covariance* matrix $\mathbf{C}$ consisting of the covariance between temporal variations of x at the ith grid point and variations in x at all grid points. The correlation matrix $\mathbf{R}$ may be derived from the covariance matrix $\mathbf{C}$ by dividing the elements in each row i by the standard deviation of x_i and then dividing the elements in each column j by the standard deviation of x_j. Another matrix of interest, $\mathbf{R}^*$, is found by dividing only each row i of $\mathbf{C}$ by the standard deviation of x_i. Equivalently, $\mathbf{R}^*$ equals the covariance between standardized x_i and non-standardized x_j. $\mathbf{R}^*$ gives the value of x_j expected when x_i has a value of 1 standard deviation based on linear regression. A row of $\mathbf{R}^*$ is called a *one-point regression map*.

C.2 INTERPRETATION OF ONE-POINT CORRELATION AND REGRESSION MAPS

One-point correlation and (regression) maps for all grid points exhibit strong positive values at the reference grid point and in the region surrounding it, but their strength declines more rapidly with distance in the maps for some reference grid points than in the maps for others. In addition, maps for some grid points exhibit dipole or wavelike patterns with significant (usually negative) correlations between

fluctuations at the reference grid point and fluctuations at distant grid points, referred to as *teleconnections*.

Grid points with strong one-point correlation (or regression) patterns are considered to have high *teleconnectivity*: that is, correlations with grid points beyond the range of the positive values surrounding the reference grid point owing to the spatial "redness" of the field. Two widely used metrics for assessing the teleconnectivity are

- The absolute value of the strongest negative correlation on its one-point correlation map, which corresponds to the strongest negative element r_{ij} in its row of the correlation matrix $\mathbf{R}$.[1]
- The length of a row vector in $\mathbf{R}$ (or $\mathbf{R}^*$), that is, the square root of the average of the squared grid point values on a one-point correlation (or regression) map, referred to as the *correlation teleconnectivity* or the *covariance teleconnectivity*, depending on which matrix it is based on. When averaging it is appropriate to weight the contribution from each grid point by the area that grid point represents. Sometimes it is appropriate to restrict the contributions to a subdomain. (For example, all plots in Chapters 12 and 13 use lengths based only on the extratropical grid points that reside in the same hemisphere as the reference grid point.)

The first of the above metrics depends entirely upon the existence of dipolar or wavelike patterns, whereas the second is more strongly influenced by the areal extent of the regions that have large positive or negative correlation (or covariance) with the reference grid point.

C.3 EMPIRICAL ORTHOGONAL TELECONNECTION PATTERNS

Empirical orthogonal teleconnection (EOT) patterns, designated EOT1, EOT2, etc., are an ordered set of mutually orthogonal spatial patterns very much like one-point correlation or regression maps.[2] In fact, the pattern of EOT1 is the map for the grid point that exhibits the strongest teleconnectivity and its time varying index is the standardized time series of x for that grid point. Orthogonality between successive EOTs is achieved by regressing the data matrix $\mathbf{X}$ upon each EOT index as soon as it is computed and, from that point onward, considering the teleconnectivity of the

---

[1] Wallace and Gutzler (1981).
[2] Van den Dool et al. (2000).

residual field, just as if it were the total field. As more and more EOTs are computed and the space-time variance that they explain regressed out of $\mathbf{X}$, the variance of the linearly independent residual field steadily declines. Useful attributes of EOT analysis are that it is simple to perform and requires relatively little computation, the results appear to be as robust as those derived from any other method, there are no user-specified parameters, and the patterns are less likely to be domain dependent than those derived by other means.

C.4 EMPIRICAL ORTHOGONAL FUNCTION ANALYSIS

Empirical orthogonal functions (EOFs) are an ordered set of spatial patterns that explain the maximal fraction of the variance of squared correlation summed over the elements of $\mathbf{R}$ or covariance summed over the elements of $\mathbf{C}$.[3] In this optimization, squared correlation or covariance is usually area-weighted: for example, for a regular latitude–longitude grid, elements of $\mathbf{R}$ or $\mathbf{C}$ are weighted by the square root of cosine of latitude.

The EOFs can be obtained either by performing singular value decomposition on the data matrix $\mathbf{X}$ or by diagonalizing $\mathbf{R}$ or $\mathbf{C}$ or their counterparts based on spatial, rather than temporal correlations or covariances, For large data matrices computation times may be substantial and highly dependent on the choice of analysis procedure. EOF analysis yields three products: (1) a set of eigenvalues listing the variance explained by each mode, (2) the EOFs themselves, which are spatial patterns, and (3) the corresponding time varying indices, referred to as *principal components* (PCs). By convention, the PC time series are standardized and the EOFs carry the amplitude information. The EOFs are thus identical to the patterns obtained by regressing the data matrix upon the standardized PC time series. In fact, the EOF patterns shown in the figures in this book were derived in that manner. Not only are the PCs temporally orthogonal like the time varying indices of the EOTs; the EOFs are spatially orthogonal as well. EOFs are the only class of functions that exhibit this kind of "bi-orthogonality."

Using the correlation matrix $\mathbf{R}$ as a basis for EOF analysis is equivalent to standardizing the rows of the data matrix, thereby giving all grid points equal weight in the analysis (apart from area weighting), irrespective of their temporal variance. Using the covariance matrix $\mathbf{C}$ retains the amplitude information in the data matrix. The EOFs in most contemporary studies are based on $\mathbf{C}$.

The advantage of EOFs is that at any truncation they explain more of the space/time variance of the data matrix than any other set of patterns. A disadvantage is that the higher order EOFs tend to be much more sensitive to sampling variability than the higher order EOTs and even the leading patterns can be sensitive to sampling variability when their eigenvalues are not well separated.[4] Even a pattern that is robust with respect to sampling variability may comprise elements that do not appear in combination in the one-point correlation maps for its centers of action, and are consequently difficult to interpret.[5] To ensure that EOFs are physically realistic, it is always worth comparing them with one-point correlation or regression maps for their primary centers of action.

C.5 EXTENDED EMPIRICAL ORTHOGONAL FUNCTION ANALYSIS

To perform an extended EOF analysis, the data matrix $\mathbf{X}$ is augmented. Each individual map (row vector) is expanded into a sequence of maps centered at time j and extending forward and backward k map times so that number of elements in the matrix is increased by the factor $(2k+1)$. $2k$ is referred to as the "analysis window." From that point onward, the analysis proceeds as in conventional EOF analysis, but instead of consisting of a single map, EOFs consist of a set of $2k+1$ maps, which depict the evolution of the pattern during the "window." The PCs still consist of single time series. Analogous map sequences can be generated using conventional EOF analysis, but the ones generated using EEOF analysis incorporate information on lagged covariance (or correlations) and may thus provide a more faithful representation of spatial patterns that evolve in a systematic way, and the analysis may be capable of separating phenomena whose spatial patterns are not orthogonal, but which evolve on different time-scales.

The analysis technique known as *singular spectrum analysis* SSA is a further elaboration of EEOF analysis.[6]

C.6 ROTATED EMPIRICAL ORTHOGONAL FUNCTION ANALYSIS

EOFs often can be simplified and rendered more robust by subjecting them to a rotation (i.e., by expressing them as linear combinations of the leading EOFs/PCs). The more widely used rotation protocols retain the orthogonality of the PCs while allowing the rotated patterns to be spatially correlated with one another. The most widely used protocol is *varimax rotation*, which maximizes the kurtosis of the grid point values of the rotated EOFs (REOFs). The user must specify the number of EOFs to be rotated. For truncations ~30, the hemispheric REOFs appear to be quite insensitive to the choice of cutoff. When the analysis is applied to Northern Hemisphere wintertime SLP, the REOFs resemble the EOTs.[7] They are noticeably simpler and demonstrably more robust than the corresponding EOFs, but at the expense of explaining smaller fractions of the hemispherically integrated variance.

[3] Lorenz (1956); Kutzbach (1970).
[4] North et al. (1982).
[5] Richman (1986); Dommenget and Latif (2002); Cheng et al. (1995).
[6] Vautard and Ghil (1989).
[7] Smoliak and Wallace (2015).

C.7 MAXIMAL COVARIANCE ANALYSIS (MCA)

Maximum covariance analysis is analogous in some respects to EOF analysis, but rather than identifying geographical patterns that optimally represent the temporal variability in a single field x, MCA finds pairs of patterns that optimally represent the temporal covariability *between* paired fields x and y. It involves performing singular value decomposition SVD upon on the temporal covariance matrix $\mathbf{C}_{xy}$, whose elements are $x'_i y'_j$. The paired variables x and y need not be on the same grid and $\mathbf{C}_{xy}$ need not be a square matrix.

If the data matrices $\mathbf{X}$ and $\mathbf{Y}$ were combined into a single multivariate matrix $\mathbf{Z}$, the covariance matrix $\mathbf{C}_{zz}$ would be made up of $\mathbf{C}_{xy}$, $\mathbf{C}_{xx}$ and $\mathbf{C}_{yy}$. MCA is not quite the same as multivariate EOF analysis of $\mathbf{C}_{zz}$ because it does not take into account the autocovariance structure within the individual fields that is included in $\mathbf{C}_{xy}$, (i.e., $\mathbf{C}_{xx}$ and $\mathbf{C}_{yy}$). It is optimal for explaining the *covariance between fields*, not the *combined variance* of the fields. Rather than a single matrix whose elements are the time varying coefficients for the respective modes, MCA yields a pair of coefficient matrices, one for x and one for y. The spatial patterns for the respective modes in the x field are derived by (temporally) regressing it upon the corresponding y coefficient time series and vice versa. Rather than a single set of spatial patterns in z, it yields separate sets of x and y patterns, each with its own set of time varying coefficients. MCA modes are not bi-orthogonal like EOFs: their spatial patterns are mutually orthogonal, but their coefficient time series are not.

The singular value vectors derived from MCA are the squared covariances explained by the respective modes. There are as many singular values as there are grid points in the smaller of the x and y matrices. Each singular value divided by the sum of the singular values can be interpreted as the *squared covariance fraction* explained by that mode. Another useful product of MCA is the set of correlation coefficients obtained by correlating the time series for the respective modes with one another, with or without lagging one them.

MCA can be performed upon any pair of interrelated variables. Applications include[8]

- SST with 200 hPa height or SLP to identify modes coupled atmosphere–ocean variability,
- OLR with 200 hPa height or SLP to identity convectively coupled modes of variability,
- geopotential height or wind defined over a large-scale domain with rain rate over an embedded smaller, mountainous domain to study how orography influences rain rate,
- geostrophic wind with the ageostrophic wind component to illuminate how geostrophic balance is maintained,
- the same field at different times as an approach to characterizing its temporal variability.

---

[8] Bretherton et al. (1992). The term "maximal covariance analysis" did not come into use until after this paper was published.

Appendix D Scaling and Application of the Weak Temperature Gradient Approximation

D.1 THE WEAK TEMPERATURE GRADIENT APPROXIMATION

The weak temperature gradient (WTG) approximation simplifies the representation of the thermodynamics of convectively coupled tropical motion systems. In isobaric coordinates, following an air parcel as it moves along its three-dimensional trajectory, the First Law can be expressed as[1]

$$\frac{d}{dt}\text{DSE} \equiv Q. \tag{D.1}$$

Under the WTG approximation, the horizontal advection of DSE and its Eulerian tendency are neglected, so that

$$Q \simeq -\omega\frac{\partial}{\partial p}\text{DSE}. \tag{D.2}$$

The diabatic heating rate Q is partitioned into a component associated with radiation Q_R, condensation heating $L(c-e)$, where c and e are the rates of condensation and evaporation, and a convective heating term $-\partial/\partial p\ (\overline{\omega_*\text{DSE}_*})$. The subscript $_*$ denotes in-cloud DSE fluxes, as explained in Section 5.2. Following Yanai et al.,[2] the condensation and convective heating terms, both of which involve model parameterizations, are combined into a single term $Q_c = L(c-e) - \partial/\partial p\ (\overline{\omega_*\text{DSE}_*})$ and

$$Q = Q_c + Q_R. \tag{D.3}$$

The radiative heating is made up of two components, a clear-sky component Q_{res} (we use this notation because it is computed as a residual) and a cloud-related feedback component Q_{rc}.

$$Q_R = Q_{res} + Q_{rc}. \tag{D.4}$$

The cloud-related radiative heating Q_{rc} is mainly associated with the absorption of infrared radiation in the anvils of deep convective clouds.

The cloud-radiative feedback is the component of Q_R that is linearly proportional to the convective heating rate:

$$Q_{rc} = rQ_c, \tag{D.5}$$

where r is called the cloud-radiative feedback parameter.[3] With this definition, the total cloud-related component of the diabatic heating rate is

$$(1+r)Q_c, \tag{D.6}$$

---

[1] In this Appendix, as in Section 5.2 of the text, diabatic heating rates are expressed in W kg^{-1}.
[2] Yanai et al. (1973).
[3] Peters and Bretherton (2005).

and the apparent heating rate is

$$Q = (1+r)Q_c + Q_{res}, \tag{D.7}$$

which is the relation used in the diagnostic calculation in Section 15.2. The vertical integral of Eq. (D.7) may be written as

$$\{Q\} = (1+r)LP + \{Q_{res}\} + F_{SH}, \tag{D.8}$$

where F_{SH} is the sensible heat flux through the bottom boundary. In deriving Eq. (D.8) it is assumed that the storage tendency of vapor is small, so $\{c - e\} = P$. In Section 15.1 it is shown that over the tropical oceans $r = 0.18$ and $Q_{res} \approx -120$ W m^{-2} based on ERA5 data.

D.2 SCALE ANALYSIS OF THE BASIC EQUATIONS NEAR THE EQUATOR

The conditions under which WTG balance is satisfied can be inferred from scale analysis of the basic equations. The horizontal momentum, continuity and hydrostatic balance equations are written in pressure coordinates as:

$$\frac{\partial\mathbf{V}}{\partial t} = -\mathbf{V}\cdot\nabla\mathbf{V} - \omega\frac{\partial\mathbf{V}}{\partial p} - f\mathbf{k}\times\mathbf{V} - \nabla\Phi \tag{D.9a}$$

$$\frac{\partial u}{\partial x} + \frac{\partial v}{\partial y} + \frac{\partial\omega}{\partial p} = 0 \tag{D.9b}$$

$$\frac{\partial\Phi}{\partial p} = -\frac{RT}{p}, \tag{D.9c}$$

where the field variables are as defined in Appendix E.

The momentum equation is nondimensionalized by introducing the following scales

$$\mathbf{V} = U\widehat{\mathbf{V}} \quad \omega = W\widehat{\omega} \quad p = P\widehat{p} \quad t = \tau\widehat{t} \quad \nabla = L^{-1}\widehat{\nabla}, \tag{D.10}$$

where τ and the capitalized roman letters refer to dimensional scales and the symbols with hats are nondimensional variables or operators. Scale analysis of the continuity equation reveals that $W = U\,P/L$. Substituting Eq. (D.10) into the horizontal momentum equation and rearranging the terms yields the nondimensional momentum equation

$$\text{Ro}_\tau\frac{\partial\widehat{\mathbf{V}}}{\partial\widehat{t}} = -\text{Ro}\left(\widehat{\mathbf{V}}\cdot\widehat{\nabla}\widehat{\mathbf{V}} + \widehat{\omega}\frac{\partial\widehat{\mathbf{V}}}{\partial\widehat{p}}\right) - \mathbf{k}\times\widehat{\mathbf{V}} - \mathcal{S}\,\widehat{\nabla}\widehat{\Phi}, \tag{D.11}$$

where $\mathcal{S} = \max(1, \text{Ro}_\tau)$ and

$$\text{Ro} = \frac{U}{fL} \qquad \text{Ro}_\tau = \frac{1}{f\tau} \qquad (\text{D.12})$$

are the conventional and timescale-based Rossby numbers, respectively. Depending upon how fast the waves of interest are propagating relative to the background flow, their characteristic timescale τ could be either L/U to L/C, where C is the scaling factor for their phase speed. In midlatitude synoptic-scale waves at the jet stream level, the eastward advection of the waves by the basic state flow is much larger than the westward phase propagation due to the beta effect, so that U~C. This is the justification for thinking of τ as the advective timescale L/U. But over the tropical warm pool, where the advection by the basic state flow is not as strong, τ should correspond more closely related to L/C, the time required for a wave to propagate past a fixed grid point

Equation (D.11) admits the possibility of different scalings of the geopotential. In the deep tropics, where $f \to 0$, geopotential scales as $\Phi = \widehat{\Phi}UL/\tau$ and $\text{Ro}_\tau \gg 1$ for large-scale motions.[4] In this case, $\mathcal{S} = \text{Ro}_\tau$ and the dominant balance in Eq. (D.11) is between the acceleration and the pressure gradient force, as in gravity waves. Beyond ~10° latitude, however, the geopotential scales as $\Phi = \widehat{\Phi}fUL$ in synoptic scale and lower frequency motions, and so $\text{Ro}_\tau \ll 1$ and the dominant balance in Eq. (D.11) is geostrophic.

Hydrostatic balance requires that the temperature anomalies be of order $\delta = UL/(R\tau)$. Thus the scale of the DSE fluctuations δ due to motions along a constant pressure surface is $\sim c_p\delta$. The vertical advection of DSE scales differently from the other terms in the DSE equation because it involves both the vertical temperature advection and the work that parcels do on the environment

$$\frac{d}{dp}\text{DSE} \sim \frac{\delta}{P}\frac{\partial}{\partial\widehat{p}}\widehat{\text{DSE}}, \qquad (\text{D.13})$$

where δ is the magnitude of the vertical change in DSE over the depth of the troposphere.

A characteristic scale for the phase speed of gravity waves C_g in terms of the vertical gradient in DSE is obtained by noting the relationship between the vertical gradient in dry static energy and stratification. One approach is to approximate the troposphere as two layers of equal mass, and estimate δ to be the difference in the DSE at the mid-points of the two layers, designated as P_2 and P_1. Alternatively, one can assume that the troposphere is continuously stratified with density scale height H and depth D (Gill, 1980) and take δ to be the increase in DSE from surface to tropopause. The characteristic phase speed squared of gravity waves is

$$C_g^2 = L^2/\tau_g^2 \sim \begin{cases} 1/2 \, ln(P_2/P_1)\frac{R}{c_p}\delta, & \text{two layer} \\ \frac{D}{\pi^2 H}\frac{R}{c_p}\delta, & \text{continous.} \end{cases} \qquad (\text{D.14})$$

where τ_g is the characteristic period of a gravity wave with wavelength L (see Fig. 15.3). Using δ from Fig. 5.2, $D = 16$ km, and $H = 8$ km results in characteristic phase speeds C of ~60 and 49 m/s, respectively.

With all the aforementioned definitions and scales, we obtain the following nondimensional thermodynamic equation

$$\frac{\tau_g^2}{\tau^2}\left(\frac{\partial\widehat{\text{DSE}}}{\partial\widehat{t}} + \frac{\text{Ro}}{\text{Ro}_\tau}\widehat{\mathbf{V}}\cdot\nabla\widehat{\text{DSE}}\right) + \widehat{\omega}\frac{\partial}{\partial\widehat{p}}\widehat{\text{DSE}} = \widehat{Q}, \qquad (\text{D.15})$$

where we find that the scale of the apparent heating is

$$Q = \frac{U}{L}\delta\,\widehat{Q}. \qquad (\text{D.16})$$

Examination of Eq. (D.15) reveals that the WTG approximation is applicable to tropical motion systems provided that the square of their timescale is at least an order of magnitude longer than that of gravity waves. This requirement is met if their phase speed is a factor of $\sqrt{10}$ slower or $\lesssim 15$ m s^{-1}.

D.3 SCALE ANALYSIS OF THE BASIC EQUATIONS AWAY FROM THE EQUATOR

Away from the equatorial belt ($\geq 10°$N/S), the Coriolis force becomes large enough to be of leading-order importance for synoptic scale and lower frequency motions. In this case, $\text{Ro}_\tau \ll 1$ so $\mathcal{S} = 1$ and the geopotential scales as $\Phi = \widehat{\Phi}fUL$; the dominant balance in the momentum Eq. (D.11) is geostrophic.

Proceeding as in the previous section, the nondimensional form of the thermodynamic energy Eq. (D.1) is

$$\frac{\tau_g^2}{\text{Ro}_\tau\tau^2}\left(\frac{\partial\widehat{\text{DSE}}}{\partial\widehat{t}} + \frac{\text{Ro}}{\text{Ro}_\tau}\widehat{\mathbf{V}}\cdot\nabla\widehat{\text{DSE}}\right) + \widehat{\omega}\frac{\partial\widehat{\text{DSE}}}{\partial\widehat{p}} = \widehat{Q}. \qquad (\text{D.17})$$

The inclusion of Ro_τ in the denominator of the first term represents the impact of the Coriolis force on the gravity wave adjustment process. As f increases, the gravity waves become increasingly influenced by Earth's rotation, rendering them less effective in maintaining WTG balance, while the adjustment toward geostrophic balance becomes increasingly important. Thus, the WTG approximation becomes less applicable with increasing latitude. However, there is some overlap between the domains in which geostrophic balance and WTG balance prevail. For example, very slowly evolving systems in which $\tau > \sqrt{10}\tau_g$ and $\text{Ro}_\tau \sim 0.1$ are in both geostrophic and WTG balance. An alternative way of understanding the overlap between geostrophic and WTG balance is to express the scale parameters in Eq. (D.17) in terms of the Rossby radius of deformation ($L_R = C/f$)

$$\frac{T_g^2}{\text{Ro}_\tau\tau^2} = \frac{L^2}{L_R^2}\text{Ro}_\tau. \qquad (\text{D.18})$$

Thus, WTG balance and geostrophic balance are both applicable for motion systems in which $\text{Ro}_\tau \ll 1$ and whose horizontal scale is equal to or or larger than the radius of deformation, which is on the order of 1000 km in the outer

---

[4] Charney (1963).

tropics ($\sim$20°N/S). At these latitudes synoptic scale waves are in both geostrophic and in WTG balance.

D.4 DERIVATION OF THE WTG MOISTURE EQUATION

The equation for the conservation of moisture is

$$\frac{dq}{dt} = (e - c) - \frac{\partial}{\partial p}\overline{\omega_* q_*}, \tag{D.19}$$

where q is specific humidity, e and c are the rates of evaporation and condensation per unit mass, and $\overline{\omega_* q_*}$ is the "convective" term representing the local moisture tendency due the net upward flux of water vapor in convective scale updrafts and downdrafts, which is included in the convective parameterization scheme.

Multiplying by L to express the conservation equation in the form of latent heat, we obtain

$$\frac{dLq}{dt} = -L(c - e) - \frac{\partial}{\partial p}\overline{\omega_* Lq_*}. \tag{D.20}$$

The phase change term $L(c - e)$ is equal to the condensation heating term in the thermodynamic energy equation, which is part of the Q_c term that appears in Eq. (D.3). To isolate the phase change term, we expand Q_c into its microphysical and convective components

$$Q_c = L(c - e) - \frac{\partial}{\partial p}\overline{\omega_* DSE_*} \tag{D.21}$$

and substitute $L(c-e)$ from Eq. (D.21) into Eq. (D.20), which yields

$$\frac{dLq}{dt} = -Q_c - \frac{\partial}{\partial p}(\overline{\omega_* DSE_*} + \overline{\omega_* Lq_*})$$
$$= -Q_c - \frac{\partial}{\partial p}(\overline{\omega_* MSE_*}). \tag{D.22}$$

Expanding into the local time derivative and advection terms, we obtain

$$\frac{\partial}{\partial t}Lq = -\mathbf{V} \cdot \nabla Lq - \omega\frac{\partial Lq}{\partial p} - Q_c - \frac{\partial}{\partial p}(\overline{\omega_* MSE_*}). \tag{D.23}$$

Under WTG balance, ω can be partitioned into components whose temperature tendencies balance the respective components of the diabatic heating. Noting that $Q = Q_c + Q_{rc} + Q_{res}$, we decompose the vertical velocity field into components $\omega \simeq \omega_c + \omega_{rc} + \omega_{res}$, satisfying the following balances

$$\omega_c\frac{\partial}{\partial p}DSE \simeq Q_c \; ; \qquad \omega_{rc}\frac{\partial}{\partial p}DSE \simeq Q_{rc}$$
$$\omega_{res}\frac{\partial}{\partial p}DSE \simeq Q_{res}. \tag{D.24}$$

With these definitions, Eq. (D.23) can be written as

$$\frac{\partial}{\partial t}Lq = -\mathbf{V} \cdot \nabla Lq - (\omega_{rc} + \omega_{res})\frac{\partial Lq}{\partial p}$$
$$- \omega_c\frac{\partial}{\partial p}MSE - \frac{\partial}{\partial p}\overline{\omega_* MSE_*}. \tag{D.25}$$

This expression is of limited use in evaluating the moisture budget because the sub grid-scale term $\overline{\omega_* MSE_*}$ can be estimated only indirectly in observational data and it is not available in the reanalysis products used in this book. To circumvent this problem, we use the mass-weighted vertical integral of Eq. (D.25), in which the convective scale vertical fluxes are replaced by the surface sensible plus latent energy fluxes $F_{SH} + F_{LH}$:

$$L\frac{\partial W}{\partial t} = -\{\mathbf{V} \cdot \nabla Lq\} - \left\{(\omega_{rc} + \omega_{res})\frac{\partial Lq}{\partial p}\right\}$$
$$- \left\{\omega_c\frac{\partial}{\partial p}MSE\right\} + F_{LH} + F_{SH}. \tag{D.26}$$

The conventional application of the WTG approximation neglects any potential contribution from gravity waves and temperature advection to the moisture tendency. Under circumstances in which it is not strictly applicable, as in the study of high-frequency variability, one may employ a "relaxed WTG formulation," which includes an adiabatic contribution to the vertical velocity ω_a, so that $\omega = \omega_Q + \omega_a$.[5]

[5] See Mapes (1997); Adames et al. (2021).

Appendix E Math Symbols and Abbreviations

E.1 MATH SYMBOLS AND NOTATION

Table E.1 *Math symbols, their meaning, and chapter/equation introduced*

Symbol	Meaning	Chapter, section, or equation
	Frames of reference	
x or λ	zonal coordinate	
y or ϕ	meridional coordinate	
z	vertical coordinate; height in $logp$ coordinates	
ϕ	latitude	
$\mathbf{i}, \mathbf{j}, \mathbf{k}$	zonal, meridional, vertical unit vectors	
t	time	
	State variables	
p	pressure	Eq. (1.1)
p_s	surface pressure	Sec. 3.1
T, T_s	temperature, surface temperature	Eq. (1.1), Sec. 2.6.2
θ	potential temperature	Eq. 1.1
θ_e	equivalent potential temperature	Sec. 5.2
ρ, α	density, specific volume	Eq. (1.7), Sec. 2.5
ρ_o	reference density	Eq. (10.7)
s	entropy	Sec. 2.5
Φ	geopotential	Eq. (1.8)
Z	geopotential height	Sec. 1.4.1
Ro, Ro$_\tau$	Rossby number	Eq. (D.12)
H	atmospheric scale height	Eq. (9.1)
	Wind and wind-related symbols	
u, v	the zonal and meridional wind components on Earth; the azimuthal and radial components in a tropical cyclone	
$u_g, v_g, (u_a, v_a)$	geostrophic (ageostrophic) zonal and meridional wind components	Eqs. (6.29) and (6.30)
$\mathbf{V}$, V	horizontal wind vector, speed	Eqs. (4.3), (3.12)
$\mathbf{V}_g$	geostrophic wind vector	Eq. (1.9)
v_r	radial wind component in a tropical cyclone	Sec. 10.2.7
w, ω	vertical velocity in height, $logp$, and pressure coordinates	Eqs. (10.1), (1.4), (5.15)
ψ	horizontal streamfunction for Eulerian mean meridional circulation	Sec. 6.1
χ	velocity potential	Sec. 18.1
χ	moisture weighted velocity potential	Eq. (4.12)
$\mathbf{C}$	the horizontal velocity covariance tensor	Eq. (12.7)
ζ	vertical component of relative vorticity	Eq. (3.17)
ζ_g	vertical component of geostrophic vorticity	Sec. 1.5.1, Eq. (1.10)
η	vertical component of absolute vorticity	Eq. (3.17)
ϖ	the angular velocity of a vortex	Sec. 18.1
q_g, q_g^*	quasi-geostrophic potential vorticity	Eqs. (7.28), (8.16)
$\widehat{q}$	pseudo-potential vorticity	Sec. 7.3, Eq. (7.16)

Table E.1 *(Continued)*

Symbol	Meaning	Chapter, section, or equation
$Tr(x)$	convergence of the horizontal transport of x	Eqs. (4.2), (4.3)
$K, M, N, \widehat{M}, \widehat{N}, \alpha, \Psi$	metrics related to the velocity covariance tensor	Eqs. (12.8)–(12.12)
Moisture-related symbols		
q	specific humidity	Eq. (4.7), Sec. 4.2
E	evapotranspiration rate	Eq. (4.7)
P	precipitation rate (rain rate)	Eq. (4.7)
R	river runoff rate	Ch. 4.14
W	column-integrated water vapor (precipitable water)	Eq. (4.7)
$\mathcal{W}$	vertically integrated water in soil	Eq. (4.14)
S, S^{+}, S^{-}	sources and sinks	Eq. (4.2)
Static stability		
Γ	vertical temperature gradient (lapse rate)	Eq. (1.2)
Γ_d	dry adiabatic lapse rate g/c_p	Eq. (1.2)
$\Gamma_d - \Gamma$	static stability in geometric height and $log\ p$ coordinates	Eq. (1.2)
$s = \kappa T/p - \partial T/\partial p, \sigma = sR/p$	static stability in pressure coordinates	Eqs. (1.3), (7.12)
N	buoyancy frequency	Sec. 1.3
Heating-related symbols		
$Q, \mathcal{Q}$	net diabatic heating rate	Eqs. (5.1), (7.11), (D.1)
Q_R	net radiative heating rate	Eqs. (5.5) and (D.3)
Q_c	sum of convective and condensational heating	Eq. (D.3)
Q_{rc}	cloud-related radiative heating rate	Eq. (D.4)
Q_{res}	residual (clear sky) radiative heating rate	Eqs. (15.2), (D.4)
F_{SH}	surface sensible heat flux	Eq. (5.5)
F_{LH}	net latent heat flux into a column of air	Eq. (5.7)
Energy-related symbols		
DSE	dry static energy	Eq. (5.3)
MSE	moist static energy	Eq. (5.8)
A	global averaged available potential energy per unit area	Sec. 6.1, Eq. (6.6)
K	global averaged kinetic energy per unit area	Sec. 6.1, Eq. (6.23)
$A + K$	mechanical energy	Eq. (7.1)
P	potential energy	Eq. (7.1)
I	internal energy	Eq. (7.1)
Lq	latent energy per unit mass	
C	conversion of available potential energy to kinetic energy	Eq. (6.13)
G	generation of potential energy due to diabatic heating	Eq. (6.14)
D	global average friction dissipation	Eq. (6.25)
$A_E\ K_E, C_E, D_E, G_E$	as for A, K, C, D, G, but for the eddies	Secs. 6.7 and 8.1, Eqs. (8.1)–(8.6)
A_T, K_T, C_T, D_T, G_T	as for A, K, C, D, G, but for the transients	Secs. 6.7 and 8.1, Eqs. (8.1)–(8.6)
A_Z, K_Z, C_Z, D_Z, G_Z	as for A, K, C, D, G, but for the zonal mean	Secs. 6.7 and 8.1, Eqs. (8.1)–(8.6)
A_M, K_M, C_M, D_M, G_M	as for A, K, C, D, G, but for the time mean	Secs. 6.7 and 8.1, Eqs. (8.1)–(8.6)
C_K	barotropic energy conversion	Eqs. (8.1), (12.13)
C_A	baroclinic energy conversion	Eq. (8.4)
C_E	energy conversion due to overturning circulations in eddies	Eq. (8.5)
C_Z	conversion in mean meridional circulations	Eq. (8.1)
W	work done in release of available potential energy	Eq. (6.4)
Friction-related symbols		
C_D	drag coefficient	Eq. (3.12)
$\mathcal{F}_x$	zonal stress	Eq. (3.17)
τ_x	zonal surface stress	Eq. (3.11)
$\mathcal{F}, \mathcal{F}_x$	frictional force per unit mass; zonally average zonal component of friction	Eqs. (6.22), (7.1)

Table E.1 *(Continued)*

Symbol	Meaning	Chapter, section, or equation
	Wave-related symbols	
k and l	zonal and meridional wave numbers	Fig. 3.10, Ch. 11
K	two-dimensional wavenumber	Ch. 11
K_s	stationary wavenumber	Eq. (11.3), Sec. 13.1.1
ω	frequency	Sec. 8.2.3
c	phase speed	
U	background zonal wind	Eq. (11.1)
η, h_o	surface displacement η about a resting depth h_o of an incompressible barotropic fluid	Eqs. (10.2)-(10.4)
h_e	equivalent depth	Sec. 10.2.3
C_g	characteristic gravity wave phase speed	Eq. (D.14)
$\mathbf{cg}; c_{gy}, c_{gp}$	group velocity	Eq. 8.13
A	wave activity	Sec. 8.2
$\boldsymbol{\Gamma}$	zonally averaged forcing vector; forcing vector	Eqs. (7.21), (20.4)
$\mathbf{F}$	the Eliassen–Palm flux	Eqs. (8.10), (9.5)
$F^{(y)}, F^{(z)}$	meridional and vertical components of the EP flux generalized to include gravity and inertio-gravity waves	Eq. (10.7)
$\mathbf{F}_M, \mathbf{F}_H$	components of the EP flux involving eddy transports of zonal momentum and heat in the generalized formulation	Eqs. (8.8), (8.9)
$\mathbf{W}$	eddy geopotential flux vector	Eq. (8.17)
$\mathbf{E}$	the extended Eliassen–Palm flux vector	Eq. (12.16)
$\mathcal{D}$	generation of wave activity by diabatic heating and friction in eddies	Eqs. (8.12), (8.18)
	Zonally averaged flow	
ψ and ψ_p, ψ_z and ψ_h	mass streamfunction in pressure, height and moist static energy coordinates	Eqs. (1.4), (5.11), (5.12), (5.13)
AAM	globally integrated atmospheric angular momentum	Eq. (3.6)
m	angular momentum per unit mass	Eq. (3.2)
M	angular momentum in a zonal ring of air	Eq. (3.1)
T_i	vertically integrated torques	Eq. (3.8)
$[v]^*, [\omega]^*$	transformed Eulerian mean meridional (TEM) circulation	Eqs. (8.9), (5.21), (5.23)
$[v]_B, [\omega]_B$	boundary forced component of the TEM circulation	Eq. (8.9)
ψ^*	the TEM streamfunction	Eq. (8.11)
v_S, ω_S	the Stokes drift	Eq. (5.21)
$\mathcal{G}$	meridional convergence of zonal momentum transport by eddies; zonal average flux of relative vorticity by the nondivergent barotropic eddies	Eqs. (3.20), (7.9)
$[H], \mathcal{H}$	zonally averaged meridional convergence of heat transport by eddies	Eqs. (5.19) and (7.10)
η^*	meridional displacement of air parcels in the eddies from their zonal average	Sec. 8.2
$\boldsymbol{\Sigma}$	pressure gradient force vector	Eq. (7.13)
$\mathbf{S}$	displacement vector or state vector	Eqs. (7.15), (20.2)
$\boldsymbol{\Psi}$	Eulerian mean merdional motion vector; transverse circulation in a TC	Eqs. (7.20), (20.1)
	Earth variables and constants	
Ω	angular velocity of Earth 7.292×10^{-5} s^{-1}	Secs. 1.5, 2.6.2
R_E	radius of Earth 6371 km	Eq. (1.4)
g	acceleration due to gravity 9.807 m s^{-2}	Eq. (1.2)
$f : f_o$	Coriolis parameter; midlatitude reference	Secs. 1.5, 7.5
β	meridional gradient of f 2.29×10^{-11} m^{-1} s^{-1} at the Equator	Sec. 7.5, Eq. (12.3), Fig. 10.8
β_*	meridional gradient of absolute vorticity	Ch 11, Sec. 13.1.1

Table E.1 *(Continued)*

Symbol	Meaning	Chapter, section, or equation
c_p	specific heat at constant pressure 1005 kJ kg^{-1} K^{-1}	Eq. (1.2)
c_v	specific heat at constant volume 717 J kg^{-1} K^{-1}	Sec. (5.2)
R	ideal gas constant 287 J kg^{-1} K^{-1}	Eq. (1.8)
L	latent heat of condensation 2.5×10^6 J kg^{-1}	Eq. (5.5)
H	height of terrain	Eq. (3.10)
T_E	equivalent blackbody temperature	Sec. 2.1
c_s	speed of sound	Sec. 5.2
	Statistical	
$\sigma(\ )$	standard deviation	Sec. 1.4.3
$r, \mathbf{R}$	correlation coefficient, scaled version of the covariance matrix	App. A.1.3, App. C
$\mathbf{C}$	covariance matrix	App. C
	Notation	
$[\ ]$	zonal mean	Sec. 1.4.2
$(\ )^*$	deviation from a zonal mean	Eq. (1.6)
$\overline{(\ )}$	time mean	Eq. (1.6)
$(\ )'$	deviation from a time mean	Eq. (1.6)
$\langle\ \rangle$	global average at a fixed height or pressure level	Eqs. (6.2), (A.20)
$\widehat{(\ )}$	deviation from the global mean on a fixed height or pressure surface	Eqs. (6.2), (6.10)
$\{\ \}$	mass-weighted vertical integral	Eq. (4.1)
	Units	
W	watt	
J	joule	
cm, m, km	centimeter, meter, kilometer	
s	second	
d	day	
m	month	
y, yr	year	
K	kelvin	
°C	degrees Celsius	
kg	kilogram	
Pa, hPa	Pascal (1 Newton m^{-1}), hectopascals	
ppmv	parts per million by volume	

E.2 ABBREVIATIONS

Table E.2 *Abbreviations*

AAM	atmospheric angular momentum
BAM	baroclinic annual mode
BDC	Brewer–Dobson circulation
CALIPSO	Cloud-Aerosol Lidar and Infrared Pathfinder Satellite Observation
CAPE	convective available potential energy
CFCs	chlorofluorocarbon gases
CGT	circumpolar global teleconnection (pattern)
CINH	convective inhibition
COSMIC	Constellation Observing System for Meteorology, Ionosphere, and Climate
CSF	column saturation fraction

Table E.2 *(Continued)*

CSI	conditional symmetric instability
CT	correlation teleconnectivity
CTI	cold tongue index
DJF	December–January–February average
DSE	dry static energy
ECMWF	European Center for Medium-Range Weather Forecasts
EEOF	extended empirical orthogonal function
ENSO	El Niño–Southern Oscillation
EOF	empirical orthogonal function
EOT	empirical orthogonal teleconnection (patterns)
EP	Eliassen–Palm flux
ERA	ECMWF Re-Analysis product (V5 and Interim)
ERBE	Earth Radiation Budget Experiment
GARP	Global Atmospheric Research Program
GCM	general circulation model
GFD	geophysical fluid dynamics
GFDL	NOAA Geophysical Fluid Dynamics Laboratory
GOES	NOAA Geostationary Operational Environmental Satellites
GPM	NASA Global Precipitation Mission
GPS	global positioning system
GWE	GARP Global Weather Experiment
IG	inertio-gravity (wave)
ITCZ	intertropical convergence zone
JJA	June–July–August average
KE	kinetic energy
LC	life cycle
LOD	length of day
MISO	monsoon interseasonal oscillation
MIT	Massachusetts Institute of Technology
MJO	Madden–Julian Oscillation
MLS	NASA mircowave limb sounder
MMC	mean meridional circulation
MRG	mixed Rossby–gravity wave
MSE	moist static energy
MSU	mircrowave sounding unit
NAM	Northern Hemisphere annular mode
NAO	North Atlantic Oscillation
NASA	National Aeronautics and Space Administration
NECC	North Equatorial Counter Current
NH	Northern Hemisphere
NOAA	National Oceanic and Atmospheric Administration
NSCAT	NASA Scatterometer
NSF	National Science Foundation
NWP	numerical weather prediction
OI	optimal interpolation
OLR	outgoing longwave radiation
OMI	OLR MJO index
PC	principal component
PJ	Pacific–Japan (pattern)
PNA	Pacific–North American (pattern)
PV	potential vorticity
QBO	Quasi-biennial oscillation
REOF	rotated empirical orthogonal function
RH	relative humidity
RMM	real-time multivariate MJO index
RT	regression teleconnectivity

Table E.2 *(Continued)*

SACZ	South Atlantic convergence zone
SAM	Southern Hemisphere annular mode
SAO	semi-annual oscillation
SH	Southern Hemisphere
SLP	sea level pressure
SO	Southern Oscillation
SPCZ	South Pacific convergence zone
SST	sea surface temperature
SSW	stratospheric sudden warming
SSZ	subtropical subsidence zone
TC	tropical cyclone
TEM	transformed Eulerian mean
TOA	top of atmosphere
TOMS	Total Ozone Measurement System
TRMM	NASA Tropical Rainfall Measurement Mission
TTL	tropopause transition layer
WACCM	Whole Atmosphere Community Climate Model
WLLN	World Wide Lightning Location Network
WTG	weak temperature gradient

Appendix F Extended Figure Captions

This Appendix provides extensions of captions for figures that require additional explanation with regard to data sources or methods. URLs for the various data sources used to produce the figures that appear in this book are provided on the companion web page.

Chapter 1

Figs. 1.3–1.13, 1.16–1.21, 1.24–1.34: From the ERA-Interim Reanalysis. Most of these figures are part of a mini-atlas based on the ERA-I Reanalysis. A broader selection of figures can be accessed from the companion webpage.

Figs. 1.14, 1.15: NOAA Global Precipitation Climatology Project (GPCP).

Figs. 1.22, 1.23: Constellation Observing System for Meteorology, Ionosphere, and Climate (COSMIC) radio occultation measurements of temperature and water vapor in conjunction with low Earth-orbiting GPS satellites.

Chapter 2

Figs. 2.11–2.14: The GFDL atmospheric (AM2) model coupled to a slab ocean. Initialized from a state of rest. Averages for Days 120–150.

Fig. 2.15: From the ERA-Interim Reanalysis.

Fig. 2.18: NOAA GOES 17 satellite.

Fig. 2.19: Based on hourly SLP fields from the ERA5 Reanalysis.

  The SLP field is highpass filtered to remove the more energetic day-to-day variability, which dominates regression patterns based on unfiltered data. The data are smoothed using a 5-point (i.e., 5 hour) binomial filter and the smoothed fields are subtracted from the original fields. The highpass filtering distorts the wave pattern such that a pulse morphs into a train of waves.

  To isolate the Krakatoa-like signal in the SLP field it is also necessary to filter out the diurnal cycle and its higher harmonics. For this purpose, the SLP time series at each grid point is Fourier transformed, the diurnal cycle and the neighboring harmonics on each side are removed, and similarly for the semidiurnal cycle, the diurnal cycle, etc., and then the reverse transform is performed. Removing the neighboring harmonics suppresses the diurnally forced variability whose amplitude and phase varies from one day to the next.

Chapter 3

Fig. 3.3: Steric sea level is derived from monthly, objectively analyzed subsurface temperature and salinity at 24 levels in the uppermost 1500 m of the ocean from 1945 onward by M. Ishii et al., The surface wind field is derived from the 20th Century Reanalysis.

Fig. 3.5: AAM from ERA-Interim Reanalysis, LOD from the Observatory of Paris Earth Orientation Centre.

Figs. 3.7–3.8, 3.11–3.13: From the ERA-Interim Reanalysis.

Chapter 4

Figs. 4.1–4.2: Fires based on MODIS data from the NASA Terra satellites. CO and CH_4 based on the Atmospheric InfraRed Sounder (AIRS) flown on NASA's Aqua satellite

Fig. 4.3: Meridional profiles of zonally averaged CO_2 are generated by a chemical transport model that assimilates in situ measurements, both ground-based and upper air, and makes extensive use of back-trajectories.

Fig. 4.4, bottom panel: Composite SSM/I water vapor imagery. SSM/I is a passive microwave sensor with seven channels.

Figs. 4.5–4.9: From the ERA-Interim Reanalysis.

Fig. 4.11: Lake level data from the US Geological Survey.

Fig. 4.12. NOAA National Ocean Data Centers World Ocean Atlas.

Chapter 5

Figs. 5.2: From the ERA-Interim Reanalysis.

Figs. 5.4–5.6: NASA Earth Radiation Budget Experiment (ERBE).

Fig. 5.9: NASA's Sea-viewing Wide Field-of-view Sensor (SeaWiFS) Project.

Fig. 5.10: NOAA Global Drifter Program. In order to derive fields like the one shown in this figure it is necessary to smooth the data obtained from the website.

Figs. 5.11, 5.14–5.19, 5.25: ERA-Interim Reanalysis.

Chapter 6

Fig. 6.8: The GFDL atmospheric (AM2) model coupled to a slab ocean. Initialized from a state of rest.

Chapter 7

Fig. 7.7: From the ERA-Interim Reanalysis.

Chapter 8

Fig. for Exercise 8.12: ERA5 Reanalysis.
Figs. 8.12, 8.13, 8.16, 8.17, 8.20, 8.21: From the
ERA-Interim Reanalysis.
Fig. 8.13: The vertical scaling of the EP flux vectors is in
accordance with Eq. (9.5).

Chapter 9

Fig. 9.1: From Aura Microwave Limb Sounder (MLS)
satellite measurements spanning 2005 through 2017.
Figs. 9.2, 9.3: From the ERA-Interim Reanalysis.
Fig. 9.6: (Right) adapted from a schematic created by E.
F. Danielsen that appeared in the Project Springeld
Report, U.S.Defense Atomic Support Agency, NTIS
607980 (1964).
Figs. 9.7, 9.8, From the ERA5 Reanalyses.
Fig. 9.11: The lower stratospheric channel (4) of the
NASA Microwave Sounding Unit.
Fig. 9.12: Geopotential height, temperature and diabatic
heating based on ERA-Interim Reanalysis; ozone
partial pressure from Aura Microwave Limb Sounder
(MLS) satellite measurements spanning 2004 to 2017.
Figs. 9.13–9.16: From the ERA-Interim Reanalysis.

Chapter 10

All figures in this chapter that were not previously
published are based on the ERA5 Reanalysis.
Figs. 10.1–10.7, 10.10–10.13: Based on ERA5 data.
Fig. 10.10 Middle panel (observed): Constructed by
regressing raw u_{50}, v_{50} and Φ_{50} upon a time series of
u_{50} at $(0°, 90°E)$, filtered to retain only $k = 1$,
eastward propagating fluctuations with 2 to 25 day
periods. Based on ERA Interim data from
1979–2015, all seasons.
Fig. 10.11 Top right panel (observed): Constructed by
regressing raw u_{50}, v_{50} and Φ_{50} upon a time series of
v_{50} at $(0°, 60°W)$, filtered to retain only fluctuations
with 2 to 6 day periods. Based on ERA Interim data
from 1979–2015, all seasons.
Figs. 10.6, 10.7: Based on daily mean data.
Fig. 10.12: In contrast to Fig. 2.19, which is based on
pulses in SLP, Figs. 10.12 and 10.13 are based on
"downwelling events," as justified by the following
considerations. Regions of deep convection are
characterized by concentrated buoyant updrafts, from
which downwelling gravity waves are continuously
emanating. Some mesoscale regions contain more
than their share of updrafts, while others contain less.
The grid in the ERA5 reanalysis is too coarse to
resolve the buoyant updrafts, from which the waves
emanate, but it is fine enough to capture much of this
mesoscale sampling variability in the vertical velocity
field, which is largest on the smallest resolved spatial

scales and the shortest resolved time scales. We refer
to the negative tail in the frequency distribution of
hourly (upward) vertical velocity at individual grid
points as "downwelling events," and in Figs.10.12
and 10.13 we demonstrate that they are regions in
space/time from which internal gravity waves
propagate and disperse. Before the compositing is
performed, the temperature field was highpass filtered
using a 7-point binomial filter, which is not quite as
strong as the 5-point filter used in constructing Fig.
2.19, but still removing all but the highest
frequencies. For suppressing the diurnally forced
variability, it is sufficient in this case to simply
remove the mean for each hour of the day.
Figs. 10.15–10.20: Based on ERA5 data.

Chapter 11

Fig. 11.1: From the ERA-Interim Reanalysis.
Figs. 11.2–11.13 and 11.19: Observations based in
ERA-Interim Reanalysis. Simulations based on the
Whole Atmosphere Community Climate Model
(WACCM), version 6 within the CESM release 2.1.1,
CESM2(WACCM6), run with prescribed
climatological mean SST. For documentation of the
model, see Gettelman et al. (2019).

Chapter 12

Figs. 12.1, 12.2, 12.4, 12.5, 12.11, 12.12: From the
ERA-Interim Reanalysis.
Figs. 13.37–13.39: NCAR-NCEP Reanalysis.

Chapter 13

Figs. 13.1–13.13: 13.23, 13.29–13.31, 13.36. From the
ERA-Interim Reanalysis.
Figs. 13.37–13.39: From the NCEP-NCAR Reanalysis.

Chapter 14

Fig. for Exercise 14.13 Based on the NOAA/NCDC
OISST (Optimum Interpolation Sea Surface
Temperature) dataset.
Fig. for Exercise 14.18: From the ERA-Interim
Reanalysis.
Figs. 14.1–14.4, 14.6–14.11: From the ERA-Interim
Reanalysis.

Chapter 15

Figs. 15.1, 15.2, 15.5–15.7: From the ERA5 Reanalysis.
Fig. 15.5: Basing the composite on downwelling events
is justified in the extended caption of Fig. 2.19 in this
Appendix. Data processing procedures are the same
as in Fig. 10.12.
Fig. 15.6 Based on 4 × daily ERA5 data.
Fig. 15.7 and 15.8: The various fields are regressed on
rain rate (NOAA Global Precipitation Climatology
Project (GPCP)) averaged over 20° of latitude ×
20° of longitude grid boxes centered on the equator at

10°of longitude intervals. The regression patterns for the individual grid boxes are shifted in longitude so that they are all centered on the Date Line, and then averaged to obtain a single composite pattern for each variable.

Chapter 16

Figs. 16.1, 16.3, 16.4, 16.6–16.13: From the ERA-Interim Reanalysis.
Figs. 16.2 and 16.13: Rain rate based in TRMM data.

Chapter 17

Fig. 17.1: SST time series based on the Extended Reconstructed Sea Surface Temperature (ERSST.v3b) analysis from the NOAA/Climate Prediction Center. SLP and u_{1000} from the ERA-Interim Reanalysis, tide gauge (SSH) data from Permanent Service for Mean Sea Level (PSMSL). Rain rate P: 1900–1978 based on station records from the global historical climatological network (GHCN); 1979 onward as estimated based on NOAA Interpolated Outgoing Longwave Radiation (OLR).
Figs. 17.3 and 17.4: Based on monthly, objectively analyzed subsurface temperature fields at 24 levels in the uppermost 1500 m of the ocean from 1945 onward by M. Ishii et al.
Figs. 17.7 and 17.8: From the ERA-Interim Reanalysis.
Fig. 17.9: 400 hPa vertical mass flux, 150hPa height, 1000 hPa wind vectors, SLP and rain rate based on the ERA-Interim Reanalysis for the period of record 1979–2015. SST from the Extended Reconstructed Sea Surface Temperature (ERSST.v3b). Sea surface height SSH based on satellite altimetry from Archiving, Validation, and Interpretation of Satellite Oceanographic Data (AVISO).
Figs. 17.10–17.16: From the ERA-Interim Reanalysis.
Figs. 17.17, 17.18: From the ERA-Interim Reanalysis, NOAA Interpolated Outgoing Longwave Radiation (OLR).
Fig. 17.20: From the ERA-Interim Reanalysis.
Fig. 17.21: Z150 based on the ERA-Interim Reanalysis, SST on ERSST v3b, available from the NOAA/Climate Prediction Center.

Chapter 18

Figs. 18.1–18.14: From the ERA-Interim Reanalysis.
Fig. 18.2 Rain rate based on TRMM data.
Figs. 18.12: Clouds from NASA Cloud-Aerosol Lidar and Infrared Pathfinder Satellite Observations CALIPSO, diabatic heating rate and relative humidity from the ERA-Interim Reanalysis, CO and O_3 from NASA microwave limb scanner MLS.
Fig. 18.13: Constellation Observing System for Meteorology, Ionosphere and Climatology COSMIC
Fig. 18.14 The original figure in Dima (2005) was based on the MJO indices RMM1 and RMM2 of Wheeler and Hendon (2004) and labeled accordingly. In order

to make it easy to compare with other figures presented in this chapter, we have relabeled it in terms of PC1 and PC2 of Δ_χ, which peak about 2 d after RMM1 and RMM2, respectively. See also Fig. 18.4.
Fig. 18.15: From the ERA-Interim Reanalysis.
Fig. 18.16: Temperature based on Constellation Observing System for Meteorology, Ionosphere and Climatology COSMIC. Zonal and meridional wind components and diabatic heating rate based on ERA-Interim, Reanalysis. Clouds based on NASA Cloud-Aerosol Lidar and Infrared Pathfinder Satellite Observations CALIPSO. Water vapor specific humidity based on NASA microwave limb scanner MLS.
Fig. 18.17–18.22: From the ERA-Interim Reanalysis. In Figs. 18.17 and 18.19, the MJO-related r.m.s. amplitude is estimated by (1) regressing the field in question upon PC1 and PC2 of Δ_χ, (2) squaring, adding, and square rooting grid point values of the two regression patterns.

Chapter 19

Fig. 19.1–19.2: "Brightness temperature" is computed from the Stefan–Boltzmann Law based on infrared radiation in the wavelength band around 10 microns. The data are made available through the Cloud Archive User Service Project (CLAUS).
Fig. 19.3: NOAA Interpolated Outgoing Longwave Radiation (OLR)
Figs. 19.4–19.6: NOAA Interpolated Outgoing Longwave Radiation (OLR), ERA-Interim Reanalysis.
Fig. 19.7: ERA-Interim Reanalysis.
Fig. 19.8: NOAA Interpolated Outgoing Longwave Radiation (OLR), ERA-Interim Reanalysis.
Fig. 19.9: ERA-Interim Reanalysis.
Figs. 19.11–19.14: NOAA Interpolated Outgoing Longwave Radiation (OLR), ERA-Interim Reanalysis.
Fig. 19.16: ERA-Interim Reanalysis.

Chapter 20

Fig. 20.1: From the ERA5 Reanalysis.
Fig. 20.3 From the International Best Track Archive for Climate Stewardship.
Figs. 20.4 and 20.5: The daily positions are taken at 24-hour intervals from the first time a position was recorded. The period of record is 1974–2018 and data for all calendar months included.
Fig. 20.5 The criterion for a tropical cyclone being included in the El Niño or La Niña composite is that the ENSO SST index Niño 3.4 be $> 0.7°C$ above to below normal on the day of formation.

Chapter 21

Fig. 21.1: NASA Tropical Rainfall Measurement Mission TRMM, gridded 3B42 analysis for the period of record (January 1998 to April 2015) available at 3 hourly intervals on a 0.25 degree by 2.5 degree latitude/longitude grid.

Fig. 21.2: NASA Tropical Rainfall Measurement Mission TRMMs lightning imaging sensor LIS and optical transient detector OTD instruments. The period of record is 1998–2011.

Fig. 21.3: The World Wide Lightning Location Network WWLLN, which records locations and times of lightning flashes based on analysis of data from a global network of ground stations

which sense the electromagnetic signature of sferics emitted by distant flashes.

Figs. 21.5–21.7: From the ERA5 Reanalysis.

Fig. 21.8: 50 hPa vertical velocity from the ERA5 Reanalysis. Visible imagery from Himawari satellite. Brightness temperature contours from Himawari satellite.

Fig. 21.10: 500 hPa vertical velocity from the ERA5 Reanalysis. Based on the period of record 1979–2020.

Bibliography

Adames, Á. F., D. Kim, S. K. Clark, Y. Ming, and K. Inoue, 2019: Scale analysis of moist thermodynamics in a simple model and the relationship between moisture modes and gravity waves. *Journal of the Atmospheric Sciences*, **76 (12)**, 3863–3881.

Adames, Á. F., and E. D. Maloney, 2021a: Moisture Mode Theory's Contribution to Advances in our Understanding of the Madden–Julian Oscillation and Other Tropical Disturbances. *Current Climate Change Reports*, **7 (2)**, 72–85.

Adames, Á. F., and E. D. Maloney, 2021b: Moisture mode theory's contribution to advances in our understanding of the Madden–Julian Oscillation and other tropical disturbances. *Current Climate Change Reports*, 1–14.

Adames, Á. F., S. W. Powell, F. Ahmed, V. C. Mayta, and J. D. Neelin, 2021: Tropical precipitation evolution in a buoyancy-budget framework. *Journal of the Atmospheric Sciences*, **78 (2)**, 509–528.

Adames, Á. F., and J. M. Wallace, 2014a: Three-dimensional structure and evolution of the MJO and its relation to the mean flow. *Journal of the Atmospheric Sciences*, **71 (6)**, 2007–2026.

Adames, Á. F., and J. M. Wallace, 2014b: Three-dimensional structure and evolution of the vertical velocity and divergence fields in the MJO. *Journal of the Atmospheric Sciences*, **71 (12)**, 4661–4681.

Adames, Á. F., and J. M. Wallace, 2015: Three-dimensional structure and evolution of the moisture field in the MJO. *Journal of the Atmospheric Sciences*, **72 (10)**, 3733–3754.

Adames, Á. F., and J. M. Wallace, 2017: On the tropical atmospheric signature of El Niño. *Journal of the Atmospheric Sciences*, **74 (6)**, 1923–1939.

Adames, Á. F., J. M. Wallace, and J. M. Monteiro, 2016: Seasonality of the structure and propagation characteristics of the MJO. *Journal of the Atmospheric Sciences*, **73 (9)**, 3511–3526.

Ahmed, F., J. D. Neelin, and Á. F. Adames, 2021: Quasi-equilibrium and weak temperature gradient balances in an equatorial beta-plane model. *Journal of the Atmospheric Sciences*, **78 (1)**, 209–227.

Alexander, M. A., 1992: Midlatitude atmosphere–ocean interaction during El Niño. Part I: the North Pacific Ocean. *Journal of Climate*, **5 (9)**, 944–958.

Alexeev, V., P. Langen, and J. Bates, 2005: Polar amplification of surface warming on an aquaplanet in "ghost forcing" experiments without sea ice feedbacks. *Climate Dynamics*, **24 (7)**, 655–666.

Andrews, D. G., 1983: A finite-amplitude Eliassen–Palm theorem in isentropic coordinates. *Journal of the Atmospheric Sciences*, **40**, 1877–1883.

Andrews, D. G., J. R. Holton, and C. B. Leovy, 1987: *Middle Atmosphere Dynamics*. Academic Press.

Andrews, D. G., and M. McIntyre, 1978: Generalized Eliassen–Palm and Charney-Drazin theorems for waves on axismmetric mean flows in compressible atmospheres. *Journal of the Atmospheric Sciences*, **35**, 175–185.

Andrews, D. G., and M. E. McIntyre, 1976: Planetary waves in horizontal and vertical shear: The generalized Eliassen-Palm relation and the mean zonal acceleration. *Journal of the Atmospheric Sciences*, **33**, 2031–2048.

Armour, K. C., N. Siler, A. Donohoe, and G. H. Roe, 2019: Meridional atmospheric heat transport constrained by energetics and mediated by large-scale diffusion. *Journal of Climate*, **32 (12)**, 3655–80.

Athanasiadis, P. J., J. M. Wallace, and J. J. Wettstein, 2010: Patterns of wintertime jet stream variability and their relation to the storm tracks. *Journal of the Atmospheric Sciences*, **67**, 1361–1381.

Bao, M., and J. M. Wallace, 2015: Cluster analysis of Northern Hemisphere wintertime 500-hPa flow regimes during 1920–2014. *Journal of the Atmospheric Sciences*, **72**, 3597–3608.

Barsugli, J. J., and D. S. Battisti, 1998: The basic effects of atmosphere–ocean thermal coupling on midlatitude variability. *Journal of the Atmospheric Sciences*, **55 (4)**, 477–493.

Battisti, D. S., 1988: Dynamics and thermodynamics of a warming event in a coupled tropical atmosphere–ocean model. *Journal of the Atmospheric Sciences*, **45 (20)**, 2889–2919.

Battisti, D. S., Q. Ding, and G. H. Roe, 2014: Coherent pan-Asian climatic and isotopic response to orbital forcing of tropical insolation. *Journal of Geophysical Research: Atmospheres*, **119 (21)**, 11 997–12 020.

Battisti, D. S., and A. C. Hirst, 1989: Interannual variability in a tropical atmosphere–ocean model: Influence of the basic state, ocean geometry and nonlinearity. *Journal of the Atmospheric Sciences*, **46 (12)**, 1687–1712.

Battisti, D. S., D. J. Vimont, and B. P. Kirtman, 2019: 100 years of progress in understanding the dynamics of coupled atmosphere/ocean variability. *Meteorological Monographs*, **59**, 8.1–8.57.

Berggren, R., B. Bolin, and C.-G. Rossby, 1949: An aerological study of zonal motion, its perturbations and break-down. *Tellus*, **1 (2)**, 14–37.

Berrisford, P., B. J. Hoskins, and E. Tyrlis, 2007: Blocking and Rossby wave breaking on the dynamical tropopause in the Southern Hemisphere. *Journal of the Atmospheric Sciences*, **64 (8)**, 2881–2898.

Bjerknes, J., 1966: A possible response of the atmospheric Hadley circulation to equatorial anomalies of ocean temperature. *Tellus*, **18**, 820–829.

Bjerknes, J., 1969: Atmospheric teleconnections from the equatorial Pacific. *Monthly Weather Review*, **97**, 163–172.

Blackmon, M. L., 1976: A climatological spectral study of the 500 mb geopotential height of the Northern Hemisphere. *Journal of the Atmospheric Sciences*, **33**, 1607–1623.

Blackmon, M. L., Y. Lee, and J. M. Wallace, 1984a: Horizontal structure of 500 mb height fluctuations with long, intermediate and short time scales. *Journal of the Atmospheric Sciences*, **41**, 961–980.

Blackmon, M. L., Y. Lee, J. M. Wallace, and H.-H. Hsu, 1984b: Time variation of 500 mb height fluctuations with long, intermediate and short time scales as deduced from lag-correlation statistics. *Journal of the Atmospheric Sciences*, **41**, 981–991.

Bony, S., B. Stevens, D. Coppin, T. Becker, K. A. Reed, A. Voigt, and B. Medeiros, 2016: Thermodynamic control of anvil cloud amount. *Proceedings of the National Academy of Sciences*, **113 (32)**, 8927–8932.

Boos, W. R., J. V. Hurley, and V. S. Murthy, 2014: Adiabatic westward drift of Indian monsoon depressions. *Quarterly Journal of the Royal Meteorological Society*, **141 (689)**, 1035–1048.

Bordoni, S., and T. Schneider, 2008: Monsoons as eddy-mediated regime transitions of the tropical overturning circulation. *Nature Geoscience*, **1 (8)**, 515–519.

Branstator, G., 1992: The maintenance of low-frequency atmospheric anomalies. *Journal of the Atmospheric Sciences*, **49 (20)**, 1924–1946.

Branstator, G., and H. Teng, 2017: Tropospheric waveguide teleconnections and their seasonality. *Journal of the Atmospheric Sciences*, **74 (5)**, 1513–1532.

Brayshaw, D. J., B. J. Hoskins, and M. Blackburn, 2009: The basic ingredients of the North Atlantic storm track. Part I: Land–sea contrast and orography. *Journal of the Atmospheric Sciences*, **66 (9)**, 2539–2558.

Bretherton, C. S., M. E. Peters, and L. E. Back, 2004: Relationships between water vapor path and precipitation over the tropical oceans. *Journal of Climate*, **17 (7)**, 1517–1528.

Bretherton, C. S., C. Smith, and J. M. Wallace, 1992. An intercomparison of methods for finding coupled patterns in climate data. *Journal of Climate*, **5(6)**, 541–560.

Bretherton, F. P., 1966: Critical layer instability in baroclinic flows. *Quarterly Journal of the Royal Meteorological Society*, **92 (393)**, 325–334.

Brewer, A. W., 1949: Evidence for a world circulation provided by the measurements of helium and water vapour distribution in the stratosphere. *Quarterly Journal of the Royal Meteorological Society*, **75**, 351–363.

Brier, G. W., and D.A. Bradley, 1964: The lunar synodical period and precipitation in the United States, *Journal of Atmospheric Sciences*, **21(4)**, 386–395.

Broecker, W. S., 1991: The great ocean conveyor. *Oceanography*, **4**, 79–89.

Cai, M., and M. Mak, 1990: Symbiotic relation between planetary and synoptic-scale waves. *Journal of the Atmospheric Sciences*, **47 (24)**, 2953–2968.

Camargo, S. J., and A. H. Sobel, 2005: Western North Pacific tropical cyclone intensity and ENSO. *Journal of Climate*, **18 (15)**, 2996–3006.

Cane, M. A., and E. S. Sarachik, 1976: Forced baroclinic ocean motions. 1. Linear equatorial unbounded case. *Journal of Marine Research*, **34 (4)**, 629–665.

Cane, M. A., S. E. Zebiak, and S. C. Dolan, 1986: Experimental forecasts of El Niño. *Nature*, **321 (6073)**, 827–832.

Chang, E. K., 1993: Downstream development of baroclinic waves as inferred from regression analysis. *Journal of the Atmospheric Sciences*, **50**, 2038–2053.

Chang, E. K., 2009: Diabatic and orographic forcing of northern winter stationary waves and storm tracks. *Journal of Climate*, **22 (3)**, 670–688.

Chang, P., L. Zhang, R. Saravanan, D. J. Vimont, J. C. H. Chiang, L. Ji, H. Seidel, and M. K. Tippett, 2007: Pacific Meridional Mode and El Niño-Southern Oscillation. *Geophysical Research Letters*, **34 (16)**.

Chapman, S., and R. S. Lindzen, 2012: *Atmospheric Tides: Thermal and Gravitational.* Springer Science & Business Media.

Charney, J.G., 1947: The dynamics of long waves in a baroclinic westerly current. *Journal of Meteorology*, **4**, 135–162.

Charney, J. G., 1963: A note on large-scale motions in the tropics. *Journal of the Atmospheric Sciences*, **20 (6)**, 607–609.

Charney, J. G., 1971: Geostrophic turbulence. *Journal of the Atmospheric Sciences*, **28**, 1087–1095.

Charney, J. G., and J. G. DeVore, 1979: Multiple flow equilibria in the atmosphere and blocking. *Journal of the Atmospheric Sciences*, **36 (7)**, 1205–1216.

Charney, J. G., and P. Drazin, 1961: Propagation of planetary-scale disturbances from lower into upper atmosphere. *Journal of Geophysical Research*, **66**, 83–109.

Chen, S. S. and R. A. Houze Jr., 1997: Diurnal variation and life-cycle of deep convective systems over the tropical Pacific warm pool. *Quarterly Journal of the Royal Meteorological Society*, **123**, 357–388.

Chen, X., and J. M. Wallace, 2015: ENSO-like variability: 1900–2013. *Journal of Climate*, **28 (24)**, 9623–9641.

Cheng, X., G. Nitsche, and J. M. Wallace, 1995: Robustness of low-frequency circulation patterns derived from EOF and rotated EOF analyses. *Journal of Climate*, **8 (6)**, 1709–1713.

Chiang, J. C., W. Kong, C. Wu, and D. S. Battisti, 2020: Origins of East Asian summer monsoon seasonality. *Journal of Climate*, **33 (18)**, 7945–7965.

Colle, B. A., and C. F. Mass, 1995: The structure and evolution of cold surges east of the Rocky Mountains. *Monthly Weather Review*, **123**, 2577–2610.

Czaja, A., and J. Marshall, 2006: The partitioning of poleward heat transport between the atmosphere and ocean. *Journal of the Atmospheric Sciences*, **63**, 1498–1511.

Deser, C., and J. M. Wallace, 1987: El Niño events and their relation to the Southern Oscillation: 1925–1986. *Journal of Geophysical Research: Oceans*, **92 (C13)**, 14 189–14 196.

Diallo, M., B. Legras, E. Ray, A. Engel, and J. A. Añel, 2017: Global distribution of CO_2 in the upper troposphere and stratosphere. *Atmospheric Chemistry and Physics*, **17 (6)**, 3861.

Dickey, J., S. Marcus, R. Hide, T. Eubanks, and D. Boggs, 1994: Angular momentum exchange among the solid Earth, atmosphere, and oceans: A case study of the 1982–1983 El Niño event. *Journal of Geophysical Research: Solid Earth*, **99 (B12)**, 23 921–23 937.

Dickinson, R. E., 1968: Planetary Rossby waves propagating vertically through weak westerly wind wave guides. *Journal of the Atmospheric Sciences*, **25**, 984–1002.

Dima, I. M., 2005: An observational study of the tropical tropospheric circulation. Ph.D. thesis, University of Washington.

Dima, I. M., and J. M. Wallace, 2003: On the seasonality of the Hadley cell. *Journal of the Atmospheric Sciences*, **60 (12)**, 1522–1527.

Dima, I. M., and J. M. Wallace, 2007: Structure of the annual-mean equatorial planetary waves in the ERA-40 reanalyses. *Journal of the Atmospheric Sciences*, **64**, 2862–2880.

Dommenget, D., and M. Latif, 2002: A cautionary note on the interpretation of EOFs. *Journal of Climate*, **15 (2)**, 216–225.

Donohoe, A., and D. S. Battisti, 2011: Atmospheric and surface contributions to planetary albedo. *Journal of Climate*, **24 (16)**, 4402–4418.

Dunkerton, T. J., 1978: On the mean meridional mass motions of the stratosphere and mesosphere. *Journal of the Atmospheric Sciences*, **35**, 2325–2333.

Dunkerton, T. J., M. T. Montgomery, and Z. Wang, 2009: Tropical cyclogenesis in a tropical wave critical layer: Easterly waves. *Atmospheric Chemistry and Physics Discussions*, **8**, 11 149–11 292.

Dutton, J. A., and D. R. Johnson, 1967: *The Theory of Available Potential Energy and a Variational Approach to Atmospheric Energetics*, Vol. 12. Academic Press New York, 331 pp.

Eady, E., 1949: Long waves and cyclone waves. *Tellus*, **1**, 33–52.

Eady, E., 1950: The cause of the general circulation in the atmosphere. *Centennial Proceedings Royal Meteorological Society*, **76**, 156–172.

Edmon, H. J., B. J. Hoskins, and M. E. McIntyre, 1980: Eliassen-Palm cross sections for the troposphere. *Journal of the Atmospheric Sciences*, **37**, 2600–2616.

Eliassen, A., 1951: Slow thermally or frictionally controlled meridional circulation in a circular vortex. *Astrophysica Norvegica*, **5**, 19.

Eliassen, A., and E. Palm, 1961: On the transfer of energy in stationary mountain waves. *Geofysiske Publikasjoner*, **22**, 1–23.

Emanuel, K., 2018: 100 years of progress in tropical cyclone research. *Meteorological Monographs*, **59**, 15.1 – 15.68.

Emanuel, K., and D. S. Nolan, 2004: Tropical cyclone activity and the global climate system. *26th Conference on Hurricanes and Tropical Meteorolgy*, 240–241.

Emanuel, K. A., 1986: An air–sea interaction theory for tropical cyclones. Part I: Steady-state maintenance. *Journal of the Atmospheric Sciences*, **43 (6)**, 585–605.

Exner, F. M., 1913: Über oszillatorische Strömungen der Luft. *Ann. Hydr.*, **41**, 145–150.

Ferrel, W., 1856: An essay on the winds and currents of ocean. *Nashville Journal of Medicine and Surgery*, **11**.

Findlaterr, J. 1969: A major low-level air current near the Indian Ocean during the northern summer. *Quarterly Journal of the Royal Meteorological Society*, **95 (404)**, 362–380.

Flasar, F. M., and Coauthors, 2004: An intense stratospheric jet on Jupiter. *Nature*, **427 (6970)**, 132–135.

Franzke, C., S. B. Feldstein, and S. Lee, 2011: Synoptic analysis of the Pacific–North American teleconnection pattern. *Quarterly Journal of the Royal Meteorological Society*, **137 (655)**, 329–346.

Friedson, A. J., 1999: New observations and modelling of a QBO-like oscillation in Jupiter's stratosphere. *Icarus*, **137 (1)**, 34–55.

Fueglistaler, S., H. Wernli, and T. Peter, 2004: Tropical troposphere-to-stratosphere transport inferred from trajectory calculations. *Journal of Geophysical Research: Atmospheres*, **109 (D3)**.

Fultz, D., R. R. Long, G. V. Owens, W. Bohan, R. Kaylor, and J. Weil, 1959: Studies of thermal convection in a rotating cylinder with some implications for large-scale atmospheric motions. *Studies of Thermal Convection in a Rotating Cylinder with Some Implications for Large-Scale Atmospheric Motions*, Springer, 1–104.

Gadgil, S., P. V. Joseph, and N. V. Joshi, 1984: Ocean–atmosphere coupling over monsoon regions. *Nature*, **312 (5990)**, 141–143.

Gallimore, R. G., and D. R. Johnson, 1981: A numerical diagnostic model of the zonally averaged circulation in isentropic coordinates. *Journal of the Atmospheric Sciences*, **38**, 1870–1890.

Garfinkel, C. I., T. A. Shaw, D. L. Hartmann, and D. W. Waugh, 2012: Does the Holton–Tan mechanism explain how the Quasi-Biennial Oscillation modulates the Arctic polar vortex? *Journal of the Atmospheric Sciences*, **69 (5)**, 1713–1733.

Garfinkel, C. I., I. White, E. P. Gerber, M. Jucker, and M. Erez, 2020: The building blocks of Northern Hemisphere wintertime stationary waves. *Journal of Climate*, **33 (13)**, 5611–5633.

Garreaud, R. D., 2001: Subtropical cold surges: Regional aspects and global distribution. *International Journal of Climatology*, **21**, 1181–1197.

Geen, R., S. Bordoni, D. S. Battisti, and K. Hui, 2020: Monsoons, ITCZs, and the concept of the global monsoon. *Reviews of Geophysics*, **58 (4)**, e2020RG000 700.

Gerber, E. P., and D. W. Thompson, 2017: What makes an annular mode "annular"? *Journal of the Atmospheric Sciences*, **74 (2)**, 317–332.

Gerber, E. P., and G. K. Vallis, 2005: A stochastic model for the spatial structure of annular patterns of variability and the North Atlantic Oscillation. *Journal of Climate*, **18**, 2102–2118.

Gettelman, A., and Coauthors, 2019: The whole atmosphere community climate model version 6 (WACCM6). *Journal of Geophysical Research: Atmospheres*, **124 (23)**, 12 380–12 403.

Gill, A. E., 1980: Some simple solutions for heat-induced tropical circulation. *Royal Meteorological Society, Quarterly Journal*, **106**, 447–462.

Gillett, N. P., and D. W. Thompson, 2003: Simulation of recent Southern Hemisphere climate change. *Science*, **302 (5643)**, 273–275.

Gleeson, T., K. M. Befus, S. Jasechko, E. Luijendijk, and M. B. Cardenas, 2016: The global volume and distribution of modern groundwater. *Nature Geoscience*, **9 (2)**, 161–167.

Gonzalez, A. O., and X. Jiang, 2019: Distinct propagation characteristics of intraseasonal variability over the tropical west Pacific. *Journal of Geophysical Research: Atmospheres*, **124 (10)**, 5332–5351.

Goody, R. M., and J. C. G. Walker, 1972: *Atmospheres: Foundations of Earth Science Series*. Prentice-Hall.

Graham, N. E., and T. P. Barnett, 1987: Sea surface temperature, surface wind divergence, and convection over tropical oceans. *Science*, **238 (4827)**, 657–659.

Gray, W. M., 1968: Global view of the origin of tropical disturbances and storms. *Monthly Weather Review*, **96 (10)**, 669–700.

Green, J. S. A., 1970: Transfer properties of the large-scale eddies and the general circulation of the atmosphere. *Quarterly Journal of the Royal Meteorological Society*, **96 (408)**, 157–185.

Green, J. S. A., 1977: The weather during July 1976: Some dynamical considerations of the drought. *Weather*, **32 (4)**, 120–126.

Grise, K. M., and D. W. Thompson, 2012: Equatorial planetary waves and their signature in atmospheric variability. *Journal of the Atmospheric Sciences*, **69 (3)**, 857–874.

Guo, J., and Z. Kuang, 2019: Spatial and temporal characteristics of asymmetries in tropical cyclones. *Geophysical Research Letters*, **46 (13)**, 7769–7779.

Hack, J. J., and W. H. Schubert, 1986: Nonlinear response of atmospheric vortices to heating by organized cumulus convection. *Journal of Atmospheric Sciences*, **43 (15)**, 1559–1573.

Hadley, G., 1735: VI. Concerning the cause of the general trade-winds. *Philosophical Transactions of the Royal Society of London*, **39 (437)**, 58–62.

Haertel, P. T., and G. N. Kiladis, 2004: Dynamics of 2-day equatorial waves. *Journal of the Atmospheric Sciences*, **61 (22)**, 2707–2721.

Haines, K., and J. Marshall, 1987: Eddy-forced coherent structures as a prototype of atmospheric blocking. *Royal Meteorological Society, Quarterly Journal*, **113**, 681–704.

Halpern, D., R. A. Knox, and D. S. Luther, 1988: Observations of 20-day period meridional current oscillations in the upper ocean along the Pacific equator. *Journal of Physical Oceanography*, **18 (11)**, 1514 – 1534.

Hannachi, A., and W. Iqbal, 2019: Bimodality of hemispheric winter atmospheric variability via average flow tendencies and kernel EOFs. *Tellus A: Dynamic Meteorology and Oceanography*, **71 (1)**, 1633 847.

Hartmann, D. L., H. H. Hendon, and R. A. Houze Jr, 1984: Some implications of the mesoscale circulations in tropical cloud clusters for large-scale dynamics and climate. *Journal of the Atmospheric Sciences*, **41 (1)**, 113–121.

Hartmann, D. L., and F. Lo, 1998: Wave-driven zonal flow vacillation in the Southern Hemisphere. *Journal of the Atmospheric Sciences*, **55**, 1303–1315.

Hartmann, D. L., and M. L. Michelsen, 1989: Intraseasonal periodicities in Indian rainfall. *Journal of the Atmospheric Sciences*, **46 (18)**, 2838–2862.

Hartmann, D. L., V. Ramanathan, A. Berroir, and G. E. Hunt, 1986: Earth radiation budget data and climate research. *Reviews of Geophysics*, **24 (2)**, 439–468.

Hartmann, D. L., and P. Zuercher, 1998: Response of baroclinic life cycles to barotropic shear. *Journal of the Atmospheric Sciences*, **55**, 297–313.

Haurwitz, B., 1941: *Dynamic Meteorology*. McGraw-Hill Book Company, Inc., 365 pp.

Hayashi, Y., 1974: Spectral analysis of tropical disturbances appearing in a GFDL general circulation model. *Journal of Atmospheric Sciences*, **31 (1)**, 180–218.

Haynes, P. H. and M. E. McIntyre, 1990: On the conservation and impermeability theorems for potential vorticity. *Journal of Atmospheric Science*, **47**, 2021–2030.

Haynes, P. H., M. E. McIntyre, T. G. Shepherd, C. J. Marks, and K. P. Shine, 1991: On the "Downward Control" of extratropical diabatic circulations by eddy-induced mean zonal forces. *Journal of the Atmospheric Sciences*, **48 (4)**, 651–678.

Heckley, W. A., and A. E. Gill, 1984: Some simple analytic solutions to the problem of forced equatorial long waves. *Quarterly Journal of the Royal Meteorological Society*, **110 (463)**, 203–217.

Held, I. M., 1975: Momentum transport by quasi-geostrophic eddies. *Journal of the Atmospheric Sciences*, **32**, 1494–1497.

Held, I. M., 1983: Stationary and quasi-stationary eddies in the extratropical troposphere: Theory. *Large-scale Dynamical Processes in the Atmosphere*, Academic Press, 127–168.

Held, I. M., and A. Y. Hou, 1980: Nonlinear axially symmetric circulations in a nearly inviscid atmosphere. *Journal of the Atmospheric Sciences*, **37 (3)**, 515–533.

Held, I. M. and T. Schneider, 1999: The surface branch of the zonally averaged mass transport circulation in the troposphere. *Journal of Atmospheric Science*, **56**, 1688–1697.

Held, I. M., M. Ting, and H. Wang, 2002: Northern winter stationary waves: Theory and modeling. *Journal of Climate*, **15**, 2125–2144.

Henderson, S. A., E. D. Maloney, and E. A. Barnes, 2016: The influence of the Madden–Julian Oscillation on Northern Hemisphere winter blocking. *Journal of Climate*, **29 (12)**, 4597–4616.

Hersbach, H., and Coauthors, 2020: The ERA5 global reanalysis. *Quarterly Journal of the Royal Meteorological Society*, **146 (730)**, 1999–2049.

Highwood, E. J., and B. J. Hoskins, 1998: The tropical tropopause. *Quarterly Journal of the Royal Meteorological Society*, **124 (549, A)**, 1579–1604.

Hitchman, M. H., C. B. Leovy, J. C. Gille, and P. L. Bailey, 1987: Quasi-stationary zonally asymmetric circulations in the equatorial lower mesosphere. *Journal of the Atmospheric Sciences*, **44 (16)**, 2219–2236.

Hoerling, M. P., A. Kumar, and T. Y. Xu, 2001: Robustness of the nonlinear atmospheric response to opposite phases of ENSO. *Journal of Climate*, **14**, 1277–1293.

Holton, J. R., 1965: The influence of viscous boundary layers on transient motions in a stratified rotating fluid. Part I. *Journal of the Atmospheric Sciences*, **22**, 402–411.

Holton, J. R., 1972: Waves in the equatorial stratosphere generated by tropospheric heat sources. *Journal of the Atmospheric Sciences*, **29 (2)**, 368–375.

Holton, J. R., 1987: Issues in atmospheric and oceanic modeling. *Dynamics of Atmospheres and Oceans*, **11**, 204–206.

Holton, J. R., and T. Dunkerton, 1978: On the role of wave transience and dissipation in stratospheric mean flow vacillations. *Journal of the Atmospheric Sciences*, **35**, 740–744.

Holton, J. R., and G. J. Hakim, 2012: *An Introduction to Dynamic Meteorology*, Vol. 88. Academic Press.

Holton, J. R., and R. S. Lindzen, 1972: An updated theory for the quasi-biennial cycle of the tropical stratosphere. *Journal of the Atmospheric Sciences*, **29**, 1076–1080.

Holton, J. R., and C. Mass, 1976: Stratospheric vacillation cycles. *Journal of the Atmospheric Sciences*, **33**, 2218–2225.

Holton, J. R., and H.-C. Tan, 1980: The influence of the equatorial Quasi-Biennial Oscillation on the global circulation at 50 mb. *Journal of the Atmospheric Sciences*, **37 (10)**, 2200–2208.

Horel, J. D., and J. M. Wallace, 1981: Planetary-scale atmospheric phenomena associated with the Southern Oscillation. *Monthly Weather Review*, **109 (4)**, 813–829.

Hoskins, B. J., and K. I. Hodges, 2002: New perspectives on the Northern Hemisphere winter storm tracks. *Journal of the Atmospheric Sciences*, **59**, 1041–1061.

Hoskins, B. J., and I. N. James, 2014: *Fluid Dynamics of the Mid-Latitude Atmosphere*. John Wiley & Sons, 408 pp.

Hoskins, B. J., I. N. James, and G. H. White, 1983: The shape, propagation and mean-flow interaction of large-scale weather systems. *Journal of the Atmospheric Sciences*, **40**, 1595–1612.

Hoskins, B. J., and D. J. Karoly, 1981: The steady linear response of a spherical atmosphere to thermal and orographic forcing. *Journal of the Atmospheric Sciences*, **38**, 1179–1196.

Hoskins, B. J., M. E. McIntyre, and A. W. Robertson, 1985: On the use and significance of isentropic potential vorticity maps. *Quarterly Journal of the Royal Meteorological Society*, **111 (470)**, 877–946.

Hoskins, B. J., A. J. Simmons, and D. G. Andrews, 1977: Energy dispersion in a barotropic atmosphere. *Quarterly Journal of the Royal Meteorological Society*, **103**, 553–567.

Hoskins, B. J., G.-Y. Yang, and R. M. Fonseca, 2020: The detailed dynamics of the June–August Hadley cell. *Quarterly Journal of the Royal Meteorological Society*, **146 (727)**, 557–575.

Houze, R. A., 2010: Clouds in tropical cyclones. *Monthly Weather Review*, **138 (2)**, 293–344.

Huaman, L., C. Schumacher, and G. N. Kiladis, 2020: Eastward-propagating disturbances in the tropical Pacific. *Monthly Weather Review*, **148 (9)**, 3713–3728.

Imbrie, J., and K. P. Imbrie, 1986: *Ice Ages: Solving the Mystery*. Harvard University Press.

Iskenderian, H., and D. A. Salstein, 1998: Regional sources of mountain torque variability and high-frequency fluctuations in atmospheric angular momentum. *Monthly Weather Review*, **126**, 1681–1694.

Ivanov, M. A., and S. N. Evtimov, 2014: Seasonality in the biplot of Northern Hemisphere temperature anomalies. *Quarterly Journal of the Royal Meteorological Society*, **140 (685)**, 2650–2657.

Jaeglé, L., R. Wood, and K. Wargan, 2017: Multiyear composite view of ozone enhancements and stratosphere-to-troposphere transport in dry intrusions of Northern Hemisphere extratropical cyclones. *Journal of Geophysical Research: Atmospheres*, **122 (24)**, 13436–13457.

James, I., 1995: *Introduction to Circulating Atmospheres* (Cambridge Atmospheric and Space Science Series). Cambridge: Cambridge University Press.

Jeffreys, H., 1926: On the dynamics of geostrophic winds. *Quarterly Journal of the Royal Meteorological Society*, **52**, 85–104.

Jiang, X., and Coauthors, 2020: Fifty years of research on the Madden-Julian Oscillation: Recent progress, challenges, and perspectives. *Journal of Geophysical Research: Atmospheres*, **125 (17)**, e2019JD030911.

Johnson, D. R., 1970: The available potential energy of storms. *Journal of the Atmospheric Sciences*, **27 (5)**, 727–741.

Jucker, M., and E. P. Gerber, 2017: Untangling the annual cycle of the tropical tropopause layer with an idealized moist model. *Journal of Climate*, **30 (18)**, 7339–7358.

Kao, S. K., 1968: Governing equations and spectra for atmospheric motion and transports in frequency, wave-number space. *Journal of the Atmospheric Sciences*, **25**, 32–38.

Kao, S. K., and C. N. Chi, 1978: Mechanism for the growth and decay of long- and synoptic-scale waves in the mid-troposphere. *Journal of the Atmospheric Sciences*, **35**, 1375–1387.

Keller, J. H., and Coauthors, 2019: The extratropical transition of tropical cyclones. Part II: Interaction with the midlatitude flow, downstream impacts, and implications for predictability. *Monthly Weather Review*, **147 (4)**, 1077–1106.

Khouakhi, A., G. Villarini, and G. A. Vecchi, 2017: Contribution of tropical cyclones to rainfall at the global scale. *Journal of Climate*, **30 (1)**, 359–372.

Kida, H., 1977: A numerical investigation of the atmospheric general circulation and stratospheric-tropospheric mass exchange: II. Lagrangian motion of the atmosphere. *Journal of the Meteorological Society of Japan. Series II*, **55 (1)**, 71–88.

Kidson, J. W., 1988a: Indices of the Southern Hemisphere zonal wind. *Journal of Climate*, **1 (2)**, 183–194.

Kidson, J. W., 1988b: Interannual variations in the Southern Hemisphere circulation. *Journal of Climate*, **1 (12)**, 1177–1198.

Kiladis, G. N., 1998: Observations of Rossby waves linked to convection over the eastern tropical Pacific. *Journal of the Atmospheric Sciences*, **55 (3)**, 321–339.

Kiladis, G. N., J. Dias, K. H. Straub, M. C. Wheeler, S. N. Tulich, K. Kikuchi, K. M. Weickmann, and M. J. Ventrice, 2014: A comparison of OLR and circulation-based indices for tracking the MJO. *Monthly Weather Review*, **142 (5)**, 1697–1715.

Kiladis, G. N., C. D. Thorncroft, and N. M. J. Hall, 2006: Three-dimensional structure and dynamics of African easterly waves. Part I: observations. *Journal of the Atmospheric Sciences*, **63 (9)**, 2212–2230.

Kiladis, G. N., M. C. Wheeler, P. T. Haertel, K. H. Straub, and P. E. Roundy, 2009: Convectively coupled equatorial waves. *Reviews of Geophysics*, **47 (2)**.

King, M. J., M. C. Wheeler, and T. P. Lane, 2015: Association of convection with the 5-day Rossby–Haurwitz wave. *Journal of the Atmospheric Sciences*, **72 (9)**, 3309–3321.

Klein, W., 1957: Principal tracks and mean frequencies of cyclones and anticyclones in the Northern Hemisphere. Research Paper 40, U.S. Weather Bureau.

Klein, W. H., 1951: A hemispheric study of daily pressure variability at sea level and aloft. *Journal of Meteorology*, **8 (5)**, 332–346.

Knapp, K. R., H. J. Diamond, J. P. Kossin, M. C. Kruk, C. Schreck, and Coauthors, 2018: International best track archive for climate stewardship (IBTrACS) project, version 4. *NOAA National Centers for Environmental Information*.

Knapp, K. R., M. C. Kruk, D. H. Levinson, H. J. Diamond, and C. J. Neumann, 2010: The international best track archive for climate stewardship (IBTrACS) unifying tropical cyclone data. *Bulletin of the American Meteorological Society*, **91 (3)**, 363–376.

Kohonen, T., 1990: The self-organizing map. *Proceedings of the IEEE*, **78 (9)**, 1464–1480.

Kohyama, T., and J. M. Wallace, 2016: Rainfall variations induced by the lunar gravitational atmospheric tide and their implications for the relationship between tropical rainfall and humidity. *Geophysical Research Letters*, **43 (2)**, 918–923.

Kolmogorov, A. N., 1962: A refinement of previous hypotheses concerning the local structure of turbulence in a viscous incompressible fluid at high Reynolds number. *Journal of Fluid Mechanics*, **13 (1)**, 82–85.

Kosaka, Y., and H. Nakamura, 2006: Structure and dynamics of the summertime Pacific–Japan teleconnection pattern. *Quarterly Journal of the Royal Meteorological Society*, **132 (619)**, 2009–2030.

Kowaleski, A. M., and J. L. Evans, 2015: Thermodynamic observations and flux calculations of the tropical cyclone surface layer within the context of potential intensity. *Weather and Forecasting*, **30 (10)**, 1303–1320.

Kraichnan, R. H., 1959: The structure of isotropic turbulence at very high Reynolds numbers. *Journal of Fluid Mechanics*, **5 (4)**, 497–543.

Kuai, L., R.-L. Shia, X. Jiang, K.-K. Tung, and Y. L. Yung, 2009: Nonstationary synchronization of equatorial QBO with SAO in observations and a model. *Journal of the Atmospheric Sciences*, **66 (6)**, 1654–1664.

Kump, L. R., J. F. Kasting, R. G. Crane, and Coauthors, 2004: *The Earth System*, Vol. 432. Pearson Prentice Hall.

Kung, E. C., 1966: Large-scale balance of kinetic energy in the atmosphere. *Monthly Weather Review*, **94**, 627–640.

Kung, E. C., and P. J. Smith, 1974: Problems of large-scale kinetic energy balance—a diagnostic analysis in GARP. *Bulletin of the American Meteorological Society*, **55 (7)**, 768–778.

Kuo, H.-L., 1951: Dynamical aspects of the general circulation and the stability of zonal flow. *Tellus*, **3 (4)**, 268–284.

Kutzbach, J. E., 1970: Large-scale features of monthly mean Northern Hemisphere anomaly maps of sea-level pressure. *Monthly Weather Review*, **98**, 708–716.

Lambeck, K., 2005: *The Earth's Variable Rotation: Geophysical Causes and Consequences*. Cambridge University Press.

Large, W. G., and S. Pond, 1981: Open ocean momentum flux measurements in moderate to strong winds. *Journal of Physical Oceanography*, **11**, 324–336.

Lau, N.-C., 1988: Variability of the observed midlatitude storm tracks in relation to low-frequency changes in the circulation pattern. *Journal of the Atmospheric Sciences*, **45 (19)**, 2718–2743.

Lau, N.-C., and E. O. Holopainen, 1984: Transient eddy forcing of the time-mean flow as identified by geopotential tendencies. *Journal of the Atmospheric Sciences*, **41**, 313–328.

Lau, N.-C., A. Leetmaa, and M. J. Nath, 2006: Attribution of atmospheric variations in the 1997–2003 period to SST anomalies in the Pacific and Indian Ocean basins. *Journal of Climate*, **19 (15)**, 3607–3628.

Lau, N.-C., A. Leetmaa, M. J. Nath, and H.-L. Wang, 2005: Influences of ENSO-induced Indo–western Pacific SST anomalies on extratropical atmospheric variability during the boreal summer. *Journal of Climate*, **18 (15)**, 2922–2942.

Lau, N.-C., and M. J. Nath, 1996: The role of the "atmospheric bridge" in linking tropical Pacific ENSO events to extratropical SST anomalies. *Journal of Climate*, **9 (9)**, 2036–2057.

Lau, N.-C., and J. M. Wallace, 1979: On the distribution of horizontal transports by transient eddies in the Northern Hemisphere wintertime circulation. *Journal of the Atmospheric Sciences*, **36**, 1844–1861.

Lee, S., 1999: Why are the climatological zonal winds easterly in the equatorial upper troposphere? *Journal of the Atmospheric Sciences*, **56**, 1353–1363.

Lee, S., and I. M. Held, 1993: Baroclinic wave packets in models and observations. *Journal of the Atmospheric Sciences*, **50**, 1413–1428.

Thompson, D. W. and E. A. Barnes, 2014: Periodic variability in the large-scale Southern Hemisphere atmospheric circulation. *Science*, **343**, 641–645.

Li, L., A. P. Ingersoll, X. Jiang, D. Feldman, and Y. L. Yung, 2007: Lorenz energy cycle of the global atmosphere based on reanalysis datasets. *Geophysical Research Letters*, **34 (16)**.

Li, T., L. Wang, M. Peng, B. Wang, C. Zhang, W. Lau, and H.-C. Kuo, 2018: A paper on the tropical intraseasonal oscillation published in 1963 in a Chinese journal. *Bulletin of the American Meteorological Society*, **99 (9)**, 1765–1779.

Lim, H., and C. P. Chang, 1983: Dynamics of teleconnections and Walker circulations forced by equatorial heating. *Journal of the Atmospheric Sciences*, **40 (8)**, 1897–1915.

Lin, L., Q. Fu, H. Zhang, J. Su, Q. Yang, and Z. Sun, 2013: Upward mass fluxes in tropical upper troposphere and lower stratosphere derived from radiative transfer calculations. *Journal of Quantitative Spectroscopy and Radiative Transfer*, **117**, 114–122.

Lindzen, R. S., and J. R. Holton, 1968: A theory of the Quasi-Biennial Oscillation. *Journal of the Atmospheric Sciences*, **25**, 1095–1107.

Lindzen, R. S., and A. V. Hou, 1988: Hadley circulations for zonally averaged heating centered off the equator. *Journal of the Atmospheric Sciences*, **45 (17)**, 2416–2427.

Lindzen, R. S., and S. Nigam, 1987: On the role of sea surface temperature gradients in forcing low-level winds and convergence in the tropics. *Journal of the Atmospheric Sciences*, **44 (17)**, 2418–2436.

Liu, X., D. S. Battisti, and G. H. Roe, 2017: The effect of cloud cover on the meridional heat transport: Lessons from variable rotation experiments. *Journal of Climate*, **30 (18)**, 7465–7479.

Lorenz, D. J., and D. L. Hartmann, 2001: Eddy-zonal flow feedback in the Southern Hemisphere. *Journal of the Atmospheric Sciences*, **58 (21)**, 3312–3327.

Lorenz, D. J., and D. L. Hartmann, 2003: Eddy-zonal flow feedback in the Northern Hemisphere winter. *Journal of Climate*, **16**, 1212–1227.

Lorenz, E. N., 1953: A multiple index notation for describing atmospheric transport processes. *Geophysical Research Papers*, **53**, 100–110.

Lorenz, E. N., 1955: Available potential energy and the maintenance of the general circulation. *Tellus*, **7**, 157–167.

Lorenz, E. N., 1956: Empirical orthogonal functions and statistical weather prediction. Sci. Rep. No. 1, *Statistical Forecasting Project*, Massachusetts Institute of Technology, Department of Meteorology, 48, 1.

Lorenz, E. N., 1963: Deterministic nonperiodic flow. *Journal of the Atmospheric Sciences*, **20 (2)**, 130–141.

Lorenz, E. N., 1967: *The Nature and Theory of the General Circulation of the Atmosphere*, Vol. 218. World Meteorological Organization, Geneva.

Lübbecke, J. F., and M. J. McPhaden, 2017: Symmetry of the Atlantic Niño mode. *Geophysical Research Letters*, **44 (2)**, 965–973.

Lumpkin, R., and G. C. Johnson, 2013: Global ocean surface velocities from drifters: Mean, variance, El Niño–Southern Oscillation response, and seasonal cycle. *Journal of Geophysical Research: Oceans*, **118**, 2992–3006.

Madden, R. A., and P. R. Julian, 1971: Detection of a 40–50 day oscillation in the zonal wind in the tropical Pacific. *Journal of the Atmospheric Sciences*, **28 (5)**, 702–708.

Madden, R. A., and P. R. Julian, 1972: Description of global-scale circulation cells in the tropics with a 40–50 day period. *Journal of the Atmospheric Sciences*, **29 (6)**, 1109–1123.

Madden, R. A., and P. Speth, 1995: Estimates of atmospheric angular momentum, friction, and mountain torques during 1987–1988. *Journal of the Atmospheric Sciences*, **52 (21)**, 3681–3694.

Mak, M.-K., 1969: Laterally driven stochastic motions in the tropics. *Journal of the Atmospheric Sciences*, **26 (1)**, 41–64.

Maloney, E. D., and D. L. Hartmann, 2001: The Madden–Julian oscillation, barotropic dynamics, and North Pacific tropical cyclone formation. Part I: Observations. *Journal of the Atmospheric Sciences*, **58 (17)**, 2545–2558.

Manabe, S., J. Smagorinsky, and R. F. Strickler, 1965: Simulated climatology of a general circulation model with a hydrologic cycle. *Monthly Weather Review*, **93**, 769–798.

Manabe, S., and R. F. Strickler, 1964: Thermal equilibrium of the atmosphere with a convective adjustment. *Journal of the Atmospheric Sciences*, **21 (4)**, 361–385.

Manabe, S., and T. B. Terpstra, 1974: The effects of mountains on the general circulation of the atmosphere as identified by numerical experiments. *Journal of the Atmospheric Sciences*, **31 (1)**, 3–42.

Mapes, B., S. Tulich, J. Lin, and P. Zuidema, 2006: The mesoscale convection life cycle: Building block or prototype for large-scale tropical waves? *Dynamics of Atmospheres and Oceans*, **42 (1-4)**, 3–29.

Mapes, B. E., 1997: *Mutual Adjustment of Mass Flux and Stratification Profiles*, 399–411. Springer.

Mapes, B. E., E. S. Chung, W. M. Hannah, H. Masunaga, A. J. Wimmers, and C. S. Velden, 2018: The meandering margin of the meteorological moist tropics. *Geophysical Research Letters*, **45 (2)**, 1177–1184.

Mapes, B. E., T. T. Warner, and M. Xu, 2003: Diurnal patterns of rainfall in northwestern South America. Part III: Diurnal gravity waves and nocturnal convection offshore. *Monthly Weather Review*, **131 (5)**, 830–844.

Marengo, J. A., W. R. Soares, C. Saulo, and M. Nicolini, 2004: Climatology of the low-level jet east of the Andes as derived from the NCEP-NCAR reanalyses: Characteristics and temporal variability. *Journal of Climate*, **17**, 2261–2280.

Martius, O., C. Schwierz, and H. Davies, 2007: Breaking waves at the tropopause in the wintertime Northern Hemisphere: Climatological analyses of the orientation and the theoretical LC1/2 classification. *Journal of the Atmospheric Sciences*, **64 (7)**, 2576–2592.

Matsuno, T., 1966: Quasi-geostrophic motions in the equatorial area. *Journal of the Meteorological Society of Japan*, **44**, 25–43.

Maximenko, N. A., O. V. Melnichenko, P. P. Niiler, and H. Sasaki, 2008: Stationary mesoscale jet-like features in the ocean, *Geophysical Research Letters*, **35**, L08603.

Mayta, V. C., and Á. F. Adames, 2021: 2-day westward-propagating inertio-gravity waves during GoAmazon. *Journal of the Atmospheric Sciences*, **78 (11)**, 3727–3743.

Minobe, S., A. Kuwano-Yoshida, N. Komori, S.-P. Xie, and R. J. Small, 2008: Influence of the Gulf Stream on the troposphere. *Nature*, **452**, 206–209.

Mitchell, T. P., and J. M. Wallace, 1992: The annual cycle in equatorial convection and sea surface temperature. *Journal of Climate*, **5 (10)**, 1140–1156.

Miyasaka, T., and H. Nakamura, 2005: Structure and formation mechanisms of the Northern Hemisphere summertime subtropical highs. *Journal of Climate*, **18 (23)**, 5046–5065.

Monteiro, J. M., Á. F. Adames, J. M. Wallace, and J. S. Sukhatme, 2014: Interpreting the upper level structure of the Madden-Julian Oscillation. *Geophysical Research Letters*, **41 (24)**, 9158–9165.

Mori, M., M. Watanabe, H. Shiogama, J. Inoue, and M. Kimoto, 2014: Robust Arctic sea-ice influence on the frequent Eurasian cold winters in past decades. *Nature Geoscience*, **7 (12)**, 869–873.

Mote, P. W., and Coauthors, 1996: An atmospheric tape recorder: The imprint of tropical tropopause temperatures on stratospheric water vapor. *Journal of Geophysical Research: Atmospheres*, **101 (D2)**, 3989–4006.

Mullen, S. L., 1986: The local balances of vorticity and heat for blocking anticyclones in a spectral general circulation model. *Journal of the Atmospheric Sciences*, **43 (13)**, 1406–1441.

Murgatroyd, R. J., and F. Singleton, 1961: Possible meridional circulations in the stratosphere and mesosphere. *Quarterly Journal of the Royal Meteorological Society*, **87 (372)**, 125–135.

Nakamura, H., M. Nakamura, and J. L. Anderson, 1997: The role of high- and low-frequency dynamics in blocking formation. *Monthly Weather Review*, **125 (9)**, 2074–2093.

Nakamura, H., and J. M. Wallace, 1993: Synoptic behavior of baroclinic eddies during the blocking onset. *Monthly Weather Review*, **121 (7)**, 1892–1903.

Nakamura, N., J. Falk, and S. W. Lubis, 2020: Why are Stratospheric Sudden Warmings sudden (and intermittent)? *Journal of the Atmospheric Sciences*, **77 (3)**, 943–964.

Nakamura, N., and D. Zhu, 2010: Finite-amplitude wave activity and diffusive flux of potential vorticity in eddy–mean flow interaction. *Journal of the Atmospheric Sciences*, **67 (9)**, 2701–2716.

Narsey, S., M. J. Reeder, D. Ackerley, and C. Jakob, 2017: A midlatitude influence on Australian monsoon bursts. *Journal of Climate*, **30 (14)**, 5377–5393.

Navato, A. R., R. E. Newell, J. C. Hsiung, C. B. Billing Jr, and B. C. Weare, 1981: Tropospheric mean temperature and its relationship to the oceans and atmospheric aerosols. *Monthly Weather Review*, **109 (2)**, 244–254.

Neelin, J. D., D. S. Battisti, A. C. Hirst, F.-F. Jin, Y. Wakata, T. Yamagata, and S. E. Zebiak, 1998: ENSO theory. *Journal of Geophysical Research-Oceans*, **103 (C7)**, 14 261–14 290.

Newell, R. E., 1963: Transfer through the tropopause and within the stratosphere. *Quarterly Journal of the Royal Meteorological Society*, **89 (380)**, 167–204.

Newell, R. E., and B. C. Weare, 1976: Factors governing tropospheric mean temperature. *Science*, **194 (4272)**, 1413–1414.

Nitta, T., 1987: Convective activities in the tropical western Pacific and their impact on the Northern Hemisphere summer circulation. *Journal of the Meteorological Society of Japan. Ser. II*, **65 (3)**, 373–390.

Nolan, D. S., Y. Moon, and D. P. Stern, 2007: Tropical cyclone intensification from asymmetric convection: Energetics and efficiency. *Journal of the Atmospheric Sciences*, **64 (10)**, 3377–3405.

North, G. R., T. L. Bell, R. F. Cahalan, and F. J. Moeng, 1982: Sampling errors in the estimation of empirical orthogonal functions. *Monthly Weather Review*, **110 (7)**, 699–706.

Obasi, G. O. P., 1963: Poleward flux of atmospheric angular momentum in the Southern Hemisphere. *Journal of the Atmospheric Sciences*, **20**, 516–528.

Okumura, Y. M., and C. Deser, 2010: Asymmetry in the duration of El Niño and La Niña. *Journal of Climate*, **23 (21)**, 5826–5843.

Oort, A. H., 1964: On estimates of the atmospheric energy cycle. *Monthly Weather Review*, **92**, 483–493.

Oort, A. H., and T. H. Vonder Haar, 1976: On the observed annual cycle in the ocean-atmosphere heat balance over the Northern Hemisphere. *Journal of Physical Oceanography*, **6**, 781–800.

Pahlavan, H. A., Q. Fu, J. M. Wallace, and G. N. Kiladis, 2021a: Revisiting the Quasi-Biennial Oscillation as seen in ERA5. Part I: Description and momentum budget. *Journal of the Atmospheric Sciences*, **78 (3)**, 673 – 691.

Pahlavan, H. A., J. M. Wallace, Q. Fu, and G. N. Kiladis, 2021b: Revisiting the Quasi-Biennial Oscillation as seen in ERA5. Part II: Evaluation of waves and wave forcing. *Journal of the Atmospheric Sciences*, **78 (3)**, 693 – 707.

Pahlavan, H. A., J. M. Wallace, and Q. Fu, 2022: Characteristics of tropical convective gravity waves resolved by ERA5 Reanalysis. *Journal of the Atmospheric Sciences* (published online ahead of print 2022). Retrieved from https://journals.ametsoc.org/view/journals/atsc/aop/JAS-D-22-0057.1/JAS-D-22-0057.1.xml

Palmén, E., 1949: Meridional circulations and the transfer of angular momentum in the atmosphere. *Journal of Atmospheric Sciences*, **6**, 429–430.

Palmén, E., 1956: Formation and development of tropical cyclones. *Proceedings of Tropical Cyclone symposium*, Brisbane, 213–231.

Palmén, E., and C. W. Newton 1969: *Synoptic Meteorology: Atmospheric Circulation Systems*. New York: Academic Press.

Palmer, T. N., G. J. Shutts, and R. Swinbank, 1986: Alleviation of a systematic westerly bias in general circulation and numerical weather prediction models through an orographic gravity wave drag parametrization. *Quarterly Journal of the Royal Meteorological Society*, **112 (474)**, 1001–1039.

Park, M., W. J. Randel, L. K. Emmons, and N. J. Livesey, 2009: Transport pathways of carbon monoxide in the Asian summer monsoon diagnosed from model of ozone and related tracers (MOZART). *Journal of Geophysical Research: Atmospheres*, **114 (D8)**.

Pelly, J. L., and B. J. Hoskins, 2003: A new perspective on blocking. *Journal of the Atmospheric Sciences*, **60**, 743–755.

Peters, M. E., and C. S. Bretherton, 2005: A simplified model of the Walker circulation with an interactive ocean mixed layer and cloud-radiative feedbacks. *Journal of Climate*, **18 (20)**, 4216–4234.

Pfeffer, R. L., 1987: Comparison of conventional and transformed Eulerian diagnostics in the troposphere. *Quarterly Journal of the Royal Meteorological Society*, **113 (475)**, 237–254.

Pfeffer, R. L., and M. Challa, 1981: A numerical study of the role of eddy fluxes of momentum in the development of Atlantic hurricanes. *Journal of the Atmospheric Sciences*, **38 (11)**, 2393–2398.

Phillips, N. A., 1956: The general circulation of the atmosphere: A numerical experiment. *Quarterly Journal of the Royal Meteorological Society*, **82**, 123–164.

Pierrehumbert, R. T., 1995: Thermostats, radiator fins, and the local runaway greenhouse. *Journal of the Atmospheric Sciences*, **52 (10)**, 1784–1806.

Pierrehumbert, R. T., 2010: *Principles of Planetary Climate*. Cambridge University Press.

Ploeger, F., and T. Birner, 2016: Seasonal and inter-annual variability of lower stratospheric age of air spectra. *Atmospheric Chemistry and Physics*, **16 (15)**, 10 195–10 213.

Plumb, R. A., 1983: A new look at the energy cycle. *Journal of the Atmospheric Sciences*, **40 (7)**, 1669–1688.

Plumb, R. A., 1985: On the three-dimensional propagation of stationary waves. *Journal of the Atmospheric Sciences*, **42**, 217–229.

Plumb, R. A., 1986: Three-dimensional propagation of transient quasi-geostrophic eddies and its relationship with the eddy forcing of the time-mean flow. *Journal of the Atmospheric Sciences*, **43**, 1657–1678.

Plumb, R. A., and A. D. McEwan, 1978: The instability of a forced standing wave in a viscous stratified fluid: A laboratory analogue of the Quasi-Biennial Oscillation. *Journal of Atmospheric Sciences*, **35**, 1827–1839.

Podglajen, A., A. Hertzog, R. Plougonven, and B. Legras, 2020: Lagrangian gravity wave spectra in the lower stratosphere of current (re)analyses. *Atmospheric Chemistry and Physics*, **20 (15)**, 9331–9350.

Priestly, C. H. B., 1949: Heat transport and zonal stress between latitudes. *Quarterly Journal of the Royal Meteorological Society*, **75**, 28–40.

Ralph, F. M., P. J. Neiman, G. A. Wick, S. I. Gutman, M. D. Dettinger, D. R. Cayan, and A. B. White, 2006: Flooding on California's Russian River: Role of atmospheric rivers. *Geophysical Research Letters*, **33 (13)**.

Randall, D. 2015: *An Introduction to the Global Circulation of the Atmosphere*. Princeton University Press.

Randel, W. J., and F. Wu, 2015: Variability of zonal mean tropical temperatures derived from a decade of GPS radio occultation data. *Journal of the Atmospheric Sciences*, **72 (3)**, 1261–1275.

Rasmusson, E. M., 1968: Atmospheric water vapor transport and the water balance of North America: II. Large-scale water balance investigations. *Monthly Weather Review*, **96**, 720–734.

Rasmusson, E. M., and T. H. Carpenter, 1982: Variations in tropical sea surface temperature and surface wind fields associated with the Southern Oscillation/El Niño. *Monthly Weather Review*, **110 (5)**, 354–384.

Rattray, M., and R. L. Charnell, 1966: Quasigeostrophic free oscillations in enclosed basins. Tech. Rep. 162 - 167, University of Washington.

Reed, R. J., W. J. Campbell, L. A. Rasmussen, and D. G. Rogers, 1961: Evidence of a downward-propagating, annual wind reversal in the equatorial stratosphere. *Journal of Geophysical Research*, **66 (3)**, 813–818.

Reed, R. J., D. C. Norquist, and E. E. Recker, 1977: The structure and properties of African wave disturbances as observed during Phase III of GATE. *Monthly Weather Review*, **105 (3)**, 317–333.

Reed, R. J., and C. L. Vicek, 1969: The annual temperature variation in the lower tropical stratosphere. *Journal of the Atmospheric Sciences*, **26 (1)**, 163–167.

Rennert, K. J., and J. M. Wallace, 2009: Cross-frequency coupling, skewness, and blocking in the Northern Hemisphere winter circulation. *Journal of Climate*, **22**, 5650–5666.

Rex, D. F., 1950a: Blocking action in the middle troposphere and its effect upon regional climate. Part I: An aerological study of blocking action. *Tellus*, **2 (4)**, 196–207.

Rex, D. F., 1950b: Blocking action in the middle troposphere and its effect upon regional climate. Part II: The climatology of blocking action. *Tellus*, **2 (4)**, 275–301.

Richman, M. B., 1986: Rotation of principal components. *Journal of Climatology*, **6**, 293–335.

Richter, J. H., J. A. Anstey, N. Butchart, Y. Kawatani, G. A. Meehl, S. Osprey, and I. R. Simpson, 2020: Progress in simulating the quasi-biennial oscillation in CMIP models. *Journal of Geophysical Research: Atmospheres*, **125 (8)**, e2019JD032 362.

Riehl, H., and J. S. Malkus, 1958: On the heat balance in the equatorial trough zone. Geophysica (Helsinki), 6, 503–538., and J. Simpson, 1979: The heat balance of the equatorial trough zone, revisited. *Contributions to Atmospheric Physics*, **52**, 287–304.

Roberts, C. D., F. Vitart, M. A. Balmaseda, and F. Molteni, 2020: The time-scale-dependent response of the wintertime North Atlantic to increased ocean model resolution in a coupled forecast model. *Journal of Climate*, **33 (9)**, 3663 – 3689.

Rodwell, M. J., and B. J. Hoskins, 2001: Subtropical anticyclones and summer monsoons. *Journal of Climate*, **14 (15)**, 3192–3211.

Rogers, R. F., and coauthors, 2017: Rewriting the tropical record books: The extraordinary intensification of Hurricane Patricia (2015). *Bulletin of the American Meteorological Society*, **98 (10)**, 2091–2112.

Rosen, R. D., and D. A. Salstein, 1983: Variations in atmospheric angular momentum on global and regional scales and the length of day. *Journal of Geophysical Research: Oceans (1978–2012)*, **88**, 5451–5470.

Rosen, R. D., D. A. Salstein, T. M. Eubanks, J. O. Dickey, and J. A. Steppe, 1984: An El Niño signal in atmospheric angular momentum and Earth rotation. *Science*, **225 (4660)**, 411–414.

Rosenthal, S. L., 1965: Some preliminary theoretical considerations of tropospheric wave motions in equatorial latitudes. *Monthly Weather Review*, **93**, 605–612.

Rossby, C.-G., 1939: Relation between variations in the intensity of the zonal circulation of the atmosphere and the displacements of the semi-permanent centers of action. *Journal of Marine Research*, **2**, 38–55.

Rossby, C.-G., and V. P. Starr, 1949: Interpretations of the angular-momentum principle as applied to the general circulation of the atmosphere. *Journal of Atmospheric Sciences*, **6 (4)**, 288.

Ruppert, J. H., X. Chen, and F. Zhang, 2020: Convectively forced diurnal gravity waves in the maritime continent. *Journal of the Atmospheric Sciences*, **77 (3)**, 1119–1136.

Sakazaki, T., 2021: Tropical rainfall variability accompanying global normal mode oscillations. *Journal of the Atmospheric Sciences*, **78 (4)**, 1295–1316.

Sakazaki, T., and K. Hamilton, 2020: An array of ringing global free modes discovered in tropical surface pressure data. *Journal of the Atmospheric Sciences*, **77 (7)**, 2519–2539.

Salby, M. L., and R. R. Garcia, 1987: Transient response to localized episodic heating in the tropics. Part I: Excitation and short-time near-field behavior. *Journal of the Atmospheric Sciences*, **44 (2)**, 458–498.

Saltzman, B., 1970: Large-scale atmospheric energetics in the wavenumber domain. *Reviews of Geophysics*, **8**, 289–302.

Sardeshmukh, P. D., and B. J. Hoskins, 1988: The generation of global rotational flow by steady idealized tropical divergence. *Journal of the Atmospheric Sciences*, **45 (7)**, 1228–1251.

Schneider, E. K., 1977: Axially symmetric steady-state models of the basic state for instability and climate studies. Part II. Nonlinear calculations. *Journal of the Atmospheric Sciences*, **34 (2)**, 280–296.

Schneider, E. K., and R. S. Lindzen, 1977: Axially symmetric steady-state models of the basic state for instability and climate studies. Part I. Linearized calculations. *Journal of the Atmospheric Sciences*, **34 (2)**, 263–279.

Schneider, T., I. M. Held, and S. T. Garner, 2003: Boundary effects in potential vorticity dynamics. *Journal of the Atmospheric Sciences*, **60 (8)**, 1024–1040.

Schopf, P. S., and M. J. Suarez, 1988: Vacillations in a coupled ocean-atmosphere model. *Journal of the Atmospheric Sciences*, **45**, 549–566.

Shapiro, M. A., 1981: Frontogenesis and geostrophically forced secondary circulations in the vicinity of jet stream-frontal zone systems. *Journal of the Atmospheric Sciences*, **38 (5)**, 954–973.

Shutts, G. J., 1983: The propagation of eddies in diffluent jet-streams: Eddy vorticity forcing of 'blocking' flow fields. *Quarterly Journal of the Royal Meteorological Society*, **109 (462)**, 737–761.

Shutts, G. J., 1986: A case study of eddy forcing during an Atlantic blocking episode. *Advances in Geophysics*, **29**, 135–162.

Simmons, A. J., and B. J. Hoskins, 1978: The life cycles of some nonlinear baroclinic waves. *Journal of the Atmospheric Sciences*, **35**, 414–432.

Simmons, A. J., J. M. Wallace, and G. W. Branstator, 1983: Barotropic wave propagation and instability, and atmospheric teleconnection patterns. *Journal of the Atmospheric Sciences*, **40**, 1363–1392.

Smagorinsky, J., S. Manabe, and J. L. Holloway Jr, 1965: Numerical results from a nine-level general circulation model of the atmosphere. *Monthly Weather Review*, **93**, 727–768.

Smith, S. D., 1980: Wind stress and heat flux over the ocean in gale force winds. *Journal of Physical Oceanography*, **10**, 709–726.

Smoliak, B. V., and J. M. Wallace, 2015: On the leading patterns of Northern Hemisphere sea level pressure variability. *Journal of the Atmospheric Sciences*, **72**, 3469–3486.

Sobel, A. H., and C. S. Bretherton, 1999: Development of synoptic-scale disturbances over the summertime tropical northwest Pacific. *Journal of the Atmospheric Sciences*, **56 (17)**, 3106–3127.

Sobel, A. H., and C. S. Bretherton, 2000: Modeling tropical precipitation in a single column. *Journal of Climate*, **13 (24)**, 4378–4392.

Sobel, A. H., J. Nilsson, and L. M. Polvani, 2001: The weak temperature gradient approximation and balanced tropical moisture waves. *Journal of the Atmospheric Sciences*, **58 (23)**, 3650–3665.

Starr, V. P., 1948: An essay on the general circulation of the Earth's Atmosphere. *Journal of Meteorology*, **5**, 39–43.

Starr, V. P., 1949: Meridional circulations and the transfer of angular momentum in the atmosphere–Reply. *Journal of Metorology*, **6**, 430.

Starr, V. P., 1968: *Physics of Negative Viscosity Phenomena*. McGraw-Hill.

Starr, V. P., J. P. Peixoto, and G. C. Livadas, 1958: On the meridional flux of water vapor in the Northern Hemisphere. *Geofisica pura e applicata*, **39**, 174–185.

Starr, V. P., and R. M. White, 1951: A hemispherical study of the atmospheric angular-momentum balance. *Quarterly Journal of the Royal Meteorological Society*, **77**, 215–225.

Starr, V. P., and R. M. White, 1955: Direct measurement of the hemispheric poleward flux of water vapor. *Journal of Marine Research*, **14**, 217–225.

Stensrud, D. J., 1996: Importance of low-level jets to climate: A review. *Journal of Climate*, **9**, 1698–1711.

Stephan, C. C., and A. Mariaccia, 2021: The signature of the tropospheric gravity wave background in observed mesoscale motion. *Weather and Climate Dynamics*, **2 (2)**, 359–372.

Stephenson, D. B., H. Wanner, S. Brönnimann, and J. Luterbacher, 2003: The history of scientific research on the North Atlantic Oscillation. *Geophysical Monograph 134*, J. W. Hurrell, Y. Kushnir, G. Ottersen, and M. Visbeck, Eds., American Geophysical Union, 37–50.

Stokes, G. G., 1847: On the theory of oscillatory waves. *Transactions of the Cambridge Philosophical Society*, **8**, 441–473.

Stone, P. H., 1978: Baroclinic adjustment. *Journal of the Atmospheric Sciences*, **35 (4)**, 561–571.

Strachey, R., 1888: On the air waves and sounds caused by the eruption of Krakatoa in August, 1883. In *"The eruption of Krakatoa and subsequent phenomena", Report of the Krakatoa Committee of the Royal Society,* Ed. G.J. Symons, Trübner and Co., London.

Straub, K. H., and G. N. Kiladis, 2002: Observations of a convectively coupled Kelvin Wave in the eastern Pacific ITCZ. *Journal of the Atmospheric Sciences*, **59 (1)**, 30 – 53.

Symons, G. J., 1888: *The Eruption of Krakatoa, and Subsequent Phenomena: Report of the Krakatoa Committee of the Royal Society*. Trübner London.

Takahashi, K., and D. S. Battisti, 2007: Processes controlling the mean tropical Pacific precipitation pattern. Part I: The Andes and the eastern Pacific ITCZ. *Journal of Climate*, **20 (14)**, 3434–3451.

Takahashi, K., A. Montecinos, K. Goubanova, and B. Dewitte, 2011: ENSO regimes: Reinterpreting the canonical and Modoki El Niño. *Geophysical Research Letters*, **38 (10)**.

Takaya, K., and H. Nakamura, 2001: A formulation of a phase-independent wave-activity flux for stationary and migratory quasigeostrophic eddies on a zonally varying basic flow. *Journal of the Atmospheric Sciences*, **58 (6)**, 608–627.

Takayabu, Y. N., 1994: Large-scale cloud disturbances associated with equatorial waves. Part II: Westward propagating inertio-gravity waves. *Journal of the Meteorological Society of Japan*, **72**, 451–465.

Takayabu, Y. N., and T. Nitta, 1993: 3-5 day-period disturbances coupled with convection over the tropical Pacific Ocean. *Journal of the Meteorological Society of Japan. Ser. II*, **71 (2)**, 221–246.

Tanaka, S., K. Nishii, and H. Nakamura, 2016: Vertical structure and energetics of the western Pacific teleconnection pattern. *Journal of Climate*, **29 (18)**, 6597–6616.

Taylor, G. I., 1929: Waves and tides in the atmosphere. *Proceedings of the Royal Society of London. Series A, Containing Papers of a Mathematical and Physical Character*, **126 (800)**, 169–183.

Thompson, C. J., and D. S. Battisti, 2000: A linear stochastic dynamical model of ENSO. Part I: Model development. *Journal of Climate*, **13 (15)**, 2818–2832.

Thompson, C. J., and D. S. Battisti, 2001: A linear stochastic dynamical model of ENSO. Part II: Analysis. *Journal of Climate*, **14 (4)**, 445–466.

Thompson, D. W. and E. A. Barnes, 2014: Periodic variability in the large-scale Southern Hemisphere atmospheric circulation. *Science*, **343**, 641–645.

Thompson, D. W., M. P. Baldwin, and J. M. Wallace, 2002: Stratospheric connection to Northern Hemisphere wintertime weather: Implications for prediction. *Journal of Climate*, **15 (12)**, 1421–1428.

Thompson, D. W. J., S. Solomon, P. J. Kushner, M. H. England, K. M. Grise, and D. J. Karoly, 2011: Signatures of the Antarctic ozone hole in Southern Hemisphere surface climate change. *National Geoscience*, **4**, 741–749.

Thorncroft, C. D., B. J. Hoskins, and M. E. McIntyre, 1993: Two paradigms of baroclinic-wave life-cycle behaviour. *Quarterly Journal of the Royal Meteorological Society*, **119**, 17–55.

Tomas, R. A., and P. J. Webster, 1997: The role of inertial instability in determining the location and strength of near-equatorial convection. *Quarterly Journal of the Royal Meteorological Society*, **123 (542)**, 1445–1482.

Townsend, R. D., and D. R. Johnson, 1985: A diagnostic study of the isentropic zonally averaged mass circulation during the first GARP global experiment. *Journal of the Atmospheric Sciences*, **42**, 1565–1579.

Trenberth, K. E., 1986: An assessment of the impact of transient eddies on the zonal flow during a blocking episode using localized Eliassen–Palm flux diagnostics. *Journal of the Atmospheric Sciences*, **43**, 2070–2087.

Trenberth, K. E., and J. M. Caron, 2001: Estimates of meridional atmosphere and ocean heat transports. *Journal of Climate*, **14 (16)**, 3433–3443.

Trenberth, K. E., and J. T. Fasullo, 2012: Climate extremes and climate change: The Russian heat wave and other climate extremes of 2010. *Journal of Geophysical Research-Atmospheres*, **117**.

Trenberth, K. E., and D. P. Stepaniak, 2003: Covariability of components of poleward atmospheric energy transports on seasonal and interannual timescales. *Journal of Climate*, **16**, 3691–3705.

Tulich, S. N., and G. N. Kiladis, 2012: Squall lines and convectively coupled gravity waves in the tropics: Why do most cloud systems propagate westward? *Journal of the Atmospheric Sciences*, **69 (10)**, 2995–3012.

Vallis, G. K., 2017: *Atmospheric and Oceanic Fluid Dynamics*. Cambridge University Press.

Van den Dool, H. M., S. Saha, and Å. Johansson, 2000: Empirical orthogonal teleconnections. *Journal of Climate*, **13**, 1421–1435.

van Loon, H., and J. C. Rogers, 1978: The seesaw in winter temperatures between Greenland and northern Europe. Part I: General description. *Monthly Weather Review*, **106**, 296–310.

Vautard, R. and M. Ghil, 1989: Singular spectrum analysis in nonlinear dynamics, with applications to paleoclimatic time series, *Physica D: Nonlinear Phenomena*, **35**, 395–424.

Venugopal, V., and J. M. Wallace, 2016: Climatology of contribution-weighted tropical rain rates based on TRMM 3B42. *Geophysical Research Letters*, **43 (19)**, 10439–10447.

Veryard, R. G., and R. A. Ebdon, 1961: Fluctuations in tropical stratospheric winds. *Meteorological Magazine*, **90**, 125–143.

Vimont, D. J., J. M. Wallace, and D. S. Battisti, 2003: The seasonal footprinting mechanism in the Pacific: Implications for ENSO. *Journal of Climate*, **16**, 2668–2675.

Virts, K. S., and J. M. Wallace, 2014: Observations of temperature, wind, cirrus, and trace gases in the tropical tropopause transition layer during the MJO. *Journal of the Atmospheric Sciences*, **71 (3)**, 1143–1157.

Virts, K. S., J. M. Wallace, M. L. Hutchins, and R. H. Holzworth, 2013: Highlights of a new ground-based, hourly global lightning climatology. *Bulletin of the American Meteorological Society*, **94 (9)**, 1381–1391.

Visbeck, M., E. Chassignet, R. Curry, T. Delworth, R. Dickson, and G. Krahmann, 2003: The North Atlantic Oscillation: Climatic significance and environmental impact. *Geophysical Monographs*, **134**.

Walker, G. T., and E. W. Bliss, 1932: World Weather V. Mem. *Royal Meteorological Society*, **4**, 53–84.

Wallace, J. M., and D. S. Gutzler, 1981: Teleconnections in the geopotential height field during the Northern Hemisphere winter. *Monthly Weather Review*, **109**, 784–812.

Wallace, J. M., and P. V. Hobbs, 2006: *Atmospheric Science: An Introductory Survey*, Vol. 92. Academic Press, 504 pp.

Wallace, J. M., and V. E. Kousky, 1968: On the relation beween Kelvin waves and the Quasi-Biennial Oscallation. *Journal of the Meteorological Society of Japan*, **47**, 496–502.

Wallace, J. M., and N.-C. Lau, 1985: On the role of barotropic energy conversions in the general circulation. *Advances in Geophysics*, Vol. 28, Elsevier, 33–74.

Wang, L., and N. Nakamura, 2015: Covariation of finite-amplitude wave activity and the zonal mean flow in the midlatitude troposphere: 1. Theory and application to the Southern Hemisphere summer. *Geophysical Research Letters*, **42**, 8192–8200.

Wang, M., Y. Du, B. Qiu, X. Cheng, Y. Luo, X. Chen, and M. Feng, 2017: Mechanism of seasonal eddy kinetic energy variability in the eastern equatorial Pacific Ocean. *Journal of Geophysical Research: Oceans*, **122 (4)**, 3240–3252.

Weare, B. C., A. R. Navato, and R. E. Newell, 1976: Empirical orthogonal analysis of Pacific sea surface temperatures. *Journal of Physical Oceanography*, **6 (5)**, 671–678.

Webster, P. J., 2020: *Dynamics of The Tropical Atmosphere and Oceans*. John Wiley and Sons, Ltd, 501 pp.

Webster, P. J., and J. R. Holton, 1982: Cross-equatorial response to middle-latitude forcing in a zonally varying basic state. *Journal of the Atmospheric Sciences*, **39**, 722–733.

Weickmann, K. M., G. N. Kiladis, and P. D. Sardeshmukh, 1997: The dynamics of intraseasonal atmospheric angular momentum oscillations. *Journal of the Atmospheric Sciences*, **54 (11)**, 1445–1461.

Wettstein, J. J., and J. M. Wallace, 2010: Observed patterns of month-to-month storm-track variability and their relationship to the background flow. *Journal of the Atmospheric Sciences*, **67 (5)**, 1420–1437.

Wheeler, M., and G. N. Kiladis, 1999: Convectively-coupled equatorial waves: Analysis of clouds in the wavenumber-frequency domain. *Journal of Atmospheric Sciences*, **56**, 374–399.

Wheeler, M. C., and H. H. Hendon, 2004: An all-season real-time multivariate MJO index: Development of an index for monitoring and prediction. *Monthly Weather Review*, **132 (8)**, 1917–1932.

Wheeler, M. C., G. N. Kiladis, and P. J. Webster, 2000: Large-scale dynamical fields associated with convectively coupled equatorial waves. *Journal of the Atmospheric Sciences*, **57 (5)**, 613–640.

White, R. H., J. M. Wallace, and D. S. Battisti, 2021: Revisiting mountains and atmospheric circulation: Zonal mean flow and stationary waves. *Journal of the Atmospheric Sciences*, **78**, 2221–2235.

White, R. M., 1951: The meridional eddy flux of energy. *Quarterly Journal of the Royal Meteorological Society*, **77 (332)**, 188–199.

Widger, W. K., 1949: A study of the flow of angular momentum in the atmosphere. *Journal of Meteorology*, **6**, 292–299.

Wijffels, S. E., R. W. Schmitt, H. L. Bryden, and A. Stigebrandt, 1992: Transport of freshwater by the oceans. *Journal of Physical Oceanography*, **22**, 155–162.

Williams, G. P., 1988: The dynamical range of global circulations— II. *Climate Dynamics*, **3 (2)**, 45–84.

Wing, A. A., K. Emanuel, C. E. Holloway, and C. Muller, 2017: Convective self-aggregation in numerical simulations: A review. *Shallow Clouds, Water Vapor, Circulation, and Climate Sensitivity*, 1–25.

Wu, G., Y. Liu, B. He, Q. Bao, A. Duan, and F.-F. Jin, 2012: Thermal controls on the Asian summer monsoon. *Scientific Reports*, **2 (1)**, 1–7.

Wu, Z., E. S. Sarachik, and D. S. Battisti, 2000: Vertical structure of convective heating and the three-dimensional structure of the forced circulation on an equatorial beta plane. *Journal of the Atmospheric Sciences*, **57 (13)**, 2169–2187.

Wyrtki, K., 1975: El Niño: The dynamic response of the equatorial Pacific Ocean to atmospheric forcing. *Journal of Physical Oceanography*, **5**, 572–584.

Xu, W., S. A. Rutledge, and W. Zhang, 2017: Relationships between total lightning, deep convection, and tropical cyclone intensity change. *Journal of Geophysical Research: Atmospheres*, **122 (13)**, 7047–7063.

Yanai, M., S. Esbensen, and J.-H. Chu, 1973: Determination of bulk properties of tropical cloud clusters from large-scale heat and moisture budgets. *Journal of the Atmospheric Sciences*, **30 (4)**, 611–627.

Yanai, M., and T. Maruyama, 1966: Stratospheric wave disturbances propagating over the equatorial Pacific. *Journal of the Meteorological Society of Japan. Ser. II*, **44 (5)**, 291–294.

Yulaeva, E., J. R. Holton, and J. M. Wallace, 1994: On the cause of the annual cycle in tropical lower-stratospheric temperatures. *Journal of the Atmospheric Sciences*, **51 (2)**, 169–174.

Žagar, N., D. Jelić, M. Blaauw, and P. Bechtold, 2017: Energy spectra and inertia-gravity waves in global analyses. *Journal of the Atmospheric Sciences*, **74 (8)**, 2447–2466.

Zhang, C., M. McGauley, and N. A. Bond, 2004: Shallow meridional circulation in the tropical eastern Pacific. *Journal of Climate*, **17 (1)**, 133–139.

Zhang, J. A., F. D. Marks, M. T. Montgomery, and S. Lorsolo, 2010: An estimation of turbulent characteristics in the low-level region of intense Hurricanes Allen (1980) and Hugo (1989). *Monthly Weather Review*, **139 (5)**, 1447–1462.

Zhang, Y., J. M. Wallace, and D. S. Battisti, 1997: ENSO-like interdecadal variability: 1900–93. *Journal of Climate*, **10 (5)**, 1004–1020.

Zhou, T., M. A. Geller, and K. Hamilton, 2006: The roles of the Hadley circulation and downward control in tropical upwelling. *Journal of the Atmospheric Sciences*, **63 (11)**, 2740–2757.

Index

Acoustic Waves, 30, 31, 50, *357*, 361
Adiabatic Lapse Rate, 33
   dry, definition of, 21
Adiabatic Temperature Tendency, 113, *124*, 141, *161*
Aerosols
   radioactive, 6, 162
   residence time in
      stratosphere, 162
      troposphere, 162
   volcanic, 167
Ageostrophic Horizontal Wind Component, 128, 162, *214, 222,* 222, 265
Ageostrophic Mean Meridional Circulation, 121, 123, 129, 205
Ageostrophic Wind Component, *113*, 114
Air Parcel, Trajectories in
   baroclinic waves, 136, *137*
   Brewer–Dobson circulation, *161*, 160–162
   gravity waves, 183
   Lagrangian mean meridional circulation, 6, 24, 86, *161*, 160–162
   MRG waves, 180, 183
   stratospheric intrusions, 162, *163*
   the TTL, 159–160, 162
   tropical cyclones, 344, 346, 348
Albedo, 89, 90
   of Earth, 86
Aleutian Low, *see* Climatology, Features of
Andes, 200, 201
   as waveguide for shallow waves, 29, *227*
   impact on the Pacific ITCZ, 290
Angular Momentum, *see* Atmospheric Angular Momentum
Angular Momentum In Tropical Cyclones, 65, 345, 346, 349
Anisotropy of the Eddies and Transients, 214, 215, 238
   coefficient of, 219, *220*
   definition of, 219–220
   energetic implications, 220
   frequency dependence of, 217, 221
Antipodal Point, 48, *49*, 49, *51*, 203
Aquaplanet, xiii, 75, 92, 265, 355
Aquaplanet Simulations, 41, 43, 46, 75, 150, 213, 244
Aquifers, 79
Arctic Oscillation, *see* Northern Hemisphere Annular Mode (NAM)
Arctic Sea Ice, 82, 243
Atlantic Meridional Mode, 310

Atmosphere–Ocean Interactions
   in aquaplanet models with slab ocean, *37*, 37–38
   in ENSO, *296*, **295–299**
   in SSZs, 264–265
   in the Atlantic Meridional Mode, 310
   in the Pacific cold tongue, *296*, 296–297, 307
   in the Pacific Merional Mode, 310
   in Tropic World, 37–38
Atmospheric Angular Momentum (AAM)
   and length of day (LOD), 60, 63
   climatology of, *82*, 83
   conservation equation for, **57–60**
   countergradient transport of, 133
   definition of, 57
   global distribution of, 43, *81*, 83
   meridional and vertical distribution of, *82*
   meridional transport of, 3, 57, **62–65**, 286
   per unit mass, 57
   sources and sinks of, **58–60**, 346
   transport by meridional circulation, **66–67**
   variations in
      ENSO-related, 307
      MJO-related, 323
      QBO-related, 175
      seasonal, 60
   vertical transport of, 66, **66–67**
Atmospheric Tides
   gravitationally driven (lunar), *356*, 355–356
   thermally (solar) driven, 41, 46, *356*, 355–356
Available Potential Energy, 55, **106–110**
   approximate expressions for, 107, 108
   conversion to kinetic energy, 108–113
      partition into eddy and zonal mean, 114–115
   definition of, 106–107, 114
   generation and dissipation of, 115, 135–136
      in stationary waves, 198
   quantitative estimates of, *111, 135, 136*
   relation to mechanical energy, 106
   zonal vs. eddy, 114–115

Balance, *see also* individual listings for Hydrostatic, Geostrophic, Gradient Wind and Weak Temperature Balance
Balance Requirement
   definition of, 55
   for
      angular momentum, 71, 119
      mechanical energy, 106

      total energy, 106
      water vapor, 78
   limitations of arguments based on, 79, 119
Baroclinic Annular Modes, **153–154**
Baroclinic Instability
   discovery of, 128
   occurrence of in spin-up experiments, 42–43, 75, 116
   role in general circulation, 136
   theory of, 101, 147
Baroclinic Waves
   angular momentum transport by, 65, 151
   critical level for, 148
   dependence of structure and amplitude structure on zonal wavenumber, 137
   dispersion, 221, 223–226
   energetics, 137–139, 143, 221, 223
   EP fluxes in, 143–146
   growth rate, 116
   horizontal scale, 42, 115, 116, *117*
   life cycle LC1 and LC2, **137–140**, 146, 341–343
      conditions under which one or the other is favored, 139
      energetics, *138*, 137–138, *138*, 139
      morphology, *138*, 138, *139*, 139
      relation to barotropic annular mode, 151–152, 239
   life cycle of baroclinic waves in general, 42, *140, 143*, 143–144, 198, 223
   phase speed, 86, 144, 193
   poleward heat transport by, 85, 96, 101
   role in general circulation, 42–43
   role in mechanical energy spectrum, 116, 361
   structure and evolution of, **227–228**
Barotropic Annular Modes, **150–153**
Barotropic Instability, 134, 135
Barotropic Model, *see also* Models
   simulated response to NH orography, 203, *204, 205,* 232–233, *238, 239*
Beta Drift, 339
Beta Effect, 245
Beta-Plane
   equatorial, 172, 177, 277, 330, 357
   midlatitude, 129
Bjerknes Feedback, *see* Feedback(s)
Blocking, 191, **247–249**
   climatology of, 249, *250*, 254
   mechanisms, 247–249, *250*
   relation to
      favored flow configurations, *251, 252,* 250–252, *253*